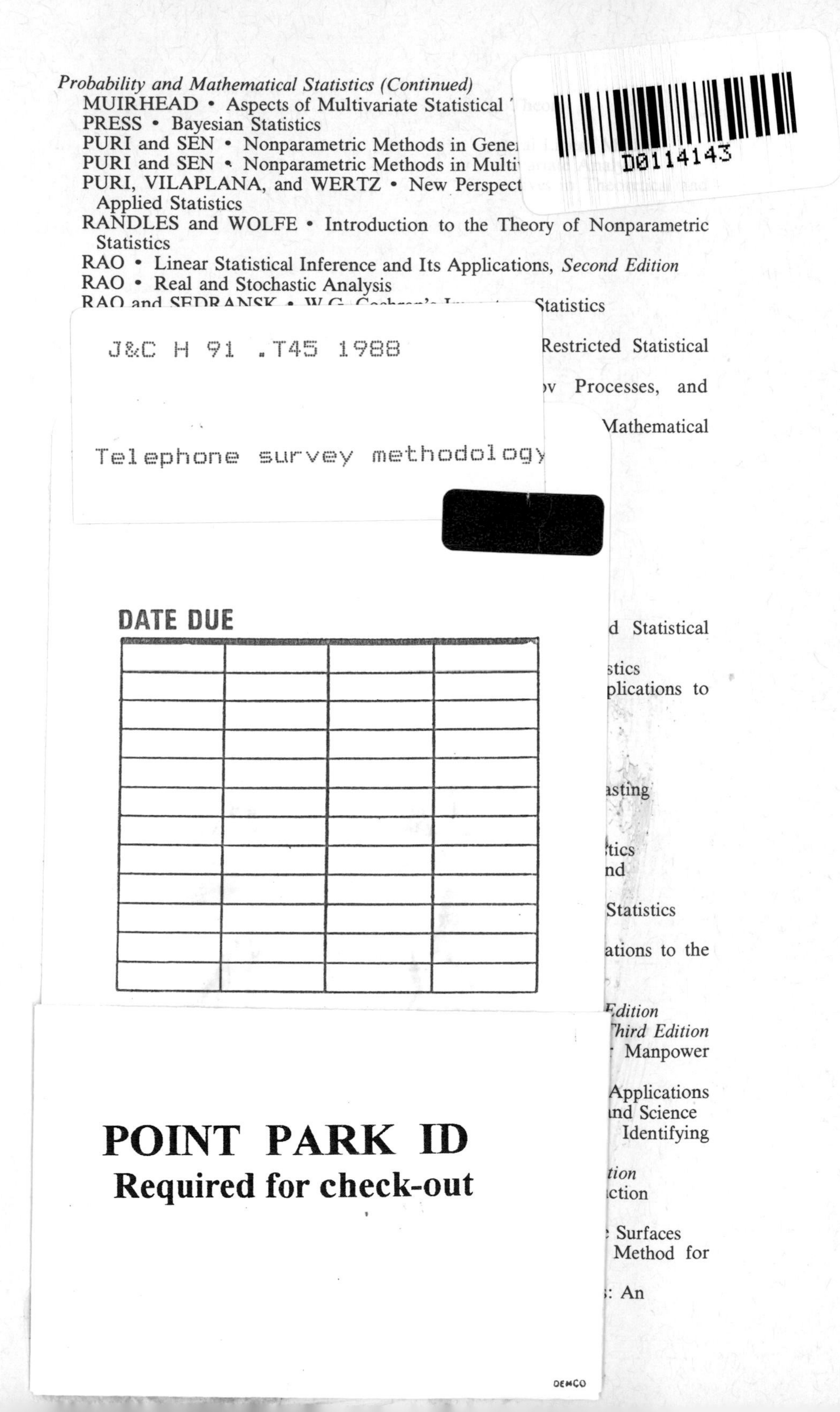

Probability and Mathematical Statistics (Continued)

MUIRHEAD • Aspects of Multivariate Statistical

PRESS • Bayesian Statistics

PURI and SEN • Nonparametric Methods in Gene

PURI and SEN • Nonparametric Methods in Multi

PURI, VILAPLANA, and WERTZ • New Perspect

Applied Statistics

RANDLES and WOLFE • Introduction to the Theory of Nonparametric Statistics

RAO • Linear Statistical Inference and Its Applications, *Second Edition*

RAO • Real and Stochastic Analysis

RAO and SEDRANSK • W.G. Cochran's I… Statistics

Restricted Statistical

…v Processes, and

Mathematical

d Statistical

stics

plications to

asting

tics

nd

Statistics

ations to the

Edition

Third Edition

Manpower

Applications

and Science

Identifying

tion

ction

Surfaces

Method for

: An

Applied Probability and Statistics (Continued)

BROWN and HOLLANDER • Statistics: A Biomedical Introduction
BUNKE and BUNKE • Statistical Inference in Linear Models, Volume I
CHAMBERS • Computational Methods for Data Analysis
CHATTERJEE and HADI • Sensitivity Analysis in Linear Regression
CHATTERJEE and PRICE • Regression Analysis by Example
CHOW • Econometric Analysis by Control Methods
CLARKE and DISNEY • Probability and Random Processes: A First Course with Applications, *Second Edition*
COCHRAN • Sampling Techniques, *Third Edition*
COCHRAN and COX • Experimental Designs, *Second Edition*
CONOVER • Practical Nonparametric Statistics, *Second Edition*
CONOVER and IMAN • Introduction to Modern Business Statistics
CORNELL • Experiments with Mixtures: Designs, Models and The Analysis of Mixture Data
COX • Planning of Experiments
COX • A Handbook of Introductory Statistical Methods
DANIEL • Biostatistics: A Foundation for Analysis in the Health Sciences, *Fourth Edition*
DANIEL • Applications of Statistics to Industrial Experimentation
DANIEL and WOOD • Fitting Equations to Data: Computer Analysis of Multifactor Data, *Second Edition*
DAVID • Order Statistics, *Second Edition*
DAVISON • Multidimensional Scaling
DEGROOT, FIENBERG and KADANE • Statistics and the Law
DEMING • Sample Design in Business Research
DILLON and GOLDSTEIN • Multivariate Analysis: Methods and Applications
DODGE • Analysis of Experiments with Missing Data
DODGE and ROMIG • Sampling Inspection Tables, *Second Edition*
DOWDY and WEARDEN • Statistics for Research
DRAPER and SMITH • Applied Regression Analysis, *Second Edition*
DUNN • Basic Statistics: A Primer for the Biomedical Sciences, *Second Edition*
DUNN and CLARK • Applied Statistics: Analysis of Variance and Regression, *Second Edition*
ELANDT-JOHNSON and JOHNSON • Survival Models and Data Analysis
FLEISS • Statistical Methods for Rates and Proportions, *Second Edition*
FLEISS • The Design and Analysis of Clinical Experiments
FLURY • Common Principal Components and Related Multivariate Models
FOX • Linear Statistical Models and Related Methods
FRANKEN, KÖNIG, ARNDT, and SCHMIDT • Queues and Point Processes
GALLANT • Nonlinear Statistical Models
GIBBONS, OLKIN, and SOBEL • Selecting and Ordering Populations: A New Statistical Methodology
GNANADESIKAN • Methods for Statistical Data Analysis of Multivariate Observations
GREENBERG and WEBSTER • Advanced Econometrics: A Bridge to the Literature
GROSS and HARRIS • Fundamentals of Queueing Theory, *Second Edition*
GROVES, BIEMER, LYBERG, MASSEY, NICHOLLS, and WAKSBERG • Telephone Survey Methodology
GUPTA and PANCHAPAKESAN • Multiple Decision Procedures: Theory and Methodology of Selecting and Ranking Populations
GUTTMAN, WILKS, and HUNTER • Introductory Engineering Statistics, *Third Edition*
HAHN and SHAPIRO • Statistical Models in Engineering
HALD • Statistical Tables and Formulas
HALD • Statistical Theory with Engineering Applications
HAND • Discrimination and Classification

TELEPHONE SURVEY METHODOLOGY

Telephone Survey Methodology

Edited by

Robert M. Groves
The University of Michigan

Paul P. Biemer
New Mexico State University

Lars E. Lyberg
Statistics Sweden

James T. Massey
National Center for Health Statistics

William L. Nicholls II
U.S. Bureau of the Census

Joseph Waksberg
Westat, Inc.

JOHN WILEY & SONS
New York • Chichester • Brisbane • Toronto • Singapore

Library of Congress Cataloging-in-Publication Data:
Telephone survey methodology.

(Wiley series in probability and mathematical
statistics. Applied probability and statistics,
ISSN 0271-6356)
Papers from a conference held Nov. 8-11, 1987 at
Charlotte, North Carolina and sponsored by the American
Association for Public Opinion Research, the American
Statistical Association, and the International Associa-
tion of Survey Statisticians.
Bibliography: p.
1. Telephone surveys—Methodology—Congresses.
2. Telephone surveys—Random digit dialing—Congresses.
3. Social sciences—Research—Congresses. I. Groves,
Robert M. II. American Association for Public Opinion
Research. III. American Statistical Association.
IV. International Association of Survey Statisticians.
V. Series.
H91.T45 1988 001.4'33 88-20559
ISBN 0-471-62218-4

Printed in the United States of America

10 9 8 7 6 5 4 3 2 1

CONTRIBUTORS

Charles H. Alexander
U. S. Bureau of the Census
Washington, D.C.

Reginald P. Baker
National Opinion Research Center
Chicago, Illinois

Robert T. Bass
National Agricultural Statistics Service
U. S. Department of Agriculture
Washington, D.C.

Sandra H. Berry
The Rand Corporation
Santa Monica, California

Paul P. Biemer
Department of Experimental Statistics
New Mexico State University
Las Cruces, New Mexico

George F. Bishop
Institute for Policy Research
University of Cincinnati
Cincinnati, Ohio

Norah Blackshaw
Office of Population Censuses and Surveys
London, England

Steven L. Botman
National Center for Health Statistics
Hyattsville, Maryland

G. J. Burkheimer
Research Triangle Institute
Research Triangle Park, North Carolina

Charles Cannell
Survey Research Center
The University of Michigan
Ann Arbor, Michigan

Robert J. Casady
U. S. Bureau of Labor Statistics
Washington, D.C.

Gary Catlin
Statistics Canada
Ottawa, Ontario, Canada

G. Hussain Choudhry
Statistics Canada
Ottawa, Ontario, Canada

Martin Collins
Social and Community Planning Research
London, England

Edith D. de Leeuw
University of Amsterdam and Free University
Amsterdam, The Netherlands

Don A. Dillman
Washington State University
Pullman, Washington

J. Douglas Drew
Statistics Canada
Ottawa, Ontario, Canada

Robert M. Groves
Survey Research Center
The University of Michigan
Ann Arbor, Michigan

Hans-Juergen Hippler
Center for Surveys, Methods, and Analysis
Mannheim, Federal Republic of Germany

Carol C. House
National Agricultural Statistics Service
U. S. Department of Agriculture
Washington, D.C.

Lecily A. Hunter
Statistics Canada
Ottawa, Ontario, Canada

F. W. Immerman
Research Triangle Institute
Research Triangle Park, North Carolina

Susan Ingram
Statistics Canada
Ottawa, Ontario, Canada

Eszter Körmendi
The Danish National Institute of Social Research
Copenhagen, Denmark

Geoff Lee
Australian Bureau of Statistics
Belconnen, Australia

William L. Lefes
National Opinion Research Center
Chicago, Illinois

James M. Lepkowski
Survey Research Center
The University of Michigan
Ann Arbor, Michigan

J. R. Levinsohn
Research Triangle Institute
Research Triangle Park, North Carolina

Lars E. Lyberg
Statistics Sweden
Stockholm, Sweden

David Maklan
Westat, Inc.
Rockville, Maryland

R. E. Mason
Research Triangle Institute
Research Triangle Park, North Carolina

James T. Massey
National Center for Health Statistics
Hyattsville, Maryland

Leyla Mohadjer
Westat, Inc.
Hyattsville, Maryland

William L. Nicholls II
U. S. Bureau of the Census
Washington, D.C.

Diane O'Rourke
Survey Research Laboratory
The University of Illinois
Urbana, Illinois

Lois Oksenberg
Survey Research Center
The University of Michigan
Ann Arbor, Michigan

Norbert Schwarz
Center for Surveys, Methods, and Analysis
Mannheim, Federal Republic of Germany

Janice Sebold
U. S. Bureau of the Census
Washington, D.C.

Monroe G. Sirken
National Center for Health Statistics
Hyattsville, Maryland

Lynne Stokes
Management Science and Information Systems
The University of Texas at Austin
Austin, Texas

Fritz Strack
University of Mannheim
Mannheim, Federal Republic of Germany

Wendy Sykes
Social and Community Planning Research
London, England

John Tarnai
Washington State University
Pullman, Washington

Owen T. Thornberry, Jr.
National Center for Health Statistics
Hyattsville, Maryland

Robert D. Tortora
National Agricultural Statistics Service
U. S. Department of Agriculture
Washington, D.C.

Dennis Trewin
Australian Bureau of Statistics
Belconnen, Australia

Johannes van der Zouwen
University of Amsterdam and Free University
Amsterdam, The Netherlands

Joseph Waksberg
Westat, Inc.
Rockville, Maryland

Michael F. Weeks
Research Triangle Institute
Research Triangle Park, North Carolina

Paul Wilson
Office of Population Censuses and Surveys
London, England

Ming-Yih Yeh
Management Science and Information Systems
The University of Texas at Austin
Austin, Texas

PREFACE

Sometimes the story behind the production of a book can be as informative as an overview of its contents. We think this is one of those times.

In summer 1985, Graham Kalton, then chair of the Survey Research Methods Section of the American Statistical Association, approached Robert Groves about his organizing a small conference of researchers involved in telephone surveys. Groves resisted. He and others involved in research on telephone surveys believed that a book-length treatment of the latest research findings in the field was needed more than a conference. Research on telephone surveys was scattered in journals across the various social science and statistical disciplines and in many unpublished memoranda in government statistical agencies. A book was needed, if only to reduce the time spent in telephone calls passing on this knowledge to others seeking information on the field.

In fall 1985, the idea of the book and the idea of the conference were joined. The conference would be used to assemble the worldwide network of researchers working in the area and to encourage the presentation of the latest work, but a parallel effort to write a comprehensive volume, with chapter drafts presented at the conference, would also be mounted. Indeed, it was believed the book would profit from such a strategy. Knowing that research on the methodology was spreading over disciplines and continents, it was clear that a more complete and comprehensive book would result from having many authors contributing rather than one or a few. In late fall 1985, Groves sent an outline of the book to several researchers for comments. In early 1986, an organizing committee was formed, consisting of: Robert Groves, The University of Michigan, as chair; Paul Biemer, New Mexico State University; Lars Lyberg, Statistics Sweden; James Massey, U.S. National Center for Health Statistics; William Nicholls, U.S. Bureau of the Census; and Joseph Waksberg, Westat, Inc., Rockville, Md. All had extensive experience in telephone survey methods.

The committee agreed that the book and the conference should be sponsored jointly by professional associations whose members were involved in the field of survey research. The American Association for Public Opinion Research (AAPOR), the American Statistical Association (ASA), and the International Association of Survey Statisticians (IASS) were approached in March 1986, to seek their sponsorship for a fall 1987 conference.

Everything worked according to plan. The three associations approved the support of the conference. They also agreed to provide financial support in the event of a deficit after the conference. This was the first time the three organizations had collaborated in such an effort.

However, no financial support was sought directly from these associations. Instead, committee members began a fund-raising campaign among several foundations, commercial and academic survey organizations, and government statistical agencies around the world to raise the money needed to support the travel of authors to the conference and for editing of the book manuscript. Two foundations and many research organizations contributed funds:

Alfred P. Sloan Foundation
Australian Bureau of Statistics
Central Statistical Office of Finland
Dentsu Research, Ltd., Tokyo
Dun and Bradstreet
Istituto Centrale di Statistica, Italy
Market Facts, Inc.
National Broadcasting Company
National Center for Health Statistics
National Science Foundation
Nielsen Media Research
Office of Population Censuses and Surveys, United Kingdom
Rand Corporation
Research Triangle Institute
Response Analysis Corporation
Statistics Canada
Statistics Sweden
Survey Research Center, The University of Michigan
Survey Sampling, Inc.
U.S Bureau of Labor Statistics
U.S. Bureau of the Census
U.S. Center for Educational Statistics
U.S. Centers for Disease Control
U.S. Department of Agriculture
U.S. Food and Drug Administration
Westat, Inc.

In spring 1986, the editors made contact with researchers throughout the world then engaged in research on telephone survey methodology. The book began to take shape. The committee decided that the book should include discussions of statistical issues related to telephone survey design and descriptions of current practice in survey implementation. Special attention would be given to research that provided information about improving the quality of telephone surveys or research which had indicated weaknesses in the method; hence, sections on coverage error, sampling error, nonresponse error, and measurement error were sought. In addition, sections on administrative issues and the use of computer-assisted telephone interviewing systems were needed. Each committee member undertook the task of being a section editor. Groves contacted Beatrice Shube at John Wiley & Sons about publishing the book. She expressed interest.

The committee sought proposals for papers from researchers known to be working in the field, and it also publicized its search for new work in newsletters around the world. These papers were to be presented as invited papers at the conference and as chapters in the book. The committee reviewed proposals for chapters to the book in two fall 1986 meetings. By early December 1986, the contents of the book was nearly complete. Unfortunately many good proposals had to be rejected because of space limitations. The committee then encouraged the presentation of some of those as contributed papers at the conference. In addition an announcement seeking other contributed papers for conference presentation was printed in various newsletters, listing a May 1, 1987, deadline for abstracts.

Based on previous conferences on restricted survey research topics, the committee believed that the conference might attract 150 to 200 persons. It organized the conference so that all participants would be at the same hotel, eat together, and relax together with ample time for informal interaction. The conference site chosen was Charlotte, North Carolina, which offered fine conference facilities and easy airplane access. The conference was set for November 8–11, 1987.

Drafts of the invited papers were obtained during the spring and summer of 1987. Section editors reviewed them and returned comments to authors. Authors were given drafts of other chapters in the same section of the volume and encouraged to cite other work in the volume. The day before the conference the authors and section editors met to review the manuscripts, decide on common terminology, identify holes in the treatment, and comment on each other's work. All authors agreed to send final manuscripts (and when possible floppy disks of text) to editors by January 31, 1988. The several reviews of draft manuscripts allowed the editors to shape sections to maximize the coverage of the particular field to which they were assigned. It led to greater coordination among chapters within each section of the volume. It also produced much more polished presentations at the conference.

Instead of the anticipated 150 to 200 conference participants, over 400 attended. They came from Australia, Canada, Denmark, Finland, France, Germany, Great Britain, Italy, the Netherlands, Sweden, the United States, the U.S.S.R., and the Virgin Islands. The presentations stimulated unusually active discussions, and individual conversations relevant to presentations spilled over to the break sessions in the lobby of the hotel. The group dinners and social hour discussion continued the exchange of views. Ideas for collaborative research and ideas for changes in survey procedures were exchanged. In short, the conference was a success.

During the conference, the committee was assisted by six graduate student fellows, selected from applicants worldwide to receive financial support to attend the conference. They were Teodora Amoloza, University of Nebraska; John Brehm, The University of Michigan; Rachel Caspar, The University of Michigan; Gosta Forsman, University of Lund, Sweden; Rose Krebill-Prather, Washington State University; and Karl Landis, The University of Michigan. The fellows took careful notes during discussions between editors and authors and during floor discussions after the papers were presented. These were entered into microcomputer files within hours and distributed to editors and authors to document ideas useful for revisions of the papers.

The organizing committee believed that several of the contributed papers presented at the conference deserved wider dissemination. After considering several other journals, the committee agreed to collaborate with the *Journal of Official Statistics* in facilitating the publication of the best of the contributed papers at the conference. Lars Lyberg, as editor of the journal, took on that task and arranged for an issue of the journal devoted to telephone survey methodology, to be published in early January 1989 (see *Journal of Official Statistics*, vol. 4, no. 4, 1989).

All the final drafts of chapters were received on the January 31, 1988, date or shortly thereafter. Yvonne Gillies and Gail Arnold put them into a common word processing system. With the assistance of Beatrice Shube at John Wiley & Sons, the book format was chosen and camera ready copy drafts were prepared. The editors met at the end of March 1988 for final editing. Authors were given one more chance to check the copy. A copy editor reviewed the manuscript. Final changes were made, and the book began production.

This volume is divided into sections on coverage of the household population by telephones (edited by James Massey), sample designs (edited by Joseph Waksberg), nonresponse in telephone surveys (edited by Robert Groves), data quality issues (edited by Paul Biemer), computer-assisted telephone interviewing (edited by William Nicholls), and survey administration (edited by Lars Lyberg). Each of the sections begins with an introduction. Sometimes (as in the case of the topics of sampling and nonresponse) this is a full review paper; for other sections a short statement of the general issues in the area, written by the section editor, is presented,

This volume assumes an audience of survey researchers and graduate students with prior training in survey research. It is suitable for a survey researcher contemplating a first use of telephone surveys as a mode of data collection, after experience with some other technique. It also describes the state of the art in the field, and therefore would be useful to researchers already actively using the method. Some chapters of the book address statistical issues and utilize mathematical notation necessary to explicate the topic. Most chapters present quantitative information justifying the conclusions of the authors. This is not a "how to do it" book, but it does contain some descriptions of procedures currently used in different survey organizations. It focuses on the use of the telephone in household surveys, not surveys of businesses or organizations. Finally, it deals with surveys that are designed initially to use the telephone mode. It ignores the large number of ongoing panel surveys which start out as face to face surveys and then, over time, begin to use telephone interviewing as a followup mode. The last section of the volume is a comprehensive listing of the literature in telephone survey methodology. In addition, this bibliography contains all references made in individual chapters in the book, whether or not they are relevant to telephone surveys.

This book would not have been written without the financial support of the organizations listed and the moral support of the three sponsoring professional organizations. The financial contributors deserve special thanks. Their gifts were made with no *quid pro quo* conditions and with honest concern for excellence in research methods. The royalties from this volume will be shared by the professional associations. Our employing organizations also deserve great appreciation for supporting our activities in assembling the book: the Survey Research Center at The University of Michigan, New Mexico State University, Statistics Sweden, the National Center for Health Statistics, the U.S. Bureau of the Census, and Westat, Inc.

CONTENTS

SECTION F ADMINISTRATION OF TELEPHONE SURVEYS

TELEPHONE SURVEY METHODOLOGY

SECTION A

COVERAGE OF THE HOUSEHOLD POPULATION BY TELEPHONES

CHAPTER 1

AN OVERVIEW OF TELEPHONE COVERAGE

James T. Massey
National Center for Health Statistics

It has been more than fifty years since a *Literary Digest* survey based on telephone directory listings predicted a landslide victory for Landon over Roosevelt in the 1936 U.S. presidential election. Noncoverage of the population by telephone was cited as an important reason for the erroneous prediction (Katz and Cantril, 1937). In 1936, it was estimated that 35 percent of the households in the United States had telephones. It took nearly forty years for telephone surveys to begin to receive acceptance among U.S. social scientists as a legitimate sole source data collection method. Two of the reasons for the renewed interest in telephone surveys during the past 15 years is the lower cost of data collection by telephone compared to face to face interviews and the increased coverage of the population by telephone. By 1960, approximately 75 percent of U.S. households had telephones (U.S. Bureau of the Census, 1965), and the coverage rate increased to 93 percent by 1986 (Chapter 3). The telephone coverage rate in a number of other countries has also increased dramatically over the past several decades, which has led to greater worldwide interest in telephone surveys.

Even with the increases in the number of households with telephones, the effect on survey estimates of excluding parts of the population in telephone surveys remains an important concern among survey researchers. This chapter summarizes some of the key methodological issues related to telephone coverage.

1. DEFINITIONS

There are several terms used in the literature to define the number of households with and without telephones. Terms such as "undercoverage," "noncoverage," "telephone access," "telephone penetration," "telephone ownership," and "telephone subscribers" have all been used to refer to the number of households with and without telephones. For this volume we have adopted the following set of terminology. The terms "telephone households" and "nontelephone households" are used to identify households with and without telephones, respectively. The term "noncoverage" refers to the households missing from a telephone survey sampling frame. This includes both nontelephone households and telephone households which are omitted from the telephone sampling frame. "Coverage rate" is defined as the percent of *all* households included in the telephone sampling frame. It is also used to refer to the percent of all households with telephones when the general characteristics of the population are being discussed. The term "undercoverage" is defined as the number of persons in telephone households who are not enumerated in sample households in a telephone survey. "Telephone access" refers to the households that can be reached by telephone, including those that do not have a telephone in their dwelling unit. "Noncoverage bias" is defined as the difference between the expected value of a given characteristic for the covered telephone survey population and the expected value of the characteristic for the total household population. The value of the characteristic for the total population is often estimated by the expected value of the characteristic in the total population.

2. COMPLETENESS OF SAMPLING FRAME

The completeness of the sampling frame is an important sample design consideration regardless of whether a survey is conducted by telephone, mail, or face to face. For large scale national face to face and mail surveys samples are often selected from geographical area frames or directory listings of household addresses. The completeness of these sampling frames depends upon both the accuracy of the frame at time of construction and the maintenance of the sampling frame over time. The sampling frames for telephone surveys often exhibit the same types of coverage errors as the sampling frames for other modes of data collection. In addition, the telephone survey sampling frames have several unique coverage problems.

For telephone surveys, an important coverage issue that applies to all sampling frames is the noncoverage of nontelephone households. Chapters 2 and 3 in this section are devoted primarily to this topic. In Chapter 2, Trewin and Lee for the first time present information on telephone coverage for countries throughout the world. Estimates of the percentage of households with telephones are shown for 22 countries along with the known characteristics of the noncovered subgroups of the population. There is a great deal of variability in the percentage of nontelephone households across countries. While the low coverage rates for some countries make a telephone survey of the population technically infeasible, there are now a number of countries where telephone surveys cannot be rejected based only on the percent of households without telephones. In Chapter 3, Thornberry and Massey analyze trends in telephone coverage in the United States across time and subgroups of the population.

In Chapter 5 of this volume, Lepkowski presents a description of three types of sampling frames commonly used for telephone surveys and discusses the sampling methods associated with them. The first sampling frame consists of a list of all possible telephone numbers and is the frame usually associated with random digit dialing (RDD). If this type of frame is complete and up to date, nontelephone households represent the only noncovered segment of the population. Omitted exchanges and out of date frames produce additional noncoverage. Another commonly used sampling frame for telephone surveys is a directory frame based on listings from telephone directories. This type of frame is generally used for small-area surveys because it also contains address information. The noncoverage of households for this type of frame can vary significantly over areas and is often unacceptable in large urban areas in the United States where the number of unlisted numbers may be greater than the number of listed numbers. A third sampling frame described by Lepkowski consists of commercially maintained directory frames. These frames are sometimes constructed from combinations of list frames such as telephone directories, city directories, and automobile registration files. These frames generally contain information which can be used to construct restricted sampling frames for subsegments of the population. The advantages and disadvantages of all of the telephone sampling frames with respect to coverage and other issues are discussed in Chapter 5.

A variation of the sampling frame of all possible telephone numbers used in RDD surveys is worth discussing with respect to coverage of telephone households. Because of the inefficiency of selecting a simple random sample of all telephone numbers, several methods have been developed to improve sampling efficiency (see Chapter 5, this volume). One method that has been used by several commercial companies is the elimination of nonworking numbers or blocks of nonworking numbers from the sampling frame. While this method improves the sampling efficiency of the frame, the likelihood of missed telephone households also increases.

3. WITHIN-HOUSEHOLD COVERAGE

Another coverage issue that is common to all household surveys of the population is the failure to enumerate persons within sample households. In face to face sample surveys conducted by the U.S. Bureau of the Census the underenumeration of persons within households is usually between 5 and 10 percent. The underenumeration varies considerably across subdomains and is sometimes as high as 50 percent for some subdomains. For telephone surveys the possibility of an even greater amount of underenumeration within telephone households exists because of the lack of visual cues and the inability in some cases to accurately link a telephone with a specific dwelling unit. In face to face surveys the interviewer can often observe separate living quarters and make finer distinctions in household membership. In Chapter 4, Maklan and Waksberg address the issue of how well RDD surveys enumerate persons within households relative to face to face surveys.

4. DIFFERENTIALS IN COVERAGE

Although the overall telephone coverage has increased dramatically throughout the world in the past 25 years, the telephone coverage for most countries is not uniformly distributed over the population. Because surveys often focus on subgroups of the population or make separate estimates for subgroups, noncoverage of the social, demographic, and economic subdomains is important for designing and analyzing the results from telephone surveys. In Chapters 2 and 3 the variation in telephone coverage across subdomains is presented along with trends among the subdomain coverage rates over time. In most countries the single most important explanatory variable for differences in telephone coverage is household income. Other variables, such as employment status, education, marital status, and race, correlated with telephone coverage, are also correlated with income.

5. NONCOVERAGE BIAS

Noncoverage bias is a function of the magnitude of the noncoverage of a telephone survey frame and the difference in characteristics between the covered and noncovered populations. While the potential for a large noncoverage bias decreases as the percentage of telephone households increases, large differences between the telephone and nontelephone household populations can result in significant noncoverage biases.

Surveys whose main variables of interest are related to income, occupation, education, or other variables correlated with telephone coverage are very likely to have significant noncoverage biases. In Chapter 2, Trewin and Lee summarize the characteristics of the nontelephone households for 22 countries and describe a number of noncoverage bias studies that have been conducted throughout the world. In Chapter 3, Thornberry and Massey estimate the noncoverage bias for a number of health characteristics by comparing the health statistics for telephone households with the health statistics for all households. These estimates are based on information collected in a national face to face survey.

Several methods have been described in the telephone survey literature to correct or adjust for the noncoverage bias associated with telephone surveys. The use of dual frame designs is one approach to compensate for noncoverage. Lepkowski (this volume, Chapter 5) discusses the use of dual RDD and directory telephone sampling frames to compensate for the noncoverage of the telephone household population by the directory sampling frames. Dual frame, mixed mode surveys have been proposed as a solution for the noncoverage of households without telephones. In this type of design a face to face household sampling frame is combined with a telephone sampling frame. Sirken and Casady (this volume, Chapter 11) examine the effect of response rates on optimal allocation in the face to face and the telephone dual frame design. Additional research on dual frame, mixed mode surveys has been done by Biemer (1983), Casady et al. (1981), Groves and Lepkowski (1985), Lepkowski and Groves (1984), and Whitmore et al. (1983).

Another approach used to decrease the effects of the noncoverage is post survey weighting adjustments. Massey and Botman (this volume, Chapter 9) discuss several post survey weighting adjustments for RDD surveys and evaluate the impact of these adjustments on the survey estimates. Thornberry and Massey (1978), Banks (1983), and Banks and Anderson (1982), each made post survey weighting adjustments to the telephone household population in a face to face survey to evaluate the potential effects of the adjustments. The adjustments were generally consistent with the direction of the noncoverage bias, although they appeared to only partially adjust for the noncoverage bias.

6. SUMMARY

During the past two decades dramatic increases have occurred in the number of households with telephones. The increased coverage is one reason for an increased interest in telephone surveys. This has led to the

development of improved sampling frames and sampling methods to help minimize the noncoverage bias due to the exclusion of nontelephone households. Even with these advances there is still a need to be aware of the potential for noncoverage biases, especially among the subdomains of the population with lower coverage rates.

CHAPTER 2

INTERNATIONAL COMPARISONS OF TELEPHONE COVERAGE

Dennis Trewin and Geoff Lee[1]
Australian Bureau of Statistics

Although commercial survey researchers embraced the telephone methodology many years ago in a number of countries, government and academic researchers delayed its use, primarily because of concerns about coverage of the household population by telephones. If the population of households without telephones is large *and* distinctive in its characteristics, telephone surveys can provide misleading indicators. As the costs of face to face interviewing have risen around the world, pressures to investigate the coverage error of telephone surveys have increased.

Every country in the world has a unique story about the development of its national telephone system. Some are stories of the independent development of individual companies, each serving a small number of

---

[1] The data on coverage rates in this chapter were provided by statisticians from many countries who replied to our letters. We are very grateful for their contributions: Australian Bureau of Statistics; Dr. Klein, Österreichisches Statistisches Zentralamt, Austria; G.J. Brackstone, Statistics Canada; Catlin et al. (1984), Statistics Canada; Drew and Jaworski (1987), Statistics Canada; Eszter Kormendi, Danish Institute of Social Research; Pentti Pietila, Central Statistical Office of Finland; G. Theodore, Institut National de la Statistique et des Etudes Economiques, France; Jürgen Friedrichs, University of Hamburg; Elias Spanoudakis, National Statistical Service of Greece; Richard Butler, Hong Kong Census and Statistics Department; Vera Nyitrai, Hungarian Central Statistical Office; Vincent Healy, Telecom Eireann, Ireland; Malka Kantorowitz, Israel Central Bureau of Statistics; Guido Rey, Istituto Centrale di Statistica, Italy; Victor Manuel Navarrete Ruiz, Instituto Nacional de Estadistica, Mexico; van Bastelaer and Leenders (1985), Netherlands; Paul Brown, New Zealand Department of Statistics; Tove L. Mordal, Norwegian Central Bureau of Statistics; Jan Kordos, Central Statistical Office, Poland; Carmen Arribas, Instituto Nacional de Estadistica, Spain; Hans Nasholm, Statistics Sweden; Wendy Sykes and Martin Collins, Social and Community Planning Research Survey Methods Centre, United Kingdom; Kevin J. McCormick, United States Bureau of the Census; and Owen Thornberry and James Massey, United States National Center for Health Statistics.

communities. Others are stories of government sponsored agencies devoted to providing service to the entire country. Some small countries have relatively simple problems of installation. Countries with large sparsely settled and inhospitable terrain face enormous obstacles to providing universal telephone service. Further, some countries have adopted charging rates which make local calls free (after a monthly service charge). Others charge each call, proportionate to the length of the call. Similarly, long distance calls are charged with different procedures. All of these differences across countries affect the rate at which the household population is covered by telephones. The main purpose of this chapter is to make international comparisons of telephone coverage and the characteristics of nontelephone households.

1. COMPARISONS OF TELEPHONE COVERAGE RATES

Figure 2.1 provides a comparison of the telephone noncoverage rates for households for 22 countries for which information was obtained. The coverage rate varies considerably among the countries, although not always directly in proportion to per capita income. Sweden has the highest coverage rate (99 percent of all households), and Hungary, Mexico, and Poland have the smallest (16 percent) among the countries represented in the study. Clearly, if other countries had been included in the study, even lower rates of coverage would have been discovered. Figure 2.1 shows that half of the 22 countries have rates above 90 percent. Then there is a cluster of eight countries with rates between 40 and 80 percent. Finally, there are three countries with rates less than 20 percent. There are often local factors that can result in coverage rates being higher or lower than expected. For example, the rate in Canada is very high — the costs of telephoning are relatively low and, in particular, local calls are free. Local calls are also free in New Zealand. In the United Kingdom, the coverage rate is lower than expected. "For a number of reasons Britain has lagged behind other developed countries in telephone penetration. Smaller distances, less population mobility, an excellent postal service, effective rationing through underinvestment in exchange equipment and a failure — by government and people — to see a social advantage in telephone ownership all contributed to slow growth" (Collins and Sykes, 1987). This situation is changing and telephone ownership in the United Kingdom is growing at a fast rate. In short, the causes of differential coverage may be both economic (costs of telephone service) and social (perceived lack of need for telephone communication).

Unfortunately, some caution must be exercised in reading Figure 2.1. First, the information pertains to different time periods across countries.

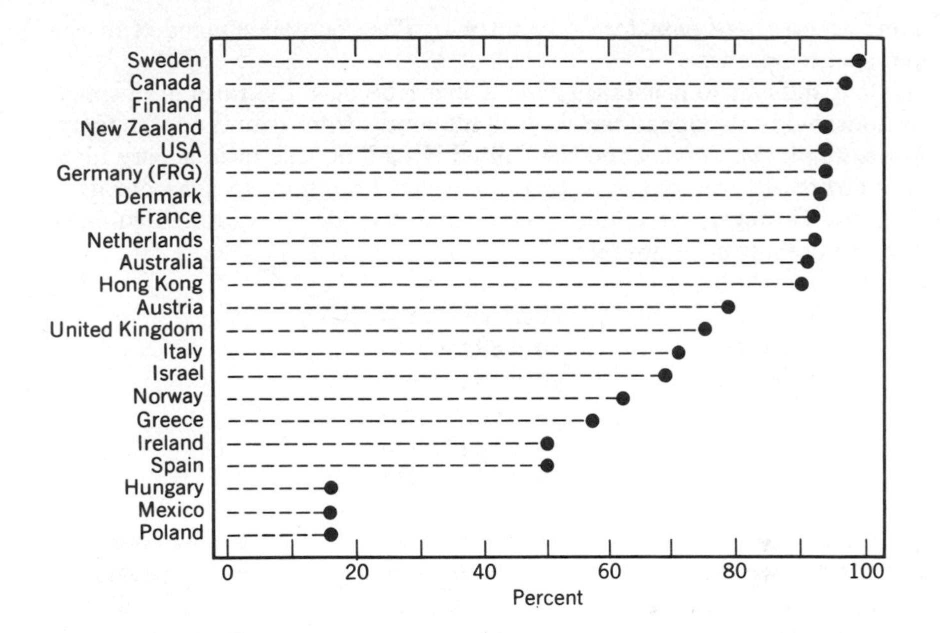

Figure 2.1 Telephone Coverage Rates for Households.

This was an inevitable result of different countries' schedules of collecting and publishing such information. For example, the data from Norway are seven years out of date and telephone coverage has grown considerably in recent years. This is one possible explanation for the surprisingly low figure for Norway. Second, the precise definition of a telephone household may differ slightly across countries. We believe that the definition used most often concerns the existence of a telephone instrument within the housing unit itself. However, some households may be labelled as telephone households if a telephone is available to them (i.e., they receive telephone calls on the instrument routinely), even if it may not be physically located within the home.

The data in Figure 2.1 came from special census or survey based studies of telephone coverage. They are estimates of the percentage of households with telephones. Another source of data is relevant to our interest in telephone coverage because it provides information on additional countries beyond the 22 described above. The United Nations (1986) publishes statistics on the number of telephones per 100 inhabitants. These statistics include public and private telephones and those used by businesses. Hence, the information can only be used to give a general guide to telephone coverage rates of households. Table 1 summarizes these data for 64 countries. The countries included in the survey are in italics.

It is difficult to generalize from Table 1 because the ratio of business to household telephones can vary significantly from country to country. For example, in Norway and the United Kingdom, this ratio is very high. In addition, differences in average household sizes make generalization difficult. However, the following categories provide a very rough guide to likely household coverage rates.

Class I	90 percent or greater
Class II	80 – 89 percent
Class III	70 – 79 percent
Class IV	50 – 69 percent
Class V	25 – 49 percent
Class VI	10 – 24 percent

Clearly, telephone surveys are going to be of most interest to countries in the first two classes. There are 21 countries in these two classes. For countries in the third and fourth classes, telephone surveys may still be of interest, either as part of a dual frame, mixed mode approach (see this volume, Chapter 11) or as a method of data collection after face to face contact. In some of these countries, the coverage rate is quite high in urban areas or large cities, making telephone interviewing a viable alternative for at least these areas.

The growth in telephone coverage of the countries in our survey may be of particular interest. Figure 2.2 shows the increase in the number of telephones per 100 inhabitants for the period 1975 (the left-hand side of each bar in the graph) through 1984 (the right-hand side of each bar). The countries are shown in the graph in descending order of coverage increase (absolute increase). Telephone coverage has been increasing in all countries and rapidly in some.

Telephone interviewing may become feasible for more countries in the future. Countries where the number of telephones per inhabitant has more doubled over the 1975–1984 period include Bahrain, Bulgaria, Cyprus, France, Malta, Qatar, Saudi Arabia, Singapore, and Yugoslavia. Other countries of rapid growth include Austria, Denmark, Germany (Federal Republic), Hong Kong, Norway, and United Kingdom.

Table 1. Classification of 64 Countries by Telephones per 100 Persons and Household Coverage by Telephones

No. of Telephones Per 100 Persons	Household Coverage Class	Countries[a]
More than 60	I (90%+)	Bermuda, *Canada*, Cayman Islands, *Denmark*, *France*, Liechtenstein, Monaco, *New Zealand*, *Norway*, Saint Pierre, *Sweden*, Switzerland, *USA*.
50 – 60	II (80%–89%)	*Australia*, *Finland*, *Germany (Federal Republic)*, Iceland, Japan, *Netherlands*, San Marino, *United Kingdom*.
40 – 50	III (70%–79%)	*Austria*, Bahamas, Belgium, *Hong Kong*, *Italy*, Luxembourg, Virgin Islands.
30 – 40	IV (50%–69%)	Barbados, French Guiana, *Greece*, *Israel*, Malta, Qatar, Singapore, *Spain*.
20 – 30	V (25%–49%)	Bahrain, Bulgaria, Cyprus, Czechoslovakia, Germany (Democratic Republic), *Ireland*, Martinique, *Mexico*, Netherlands Antilles, New Caledonia, Puerto Rico, United Arab Emirates.
10 – 20	VI (10%–24%)	Argentina, Costa Rica, French Polynesia, *Hungary*, Korea, Kuwait, Panama, *Poland*, Portugal, Reunion, Saudi Arabia, Seychelles, South Africa, USSR, Uruguay, Yugoslavia.

[a]Italicized names are countries for which household coverage estimates appear in Table 2.

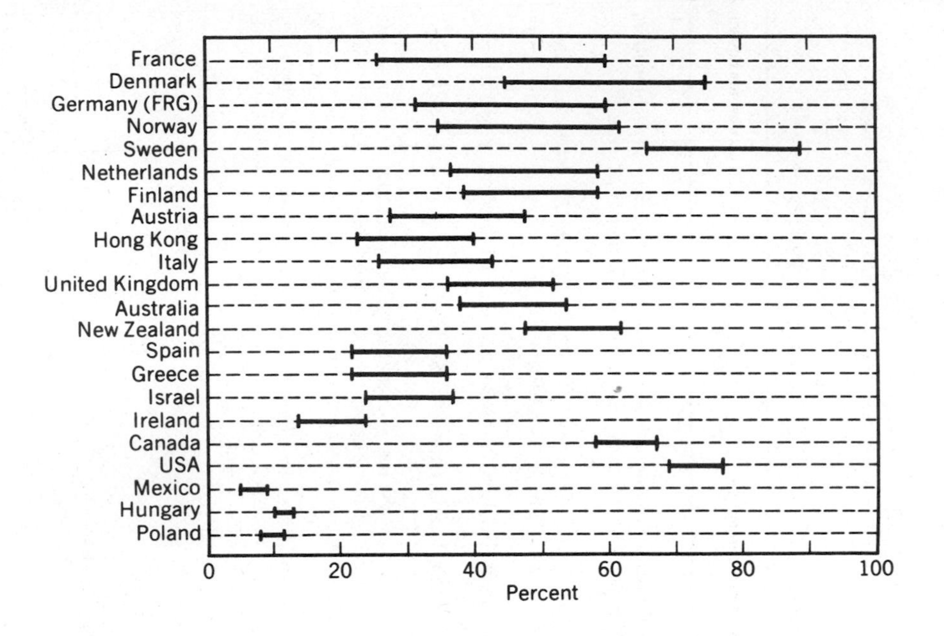

Figure 2.2 Changes in number of telephones per 100 persons, 1975–1984, ordered according to magnitude of absolute increase in numbers of telephones.

2. CHARACTERISTICS OF NONTELEPHONE HOUSEHOLDS

Table 2 contains information on the 22 countries in our survey about characteristics of households without telephones. Although the type of information available varied, some commentary on geographical differences in coverage and socioeconomic characteristics of persons without telephones in their homes are presented for most of the countries. Table 2 shows a remarkable number of common elements among the characteristics of nontelephone households.

Geographic Characteristics. The noncoverage usually varied considerably across regions, generally being lowest for the most remote regions. In most countries noncoverage was highest for rural localities and lowest for large cities. Clearly, the cost of installing and maintaining switching systems and wiring in remote areas retards the introduction of telephone service. However, in some of the Scandinavian countries, the opposite appeared to be the case. Coverage was slightly higher outside the major cities.

Size of Households. Noncoverage was almost universally higher for single person households and the very largest households (e.g., more than

Table 2. Household Telephone Coverage Rate, Year of Estimate, and Characteristics of Nontelephone Households for 22 Countries

Country	Coverage Rate	Year	Geographic Characteristics of Nontelephone Households	Other Characteristics of Nontelephone Households
Australia	91	1986	At the state level, noncoverage varies from 6 percent to 12 percent. It is about 7 percentage points higher outside the capital cities.	Noncoverage is higher for rented houses, for households where the head is unemployed, young or with low income, and for single person households.
Austria	79	1984	There are no major regional differences in noncoverage; Vienna and other towns have lower noncoverage than rural areas.	Noncoverage is very high (46 percent) for unskilled workers, slightly above average for pensioners and skilled workers, and low for the self-employed (7 percent).
Canada	97	1986	At the province level, noncoverage ranges from 2.0 percent to 6.0 percent. Noncoverage is slightly higher outside the major urban areas.	Noncoverage is higher for rented houses, single person households, low income households, and those with unemployed persons.
Denmark	93	1986	No great difference between regions but noncoverage is lower for rural communities.	Noncoverage decreases with age of head of household. It is significantly higher for unmarried and divorced people.
Finland	94	1986	Noncoverage ranges from 12 percent to a low of 3 percent in Helsinki.	Noncoverage is higher for manual workers, pensioners, and those with low income.
France	92	1985	No great difference between regions but noncoverage is slightly higher for rural communities.	Noncoverage is higher for agriculture workers, where the age of the head of household is low, and low income households. Average household size is smaller.
Germany (Federal Republic)	94	1985	Noncoverage is higher for rural areas.	Noncoverage is higher for pensioners, those receiving public assistance, and those with low income.
Greece	57	1981/82	Noncoverage varies from 70 percent in rural areas to 30 percent in urban areas.	Varies from 78 percent in lowest of 8 income groups to 20 percent in highest income group.
Hong Kong	90	1987	Not relevant.	Highest for low income groups.

Table 2 (Continued).

Country	Coverage Rate	Year	Geographic Characteristics of Nontelephone Households	Other Characteristics of Nontelephone Households
Hungary	16	1985	Varies considerably between urban and rural areas with very low coverage in rural areas.	Noncoverage is considerably lower for households with non-manual workers (current and retired). It is also marginally lower for economically inactive households. Noncoverage decreases with income and age of head of household.
Ireland	50	1987	Noncoverage is higher in the more remote regions on the Western Seaboard.	Not known.
Israel	69	1983	Noncoverage varies considerably by region from 16 percent to 93 percent. It is 17 percent in large cities but 47 percent in rural localities.	Noncoverage is high for very small and very large households; it declines as the age of household increases; it is higher for low income groups and unemployed persons.
Italy	71	1986	Noncoverage rates vary from 15 percent to 49 percent depending on urbanization of regions.	Not known.
Mexico	16	1980	Noncoverage rates vary from 99 percent in rural localities to 72 percent in large cities.	Not available.
Netherlands	92	1984	Noncoverage rates vary from 5 percent to 19 percent across regions.	Noncoverage was highest for households with a young head, single person and large households, unemployed and students, low income earners, and skilled/unskilled labourers.
New Zealand	94	1985/86	Not available.	Not available.

Table 2 (Continued).

Country	Coverage Rate	Year	Geographic Characteristics of Nontelephone Households	Other Characteristics of Nontelephone Households
Norway	62[a]	1980	Not a great deal of variation across regions. Little difference between urban and rural areas.	Average household size is smaller. Noncoverage is highest for the low income groups except for those with little or no earned income.
Poland	16	1986	Noncoverage rates vary from 96 percent in rural areas to 68 percent in large cities.	Not known.
Spain	50	1980/81	Noncoverage rates vary across regions from 28 percent to 82 percent. Noncoverage is 16 percent in large cities but 75 percent in rural localities.	Noncoverage is much higher for small households. It also increases as the age of the dwelling increases.
Sweden	99	1987	No details provided.	Noncontact is more of a problem than noncoverage. Noncontact is highest for males, younger people, single people, and low income groups.
United Kingdom	75	1986	No details provided.	Noncoverage is high for very old and very young household heads, single person households, rented accommodation, low income earners, and unemployed persons.
USA	94	1986	Noncoverage varies from 4 to 21 percent across states. Noncoverage is higher in rural areas and inner cities.	Noncoverage is high for households with unemployed persons, blacks and hispanics, single person and large households, low income groups, and households with young heads.

[a]Now believed to be considerably understated.

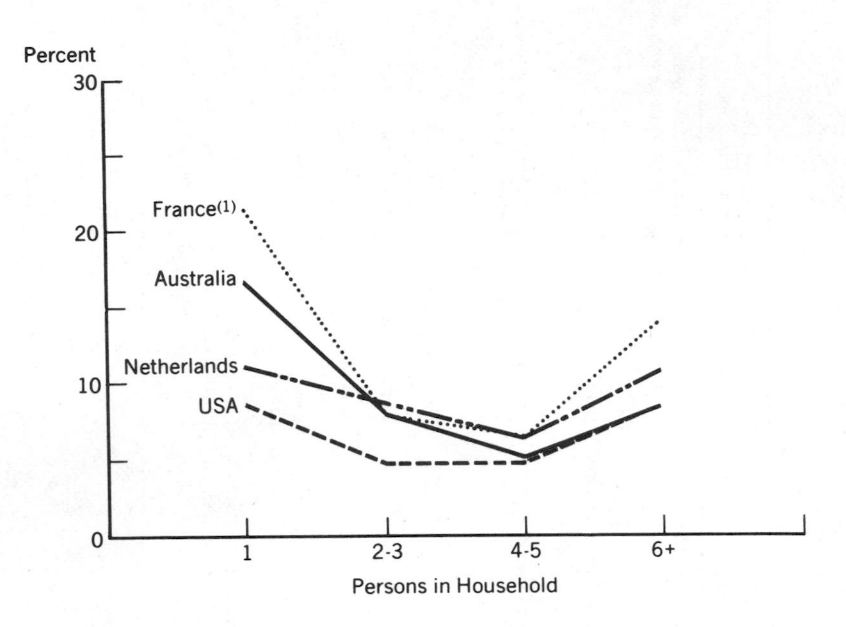

Figure 2.3 Noncoverage by household size. 1 = Different reference period than in Table 1.

six persons). As shown in Figure 2.3 for four countries which provided the relevant data, noncoverage was lower in terms of households than in persons. That is, on average, there were more persons in telephone households than nontelephone households.

Income. Not surprisingly, the average income of nontelephone households was considerably lower than that of telephone households. For seven countries, Figure 2.4 shows the differences between income groups, with the left-hand side of each bar showing noncoverage in the highest income group, and the right-hand side noncoverage for the lowest income group. The largest coverage differences by income appear in Greece and Israel. The Netherlands exhibit little difference across income groups.

Employment Status. Table 3 presents telephone noncoverage rates by employment status for four countries. Noncoverage was universally higher for unemployed heads of households and unemployed persons in general. The situation was mixed when the head of household was not in the labour force (e.g., retired). In some countries, noncoverage was slightly lower for such households.

Canada provided unemployment and labour force participation rates separately for telephone and nontelephone households. The

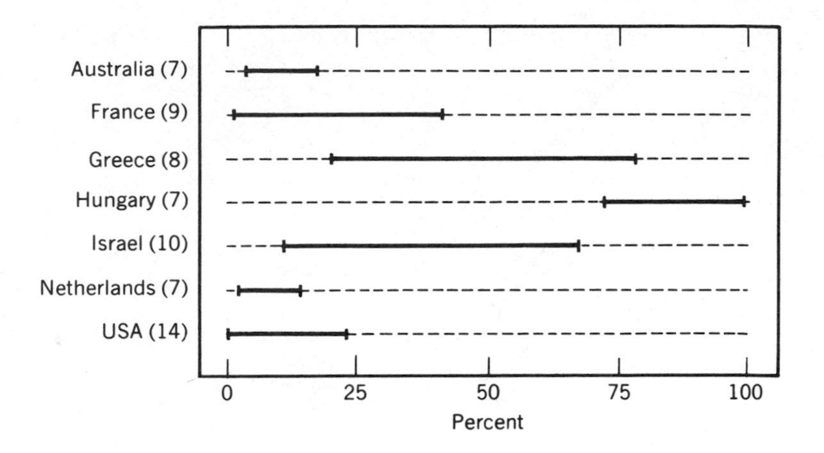

Figure 2.4 Noncoverage by income groups. Number of income groups shown in parentheses.

Table 3. Noncoverage by Employment Status of Head of Household

Country	Employed	Unemployed	Not in Labour Force	Retired
Australia	6.5	27.2	11.3[a]	b
Austria	19	32	b	25
Netherlands	7.0	19.0	b	7.5
USA	3.9	14.0	6.1[a]	b

[a] Retired are included in this group.
[b] Not available separately.

unemployment rate is 12.0 percent for telephone households compared with 32.7 percent for nontelephone households. The participation rate is 62.9 percent compared with 48.3 percent for nontelephone households.

Age of Household Head. As shown in Figure 2.5, the pattern in most countries was for noncoverage to decline as the age of the household head increased but with an increase in most, but not all, countries for

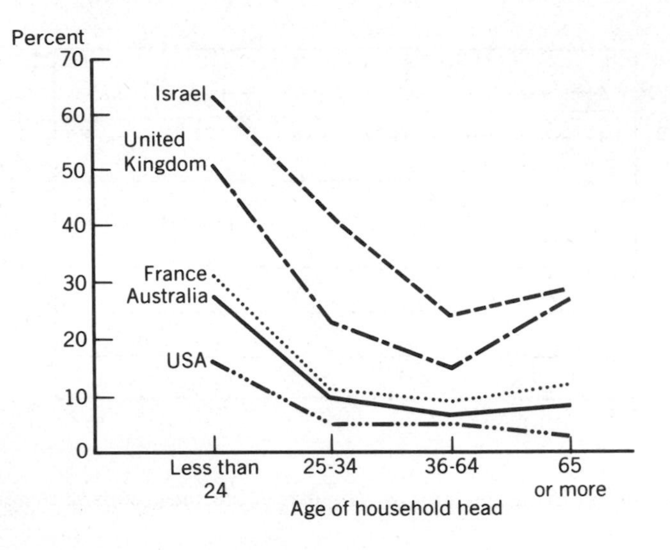

Figure 2.5 Noncoverage by age of household head.

household heads aged 65 or more. The notable exception to this is the United States, where the telephone coverage for persons over 65 years of age is lower than that for other age groups (see Chapter 3).

Type of Residence. Some countries provided details of noncoverage rates for different types of residence. The consistent pattern was for noncoverage to be significantly higher for rented accommodation than for owner-occupied accommodation. Noncoverage also increases with the age of the dwelling.

Marital Status. Noncoverage is higher for single and divorced persons than for other categories.

Occupation. Noncoverage is higher for households where the head is nonskilled.

Race and Ethnic Origin. In the United States, the noncoverage rates varied considerably by race and ethnic origin. It was 6 percent for white persons, 16 percent for black persons, and 19 percent for hispanic persons.

Almost all of the correlates of telephone ownership seem to be related either to the cost of providing telephone service to a housing unit or to the relative importance of the charges to the occupants.

The cost per telephone installed in remote areas is much higher than in cities and hence noncoverage tends to be greater there. The percentage of a household budget spent on telephone service is higher among lower income groups, the unemployed, and those with few employed members, and hence these categories cannot afford telephone service. The disproportionate tendency for the elderly to have telephones, despite their lowered economic status, may result from the value of immediate contact with relatives and others who could provide assistance to them when in need.

3. STUDIES OF BIAS

In many countries it is clear that most surveys would suffer from significant bias if only the telephone population was covered. Some countries have undertaken studies of this potential bias and the results of these studies are worth mentioning.

Australia. Jones (1982) used data from the Household Expenditure Surveys undertaken by the Australian Bureau of Statistics (ABS) during 1974–1975 and 1975–1976 to search for demographic characteristics of households that would identify subgroups of the population which have a similar degree of telephone access. The Automatic Interaction Detector (AID) technique was used and found some groups with very low telephone access (e.g., manual employee, aged under 30, living in rented accommodation), and other groups with very high telephone access (e.g., professional, managerial, or white collar worker with high income, living in own home). Jones conjectured that "poststratification" of the results of a telephone survey by the variables that were proved to be significant under the AID analysis would lead to a significant reduction in bias.

Steel and Boal (1987) have extended this analysis using more recent data from the 1984 Household Expenditure Survey, and the March 1986 Monthly Labour Force Survey undertaken by the ABS. They confirmed that noncoverage can be as high as 30 percent or more for particular subgroups of the population, even when the national average is around 9 percent. "A picture emerges of the penetration rate being low for young people, people not living in a family situation, especially those living alone, and for low income, unemployed persons." Steel and Boal considered the relative biases that would occur if a survey was restricted to the population that was accessible by telephone, and found that they ranged from −12.2 percent for the unemployment rate to +5.6 percent for the percentage of households where the income of the household head exceeded A$500 per week. Steel and Boal were not confident that poststratification by one or two variables could be used to correct for these biases, because of the variety of variables that were needed to identify specific subgroups of the population with low telephone access.

Canada. "The number of households that cannot be reached in a telephone survey is small. For many surveys of the general population excluding this group would have a negligible effect on the estimates of most household or personal characteristics. This result is very encouraging for increasing the use of telephone sampling methods as a way of reducing the cost of special surveys. The implication for ongoing surveys such as the Labour Force Survey, where very precise estimates at the national level are required, is less sure." (Catlin et al., 1984). In fact, a downward bias of three-tenths to four-tenths of a percentage point in the national estimate of unemployment would result if nontelephone households were excluded.

Also, some specific subpopulations would not be adequately covered by telephone surveys. Statistics Canada cite groups such as young males or those with education below the postsecondary level as examples of those that would not be well covered by telephone surveys.

The Netherlands. Van Bastelaer and Leenders (1985) have undertaken an extensive study of the bias that would result from a restricted sample of telephone owners. They also look at the extent to which this bias can be reduced by special weighting procedures. They conclude:

> The restriction of the observed population to the subpopulation with telephone access — in a telephone interview — can provide biased results. The bias . . . arises because the telephone access is associated with some background variables. The results of the Consumer Survey confirm the association of telephone access with age, marital status, activity status, occupation, education and income of the head of household, size of the household and province. The sex of the head of household is not associated with telephone access. The conclusion that unbiased estimates for the total target population are impossible to obtain on the basis of a restricted sample of telephone owners due to those associations, is premature.

They have looked at using variables such as province, age, marital status, and sex as weighting variables. Their study confirms a significant reduction in the bias. They have concluded that the bias is nonsignificant for the consumer confidence variables being examined, but the bias is still significant when the variables are measuring facts (e.g., car ownership).

United Kingdom. Collins and Sykes (1987) have concluded that in the United Kingdom (which has a relatively low telephone coverage rate) it will rarely be acceptable for social surveys or research aimed at public policy to be based upon the 80 percent or so of the population who have a telephone. Many of these surveys will be focused on the groups where the coverage is lowest.

United States. The bias due to noncoverage has been studied for a number of variables. Thornberry and Massey (1978) used data from the 1976 National Health Survey. They observed that there were some sociodemographic differences between the nontelephone population and the general household population. On health related variables the populations were very different. Poststratification of the telephone household estimates by income-region-color and by age-sex-color were examined and they found that ratio estimation did not always produce better estimates. The poststratification adjustments were, however, generally in the right direction and the income-region-color adjustment performed better than the age-sex-color adjustment.

Banks (1983) came to very similar conclusions using data from a 1976 survey conducted for the National Center for Health Administration Studies, and noted that the "phone coverage is not as good an indicator of nontelephone bias as commonly supposed."

A report from the U.S. Office of Management and Budget (1984) found that the bias can be significant because of the special characteristics of the population without telephones. Nonownership of the telephone is more common among various 'deprived' subgroups: blacks, lower income households, those who make above average use of medical services, or those who live in areas of high crime victimization .

In contrast, Freeman et al. (1982) found that after age-sex-county poststratification, the distributions of variables from the 1976 California Disability Survey (a random digit dialing survey) were similar to equivalent variables from the 1978 Survey of Income and Education. Although there were small but meaningful differences in disability rates for some subgroups (young persons, black persons, some levels of education) that persisted after weighting, they felt that these differences should be attributed to biases caused by nonresponse and undercoverage of one and two person households, rather than to noncoverage.

Buss (1979) in a survey of high school dropouts, and Wiseman and McDonald (1979) in an analysis of several consumer telephone surveys, found that the biases due to unanswered numbers and nonresponse overshadowed their problems with noncoverage.

4. CONCLUSION

This study is a report on telephone coverage for several countries throughout the world. It identifies similarities in correlates of telephone ownership across several of the countries, many of them related to the socioeconomic status of the household and the relative cost of delivering telephone service to the household. There are clearly, however, larger issues of societal differences in patterns of communication, differences in spatial distances among relatives and friends, and government policy affecting telephone equipment installation. The latter no doubt explains rapid gains in telephone coverage that several countries have experienced in the last few years.

In speculating about coverage error in telephone surveys, it appears likely that as the noncoverage rate declines in countries, those remaining without telephones are increasingly distinctive. That is, the differences in characteristics of those covered and those not covered are larger in high coverage countries than in low coverage countries. Since coverage error is a function of both the rate and the relative characteristics of the covered and noncovered, it is unclear whether large reductions in coverage error are associated with higher telephone coverage rates. Therefore, individual studies in each country are required to estimate coverage bias in telephone surveys associated with the nontelephone population.

CHAPTER 3

TRENDS IN UNITED STATES TELEPHONE COVERAGE ACROSS TIME AND SUBGROUPS

Owen T. Thornberry, Jr. and James T. Massey[1]
National Center for Health Statistics

During the past 25 years significant changes have occurred in the proportion of households in the United States with telephones. The trend toward nearly complete telephone coverage of households has made the telephone interview with random digit dialing (RDD) sampling an attractive and acceptable methodology for conducting surveys. Cost is low relative to the area sample/face to face interview, and data of acceptable quality can be obtained. However, there are potential sources of error that should not be ignored including bias associated with the failure to interview a representative sample of the population. While a major historical concern about the use of telephone surveys has been the potential of noncoverage bias, this source of error is now often ignored. Indeed, as Groves (1987) has suggested, "coverage error is the forgotten child among the family of errors to which surveys are subject."

This paper addresses issues related to telephone coverage of the United States population and potential bias from the exclusion of nontelephone households. There are four general areas to the investigation. First, trends over time in telephone coverage in the United States are examined. Second, the major correlates of telephone coverage and the population subdomains most affected by noncoverage are identified. Next, the issue of noncoverage bias is addressed by an assessment of estimates of health characteristics for persons in telephone

[1] The authors gratefully acknowledge the contributions of others at the National Center for Health Statistics: Vance Hudgins and Anthony Thomas for processing the multiple years of data; Cecelia Snowden and Van Parsons for conducting the SEARCH analysis and calculating the design effects for a number of statistics; and Evelyn Michaliga and Jeanenne Barry for the preparation of the manuscript.

and nontelephone households. The paper concludes with a brief discussion of the potential combined effects of noncoverage bias and nonresponse bias.

1. SOURCE OF DATA

This analysis of telephone coverage in the United States is based on data collected in the National Health Interview Survey (NHIS) from 1963 through 1986 and is an updating and expansion of previous papers by the authors (Thornberry and Massey, 1978, 1983). The NHIS, conducted by the National Center for Health Statistics (NCHS), is a continuous survey of the civilian noninstitutionalized population living in the United States. Its purpose is to provide national data on the incidence of illness and injury, the prevalence of diseases and impairments, the extent of disability, the utilization of health services, and other health related topics.

Interviews are conducted each week throughout the year in a probability sample of households. The sampling plan for the survey follows a multistage probability design that permits a continuous sampling of households. The sample is designed in such a way that the sample of households interviewed each week is representative of the target population, and that weekly samples are additive over time. The data are collected through a personal household interview conducted by Bureau of the Census interviewers. The usual annual NHIS sample consists of approximately 40,000 eligible occupied households containing around 110,000 individuals. The completion rate for the eligible households averages between 96 and 97 percent.

Beginning in 1963, each household in the NHIS sample was asked at the end of the interview to provide a telephone number. Each schedule is coded according to the telephone status of the household (telephone available number provided; telephone available number not provided; no telephone; telephone status not ascertained). Telephone status is generally ascertained for more than 99 percent of the completed interviews.

Our analysis of the telephone coverage using the NHIS is affected by both the response and coverage for the NHIS. For most of the 1963–1987 period the sampling frame for the NHIS was based on information collected in the decennial censuses, updated by a sample of building permits issued since the most recent census. These frames are estimated to undercount the number of households in the United States by approximately 1 or 2 percent. In addition, the NHIS has undercounted the estimated number of persons (Census estimates) in the United States by 4 to 7 percent since 1963. This undercount represents persons missed in the NHIS households. Although there is no direct evidence, it is generally assumed that a disproportionate number of the persons missed in the NHIS reside in households without telephones. The NHIS estimates are adjusted each quarter to the Bureau of the Census estimates of the population for 60 age-sex-race subdomains. Whether these adjustments adequately correct for persons by telephone coverage status is not known. Maklan and Waksberg address related issues in the next chapter.

Variance estimates were calculated for a number of the statistics presented in the paper, but only a few of the estimates are shown due to a limitation of space. The standard errors for selected 1985 statistics are shown in Table 2. The corresponding coefficients of variation (cv) should be a good approximation to the cv's for the other data years shown. The cv's for most of the person statistics shown in Table 2 ranged from 3 to 7 percent. When cv's were calculated for similar regional statistics shown in Table 4, most of the cv's ranged from 10 to 15 percent.

2. TIME TRENDS IN TELEPHONE COVERAGE

The estimated percent distribution of households in the United States by telephone coverage status is given in Figure 3.1 for the years 1963-1986.[2] According to these estimates derived from NHIS data, about one-fifth of U.S. households were without telephones in 1963. By the early 1970s this figure had dropped to around 10 percent and by the early 1980s to around 7 percent with no apparent further decline. The apparent slight increase in noncoverage in 1983 probably reflects the effect of higher than normal unemployment during that period. There is no evidence for 1985 or 1986 of significant increases in noncoverage related to the divestiture of AT&T.

Estimates of percent of households without telephones by selected household characteristics are shown in Table 1 for the years 1963, 1970, 1975, 1980, and 1985-1986 combined. Households in the South are consistently less likely to have telephones than are those in the other three geographic regions. In 1963, about one-third of households in the South did not have telephones — around twice the percent for each of the other three regions. This ratio appears to have remained fairly consistent throughout the 24-year period.

---

[2]Number of sample households are between 35,000 and 44,000 for each year 1963-1985, and approximately 24,000 for 1986.

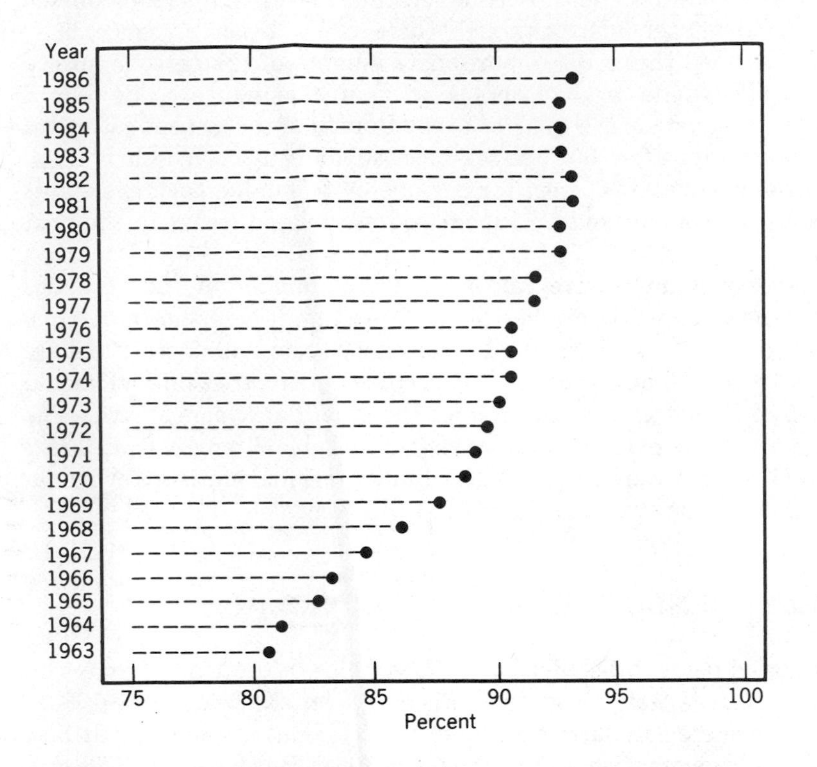

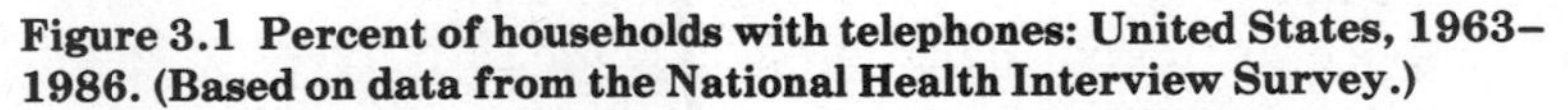
Figure 3.1 Percent of households with telephones: United States, 1963–1986. (Based on data from the National Health Interview Survey.)

Patterns of coverage by urban rural residence changed over time. In 1963 households classified as rural farm had higher noncoverage than those classified as urban or rural nonfarm. By 1980 the lowest noncoverage was for rural farm households. This change is a reflection of declining rural farm populations, urbanization, and technological developments resulting in access to telephone services in remote rural areas.

Throughout the period there is higher coverage in a standard metropolitan statistical area (SMSA) than in non-SMSA areas and higher coverage outside central city areas of SMSAs. One person households and households with large numbers of members are consistently more likely to be nontelephone.

Data on telephone noncoverage for selected characteristics of individuals for the five periods is shown in Table 2. The purpose of these person based statistics is to examine patterns of noncoverage by

Table 1. Percent of Households Without Telephones by Selected Characteristics: United States, 1963, 1970, 1975, 1980, and 1985–1986

Characteristic	1963	1970	1975	1980	1985–1986
All households	19.7	11.7	9.6	7.5	7.2
Region					
Northeast	13.6	7.9	7.1	5.2	4.5
North Central/Midwest	14.7	8.7	6.6	5.3	5.3
South	30.8	18.1	14.0	11.5	10.4
West	17.8	9.9	9.7	6.2	6.9
Urban rural residence					
Urban	15.6	9.6	8.3	7.0	6.4
Rural farm	32.0	15.9	7.9	4.5	4.5
Rural nonfarm	29.2	16.7	13.9	9.1	9.9
SMSA – Non-SMSA					
SMSA – central city	17.3	11.7	10.5	9.1	7.9
SMSA – not central city	11.4	6.2	5.6	4.3	4.8
Non-SMSA – nonfarm	28.8	17.2	14.1	10.4	11.0
Non-SMSA – farm	—[a]	15.9	8.6	4.8	3.3
Number of persons in household					
One	—	19.1	15.4	11.7	9.9
Two	—	10.0	7.8	5.5	5.2
Three	—	10.0	8.5	6.5	6.4
Four	—	8.7	6.7	5.8	6.1
Five	—	8.4	7.7	6.9	7.9
Six	—	10.7	10.5	8.7	10.8
Seven or more	—	17.9	14.6	12.0	16.3
Number of sample households	42,801	37,055	40,131	37,878	58,452

[a]Data not available.

sociodemographic characteristics. With a few exceptions, patterns of noncoverage are consistent throughout the 24-year period.

Noncoverage for males is consistently about 1 percent higher than for females. Throughout the period noncoverage for blacks is about three times higher than for whites; however, the total rate for other races (primarily oriental) has remained at about 11 percent since 1970. By age, the highest noncoverage for each period was for very young children and young adults. The pattern of coverage for the elderly in the noninstitutionalized population has changed somewhat. In 1963, noncoverage for persons 65 years and over was higher than for persons 35 to 64. Currently, older Americans have lower noncoverage than for any other age group. This probably is a result of the trend of geographic separation of older Americans from their adult children and the increased importance of the telephone for social and other contacts. Of the

Table 2. Percent of Persons in Nontelephone Households by Selected Characteristics: United States, 1963, 1970, 1975, 1980, and 1985–1986

Characteristic	1963	1970	1975	1980	1985-1986
All persons	19.8	11.2	9.2	7.1	7.2 (0.26)[a]
Sex					
Male	20.3	11.9	9.8	7.6	7.6 (0.28)
Female	19.2	10.5	8.6	6.6	6.7 (0.26)
Age					
Under 6 years	26.9	17.1	15.3	13.0	12.3 (0.61)
6–16 years	20.6	10.9	9.3	7.1	8.5 (0.54)
17–24 years	26.4	14.7	13.8	11.3	11.2 (0.46)
25–34 years	21.2	11.5	9.6	7.7	8.0 (0.35)
35–44 years	14.3	8.5	7.1	5.0	5.0 (0.27)
45–54 years	14.0	7.2	5.5	4.2	4.2 (0.28)
55–64 years	14.5	8.6	5.3	3.6	3.7 (0.26)
65–74 years	16.9	9.7	5.6	4.0	3.2 (0.27)
75 years and over	20.0	10.8	6.3	4.2	3.3 (0.34)
Race					
White	16.4	9.3	7.6	5.8	5.8 (0.25)
Black	46.7	25.6	20.4	16.8	15.6 (0.82)
Other	32.1	10.7	11.6	9.4	10.9 (2.06)
Marital status					
Married	16.8	9.1	7.0	4.9	5.0 (0.21)
Widowed	19.1	11.3	7.6	5.1	4.5 (0.32)
Divorced	25.7	16.2	12.9	11.2	9.8 (0.61)
Separated	41.0	26.5	25.3	18.0	18.8 (1.09)
Never married	18.7	11.6	10.2	8.6	8.1 (0.33)
Under 17 years	23.0	12.9	11.2	9.1	10.6 (NA)
Employment status					
Currently employed	16.4	8.8	6.9	5.4	5.2 (0.19)
Unemployed	32.1	16.2	16.0	13.5	16.0 (0.85)
Not in labor force	19.2	12.0	9.3	7.2	7.2 (0.47)
Number of sample persons	137,441	115,837	115,685	102,164	153,048

[a]Standard errors for 1985 statistics are shown in parentheses

countries studied by Trewin and Lee (this volume, Chapter 2) the United States is the only country where the older age group has the highest coverage.

Throughout the period, persons 17 years and over who were divorced or separated had higher noncoverage than those currently married, never married, or widowed. Unemployed persons had higher noncoverage than employed persons. The high noncoverage for unemployed persons and persons separated from their spouse is at least in part a reflection of their recent transition in lifestyle and economic flux.

Although there are some deviations in the trend of telephone coverage among demographic subdomains since 1963, the major conclusion from the results shown in Tables 1 and 2 is that the percent of households without telephones is now approximately one-fourth to one-third of what it was in 1963. With the exception of the elderly, the subdomains that had the largest noncoverage in 1963 generally have the largest noncoverage today. Figure 3.2 shows the percent of persons without telephones for selected race and age subgroups of the population.

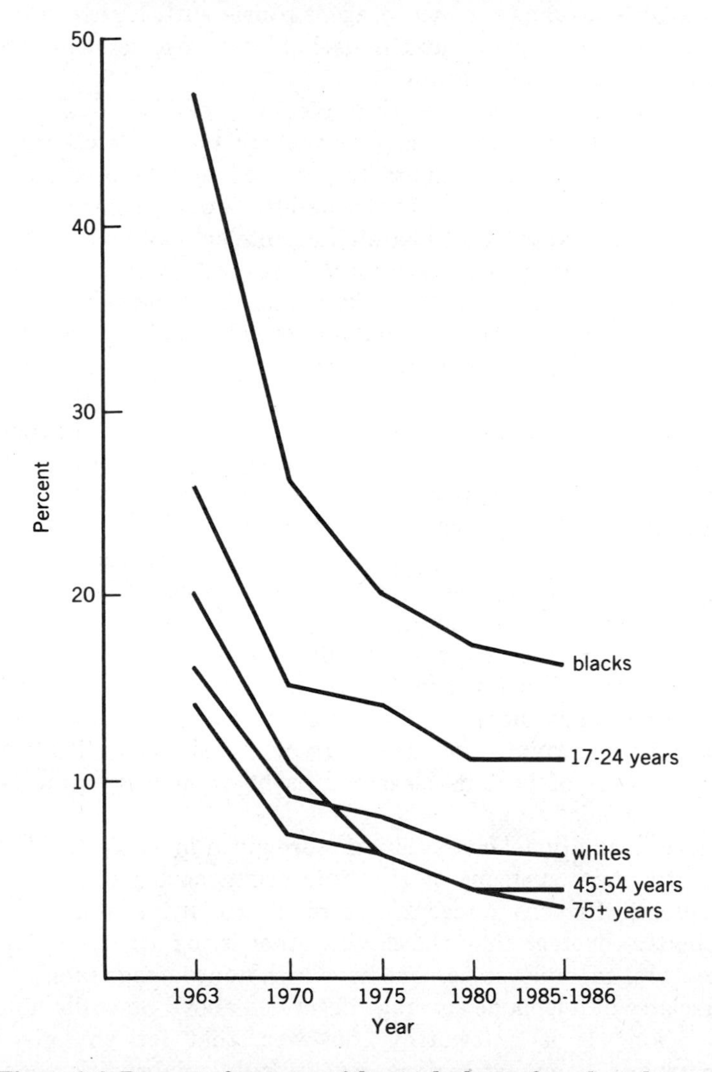

Figure 3.2 Percent of persons without telephones for selected race and age subgroups: United States, 1963–1986.

3. TELEPHONE COVERAGE: 1985–1986

The percent of households without telephone coverage by geographic region are shown in Table 3 for 1985–1986. Households in the South, with 10 percent nontelephone, are less likely to have coverage than are those in the other three regions, each with less than 7 percent noncoverage. For all household characteristics shown, noncoverage is consistently higher in the South than for the other regions and household coverage is relatively complete for the Northeast and Midwest.

Data on noncoverage for selected characteristics of individuals are presented in Table 4. Persons in households without telephone coverage are disproportionately black and under 35 years of age; 94 percent of whites are in telephone households as contrasted to 84 percent for blacks. While young persons are disproportionately in nontelephone households, there is little variation in coverage by age for those over 35 years of age (95 percent or higher). Telephone ownership is higher for persons who are widowed (96 percent) or are married and living with their spouse (95 percent), than for the never married (92 percent), divorced (90 percent), or separated (81 percent).

Telephone coverage increases with increasing family income and education (highest education of adult family member). For example, coverage increases from 71 percent for the lowest income category (less than $5,000) to 99 percent for persons in households with family incomes at $35,000 or more. Coverage for persons in families with incomes of $20,000 or more is relatively complete at 96 percent or higher. This latter income group includes almost two-thirds of the U.S. population.

Persons in the NHIS were classified as above and below the poverty threshold. This index is based on family size, number of children under 18 years of age, and family income using the 1984 poverty levels derived from the August 1985 Current Population Survey. As shown in Table 4, more than one-quarter of persons classified as below poverty were in nontelephone households.

Data for several additional characteristics are given in Table 4. With the exception of Cubans, persons of Hispanic origin have low rates of telephone coverage. Persons under 25 years of age living with both parents have better coverage than those with other living arrangements. As noted previously, unemployed persons have high noncoverage rates.

These correlates of telephone coverage described above generally hold within each region. It is noteworthy, however, that for any given subgroup a comparison among regions reveals only minor differences except for the South, where coverage is almost always lower than for the other three regions. For example, while blacks have lower coverage than whites in all four regions, only 80 percent of blacks in the South reside in

Table 3. Percent of Households Without Telephones by Selected Characteristics and Geographic Region: United States, 1985–1986

Characteristic	All Regions	North-east	Mid-west	South	West
All households	7.2	4.5	5.3	10.4	6.9
MSA – Non-MSA					
MSA – central city	7.9	7.0	7.3	9.6	6.8
MSA – not central city	4.8	2.9	2.8	7.7	5.2
Non-MSA	10.6	5.1	6.7	14.3	12.0
Geographic distribution					
MSA					
1,000,000 or more	5.5	4.9	4.1	7.6	5.4
250,000 – 999,999	6.7	3.9	5.8	9.1	6.7
Under 250,000	7.2	3.0	4.4	10.6	7.8
Other urban areas	10.0	6.8	6.8	14.3	8.3
Rural areas (except in MSA)	11.0	4.2	6.6	14.4	17.5
Number of persons in household					
One	9.9	7.2	8.4	12.5	10.1
Two	5.2	2.9	4.0	7.6	5.0
Three	6.4	3.9	4.4	9.7	5.5
Four	6.1	3.3	3.4	10.2	5.5
Five	7.9	4.4	4.8	13.5	6.9
Six	10.8	6.1	6.8	17.2	9.9
Seven or more	16.3	6.4	10.4	24.9	15.0
Number of sample households	58,452	12,395	14,727	19,689	11,641

telephone households, as contrasted to about 90 percent for each of the other three regions. Similarly, while telephone coverage increases with increasing family income within each region, the level of coverage in the South is lower for each family income category than in the other three regions.

4. SOCIODEMOGRAPHIC VARIABLES RELATED TO TELEPHONE COVERAGE

An investigation of the sociodemographic variables most related to telephone noncoverage was conducted using a SEARCH analysis (Sonquist, Baker, and Morgan, 1973). The SEARCH program uses a sequential one-way analysis of variance procedure to systematically identify the independent variables which explain the highest percent of the total variation associated with a dependent variable. For our analysis the dependent variable was the presence or absence of a telephone in a

Table 4. Percent of Persons in Nontelephone Households by Selected Characteristics and Geographic Region: United States, 1985–1986

Characteristic	All Regions	North-east	Mid-west	South	West
All persons	7.2	4.1	4.8	11.0	6.9
Sex					
Male	7.6	4.3	4.9	11.8	7.4
Female	6.7	3.9	4.7	10.2	6.4
Age					
Under 6 years	12.3	6.8	9.2	18.5	10.6
6–16 years	8.5	5.0	5.1	13.6	7.2
17–24 years	11.2	6.2	7.9	16.7	11.4
25–34 years	8.0	4.4	5.2	12.1	7.9
35–44 years	5.0	3.0	2.9	7.5	5.1
45–54 years	4.2	2.6	3.0	6.2	4.3
55–64 years	3.7	2.5	2.5	5.5	3.6
65–74 years	3.2	2.0	2.5	4.7	3.1
75 years and over	3.3	2.4	2.1	5.3	2.5
Race					
White	5.8	3.4	4.2	8.8	6.0
Black	15.6	10.5	10.1	19.9	10.1
Other	10.9	3.2	5.5	10.8	14.3
Black-nonblack					
Black	15.6	10.5	10.1	19.9	10.1
Nonblack	6.0	3.4	4.2	8.8	6.7
Marital status					
Married, spouse in household	4.8	2.4	2.8	7.8	4.7
Married, spouse not in household	16.2	10.9	8.7	20.5	21.4
Widowed	4.5	2.9	3.2	6.5	4.5
Divorced	9.8	4.8	7.9	13.5	10.0
Separated	18.8	13.5	16.6	24.8	15.3
Never married	8.1	5.1	5.9	11.8	8.2
Under 14 years	10.6	5.9	7.2	16.6	8.7
Highest education of responsible adult family member					
8 years or less	18.1	12.0	11.2	23.3	19.6
9–11 years	19.2	12.7	15.4	24.4	17.5
12 years	8.0	4.1	5.4	12.3	9.0
13–15 years	4.1	2.2	2.5	5.9	4.6
16 years	1.8	0.8	1.0	2.7	2.3
17 years or more	0.8	0.6	0.4	1.1	1.0
Family income					
Less than $5,000	29.2	16.8	22.7	36.9	31.9
$5,000 – $6,999	20.0	16.5	16.4	25.1	18.5
$7,000 – $9,999	18.1	12.7	12.3	23.6	18.7
$10,000 – $14,999	12.9	7.2	8.4	19.0	13.2
$15,000 – $19,999	7.7	3.7	4.8	11.4	8.5
$20,000 – $24,999	3.6	2.6	1.8	5.9	3.2
$25,000 – $34,999	1.8	1.5	0.8	2.5	2.4
$35,000 – $49,999	0.9	0.5	0.5	1.0	1.4
$50,000 or more	0.4	0.4	0.1	0.4	1.0
Unknown	8.0	4.3	4.7	10.8	10.1
Poverty index					
Above poverty	3.7	2.3	2.3	5.5	4.0
Below poverty	27.4	17.8	20.3	35.1	27.8
Unknown	9.4	5.2	5.7	12.7	11.2

Table 4 (Continued).

Characteristic	All Regions	North-east	Mid-west	South	West
Hispanic origin					
Puerto Rican	22.1	22.8	33.4	11.8	18.4
Cuban	4.5	2.0[a]	3.3[a]	5.1	4.7[a]
Mexican	18.1	16.7[a]	15.0	24.6	13.7
Other Latin American	12.6	14.9	3.9[a]	7.2	17.9
Other Spanish	8.9	10.2	16.1	7.1	8.1
Unknown	7.8	3.9	2.4	13.9	9.4
Not Spanish origin	6.5	3.2	4.6	10.4	5.7
Living with relative (under 25 years and never married)					
Both parents	6.3	3.2	3.5	10.7	5.9
Mother	17.7	11.0	14.4	24.7	14.3
Father	13.5	8.4	7.1	20.3	13.3
Neither parent, but other adult relative	15.5	10.3	5.6	20.9	15.6
Other	15.3	12.4	14.0	16.6	18.0
Employment status (18 years and over)					
Currently employed	5.2	2.0	3.0	8.0	5.5
Unemployed	16.0	9.4	11.8	22.9	15.8
Not in labor force	7.2	4.2	5.7	10.4	6.7
Number of sample persons	153,048	32,504	38,478	51,915	30,151

[a]Estimate based on less than 100 sample persons.

household. The independent variables used in the analysis were age, sex, race, marital status, highest education of adult, family income, size of family, employment status, region, and MSA. The SEARCH procedure not only identifies the independent variables which are most related to telephone coverage, but indicates the optimal dichotomy within each of the variables.

Figure 3.3 presents the results of the SEARCH analysis of telephone coverage using a tree diagram. The diagram indicates that family income initially explained the greatest amount of the variation associated with telephone coverage. Additional significant amounts of telephone coverage variation were explained by age, highest education of an adult household member, and family income dichotomies. The sample split into seven final groups. The telephone coverage among the subdomains represented by the independent variables ranged from 46 percent to 97 percent. Two-thirds of the sample ended up in the group with a family income of $15,000 or more and at least a high school education for some adults in the household. This was the group with 97 percent telephone coverage. The final group with the lowest coverage contained approximately 2 percent of the sample and was made up of persons under 55 years of age, family income of less than $5,000, and no one in the household with a high

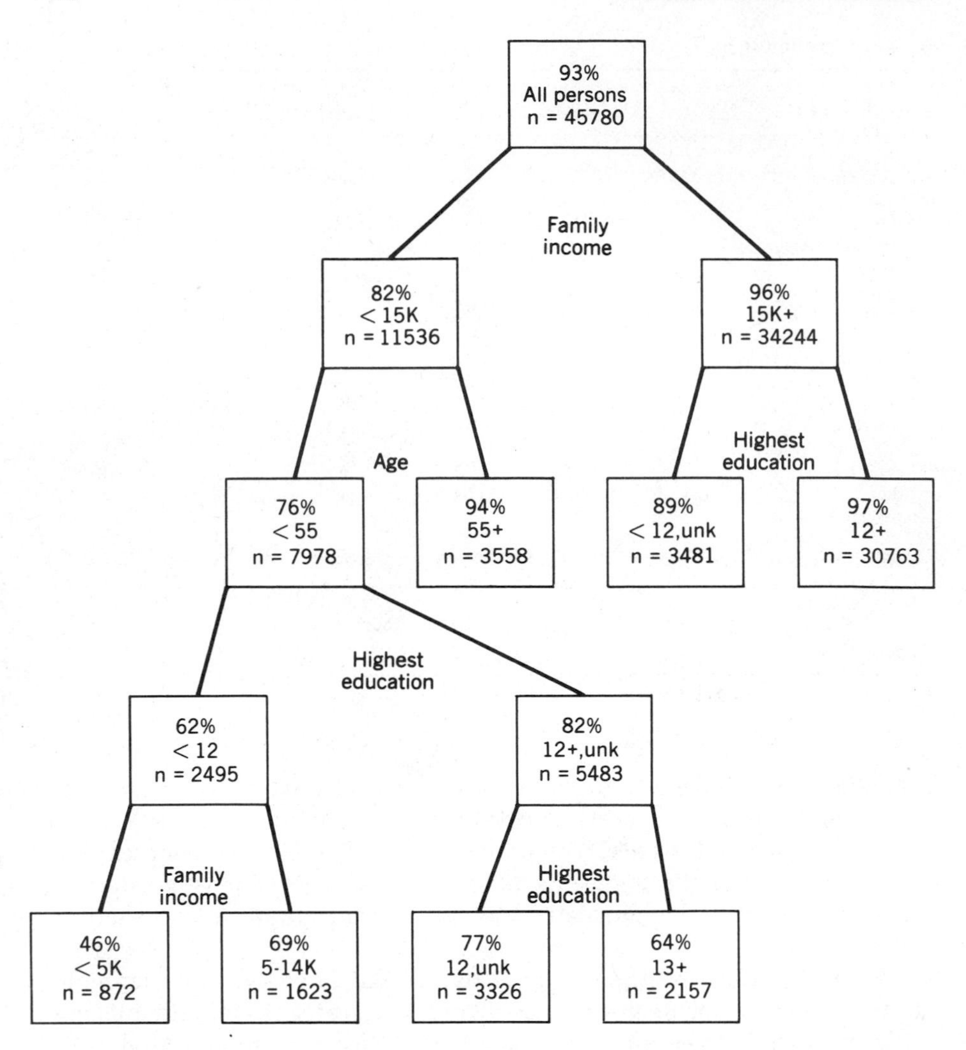

Figure 3.3 SEARCH analysis of sociodemographic characteristics associated with telephone statistics of households. Information presented in the boxes represents percent of households with telephones, distinguishing characteristic for sample persons in box, and number of sample persons falling in box (13 percent of total variation explained.)

school education. It is interesting to note that persons 55 years or older with family incomes of less than $15,000 have a telephone coverage rate of 94 percent.

The most significant finding from this analysis is the importance of family income, age, and education in predicting telephone coverage. What is surprising in the analysis is that variables such as race, marital status, geographic region, family size, and employment status never explained enough of the variation associated with telephone coverage to be shown in the final summary. Even in a more detailed analysis, race never explained as much of the coverage variation as other variables at any stage of the SEARCH analysis.

5. COVERAGE AMONG POPULATION SUBDOMAINS

Much social research and health related research in particular is focused on population subgroups with lower levels of income. A major concern with the use of RDD is the unknown bias that may result from the exclusion of nontelephone households for those populations. Generally data on coverage for population subgroups of particular interest is not available.

In the previous section, family income was established as the major correlate of telephone ownership. A next step is to assess the extent to which the relationships between telephone ownership and other sociodemographic variables diminish when family income is controlled. This will provide insights into those population subgroups most subject to bias related to telephone noncoverage. The issue thus becomes one of the degree of variation in coverage within categories of family income for selected sociodemographic variables.

An examination of a number of variables revealed, as would be expected, that telephone coverage increases with increasing income. Within specific income categories, however, the following general pattern emerges. The correlates of telephone ownership discussed previously still exist for the lower income groups, but tend to diminish in magnitude with increasing family income.

For example, telephone noncoverage by race and age for categories of family income is given in Table 5. As expected, for all age/race groups, noncoverage decreases with increasing family income. For the total population and for nonblacks, coverage is relatively complete (94 percent or higher) for all age groups with family incomes of $20,000 or more; further, coverage for persons 65 years and over is relatively complete regardless of income level, with 87 and 95 percent coverage for the lowest two income categories and 98 percent or higher for the other income groups.

While blacks still evidence higher noncoverage than nonblacks within specific age/income categories, the magnitude of the difference is diminished. Coverage for blacks is relatively complete for all age groups with family incomes of $25,000 or more. Similar to nonblacks, coverage

Table 5. Percent of Persons in Nontelephone Households by Race, Age, and Family Income: United States, 1985–1986

	Family Income						
Race and Age	All incomes	Less than $5,000	$5,000 $9,999	$10,000 $14,999	$15,000 $19,999	$20,000 $24,999	$25,000 or more
Total							
All ages	7.1	29.2	18.9	12.9	7.7	3.6	1.2
Under 6 years	12.0	45.6	32.0	24.1	11.4	5.4	1.4
6–16 years	8.4	40.6	25.2	16.4	8.7	4.4	1.2
17–24 years	11.0	24.1	25.7	19.5	12.2	5.9	1.9
25–44 years	6.4	35.7	23.3	15.7	8.6	3.7	1.2
45–64 years	3.8	25.5	14.7	5.7	4.2	1.7	0.7
65 years and over	3.2	12.7	5.1	2.2	1.1	0.4	0.3
Black							
All ages	16.1	36.6	23.7	16.1	11.2	9.1	2.7
Under 6 years	24.4	45.5	29.0	23.3	16.6	16.3	4.1
6–16 years	18.8	42.2	28.1	17.3	11.3	9.3	2.9
17–24 years	18.6	33.9	28.7	18.0	11.5	12.2	3.5
25–44 years	14.9	41.6	26.0	19.7	13.3	9.1	2.5
45–64 years	9.3	27.9	16.7	8.6	5.6	3.2	1.8
65 years and over	8.7	18.6	9.0	1.3	2.7	0.0[a]	1.0
Nonblack							
All ages	5.9	25.9	17.5	12.3	7.2	2.9	1.0
Under 6 years	9.9	45.7	33.3	24.3	10.6	4.0	1.2
6–16 years	6.5	39.0	23.7	16.1	8.2	3.7	1.1
17–24 years	9.9	21.6	24.8	19.7	12.3	5.0	1.7
25–44 years	5.4	32.8	22.5	15.0	7.9	3.1	1.1
45–64 years	3.3	24.6	14.2	5.2	4.0	1.6	0.7
65 years and over	2.7	11.0	4.6	2.3	1.0	0.4	0.3

[a]Estimate based on less than 100 sample persons.

for blacks 65 years and over is complete for all except the lowest income groups, ranging from 81 and 91 percent for the lowest two income groups to 97 percent or higher for family incomes of $10,000 or more. The patterns of coverage by family income described above generally hold for each geographic region as shown in Table 6. The previously noted lower coverage in the South than in the other regions generally remains within categories of family income.

Telephone coverage for persons classified as above or below poverty based on the index described previously were examined by race and age. Over one quarter of persons classified as below poverty were in nontelephone households. This level of coverage was found for all age/race groups in the below-poverty category, with the exception of the elderly (with 12 percent in nontelephone households) and persons under 6 years of age (with 37 percent in nontelephone households).

Similar data were examined by poverty status and geographic region. While the overall patterns of coverage for persons below poverty were similar within each region, there was higher noncoverage for most age/race groups in the South and West than in the Northeast and Midwest. For the South, the pattern of higher noncoverage for blacks than nonblacks among age groups generally had disappeared for the below-poverty group.

In summary, as demonstrated by this examination of selected sociodemographic characteristics of households and persons, there is considerable variation in the level of telephone coverage among population subdomains. Overall, nontelephone households represent only about 7 percent of all households, and many subgroups with relatively high levels of noncoverage constitute small proportions of the total population. Further, for surveys of many population subgroups, the coverage which can be obtained through RDD may be well within accuracy requirements. For example, telephone coverage is relatively complete for persons in middle and high income families. For older Americans, telephone coverage is relatively complete regardless of race or income. If level of coverage were the only consideration, it would be difficult to fault the use of RDD in planning a survey of the elderly. On the other hand, some population subgroups exhibit very high levels of noncoverage. For example, an RDD survey designed to obtain data for preschool age children would exclude the following from the sample frame: 12 percent overall; one-quarter of all black children; almost one-third of

Table 6. Percent of Persons in Nontelephone Households by Region, Age, Race, and Family Income: United States, 1985–1986

Region, Age, Race	All Incomes	Family Income: Less than $5,000	$5,000 $9,999	$10,000 $14,999	$15,000 $19,999	$20,000 $24,999	$25,000 or More
Northeast							
Total	4.0	16.8	14.4	7.2	3.7	2.6	0.9
Age							
Under 6 years	6.6	25.4	27.2	16.0	7.2	3.4	0.9
6–16 years	4.7	22.9	21.5	8.5	2.1	3.0	1.1
17–24 years	6.6	15.1	20.3	10.2	8.3	3.9	1.6
25–44 years	3.5	20.3	18.4	10.0	4.3	2.5	0.8
45–64 years	2.7	18.0	13.0	4.8	2.1	2.2	0.5
65 years and over	2.2	8.7	4.0	2.1	0.6	0.6	0.2
Race							
Black	11.1	23.4	22.8	10.4	7.6	5.8	4.6
Nonblack	3.3	14.9	12.8	6.8	3.3	2.3	0.6
Midwest							
Total	4.8	22.7	14.0	8.4	4.8	1.8	0.5
Age							
Under 6 years	9.1	35.1	27.8	16.5	6.4	1.7	0.4
6–16 years	5.1	25.6	17.3	12.0	5.1	1.8	0.8
17–24 years	7.9	22.2	20.4	12.9	6.5	2.9	0.5
25–44 years	4.2	26.5	15.8	10.3	5.4	2.0	0.5
45–64 years	2.5	16.5	9.3	2.8	4.1	1.5	0.4
65 years and over	2.3	8.5	4.3	0.8	1.4	0.5	0.4
Race							
Black	10.5	25.8	13.9	8.8	5.8	2.9	1.1
Nonblack	4.2	21.4	14.0	8.4	4.7	1.7	0.5

Table 6 (Continued).

		Family Income					
Region, Age, Race	All Incomes	Less than $5,000	$5,000 $9,999	$10,000 $14,999	$15,000 $19,999	$20,000 $24,999	$25,000 or More
South							
Total	11.0	36.9	24.2	19.0	11.4	5.9	1.5
Age							
Under 6 years	18.6	59.0	41.1	34.1	17.2	8.4	1.7
6–16 years	13.9	51.7	32.2	23.3	13.6	8.2	1.6
17–24 years	16.4	28.2	32.5	28.5	16.2	9.3	2.8
25–44 years	9.9	49.0	30.2	22.6	12.9	6.3	1.5
45–64 years	5.9	31.6	18.2	7.2	4.8	2.0	0.9
65 years and over	4.9	15.5	6.4	3.3	1.6	0.3	0.1
Race							
Black	20.6	44.2	28.3	19.7	14.4	12.0	2.8
Nonblack	8.7	31.9	22.1	18.8	10.6	4.6	1.3
West							
Total	6.5	31.9	18.6	13.2	8.5	3.2	1.7
Age							
Under 6 years	9.9	43.4	24.1	21.0	11.4	6.8	2.4
6–16 years	7.0	40.5	23.9	15.0	9.9	3.9	1.5
17–24 years	10.8	28.7	22.6	18.8	13.9	6.1	2.5
25–44 years	6.4	33.7	23.1	15.2	8.9	3.3	1.8
45–64 years	3.7	34.0	14.6	7.7	5.4	0.9	1.1
65 years and over	2.5	15.3	5.3	2.8	0.6	0.0	0.6
Race							
Black	9.9	35.3	15.5	23.4	5.7	11.9	1.4
Nonblack	6.4	31.7	18.9	12.7	8.6	2.9	1.7

those living in poverty nationwide; and about half of those living in poverty in the South.[3]

6. BIAS FROM THE EXCLUSION OF NONTELEPHONE HOUSEHOLDS

In previous sections the nature and direction of bias in sociodemographic characteristics of the population resulting from the exclusion of nontelephone households were examined. What remain to be investigated are differences between telephone and nontelephone populations in dependent variables or study characteristics and the extent to which the exclusion of nontelephone households would introduce bias into national estimates of these characteristics. If it can be demonstrated that there are few telephone-nontelephone differences in estimates of a variety of dependent variables within most population subdomains, one would be less concerned about potential bias from the exclusion of nontelephone populations. The examples that are provided here are derived from the NHIS and focus on several health or health related variables.

Table 7 provides estimates of selected health characteristics and use of health services for persons in telephone and nontelephone households and for the total U.S. population. In general, the observed differences in the telephone-nontelephone estimates are of the type and in the direction expected, given that the nontelephone population consists disproportionately of persons in low income families. Relative to the population with telephone coverage, the nontelephone population is more likely to be limited in activity because of chronic health conditions, to have higher rates of restricted activity and bed disability days, and to have lower rates of utilization of health services with the exception of hospitalizations. The most pronounced difference relates to the

---

[3]A possible method for reduction of household undercoverage for RDD surveys is to contact members of nontelephone households through telephones in other housing units. The potential effectiveness of this approach can be estimated with data on access to and use of nonbusiness telephones by members of nontelephone households. In 1982 questions were added to the face to face NHIS to determine access to a telephone for each sample household. The results of that study suggest that only 36 percent of nontelephone households could be contacted by telephones in other housing units. That is, about two-thirds of nontelephone households were reported as not having access to a telephone in another housing unit. Further, of those households with access, 40 percent reported access to a telephone in a nonadjacent building. The conclusion of the study was that total coverage of households in RDD surveys would be increased by less than one percent through procedures designed to contact nontelephone households on telephones in other housing units. Thus, since a significant reduction in potential bias due to undercoverage would not be expected, the required expenditure of resources does not appear warranted.

Table 7. Selected Health Characteristics by Telephone Coverage: United States, 1985–1986

Health Characteristic	Nontelephone Households	Telephone Households	All Households
		Per Person Per year	
Disability Days			
Restricted activity days	16.2	14.9	15.0
Bed disability days	7.7	6.2	6.3
Work loss days	5.1	5.3	5.3
		Percent With	
Chronic Conditions			
Limitation in activity[a]	15.2	13.9	14.0
Limitation in major activity[a]	11.2	9.3	9.5
		Per Person Per Year	
Use of Health Services			
Physician visits	4.3	5.4	5.3
Dentist visits	0.8	1.4	1.3
		Per 100 Persons Per Year	
Hospitalizations			
Hospital discharges (excluding deliveries)	11.1	10.6	10.6

[a]Due to chronic conditions.

utilization of dental services, which is the health characteristic most highly correlated with family income. Thus a survey based on telephone households would overestimate use of health services and underestimate limitation of activity, unless compensatory adjustments were made. However, it should be noted that as was true for the sociodemographic characteristics, the differences between the values for the telephone population and the total population are generally small.

Table 8 provides data on health insurance coverage by type of coverage for persons under 65 years of age and 65 years and over. For the under 65 population, persons in nontelephone households, as contrasted to those in telephone households, are much less likely to have private health insurance (35 versus 80 percent), more likely to be on public assistance (23 versus 5 percent), and less likely to have any type of coverage (61 versus 87 percent). Similarly for the 65 years and over population, those in nontelephone households are less likely to have Medicare coverage (87 versus 95 percent) or supplemental private health insurance (28 versus 77 percent) and more likely to be on public assistance (25 versus 6 percent). It is noteworthy that while the nontelephone

Table 8. Percent of Persons With Health Insurance Coverage by Type of Coverage and Age by Telephone Coverage: United States, 1986

Type of Coverage	Nontelephone Households	Telephone Households	All Households
Persons under 65 years			
Private health insurance	35.1	80.2	76.9
Medicare	1.4	1.2	1.2
Public assistance	22.6	4.5	5.8
Military-Veterans Administration	1.4	2.9	2.8
All types of coverage	61.1	87.0	85.0
Persons 65 years and over			
Private health insurance	28.0	77.0	75.3
Medicare	87.3	95.3	95.0
Public assistance	25.3	6.0	6.6
Military-Veterans Administration	7.8	4.9	5.0
All types of coverage	95.4	99.4	99.3

population constitutes less than 10 percent of the under 65 population, their exclusion would raise the estimate of private health insurance coverage by 3 percent — from 77 to 80 percent.

Data on health practices and knowledge of health practices are presented in Table 9. Generally, persons in nontelephone households have poorer health practices and lower knowledge of correct health practices than those in telephone households. For example, a significantly higher percentage of persons in nontelephone households smoke cigarettes (50 versus 29) and a significantly lower percentage exercise regularly (32 versus 41), wear seatbelts regularly (22 versus 37), and have a working smoke detector in their home (35 versus 62). On knowledge items they are less likely to know the value of fluoridated water and dental sealants and less likely to have heard of fetal alcohol syndrome. Examples related to child health suggest that expectant women in nontelephone households are more likely to smoke cigarettes and that children are less likely to have been breastfed. All of these results are consistent with differences between persons in lower versus higher income families and persons with lower versus higher education.

With few exceptions, the relationships described above for the total population exist within each region and among age groups. For example, within age groups, persons in nontelephone households have lower rates

Table 9. Percent of Persons With Selected Health Practices or Knowledge of Health Practices by Telephone Coverage: United States, 1985

Selected Health Practice Characteristic	Nontelephone Households	Telephone Households	All Households
Health practices			
20% or more overweight[a]	23.7	24.0	24.0
Exercise regularly[a]	32.4	40.5	40.0
Currently smoke cigarettes[a]	49.6	28.8	30.1
Wear seatbelts[a]	21.6	36.6	35.8
Smoke detector in home[a]	35.0	61.8	60.2
Mental stress in current job[b]	7.6	17.0	16.6
Knowledge of health practices			
Drinking fluoridated water helps prevent tooth decay[a]	71.2	78.8	78.2
Dental sealants help prevent tooth decay[a]	8.3	18.4	17.7
Heard of fetal alcohol syndrome[a]	40.0	56.9	55.6
Child health			
Smoked cigarettes 12 months preceding birth of child[c]	43.2	30.2	31.8
Ever breastfed[d]	32.7	56.4	53.6
Breastfed 6 months or more[e]	9.5	25.1	23.3

[a]Persons 18 years of age and over.
[b]Persons 18 years of age and over in labor force.
[c]Women 18–44 years of age giving birth in past 5 years.
[d]Children under 5 years of age.
[e]Children 6 months to 4 years of age.

of utilization of physician and dental services, higher rates of activity limitation, and poorer health habits than persons with telephones.

A variety of health and health related characteristics were examined by family income. The relationships described above almost always held. For example, as shown in Table 10, for each family income category, persons in nontelephone households were significantly less likely to have private health insurance coverage than were persons in telephone households. For the lower income groups, the estimate of insurance coverage based on the telephone population differed significantly from that based on the total population. For example, for persons in families with incomes of less than $10,000, one would estimate that 38 percent had private health insurance based only on persons with telephone coverage as contrasted to 32 percent based on the total population.

Table 10. Percent of Persons Under 65 Years of Age with Private Health Insurance Coverage by Family Income and Poverty Index by Telephone Coverage: United States, 1986

Family Income and Poverty Index	Nontelephone Households	Telephone Households	All Households
Family income			
All Incomes	35.1	80.2	76.9
Less than $5,000	15.7	39.8	32.2
$5,000 – $9,999	18.0	36.1	31.7
$10,000 – $19,999	47.8	68.2	66.0
$20,000 – $34,999	74.1	88.9	88.5
$35,000 or more	83.9	94.2	94.1
Poverty index			
Below poverty	16.8	34.5	29.5
Above poverty	59.3	87.1	86.1

Table 11. Percent of Persons 18 Years of Age and Over Who Currently Smoke Cigarettes, by Family Income and Education by Telephone Coverage: United States, 1985

Family Income and Education	Nontelephone Households	Telephone Households	All Households
Family income			
Less than $10,000	49.8	28.5	32.4
$10,000 – $19,999	52.4	31.2	33.1
$20,000 – $34,999	47.8	30.4	30.7
$35,000 or more	43.5	26.0	26.1
Education			
Less than 12 years	51.5	33.1	35.4
12 years	50.0	32.5	33.5
13–15 years	43.3	26.8	27.3
16 years or more	42.3	18.1	18.4

As a final example, data on cigarette smoking by income and by education are given in Table 11. Within each income or education subgroup, the estimates based on telephone and nontelephone households differ significantly and the conclusions from an analysis of the relationship between smoking and income or education may well differ.

An examination of a variety of health variables by age groups within income categories was also conducted. For the majority of age-income cells, the telephone-nontelephone differences described previously remained, although the magnitude of the difference was generally diminished.

In summary, the health and health related characteristics of persons in nontelephone households are very different from those of persons in telephone households. It is clear from the data presented that there are factors in addition to age and income which should be taken into account in an assessment of the bias introduced by the exclusion of nontelephone households. Massey and Botman (this volume, Chapter 9) examined the effect of weighting adjustments by sex, age, region, and income to reduce telephone noncoverage bias. They concluded that the adjustments only partially corrected for the noncoverage bias.

7. DISCUSSION: COVERAGE AND RESPONSE

The major focus of this chapter has been on telephone coverage and the potential bias from the exclusion of the nontelephone households. Section C of this volume addresses issues related to nonresponse. However, it is difficult to address coverage as a survey design issue independent of response issues. This is particularly true when considering the telephone mode with RDD. Lower response in RDD surveys than in face to face surveys is a common observation.

The issue of telephone coverage in RDD surveys relates to contact with a representative sample of the population. The issue of response relates to obtaining data from a representative sample of that population. The major concern about both noncoverage and nonresponse is the potential for bias. Those persons for whom data are not obtained may have characteristics very different from those for whom data are obtained, and consequently there is the concern that the survey estimates may not accurately reflect the characteristics of the target population. It is generally assumed that the higher the rates of noncoverage and nonresponse, the greater the risk of this bias.

In order to speculate on the nature of nonresponse bias in telephone surveys, it is useful to estimate response rates for various population subgroups. We are generally prevented from making this estimation because relative sizes of different demographic groups within the telephone household population are generally not known. However, in a previous paper (Thornberry and Massey, 1983) the authors attempted to derive estimated response rates for population subgroups within telephone surveys and to examine the potential combined effects of noncoverage and nonresponse on RDD interview rates.

Table 12 presents estimates of interview rates for selected demographic subdomains averaged over three national RDD telephone

Table 12. Estimated Percent of Target Population Interviewed (Response Rate x Coverage) for Selected Demographic Subdomains Averaged Across Three RDD Surveys

Subdomain	Average Response Rate	Telephone Coverage	Percent Interviewed
Sex			
Male	76	92	70
Female	79	93	73
Age			
17–24 years	75	89	67
25–44 years	84	94	79
45–64 years	77	96	74
65 years and over	62	97	60
Race			
White	78	94	73
Black	73	84	61
Education			
0–11 years	65	87	56
12 years	76	94	71
13 years or more	91	97	88
Total	77	93	72
Average sample size	6,265		

surveys either conducted or sponsored by NCHS. The estimates for population subdomains were derived by multiplying the estimated coverage for each subdomain by the average estimated response for that subdomain. Each of the three surveys used a different respondent rule to obtain information about persons who were 17 years old or older (Thornberry and Massey, 1983).

Based on the estimates in Table 12, one would conclude that although telephone coverage for persons 65 years of age or older is higher than for any other age group, the overall interview rate is lower, reflecting high nonresponse of this age group in RDD surveys. The very low overall interview rates for blacks and for persons with low levels of education reflects both low coverage and low response.

These findings and the previous detailed assessments of coverage issues provide strong evidence that the potential of having a substantial bias due to undercoverage and/or nonresponse is very real in surveys based only on telephone households. Potential biases among certain subdomains are more likely than among other subdomains and the overall coverage and response rates for the total population surveyed may be misleading.[4] This highlights the importance of subdomain weighting adjustments through poststratification or other means.

In conclusion, issues related to both coverage and response should be examined carefully in the design of RDD surveys and in the interpretation of the data obtained. The implications of high undercoverage or nonresponse overall and among subdomains depends upon the major objectives of the survey and the importance of subdomain estimates.

---

[4]These findings also provide an argument against performing trend analysis on data based on different sampling frames and collection modes. Even with estimates of characteristics for the total population, where the effect of noncoverage and nonresponse is at a minimum, trends that appear to be important may reflect nothing more than mode differences; that is, as Smith (1987) has suggested, the medium may become the message.

CHAPTER 4

WITHIN-HOUSEHOLD COVERAGE IN RDD SURVEYS

David Maklan and Joseph Waksberg[1]
Westat, Inc.

Most of the analyses of coverage in random digit dialing (RDD) surveys that have been carried out to date have tended to focus on sampling frame inadequacies, that is, the part of the target population that does not reside in telephone households. Several papers in this volume provide data on noncoverage of this type. However, there is another cause of undercoverage arising from the failure to obtain complete listings of household members from the respondents to a survey. We have not found any information on this subject in the published literature. While there have been a few experiments on alternative approaches to obtaining a household roster and, related to this, methods of subsampling persons within households (Groves and Kahn, 1979; Bercini and Massey, 1979), these reports have used response rates as the criteria for choosing among alternatives and have not examined possible effects on undercoverage.

It would be surprising if undercoverage were not present in RDD studies since it exists in face to face household surveys. The persistent undercoverage of the population in decennial censuses has been well publicized in the press, professional journals, and in Census Bureau reports and publications. Most household sample surveys do even worse than censuses in population coverage. For instance, the Census Bureau's Current Population Survey (CPS), generally considered a model of a well

[1]The authors are grateful to the U.S. Office of Smoking and Health/CHPE/CDC for permitting analysis of household screening data from its 1986 Adult Smoking Survey. The opinions expressed in this chapter are those of the authors and do not necessarily represent the opinions of OSH/CHPE/CDC. We also wish to thank Ms. Louise Hanson for providing data from the Oral and Pharyngeal Cancer Study, Ms. Marianne Cummings for her programming assistance, and Mrs. Elizabeth Gaughan and Mrs. Mary Lou Pieranunzi for their organizational and typing assistance.

administered survey with considerable resources devoted to achievement of high quality, has in recent years accounted for about 93 percent of persons 14 years old and over of the number reported in the 1980 census or in updates of the 1980 census population (Bailar, 1986). Most face to face surveys carried out by other organizations cannot achieve the CPS levels, or if they do, it is at considerable expense. With this pattern of undercoverage in carefully conducted, face to face personal interview surveys, it is reasonable to be concerned about the possibility that undercoverage might be even more severe in telephone surveys. This paper examines data which shed some light on this issue.

Unfortunately, none of the available data provided direct estimates of undercoverage such as might come from comparing the listings of household members with auxiliary and theoretically complete information on residents of the sample households. Reinterviews using more detailed and probing questionnaires are the most common source of auxiliary data, but such reinterviews do not seem to have been carried out on a sufficient scale in RDD studies to permit useful analyses. Our approach has therefore been to compare key indices of within-household coverage with those from surveys and censuses that can be considered as standards. We believe that this provides a reasonable indicator of whether RDD coverage is better or worse than coverage in face to face interviews, and of the size of the difference.

One obvious indicator of within-household coverage is a distribution of the sample by household size. Within-household misses should, of course, result in a general downward shift in household size resulting in a lower average number of persons per household. One analysis presented in this chapter thus compares the household sizes in a large RDD survey with the CPS.

All coverage studies that we have seen show that in the United States coverage problems in minority groups are much more severe than in the general population. Separate data on blacks, and possibly other minorities, are thus desirable partly because they are probably particularly sensitive indicators of undercoverage, and partly because they are often important subdomains of analysis in their own right. The other pattern of undercoverage which appears to be present in virtually all household surveys is the poorer coverage of males than of females, much of it concentrated in ages 20 to 40. Demographers have used sex ratios as clues to coverage problems in U.S. censuses. Increased levels of undercoverage can be expected to show up as higher sex ratios, and age-specific sex ratios reveal patterns among age groups.

Most of our analyses thus rely on analyses of sex ratios, more specifically on comparisons of sex ratios in several large RDD studies, with those in CPS. One reason is that the data file used for the household size distributions did not contain information on race and, as indicated above, coverage estimates by race seemed particularly important. The second reason is that sex ratios permit examination by age groups.

Household size distributions and sex ratios do not provide direct estimates of undercoverage. However, they should be reasonable indicators of broad levels of coverage and reveal whether substantial problems exist.

Comparisons between RDD and CPS (or the censuses) are appropriate because the coverage studies carried out on both the censuses and CPS show that most of the coverage losses appear to be due to an incomplete reporting of persons within households or to missing those who do not have a strong attachment to any households, particularly for the minority populations. Similarly, missed households are a minor problem in RDD surveys if there are intensive followups for the unanswered number cases since the sampling frame for RDD is complete, by definition. With intensive followup, Westat has been able to keep the unanswered number cases to under 2 or 3 percent, and less than half of them appear to be residential numbers. The coverage losses in RDD (other than those caused by exclusion of nontelephone households) thus are also primarily due to missed persons in interviewed households.

The sex ratio comparisons of RDD with CPS use a number of versions of CPS. Use of unweighted CPS data is a fairly direct guide to the level of male undercoverage in RDD relative to that in CPS. (The weighted CPS figures shown in published reports reflect the population estimates used in the poststratification that is part of the CPS estimation procedure. They thus do not rely on survey results, and weighted population figures do not provide data on CPS operations.) The CPS weights adjust for both nonresponse and undercoverage but since response rates are very high in CPS, the difference between unweighted and weighted sex ratios mostly reflects differential coverage in CPS. Comparisons of the RDD and CPS unweighted data with CPS weighted totals consequently indicate coverage differences from census level projections of the current population. Finally, it is possible to compare the RDD data to the "true" sex ratios, obtained by adjusting the weighted CPS data using estimates of the undercount in the 1980 census.

Sections 1 and 2 compare the RDD sex ratios to all three CPS and census level sources. A more limited set of comparisons is shown for the household size distributions.

In addition to the basic comparisons described above, Section 3 reports on an experiment regarding the effect on coverage of two alternative methods of screening households for particular types of persons eligible for a survey. One of Westat's RDD studies was restricted to persons 13 to 24 years of age. Independent random samples of households were selected quarterly and screened for eligible persons. Different screening questions were asked in the first two quarters of the survey, and the effect on coverage was studied.

Sampling errors are not shown in the detailed tables, but it can be noted that because of the large sample sizes they are quite low and mostly do not influence the comparisons. Exact estimates of the sampling errors have not been computed, but upper limits have been derived. If the males and females had been selected as independent samples, the standard errors of the sex ratios would have been approximately: (1) for the RDD sex ratios of the black population in Tables 1 and 2, about .04 for the sex ratio for all blacks, and .07 for each of the three broad age groups; (2) for estimates of black sex ratios from CPS in Table 2, .02 for all blacks and .03 for the three broad age groups; (3) for the RDD sex ratios in Tables 6 and 7, .008 for the total population, .014 for broad age groups, and .024 for detailed age groups; (4) for the CPS sex ratios in Table 7, .006 for the total, .010 for broad age groups, and .018 for the detailed groups. The assumption of independence does not hold with household surveys since males and females tend to be positively correlated within households. This positive correlation reduces the standard errors of sex ratios. The reduction is probably substantial; it is likely that the standard errors are not much more than half the figures shown above.

1. COVERAGE IN ORAL AND PHARYNGEAL CANCER STUDY

1.1 Description of Westat Study

Westat conducted a case control study on oral and pharyngeal cancer in 1985. The study was restricted to a limited set of geographic areas in the United States: the State of New Jersey, the Atlanta metropolitan area, the Los Angeles metropolitan area, and the suburban part of the San Francisco area. Since the areas were not selected as a random sample of the United States, differences between the survey results and the national surveys with which RDD is compared may reflect geographic differentials as well as coverage. However, the areas contain a diverse population and are spread through three of the four regions of the United States. The quality of reporting in the four areas should thus be indicative of what is likely to be found in national surveys.

The controls for the study were obtained by using RDD to screen an equal probability sample of households within the specified geographic areas. A sample of 10,922 households was designated for screening. Within each household, telephone interviewers attempted to identify and list all persons 18 years old and over and obtain race, sex, and age for each one. The screening was completed in about 80 percent of all households in the sample. There were slightly over 2,300 black persons 18 years and over in the screened households.

1.2 Face to Face Interview Data Used in Analysis

Our comparisons are with data from the March 1985 CPS. March 1985 is quite close to the period of time covered by the oral and pharyngeal cancer study, most of which was conducted during 1985. This chapter also shows CPS sex ratio data for March 1986 and for the 1983 National Health Interview Survey (NHIS). Experiences in these other sources are a guide to the stability of the sex ratios in surveys conducted through face to face interviews. It should be noted, however, that data collection for both CPS and NHIS are carried out by the Census Bureau which invests considerable resources into ensuring high quality. They are thus rather high standards of comparison. Sampling and data collection procedures for the CPS and NHIS have been described in some detail in published reports (Bureau of the Census, 1978, National Center for Health Statistics, 1985) and, therefore, are not repeated here.[2] However, there are various types of CPS estimates with which the Westat data are compared. All of the CPS data are based on special tabulations prepared by Westat for households with telephones from the CPS microfiles for March 1985 and March 1986. The 1983 NHIS data came from special tabulations of telephone households prepared by Owen Thornberry of the National Center for Health Statistics and supplied to Westat for this analysis.

The CPS unweighted data, of course, show the sex ratios for the identified population. The weighted CPS data adjust for variable sampling rates, nonresponse, and use poststratification to independent estimates of the population. They are similar to the totals shown in the published CPS reports except for the exclusion of the approximately 7 percent of the U.S. population without telephones. Since the CPS poststratification uses independent estimates of the population by race, sex, and age, the weighted sex ratios are essentially ratios of the independent estimates and not of the population as enumerated in CPS. The differences between the weighted CPS figures and either the RDD or

[2]Although most of CPS is collected by telephone, household coverage data are obtained in the first month of interview which is always conducted face to face.

unweighted CPS data are indicators of male undercoverage in the two surveys.

While close to constituting a self-weighted sample, CPS is not completely so. To make certain the varying probabilities do not seriously affect the sex ratios, we have created a "deflated weighted" CPS series. This was obtained by eliminating the poststratification factors from the weighted figures.[3] The deflated figures thus reflect the variable rates. They also include nonresponse adjustments and the CPS first stage ratio estimates, but these are unlikely to have any perceptible effect on the sex ratios. In any case, the deflated weighted ratios are very close to the unweighted ratios, and one would draw the same conclusions from either set.

The independent population figures used for poststratification in CPS (and NHIS) are projections of the 1980 Census data, and the coverage is thus at the 1980 Census level. Although census undercoverage cannot be measured exactly, we used the Bureau's best estimates of the "true" population to derive what we refer to as the "true" sex ratios. This was done by adjusting the weighted levels by the Census Bureau's estimates of undercoverage in the 1980 census (U.S. Bureau of the Census, 1985).

As noted earlier, comparisons of the RDD study with either the CPS unweighted or the deflated weighted figures show how RDD compares with a well executed face to face interview survey. Comparisons with the weighted CPS indicate how far RDD (and CPS) is from the best that can reasonably be expected. Finally, comparisons with the "true" sex ratios indicate the total undercoverage in both the RDD and CPS surveys.

We should like to reiterate that the sex ratios do not provide direct measures of undercoverage per se, but of males relative to females. However, we believe this is a reasonable guide to total undercoverage. In March 1985, the CPS ratios of the sample estimates of the number of blacks to the census level figures were .81 for males 14 years and over and .91 for females. Undercounts in the census for blacks 14 years old and over was .91 for males and .98 for females. The combined effect of the two is that total coverage of blacks 14 years old and over was 74 percent for males and 89 percent for females. The prevailing belief is that the difference in coverage between the two sexes reflects mostly a differential within household coverage.

The household size analysis discussed in Section 3 does provide direct evidence on within-household coverage but, unfortunately, separate data by race are not available.

---

[3]The poststratification factors are for March 1985 and were obtained in unpublished form from the Census Bureau.

1.3 Data Presented

Table 1 shows the number of black persons 18 years and older listed in interviewed households in the RDD Oral and Pharyngeal Cancer Study and their age-specific sex ratios. The ratio for all black persons 18 years and over is 1.31. There are some erratic patterns in the detailed age groups, possibly due to sampling variation or response rate differentials. For broad age groups, clearer patterns emerge.

Table 2 includes comparable ratios for the March 1985 CPS, using both unweighted and weighted data. Table 2 also contains the sex ratios for the two adjusted CPS data sets — the deflated weighted CPS estimates and the "true" counts.

Table 3 shows sex ratios in the March 1986 CPS, March 1985 CPS, and the 1983 NHIS. Table 3 is a guide to the stability of the sex ratios over time. The tabulations of the NHIS are based on a slightly different classification of age than were used in the CPS and the Westat RDD survey.

Table 1. Black Persons 18 Years and Older in Westat's RDD Oral and Pharyngeal Cancer Study, by Sex and Age

	Number of Persons		
Age	Females	Males	Sex Ratio
18–19	76	55	1.38
20–24	173	156	1.11
25–29	177	127	1.39
30–34	152	114	1.33
35–39	159	105	1.51
40–44	127	102	1.25
45–54	167	138	1.21
55–64	128	110	1.16
65–74	94	64	1.47
75+	51	27	1.89
18–29	426	338	1.26
30–44	438	321	1.36
45+	440	339	1.30
Total	1304	998	1.31

Table 2. Comparison of Sex Ratios in Westat's RDD Oral and Pharyngeal Cancer Study with 1985 CPS Interview Results in Telephone Households for Black Persons 18 Years and Over

Age	RDD Study	Unweighted CPS	Deflated Weighted CPS	Weighted CPS	"True" Sex Ratios
18–19	1.38	1.08	1.07.	1.14	1.12
20–24	1.11	1.24	1.30	1.11	1.04
25–29	1.39	1.34	1.34	1.17	1.06
30–34	1.33	1.50	1.53	1.27	1.14
35–39	1.51	1.44	1.48	1.35	1.16
40–44	1.25	1.54	1.58	1.32	1.14
45–54	1.21	1.36	1.38	1.29	1.13
55–64	1.16	1.34	1.38	1.30	1.24
65–74	1.47	1.58	1.60	1.44	1.40
75+	1.89	2.02	2.01	1.92	2.00
18–29	1.26	1.24	1.27	1.14	1.07
30–44	1.36	1.49	1.53	1.31	1.14
45+	1.30	1.47	1.49	1.39	1.31
Total	1.31	1.40	1.43	1.28	1.18

Source: Westat tabulation of CPS microfile.
Note: See text for method of calculating deflated weighted estimates and true counts.

1.4 Analytic Results

It is useful to start with an examination of the data in Table 3. The weighted sex ratios are relatively flat between 25 and 55 years of age at which point differential mortality starts showing an effect in higher sex ratios. The ratios are generally higher for Hispanics than for whites and other, and still higher for blacks. The patterns are quite similar in the three time periods covered. The only group for which the unweighted ratios are consistently lower than for the weighted ones is Hispanics 55 to 64 years of age. The Hispanic independent figures used as controls in the poststratification are weaker than for the other two race ethnic groups and one should not draw major conclusions from this single reversal of the common pattern.

In any case, the three years are close enough so that it does not make much difference which one is used to compare with the RDD study. We will use March 1985 since that is the period closest to the time of the RDD study.

Table 3. Sex Ratios in Telephone Households in CPS and NHIS by Age, Race, and Hispanic Origin

Race or Hispanic Origin and CPS Age	CPS, March 1986			CPS, March 1985				NHIS, 1983	
	Unweighted	Weighted	Deflated Weighted	Unweighted	Weighted	Deflated Weighted	Age	Unweighted	Weighted
Black									
18-19	1.29	1.15	1.35	1.08	1.14	1.07	17-24	1.13	1.06
20-24	1.13	1.10	1.16	1.24	1.11	1.30			
25-39	1.36	1.20	1.35	1.42	1.25	1.44	25-34	1.49	1.24
40-44	1.40	1.32	1.46	1.54	1.32	1.58	35-44	1.37	1.25
45-54	1.21	1.27	1.20	1.36	1.29	1.38	45-54	1.47	1.30
55-64	1.38	1.30	1.44	1.34	1.30	1.38	55-64	1.37	1.34
65+	1.74	1.57	1.71	1.72	1.60	1.73	65+	1.58	1.59
Hispanic[a]									
18-19	0.86	0.99	0.88	1.08	1.19	xx	xx	xx	xx
20-24	0.99	0.87	0.93	1.10	0.90	xx	xx	xx	xx
25-39	1.16	1.01	1.10	1.18	1.09	xx	xx	xx	xx
40-44	1.21	1.12	1.22	1.21	1.08	xx	xx	xx	xx
45-54	1.12	1.11	1.22	1.16	1.08	xx	xx	xx	xx
55-64	1.25	1.29	1.43	1.24	1.27	xx	xx	xx	xx
65+	1.43	1.39	1.54	1.42	1.42	xx	xx	xx	xx
Other[b]									
18-19	0.97	0.98	0.98	1.02	1.00	1.00	17-24	1.06	1.02
20-24	1.06	1.04	1.06	1.08	1.05	1.10			
25-39	1.06	1.02	1.06	1.04	1.02	1.04	25-34	1.07	1.04
40-44	1.02	1.03	1.03	1.04	1.04	1.06	35-44	1.08	1.05
45-54	1.06	1.06	1.06	1.09	1.06	1.11	45-54	1.09	1.06
55-64	1.13	1.13	1.13	1.12	1.13	1.13	55-64	1.15	1.15
65+	1.47	1.44	1.48	1.47	1.45	1.49	65+	1.43	1.46

[a]Hispanic data not available for NHIS.
[b]'Other' includes Hispanics for the NHIS.
Source: Westat tabulations of CPS public use tape and NCHS tabulations of NHIS.

Table 3 also indicates close agreement between the deflated weighted ratios and the unweighted ones in both 1985 and 1986. Apparently, the variability in the CPS sampling fractions and nonresponse adjustment have only minor effects on the sex ratios. We will therefore subsequently use the unweighted ratios and ignore the deflated weighted ones.

Table 1 shows the numbers of sample cases in the RDD study in each age and sex group. The sample sizes are rather skimpy for some ages, particularly 18 to 19, and 75 years and over. We have thus combined them into three broad age groups for which sampling errors are moderately low.

Table 2 permits comparisons between the Westat RDD sex ratios and the various types of CPS estimates. The RDD rates tend to be lower than the unweighted CPS, or the deflated weighted estimates. We draw the inference that the within-household coverage in the Westat RDD survey is at least as good, and possibly a little better than CPS. We cannot say with certainty that the RDD survey actually did better than CPS. The analysis is complicated by the fact that the RDD study was not based on a national survey, and the presence of sampling errors and of nonresponse. However, there is fairly strong evidence that it was at least as good. We should note that the four areas in the Oral and Pharyngeal Cancer study had somewhat higher sex ratios in 1980 than the United States as a whole for adult blacks in the ages with high levels of undercoverage. For blacks 15 to 59 years of age, the sex ratio was 1.16 in the four areas, slightly greater than the 1.14 in the total United States. The favorable performance of the RDD study is thus not due to the possibility that the areas had unusually low sex ratios.

We also draw the conclusion that this coverage is probably generally available in RDD surveys rather than being a unique feature of the Oral and Pharyngeal Cancer Study. The questions on household membership that were asked were fairly simple and direct and did not involve any unusual probing or other special devices. They consisted of two questions. The first was: "I would just like to ask you a few general questions about members of your household. First, how may people living in this household, including yourself, are at least 18 years old?" The second was: "And what is your age/are the ages of those adults? Please start with the oldest." There was an ending check that the number listed agreed with the answer to the first question. Most RDD surveys concerned about household membership are likely to ask similar questions, and should attain comparable coverage.

However, the fact that RDD coverage is at least as good as CPS does not imply there are no coverage problems. Table 2 also indicates that RDD does a little worse than was achieved in the 1980 census and is considerably lower than the "true" sex ratios. For all black persons 18 years and over, the RDD survey's sex ratio was 1.31 compared to the "true" ratio of 1.18. The weighted CPS figure was 1.28, only slightly better than the unweighted RDD ratio. RDD looks good when compared to CPS, but both do rather poorly.

2. COVERAGE IN ADULT SMOKING BEHAVIOR AND ATTITUDE STUDY

2.1 Description of Westat Adult Smoking Study

During the months of September 1986 through January 1987, Westat conducted a computer-assisted telephone interviewing (CATI) survey of the U.S. adult population aged 17 years and above to measure their smoking behavior, attitudes toward smoking, and knowledge of health and other factors associated with smoking. An RDD sample was selected using a modification of the Mitofsky-Waksberg sampling procedure. The sampling frame covered all telephone exchanges in the continental United States, plus Alaska and Hawaii.

A sample of 36,405 telephone numbers in 2,427 clusters was designated for screening (15 numbers per cluster of 100 numbers). These telephone numbers resulted in the identification of 21,982 households. The residential status of an additional 292 telephone numbers (0.8 percent of all numbers) could not be determined. Group quarters with more than five persons were excluded from the sample. The screening was completed in 18,756 households which constituted over 85 percent of the identified households.

The household screening instrument included a listing of all household members along with their respective sex and age. An adult household member was first asked to provide the desired data for each person living in the household. He or she was then asked about any students who usually lived there but were away from home and living in a dormitory, fraternity, or sorority house. This is a little different from the CPS residence rules, which place all college students in their parent's households. The average household size in Westat's Adult Smoking Behavior and Attitude Study can thus be expected to be slightly below that reported in the CPS. Upon completion of the listing, the number of household members, including students, was verified. The relevant screening questions are shown in Table 4.

Race data was collected only for the one household member selected to be the respondent for the detailed interview and not about all household members. In addition, 13 percent of the selected respondents in households that cooperated with the screening effort did not complete the detailed interview. Consequently, it was not possible to disentangle

Table 4. Screening Questions: Adult Smoking Behavior and Attitude Study

S7	May I have your age please? ASK IF NOT OBVIOUS: Are you male or female?
S8	May I have the ages of the other people who live in the household? ASK IF NOT VOLUNTEERED: Is the (AGE) year old male or female?
S9	Are there any students who usually live here but are away from home and living in a dormitory, fraternity or sorority house?
S10	May I have the ages of the students who are living in a dormitory, fraternity or sorority house? ASK IF NOT VOLUNTEERED: Is the (AGE) year old male or female?
S11	Including students away at college, I have recorded that there are (NUMBER) people in the household. Is this correct? IF INCORRECT: Let me check what I have recorded. INTERVIEWER READ AND UPDATE AS APPROPRIATE THE AGE AND SEX OF EACH HOUSEHOLD MEMBER. ADD OR DELETE HOUSEHOLD MEMBERS IF REQUIRED.

the effects of coverage and differential response rates at the extended interview level. Therefore, the analysis presented below is based on data from the screening instrument only and race breakdowns are not reported.

The RDD sample for the Adult Smoking Behavior and Attitude Study was not designed to yield a self-weighting sample. A fixed number of telephone numbers per cluster was used, instead of a fixed number of households, resulting in some variation in probability of selection among clusters. In addition, as is usual in RDD, households with more than one telephone number had multiple chances of selection. The data have been weighted to adjust for these differential probabilities but do not include any poststratification to independent population figures.

2.2 Data Presented

The information presented in Table 5 compares the household size data from the Westat Adult Smoking Study to that obtained from the unweighted March 1986 CPS. The table also compares the sex ratios within household size categories.

The data displayed in Table 6 shows the number of persons listed in the households screened for the Westat survey as well as age-specific sex ratios. Data about persons under 18 years of age were collected as part of the study's household screening effort (unlike the Oral and Pharyngeal Cancer Study) and are included in the table. However, in order to

Table 5. Comparison of Household Size Distribution and Sex Ratios in Westat's Adult Smoking Behavior and Attitude Study with March 1986 CPS

	Percent of Households		Sex Ratios	
Household Size	Westat RDD Study	CPS (Unweighted)	Westat RDD Study	CPS (Unweighted)
1 person	20.9	22.7	1.60	1.85
2 persons	34.1	31.7	1.10	1.10
3 persons	18.6	18.1	1.04	1.06
4 persons	16.2	16.1	0.97	1.00
5 persons	6.9	7.4	1.00	0.98
6 persons	2.1	2.5	1.05	0.99
7 persons	1.2	1.5	1.01	1.00
Total	100.0	100.0	1.07	1.09
Mean household size	2.66	2.68	—	—

Source: CPS data from Westat tabulation of telephone households from CPS public use tape.

simplify comparisons with the results derived from the Oral and Pharyngeal Cancer Study, the data for this age group have not been combined with that for household members 18 years of age or older.

The overall sex ratio is essentially the same for both the unweighted and weighted data displayed in Table 6. The pattern across the detailed age groups is considerably less erratic than was found in the Oral and Pharyngeal Cancer Study, probably because of the larger sample size. There is a slight predominance of males among the Adult Smoking Study's household members under 20 years of age, with females predominating thereafter, increasingly so among the three oldest age groups.

Table 7 compares the sex ratios in each age group computed from the Westat survey data to those derived from the March 1986 CPS. March 1986 is close to the period during which Westat conducted its survey. Both the unweighted and weighted CPS ratios are shown here.

2.3 Analytic Results

The first measure examined, average household size, permits comparison between CPS and the Westat RDD survey with respect to total coverage

Table 6. Household Members in Westat's RDD Adult Smoking Behavior and Attitude Study, by Sex and Age

	Number of Persons				Sex Ratio	
	Females		Males			
Age	Unweighted	Weighted[a]	Unweighted	Weighted[a]	Unweighted	Weighted[a]
0–17	6,015	5,692	6,289	5,909	0.96	0.96
18–19	904	808	916	828	0.99	0.97
20–24	2,158	1,987	2,151	1,974	1.00	1.01
25–29	2,199	2,105	2,126	1,997	1.03	1.05
30–34	2,206	2,106	2,036	1,944	1.08	1.08
35–39	2,193	2,054	1,998	1,882	1.10	1.09
40–44	1,805	1,617	1,690	1,535	1.07	1.05
45–54	2,755	2,449	2,577	2,300	1.07	1.06
55–64	2,491	2,289	2,233	2,044	1.12	1.12
65–74	1,996	1,837	1,636	1,486	1.22	1.24
75+	1,209	1,152	735	717	1.64	1.61
0–17	6,015	5,692	6,289	5,909	0.96	0.96
18–29	5,266	4,899	5,193	4,800	1.01	1.02
30–44	6,204	5,776	5,724	5,359	1.08	1.08
45+	8,451	7,726	7,181	6,548	1.18	1.18
Total	25,931	24,096	24,387	22,616	1.06	1.07

[a]Adjusted for probabilities of selection.

Table 7. Comparison of Sex Ratios in Westat's RDD Adult Smoking Behavior and Attitude Study With March 1986 CPS Interview Results in Telephone Households

Age	Westat RDD Study	Unweighted CPS	Weighted CPS
0-17	0.96	0.94	0.95
18-19	0.97	0.99	1.00
20-24	1.01	1.06	1.03
25-29	1.05	1.11	1.04
30-34	1.08	1.07	1.02
35-39	1.09	1.09	1.05
40-44	1.05	1.07	1.06
45-54	1.06	1.08	1.08
55-64	1.12	1.15	1.15
65-74	1.24	1.32	1.28
75+	1.61	1.81	1.76
0-17	0.96	0.94	0.95
18-29	1.02	1.07	1.03
30-44	1.08	1.08	1.04
45+	1.18	1.24	1.23
Total	1.07	1.09	1.07

Source: CPS data from Westat tabulation of telephone households from CPS public use tape.

within telephone households. Analysis of the data in Table 5 shows that although the household size distribution for the Westat RDD survey is skewed somewhat toward two person households[4], and away, most notably, from one person households, the mean household sizes are almost identical. Some of the difference shown could be attributable to the slight difference in residence rules adopted by the two surveys. The face to face CPS and the Westat RDD studies appear to be comparable in their total coverage within telephone households.

Additional information is provided by the sex ratios presented in Table 5. Again, for most household size categories the results for the two surveys are quite close. The notable exception is the one person household where the sex ratio is smaller in the RDD survey. In addition, the ratio for six person households is higher for the RDD survey. However, the latter finding is based on a rather small sample.

[4]This distribution is similar to that found by Monsees and Massey (1979).

Table 7 contains comparisons of the sex ratios from the Westat RDD survey with both the unweighted and weighted CPS estimates. Because the sample sizes for the age and sex groups presented in Table 6 are quite large, the standard errors for the sex ratios in Table 7 are fairly small, less than .02 to .03 for the detailed age groups and .01 to .02 for the broader groups in the RDD study and even smaller for CPS.

Table 7 shows that the sex ratios from the Westat RDD survey are lower than those for the unweighted CPS in most of the specific age categories and that its overall ratio is also slightly lower than the unweighted CPS ratio. Coverage in the younger age groups (under age 30) appears to be somewhat better in the RDD survey. In addition, the RDD sex ratios are very comparable to those from the weighted CPS, with identical overall ratios. However, the magnitude of the differences in the two oldest age categories (age 65 and over) between the RDD and the weighted CPS ratios probably does not imply better coverage of males, but indicates poorer response rates in the RDD survey for females in these age groups.

Taken together, these findings lend considerable support to the conclusion drawn from the Oral and Pharyngeal Cancer Study that the coverage of persons in households with telephones generally available in RDD surveys is at least as good, if not better, than that provided by CPS. While sampling errors and the uncertain effects of nonresponse prevent us from concluding that the within-household coverage in the Westat RDD study was as good as the 1980 Census, the results are encouraging in regard to the quality of RDD surveys versus face to face interviews.

The results also point to a possible explanation for the lower sex ratio in one person households shown in Table 5. Given the RDD survey's apparent lower coverage of females aged 65 and above, it is likely that differential response rates across age groups, especially within one person households, are a potentially important explanatory factor. Specifically, older persons living alone, and older women in particular, seem to be more likely to withhold cooperation than others in single person households. Hence, while the overall quality of coverage offered by RDD surveys may well be comparable or superior to CPS, telephone survey researchers may still wish to consider incorporating into their instruments and procedures additional methods for soliciting and encouraging the cooperation of older persons.

3. EXPERIMENT COMPARING SHORTCUT IDENTIFICATION OF A TARGET POPULATION WITH MORE DETAILED QUESTIONS

The ability and willingness of household respondents to supply complete household rosters may be influenced by the form of the questionnaire. Groves and Kahn (1979) describe a number of alternative approaches to identifying household members. They came to the conclusion that obtaining a full household listing was preferable to several shortcut procedures. The purpose of obtaining household members in their study was to aid in the selection of one random adult in each sample household and it is not clear whether the same conclusion would apply to other survey goals. Bercini and Massey (1979) examined whether placing the request for a household roster early in an interview creates higher nonresponse than putting it near the end.

A different issue in regard to completeness of reporting occurs when one is not interested in the entire population, but wants to study a particular subset. Screening within households can be reduced if the respondent is given a description of the target population. However, with such an approach there is a possibility that some respondents will then deny the existence of members of the target group in order to shorten or end the interview. In this section of the chapter, we describe an experiment regarding the effect on coverage of informing respondents about the target group.

Westat conducted an RDD survey in 1986 and 1987 in which the household screening attempted to identify persons 13 to 24 years of age in the sampled households. In order to reduce the response burden on the approximately 70 percent of the households without persons in this age range, it seemed desirable to use a simple question on whether the household contained members in the target ages, rather than asking for a full household roster. An experiment was therefore introduced to compare the quality of responses when the simple question was used with the quality for an alternative set of questions in which the target population was masked.

The experiment was carried out as part of an RDD survey conducted for the Department of the Army. The study was carried out continuously with a separate national sample of households introduced each month, although data were cumulated over a three month period for analysis purposes. The experiment was carried out during the last quarter of 1986 and the first quarter of 1987 thus minimizing the possible impact of seasonal effects due to lengthy absences from the household during summer vacation. The sample size was quite large, consisting of 17,258 completed screeners in the October to December 1986 quarter and 16,601 completed screeners in January to March 1987. The six monthly samples in the two quarters were approximately equal, ranging from a low of 5,228 to a maximum of 5,893. The response rates for the screeners were 84 percent for the October to December quarter, and 81 percent for the January to March quarter.

A single question about the presence of members of the target population was asked in the October to December quarter. Household respondents were asked:

> Since the survey we are conducting for the U.S. government is concerned with the career plans of young adults, we need to know how many young adults live in your household. Including anyone away on vacation, away on business or living away at school, how many young people between the ages of 13 and 24 live in your household?

In the following quarter, the target population was disguised. Respondents were asked two questions:

> We have a few questions to see if anyone in your household will be included in this survey.
>
> How many people aged 25 or older live in your household?
>
> How many people between the ages of 13 and 24 live in your household?

The reporting of household membership was virtually identical in the two quarters. The key statistic examined was the percentage of households with at least one person 13 to 24 years of age. This percentage was 30.4 in October to December and 30.1 in January to March. In fact, there was considerable uniformity among the six months. The monthly percentages are shown below.

Thus, this experiment does not support the hypothesis that respondents will attempt to reduce their involvement in surveys by denying the presence of members of the target population if the definition of the target population is made clear to the respondents at the beginning of the interview.

Month	Percent of households with members 13 to 24 years old
October	31.1
November	30.0
December	30.1
January	30.8
February	29.9
March	29.5

SECTION B

SAMPLING FOR TELEPHONE SURVEYS

CHAPTER 5

TELEPHONE SAMPLING METHODS IN THE UNITED STATES

James M. Lepkowski[1]
The University of Michigan

As the use of telephone interviewing for household surveys has grown in the last 25 years, telephone sampling methods have increased in diversity. Early telephone sample designs used telephone directories as frames for selection because directories were readily available and were thought to contain a "representative" portion of the telephone household population. When the proportion of unlisted telephone numbers was small, the representativeness argument was persuasive, especially given the convenience and cost advantages of telephone surveys relative to the personal visit mode. Unfortunately, the frequency of unlisted numbers increased to a level that raised concerns about the accuracy of telephone surveys based on directory samples.

In response to these directory sampling problems, three other approaches to telephone sampling were pursued. First, *random digit dialing* (RDD) provided coverage of both listed and unlisted telephone households by generating telephone numbers at random from the frame of all possible telephone numbers. Second, *list-assisted designs* used information in telephone directories (or other frames based on telephone directories) to generate telephone number samples that include both listed and unlisted telephone households. And third, *multiple frame sampling methods* combined directory and RDD sampling frames and sampling methods in a single design.

[1] The preparation of this manuscript was partially supported by the Bureau of the Census through Joint Statistical Agreement 87-15. I thank Joe Waksberg, Bob Mason, and Bob Casady for their careful reading of earlier drafts and for suggestions for improving the manuscript.

Despite the obvious improvements in coverage offered by RDD, list-assisted, and multiple frame methods, they have features that detract from the convenience and cost advantages of the earlier methods of conducting telephone surveys. As a result, directory sampling methods continue to be used. For example, local community telephone surveys with short survey periods often use directory samples despite the shortcomings in coverage. Thus, the evolution of telephone sampling design has expanded the range of designs in use rather than selected a single all-purpose design.

The purpose of this chapter is to review sampling methods for telephone population surveys as they have evolved from directory methods to RDD, list-assisted, and multiple frame methods. Since many of the designs were developed to address deficiencies in other methods, it is natural to compare the features of the designs. Discussions of specific sampling techniques contrast the coverage and nonresponse properties, sampling variance, and sampling and interviewing costs associated with various designs.

The review is restricted to probability sampling methods for telephone household surveys. The discussion is further limited to the U.S. telephone system and concentrates primarily on surveys that are regional or national in extent. Despite this limitation to the discussion, it should be recognized that many of the methods described can be used in surveys outside the United States, and for local or community surveys.

Before consideration of specific sampling methods, a brief review of the U.S. telephone system is given in the next section. Then sampling methods for three principal telephone sampling frames, directories (Section 2), commercial lists (Section 3), and telephone numbers (Section 4), are described. Two approaches to combining the strengths of RDD and directory based methods, list-assisted and dual frame sampling, are described in Section 5. The chapter concludes with remarks about likely future developments in telephone sample design.

1. TELEPHONE NUMBERS AND TELEPHONE SAMPLING FRAMES

The characteristics of the U.S. telephone system are well known, but for the sake of definition of terms, a brief review is given. Telephone numbers consist of 10 digits: the first three are the *area code* (there are currently 111 in the United States and another 14 in Canada), and the next three digits are the *prefix* (referred to as *central office codes* by the telephone system and colloquially as *the exchange*. Area code numbers must be different from numbers used for prefixes. More than 37,000 area code-prefix combinations are currently used in the U.S. telephone system. The last four digits or *suffix* identify a particular subscriber within an area code-prefix combination.

Geographic grouping of telephone numbers is limited. Area codes are both telephone number and geographic groupings, but prefixes do not necessarily have geographic definition. Suffixes within prefixes are assigned arbitrarily without geographic criteria. After the area code, the only other unit that routinely has geographic definition is the *exchange*, a geographic area serviced by a single telephone business office. Although the terms exchange and prefix are often used interchangeably, exchanges can contain two or more prefixes; nearly half contain only one prefix.[2] Within the exchange, there is seldom geographic assignment of prefixes. Some large exchanges do have *wire centers* which handle subsets of prefixes within defined geographic areas in the exchange, but these are exceptions rather than the rule.

Exchange and area code boundaries generally do not correspond to political or Census boundaries, except perhaps at a state level. This lack of correspondence reduces the quality of telephone-to-political-unit matches, which might be useful in expanding information on telephone units for purposes of sample selection. It also makes selection of a telephone sample from a set of political units awkward. A telephone sample of residents in a county, for instance, may require identification of all exchanges which cover or extend into the county and a screening question in the survey instrument to determine if the household is in the county and therefore eligible for the study. The screening operation will increase the costs of sample selection and may reduce response rates.

The *sampling frame*, the set of materials from which a sample of telephone households is selected, is a key feature of telephone sample designs. It can largely determine the type and feasibility of sampling methods that can be used. The frame determines the coverage of telephone households, and it also may provide auxiliary information that can be used to reduce nonresponse, a persistent problem for telephone surveys. The frame also can have a substantial impact on the cost of conducting a survey. For example, a frame that has a substantial proportion of ineligible listings (i.e., listings that are not eligible for the study) will increase the costs of identifying eligible ones.

There are three basic frames available for general population telephone household sampling: (1) directories of telephone subscribers, (2) commercial lists derived from directories, and (3) the set of all possible

---

[2]For example, among the 37,712 exchanges in the 48 coterminous states as of October 31, 1987, 15,667 or 41.5 percent had only a single prefix; all others had two or more.

telephone numbers. Other frames are available but are not widely used for general purpose telephone surveys. Each of these frames has different properties that must be considered in sampling design. Among the most important deficiencies are three frame deficiencies: (1) ineligible listings, for which there is no telephone household that is assigned to a frame element; (2) duplicate listings, for which there are two or more frame elements that identify the same telephone household; and (3) noncoverage, in which telephone households in the population do not have any frame element which can identify them. Some or all of these deficiencies are present in each frame, but the frequency varies from frame to frame.

2. DIRECTORY SAMPLING

The directory frame consists of the printed name, address, and telephone numbers of subscribers living in a geographic area covered by the directory. The characteristics of the directory as a frame are familiar, including the important directory frame problem, failure to cover unlisted or unpublished telephone households, who pay to suppress their listing, and failure to cover those subscribing after publication of the directory. However, ineligible and duplicate listings must also be contended with. Ineligible listings occur when a listed subscriber has discontinued service or when a listed number is a business or other unit that is not a household. The latter problem appears to be decreasing in frequency as more directories are listing residential and nonresidential numbers in separate sections. Duplicate listings occur when households have more than one listed telephone number or purchase listings for several members of the household.

Directory samples will produce biased estimates for the entire population of telephone households to the extent that listed and unlisted households differ with respect to the survey measures. Comparisons of listed and unlisted household characteristics show important differences for many measures (Brunner and Brunner, 1971; Fletcher and Thompson, 1974; Glasser and Metzger, 1975; Perry, 1968; Rich, 1977; Roslow and Roslow, 1972). Thus, there is a potential for substantial coverage bias in directory samples. The size of the bias will depend on both the size of the difference between listed and unlisted household characteristics and the size of the proportion not covered. The noncoverage rates vary across population characteristics. For example, the proportion unlisted is markedly lower in rural areas (Groves and Lepkowski, 1986); directory samples in rural areas may be subject to considerably less coverage bias than in urban locations.

An important feature of a sampling frame is the availability of auxiliary information to improve sample selection. Auxiliary information available in a directory is limited. The name may provide an indication of some household characteristic (e.g., ethnic group). The address information can be used to link individual listings to geographically specific data from other sources, but the alphabetic sorting of the list makes manual matching of an entire directory tedious. There is little other useful auxiliary information available; stratification or other controls cannot be used effectively in most directory sampling.

Because directory sampling is primarily a clerical task, relatively straightforward methods are used. For example, a systematic sample of directory lines can be selected from the alphabetically ordered list; simple random selection of lines is a needless complication. For a sample of desired size n_d households, a larger number of lines must be selected to account for blank listings and nonresponse. A sampling interval k is calculated as the ratio of an estimated total number of lines in the directory to the number of lines needed to achieve the desired sample of households (denoted by m_d). A random start from 1 to k is recommended. Alternatively, selection can be made by measuring total column inches and determining a column-inch interval. Selected listings are identified by measuring column-inch intervals after a random start. Templates for counting lines can also be used to speed the process and assure greater accuracy in sample selection. Stratified selection is generally not employed because of the limited auxiliary information available and the expense of reordering the list. Some geographic stratification is achieved when directories divide an area into exchanges and samples are selected systematically across exchanges.

Since the counting process can be tedious, the sample may be clustered. For example, all households identified by the selected line and the next $b - 1$ lines can be selected by increasing the interval to $m_d = k \cdot b$. One may also sample pages systematically or at random and then subsample listings within selected pages at a fixed rate usually applied systematically.

Care must be exercised to be sure that the selection procedure achieves known, nonzero chances of selection. Substituting the next listed household for a selected line that is not a household introduces unequal probabilities of selection and, if not corrected in estimation, bias. Alternatively, ineligible listings can be skipped, with an increased sample size of listings selected to account for anticipated ineligible listings. Multiple line listings must also be selected carefully to avoid bias. For example, a rule can specify that a household can be selected only if the line with the telephone number is selected; other lines are treated as ineligible listings.

For purposes of variance estimation, it is often assumed that systematic selection from the alphabetically sorted directory is equivalent to a simple random selection of listed households. Because of ineligible listings, the achieved sample size, denoted by n_d, is not a fixed quantity but a random variable. The sample mean $\bar{y} = y/n_d$, where $y = \sum y_i$ is the sample total for characteristic Y, is a ratio estimator that is technically biased. However, conditional on a fixed sample size, the usual simple random sampling variance estimate

$$\text{var}(\bar{y}) = \frac{1}{n_d(n_d)} \left(\sum y_i^2 - \frac{(\sum y_i)^2}{n_d} \right)$$

can be used. Application of this formula to clustered directory samples is likely to underestimate the variance slightly. In surveys where person level analysis is conducted, this formula will usually also understate the sampling variance.[3]

Despite coverage problems, directories remain a popular frame for local surveys because they can be obtained at low or no cost. However, there are other sampling costs arising from the clerical task of selecting listings from a directory. Interviewing costs will also be affected by the fact that some listed numbers will have been disconnected. Directory sampling may also incur additional costs to improve response rates when advance letters are mailed.

3. COMMERCIAL LIST SAMPLING

Two types of commercial list frames can be distinguished: city directories and master address lists. Both are based on telephone directories, with features added by a commercial firm to enhance their utility for telephone sampling. City directories are printed address listings for a city with a telephone number for every address obtained from a directory or from household canvassing. The city directory may include lists ordered both geographically and by telephone number.

Master address lists are nationwide in extent and are machine readable, greatly facilitating selection of widespread or geographically specified populations. Some master address lists are assembled by

[3] When one person is selected per household, the estimator for $\bar{y}$ needs to have household weights, and the variance will be increased.

collecting all telephone directories in the country, manually comparing directory and current list entries, keying differences that are found, and updating the list frame through computer processing. In one frame, the updating process is performed continually, revising the master address list within six to eight weeks of directory publication (Donnelley Marketing, 1986). Other sources are used to improve the coverage of the list. For example, computerized automobile registrations obtained from states for which they are publicly available (approximately 30 states at present) are, through a combination of computer and manual matching, used to identify unique information (e.g., unlisted households) from the automobile registration files that can be added to the list (Donnelley, 1986).

Master address lists often contain detailed geographic information for listings. For example, the zip code and Census block data are added by manually matching the listed address to post office and Census maps. Master address lists also may include household characteristics such as income and household size imputed by an algorithm based on Census block, tract, and other information. This detailed geographic and other information can facilitate stratified selection and selection of a geographically defined target population.

Commercial lists have coverage deficiencies similar to those of directories. In addition, if a firm fails to obtain a directory for a given area (particularly remote rural areas), portions of the telephone household population will not be covered. Ineligible listings may be somewhat more common in master address lists than in telephone directories because of the time required to keypunch revisions and update the list.

Auxiliary information and computerization provide additional opportunities for sample selection from commercial lists, especially master address lists. Since the address lists are typically sorted by geography (e.g., zip code and street address within zip code), systematic selection from the list will implicitly stratify the selection by geography. Stratified selection of telephone numbers is thus direct and convenient. There does not appear to be any advantage to clustered sample selection, although it could be employed.

Since commercial firms maintain their own lists, they often provide sample selection services. However, commercial firms primarily sample from the list for mass mailing firms and may not be capable of developing specialized selection procedures for a client interested in telephone samples. Selection by the commercial firm may be a benefit for survey operations lacking sampling expertise. For other operations, which have such expertise, it is often unacceptable for this critical phase of survey operations to be out of direct control. Of course, clients may purchase the list and select samples themselves, but for large populations the cost may be prohibitive since the commercial firm charges are usually based on the number of names supplied.

Depending on the requirements of the sample design, the size of the population to be surveyed, and the sort order of the list, commercial lists can have relatively high sampling costs. City directories must generally be leased, and the lease can be a large item in a small survey budget. Master address sampling costs may be based on a single charge for each selected listing. On the other hand, Groves and Lepkowski (1986) report on a pricing algorithm based on a two-part fee, one part for each record passed in the portion of the list being accessed and another part for each listing selection. They report that the combined passing and selection cost of sampling from one firm's list (i.e., the Metromail Corporation) was approximately $0.75 per sample listing for a sample of 800 listings in the state of Michigan. Since the algorithm was based on a fee for passing all records in Michigan (a fixed cost for Michigan samples), larger samples would typically be less expensive per selection. However, a national sample selected on the same basis would have been much more expensive because of the large fixed cost of passing more than 70 million records in the complete master file.

Systematic selection of lines or listings can be assumed to be equivalent to stratified random sampling in which implicit strata (i.e., groups of consecutive selections) are collapsed to form strata of several selections each. Under an assumption of random ordering of the population within these collapsed strata, the usual stratified expressions for estimates and their sampling errors can be applied. For example, under proportionate allocation (equal numbers of selections in each collapsed implicit stratum), $\bar{y} = \sum_h \sum_i y_{hi}/n_d$ where h denotes the hth stratum. The variance is estimated as

$$\text{var}(\bar{y}) = \sum_h \frac{1}{n_d(n_d - 1)} \left(\sum_i y_{hi}^2 - \frac{(\sum_i y_{hi})^2}{n_{hd}} \right),$$

where it is assumed that the stratum sample sizes n_{hd} are fixed.

A cost and error comparison of telephone directory, city directory, and master address sampling is appropriate in choosing a frame. Commercial lists allow stratified selection that can produce gains in precision compared to unstratified sample selection. However, the larger expenditure for frame acquisition and sample selection associated with a commercial list may overwhelm gains in precision due to stratification. For proportionate allocation, the gains in precision are often only modest and may not justify substantially larger sampling costs. Disproportionate optimal allocations may be employed to achieve larger gains, but multipurpose surveys will seldom find consistent gains across a reasonably large number of estimates.

The comparison of sampling variance and cost may be less important than bias considerations. Those commercial lists that supplement telephone directories with other sources (e.g., telephone numbers of persons with automobile registrations) may provide improved coverage over simple uses of directories. Where sampling variance is small relative to the size of the estimate, slight bias reduction through the use of a commercial list may outweigh the convenience of telephone directory sampling. On the other hand, for estimates for small subclasses, sampling variance will usually be fairly large, and modest bias reduction considerations are not as important.

4. RANDOM DIGIT DIALING SAMPLING

Random digit dialing (RDD) methods are based on the frame of all possible telephone numbers. The telephone number frame is commonly assembled by appending suffixes to area code-prefix combinations obtained from Bell Communications Research (BCR) for a fee ($320 in 1988). The BCR list includes an exchange name, geographic coordinates of the center of the exchange, time zone, and limited additional information for each area code-prefix combination in the United States and Canada. BCR also includes new area code-prefix combinations three months before they are added to the telephone system.

Frame problems for the telephone number frame differ from those of directory based frames. One frame problem which cannot occur with an up to date telephone number frame is noncoverage (i.e., telephone households which are assigned a number that is not in the frame). Of course, an out of date list missing new area code-prefix combinations may miss sets of 10,000 numbers in use. While the telephone number frame in principle provides complete coverage of the telephone household population, ineligible listings are a much larger problem. Nationally, more than three-quarters of all possible telephone numbers generated from the BCR list are not assigned to households (Chilton, undated; Glasser and Metzger, 1972; Groves and Kahn, 1979). Duplicate listings occur for a telephone household because more than one telephone number is assigned to the household.

Because only a limited amount of auxiliary information is available for each telephone number in this frame, additional information is often sought to improve sample design efficiency. For example, stratification of sampling units from the telephone number frame is limited to the few system variables and their derivatives provided with the BCR file. Geographic ordering with systematic selection provides implicit stratification of exchanges; counts of prefixes can be used as size measures for exchanges. Additional information may be sought from the telephone system itself by contacting business offices located within exchanges and asking for detailed information about their local system (e.g., the location of sets of numbers that are not assigned). However, telephone business offices are becoming increasingly less cooperative about releasing the status of a telephone number; they can no longer be routinely depended on to help determine the status of banks or of individual telephone numbers. Linkages to other databases, such as the Census, are another approach, but the irregular boundaries of area codes and exchanges limit the ability to create linkages. For example, Landenberger, Groves, and Lepkowski (1984) describe manual linkages in which exchange and Census maps were compared visually. The exchange maps typically provide inadequate detail for precise matching, and the level of effort needed and the quality of the linkage have been discouraging. Computer algorithm linkages may be feasible if exchange maps are digitized in a system compatible with Census map digitization. [A data file that provides information useful for stratification and geographic coding has recently become available. It is described in Mohadjer's paper (Chapter 10)].

The limited geographic content of the telephone number frame also complicates sampling a geographically defined population. Surveys using the telephone number frame must rely on respondent reports of housing unit location in order to identify the target area precisely. There is, of course, a cost associated with obtaining geographic information from respondents, and response errors lead to misclassification. Some increase in nonresponse rates may also occur when detailed geographic information is requested, especially at an early point in an interview.

Although the cost of obtaining the area code and central office code combinations from BCR is small, and four digit suffixes can be generated at random with little cost, there are other potentially large costs of using this frame. The large proportion of ineligible listings increases data collection costs since interviewer (and sometimes supervisor) time is required to determine the status of the number. Much of the development of alternative telephone sample designs has been directed at reducing the proportion of blank and other problem listings in the sample to reduce the costs of using this frame.

4.1 Random Digit Dialing Element Sampling Methods

A naive approach to random telephone number generation would be to generate 10-digit numbers at random. However, these randomly generated numbers would include numbers with area codes and prefixes that are not in use. Cooper (1964) first proposed RDD element sampling by adding four-digit suffixes to known prefixes to generate numbers for a local survey, although systematic or other generation methods could be used (Payne, 1974). In these early RDD element samples, the list of prefixes was obtained from telephone directories in the local area. Glasser and Metzger (1972) reported a national survey application in which extensive effort was devoted to obtaining a national list of area codes and prefixes.

Among telephone numbers generated by appending four-digit random numbers to known area code-prefix combinations, 75 to 80 percent are not assigned to a household. Further, a single dialing does not necessarily provide enough information to determine the status of a number. A number dialing may result in an answer and immediate identification as a household, a business, or another type of nonresidential unit. However, in RDD samples the result of dialing is, compared to directory based samples, more frequently one that requires further dialing to resolve. For example, not all nonworking numbers are answered by a recording or an operator intercept indicating that the number is not in service. Particularly in rural areas, nonworking numbers may simply switch to a ringing machine when dialed. Unproductive calls are necessary for numbers that are not in service because in many cases they ring without any answer. Ring without answer numbers may remain unanswered throughout a survey period. The survey staff may be required to contact a local business office to try to determine the household status of the number (which is less and less frequently a successful practice) or an arbitrary rule may be established which classifies persistent ring without answer numbers as nonresidential.

RDD samples also tend to have more dialings that result in wrong connections (identified only by verification of the number), fast busy signals, and silence or strange noises. Wrong connections, fast busy signals, silence, and strange noises require confirmation by redialing, often over a period of several days to be certain that the result is not just a temporary phenomenon. Since procedures must be specified for each dialing result, the telephone number frame increases the administrative and operational costs to telephone survey operations that would not be present with a frame with fewer blank listings.

Despite the large number of ineligible listings, RDD element samples may be treated as simple random samples for purposes of estimation. Of course, many more telephone numbers than the desired number of telephone households must be generated to account for ineligible listings and nonresponse. The method yields a simple random selection of telephone households, but the achieved sample size is not a fixed quantity but a random variable. A sample estimate of the form y/n is a ratio estimator, but estimation is made conditional on fixed n. In designs which might also use stratification (e.g., separate selection by state or region), the gains from stratification are likely to be diminished by the high proportion of ineligible listings [see, for example, Kish (1965), section 4.5]. Stratified sampling formulas can be used, but simple random sampling formulas will likely provide only a slight overestimate of precision at lower computing costs.

The problems that early RDD element sampling uncovered are not evenly distributed throughout the frame of telephone numbers. Rural areas have more wrong connection, ring without answer, and other problem numbers than urban areas. Exchanges in rural areas also have higher rates of ineligible numbers than urban areas, and because of the mechanical switching equipment still in use in many rural areas, frequently only banks of suffixes defined by the first of the last four digits (i.e., 1,000 banks) are used. At the same time, in urban areas prefixes assigned for business use can be identified.

With advance preparation and the cooperation of telephone business offices, banks without residential numbers and nonresidential prefixes can be eliminated prior to number generation. This is an expensive process, especially for the entire United States. Using exchanges as primary units in a multistage sample can reduce the cost of this operation, but a consideration of the cost efficiency of the multistage RDD design must include an assessment of increased variances associated with the clustered selection. Chilton Research Service eliminated 1,000 banks of nonworking numbers from their telephone number frame using such a multistage design, and increased the proportion of residential numbers in their telephone number frame to nearly 50 percent (Nelson, 1977).

Other approaches can be used to eliminate or reduce the number of banks of telephones not in service, and thereby decrease the proportion of numbers that are ineligible. For banks of 100 consecutive numbers defined by the first 8 digits of the telephone number, the number of listed numbers can be determined for every 100 banks in the survey area using reverse directories in a local area or from a commercial list for larger areas. Banks with no or only a few listed numbers are eliminated from the frame. Approximately 65 to 70 percent of the numbers in the remaining 100 banks are found to be residential (Survey Sampling, Inc., 1986). Another approach in repetitive survey operations is to use information from prior surveys to identify the assignment status of 100 or 1,000 banks

A second innovation introduced by Potthoff also addresses these problems. Suppose that at the first stage c numbers are selected at random and without replacement within a prefix area. To achieve equal probabilities of selection for households, three results of the dialing of the c primary numbers are important.

1. If all primary numbers are inauspicious, the prefix area is not subsampled further.

2. If two or more primaries are auspicious, exactly $c \cdot b$ numbers are generated at the second stage and dialed. Whether these secondary numbers are auspicious or not, no additional numbers are generated in these prefix areas. Thus, for a large proportion of prefix areas, the second stage sampling does not require replacement.

3. If only one primary number is auspicious, two sets of numbers are generated at the second stage. A total of $c(b-1)$ numbers are generated in a *nonsequential segment* and dialed just as in the prefix areas where two or more primary numbers were auspicious. An additional b numbers are then generated and dialed as part of a *sequential segment* within the same prefix area. If a sequential segment number is inauspicious when dialed, it is replaced with another randomly generated number from the same prefix area. The second stage sampling continues in the sequential segment until a total of b auspicious numbers have been generated.

Potthoff shows that this procedure yields an equal probability sample of households. For only a small number of selected prefix areas is replacement required, and then only for a small segment of numbers. The procedure also reduces ambiguities about the status of numbers dialed at the first stage and, as c increases, reduces the chances of obtaining a prefix area that will be exhausted in the second stage.

Implementation of this procedure requires knowledge about the proportion of auspicious numbers that are households in order to determine sample size. One must also be concerned about the increased administrative and training complexity of the procedure. More formal cost and error comparisons of the various approaches need to be made before the advantages of the procedure can be assessed.

There are several other features of the two-stage Mitofsky-Waksberg procedure that are of interest. A set of selected prefix areas may be used for more than one study or survey period. However, the measure of size used to select the prefix area will change over time, and subsequent second stage sampling at a later period will not necessarily be based on the correct measure of size for the prefix area. The subsequent samples will have probabilities of selection that are not equal. The extent of change that occurs in the measure of size for prefix areas has not been investigated, and the optimal length of time that prefix areas can be reused is unknown.

Deeper and more precise stratification of prefix areas is often desired. Groves and Kahn (1979) stratified prefix areas using the limited auxiliary information on the BCR tape. Manual matching of exchanges to Census units has been found to be costly and of limited value. However, auxiliary information obtained from commercial list frames can be used to provide more extensive stratification. Mohadjer (this volume, Chapter 10) provides an interesting use of commercial list information to develop more extensive auxiliary information for stratification.

The size of prefix area and the size of the subsample to be selected at the second stage are two closely related features of the two-stage design. The prefix area is commonly set at 1,000 or 100 consecutive numbers, but there is no theoretical restriction on prefix area size. Inglis, Groves, and Heeringa (1987) report that the density of household numbers does not decrease markedly when neighboring banks of selected prefix areas are examined. For prefix areas of 200 and 400 numbers, the density of household numbers remained virtually identical.

Optimal values of b for prefix areas of 100 numbers were examined in Waksberg (1978) and Groves (1978). Within-prefix-area homogeneity has been found in empirical work to be relatively low, which suggests that large values of b are appropriate. However, the costs of first stage sampling are usually small relative to the interviewing costs, and small values of b appear to be optimal. With small values of b, repeated use of the same set of prefix areas over several surveys or survey periods is often considered essential if benefits of the two-stage procedure are to be realized. Of course, the number of repeated uses depends both on the exhausted cluster problem and on the rate of change of the measures of size of prefix areas. A simultaneous investigation of these properties of the two-stage design is needed.

The two-stage design may also be modified to address some rare population sampling problems. To the extent that a rare population is clustered geographically within prefixes or banks, first- and second-stage rules may be modified to improve screening rates for subsets of households (Blair and Czaja, 1982; Inglis, Groves, and Heeringa, 1987). The modification could be extended to other units that are not households. For example, in a study of black households, the prefix area can be selected for subsampling if the sample number is a black household, and second stage sampling continues until a fixed number of black households is obtained. Czaja and Blair improved screening rates for black households from 9 percent using RDD methods to 21 percent using a two-stage procedure for black households. Waksberg (1983) notes several improvements that could be made to this approach, including improved stratification of prefix areas using auxiliary information from other sources. Inglis, Groves, and Heeringa (1987) further group prefix areas into high, medium, and low density black household strata. The black household selection rule at the first stage is not likely to yield similar gains for other populations because the geographic clustering observed among black households is not repeated among other population groups.

Estimation procedures for the two-stage designs may involve weighting to account for unequal probabilities of selection, stratified cluster variance formulas, and Taylor series approximation (or a repeated replication procedure) for variances of nonlinear estimators. Estimation procedures appropriate for the two-stage designs are given in standard textbooks on survey sampling [see, for example, chapter 6 in Kish (1965)]. The stratified cluster variance estimates for ratio means have, for the same size sample, larger variances than those obtained from a simple random sampling estimation procedure. The loss in precision for clusters that are prefix areas of 100 consecutive numbers is generally small, typically less than 20 percent (Groves, 1978). The loss is, on average, even smaller for subclass estimates. Nonetheless, the use of the appropriate sampling variance formulas is generally recommended to take account of the complexity of the design in estimation.

Despite higher proportions of household numbers obtained through the use of two-stage and other sampling procedures, RDD methods appear to be more expensive to implement than directory and commercial list sampling. Two stage procedures suffer losses in precision due to clustered sample selection. In addition, directory based methods may also provide lower nonresponse rates (and lower bias) when advance letters are used. Despite these advantages of directory based sampling methods, the coverage deficiencies of directories and commercial lists are considered to be so important that the lower efficiency and increased nonresponse rates for RDD methods are accepted.

5. LIST ASSISTED AND DUAL FRAME SAMPLING METHODS

List assisted and dual frame approaches have been used to capitalize on the complementary strengths of the directory and telephone number frames. Two types of list assisted methods are distinguished here. In the first, a sample from a list frame is used to generate a sample of telephone numbers which include unlisted as well as listed numbers. Generally, these list assisted procedures are designed to simplify sampling operations or to increase the proportion of productive dialings compared to RDD methods. In the second type of list assisted method, directory frame information is used to improve the characteristics of the telephone number frame. The directory frame information may include measures of size or related data for prefix areas as well as name and address of listed telephone subscribers selected for the RDD sample.

This classification scheme for telephone sampling designs is somewhat arbitrary, and some designs may be classified in more than one category. For example, the use of commercial lists to provide auxiliary information for stratification of prefix areas can be viewed either as a type of list assisted design in which a list provides characteristics for the telephone number frame or as a modification of the two-stage RDD design. The Sudman procedure described in Section 4.2 may also be considered to be either a list assisted design or a modification of the general two stage RDD design.

Not all list assisted methods are probability sampling procedures. Several employ devices which do not allow determination of the probability of selection for elements unless assumptions are made about the distribution of listed and unlisted telephone subscribers in the population. This feature has not detracted from the use of these designs because they still have properties that make them useful from a practical perspective.

5.1 List Assisted Sampling Based on Directory Samples

Plus digit sampling is a list assisted procedure in which a sample is selected from a directory and an integer is added to the last digit of the selected numbers. For example, in *plus one dialing*, one is added to the last digit of each listed telephone number selected from a directory. The resulting sample generally includes both listed and unlisted telephone numbers. Any digit from one to nine could be added, as could two or more digit numbers, and the digit(s) may also be randomly selected.

From a practical perspective, plus digit dialing yields a higher proportion of productive numbers than the RDD element design. For example, Landon and Banks (1977) found in a local survey that plus one dialing had a higher proportion of productive calls than an RDD element sample for the same area, but, at the same time, the plus one procedure also yielded different and apparently incorrect results. In contrast, Trendex, Inc. (1976) found in a national survey that samples obtained by addition of 10 to directory sample numbers yielded similar findings to an RDD element sample.

The plus digit approach has a number of theoretical problems. To establish that unlisted numbers have nonzero chance of being selected, it must be assumed that unlisted numbers are evenly mixed among listed numbers; this is an unlikely assumption and it is difficult to verify. Even under such an assumption, the probabilities of selection of telephone households are likely to be unequal. In general, the unknown probability feature of these designs has discouraged wider application despite their simplicity.

A closely related procedure is replacement of the last d digits (usually at least 2) of a directory sample number with randomly chosen digits. Again, it is questionable whether this procedure provides nonzero probabilities for all telephone households. The more digits randomized, the more likely that the probabilities of selection will be nonzero and nearly equal for listed and all unlisted numbers. But as d increases, the proportion of productive calls also decreases. As with the addition of a digit, the method is simple and appealing for small scale operations, but its theoretical properties detract from its utility.

Instead of generating a single number from each directory selection, Frankel and Frankel (1977) suggest a half-open interval to define a cluster of telephone numbers for dialing. In numeric order directories, a selection consists of a selected listed number and all numbers up to the next listed number. In alphabetic order directories where the next listed number is not known, the respondent at each successive number is asked whether the number is listed in the directory. If it is reported as listed, all households with numbers up to but not including that household are included in the half-open interval. This method does achieve known and nonzero probabilities of selection for all telephone households provided the effects of reporting errors about listing status and nonresponse on probabilities of selection are ignored. The clusters of numbers formed by the half-open intervals will not be of the same size: one listed number could be followed by only a few unlisted numbers, while another is followed by many. This uncontrolled variation in cluster size can introduce operational problems since the number of numbers generated for each selected listed number will vary. It can also introduce estimation difficulties since cluster sizes and the sample size will be random variables.

A final type of list assisted method is the design proposed by Sudman (1973). The procedure, originally suggested by Stock (1962), starts with a directory sample to generate the selected prefix areas. In the second stage, sampling consists of replacing the last three digits of each selected listed number with randomly generated 3 digit numbers. The second stage numbers are called until a fixed number of listed households are obtained from the prefix area. Numbers that are not listed numbers are replaced by another randomly generated number. Interviews are attempted with listed and unlisted households identified in the second stage.

The chance of selecting a prefix area is proportional to the number of listed numbers. Unlisted numbers in prefix areas with no numbers in a directory (e.g., prefix areas in central office codes opened after publication of the directory) have zero chance of selection. Determination of the listed status of a number often depends on a respondent report which can be in error (see Traugott, Groves, and Lepkowski, 1987), but a reverse telephone number order directory can eliminate this potential source of error.

Operationally, the Sudman procedure can be implemented for a local area survey using a directory sample of listings. As opposed to the Mitofsky-Waksberg approach, the Sudman procedure will have unequal size clusters of telephone households, although the variation in cluster size is not likely to be substantial. The procedure also must replace unlisted numbers. With prefix areas of 1,000 numbers, the exhausted prefix area problem is much smaller than for 100 number prefix areas; exhausted prefix areas must still have compensatory weighting to adjust for unequal probabilities of selection.

Application of the Sudman procedure in situations where there is likely to be little homogeneity among observations within clusters allows the use of simple random sampling variance formulas. However, when the selection is across widespread areas with multiple exchanges, the use of stratified cluster variance formulas and Taylor series approximations is recommended.

5.2 List Assisted Sampling Based on Telephone Number Samples

Other list assisted designs use directory information as a source of auxiliary data for sampling from the telephone number frame. The use of commercial lists to provide auxiliary data for stratifying prefix areas was discussed as a modification of the two-stage RDD design, but it is, in principle, a list assisted sample design.

Other list assisted methods in this category can be considered as modifications of RDD designs. For example, list frames can supply name and address of selected telephone households for mailing an advance letter to improve response rates (Traugott, Groves, and Lepkowski, 1987). The advance letter can be used for listed households in any type of RDD sample design.

Information about listed numbers in the population of all telephone numbers can also be used to improve telephone sampling methods. Survey Sampling, Inc., has constructed a frame of listed telephone numbers and counts of listed numbers by prefix areas of size 100 (Survey Sampling, Inc., 1986). These measures of size can be used in several ways in sample selection.

Instead of using the general two-stage RDD design with unknown measures of size equal to the number of listed numbers in each prefix area, a sample of prefix areas can be selected with probabilities proportional to the number of listed numbers (Lepkowski and Groves, 1986). At the second stage, a replacement sampling scheme can be used in which numbers are generated at random without replacement until a fixed size sample of b listed numbers is obtained within selected prefix areas.[7] Alternatively, the second stage sampling rate can be calculated with the known measure of size for the prefix area and used to obtain a sample of telephone numbers. An equal probability of selection design is obtained, and no replacement of numbers is required.

A large proportion of the prefix areas will have no listed numbers and a zero chance of being selected. However, some of these prefix areas will contain unlisted household numbers. Only 3 or 4 percent of all telephone households are found in these low density prefix areas. Nonetheless, if no attempt is made to sample households in these low density prefix areas, the design suffers from a coverage bias. A two stage RDD procedure could be used to sample households among low density prefixes, but the second stage proportions of numbers that are households are low (approximately 30 percent).

Probability proportional to size selection is actually a needless complication. A simple element sample of telephone numbers selected from the high density prefix areas will have the same proportion of numbers that are households as do the second stage numbers in the probability proportional to size design. Survey Sampling, Inc., employs an element sample of numbers selected from prefix areas with three or more listed numbers (a subset of the high density prefix areas). No sample is selected from the low density prefix areas. It is argued that the coverage bias is small because the proportion of households in low density prefix areas is small (Survey Sampling, Inc., 1986).

If one does not have access to a complete frame of listed numbers in all prefix areas, or cannot afford to purchase such counts, a two phase or double sampling method can be used (Lepkowski and Groves, 1986). In the first phase, a large sample of prefix areas is selected (with equal probabilities), and counts of unlisted numbers are obtained only for first phase sample units. For local surveys, the counts may be obtained from reverse directories ordered by telephone number. For larger areas, the

---

[7] This is equivalent to a two-stage RDD procedure with $k = 2$ and eligibility equal to listed.

counts can be purchased from a commercial firm such as Survey Sampling, Inc., or all listed numbers in the first phase prefix areas (including name and address) can be purchased from a commercial list frame firm.[8] In the second phase, a simple element sample is selected from the high density prefix areas (e.g., those with three or more listed numbers), and two-stage RDD methods can be used to provide complete coverage by sampling in the low density prefix areas. Replacement does occur in the low density prefix areas, but only a small portion of the sample will be selected from them.

5.3 Dual Frame Sampling

A more direct approach to combining the strengths of RDD and list frame sampling has been the use of dual frame methods. A sample is selected from a directory or commercial list frame, and simultaneously an RDD sample is selected from the frame of all possible telephone numbers. Numbers are called in both samples and interviews attempted in all telephone households.

There are three domains of interviewed households. Let n_d denote the number of completed interviews from the directory frame, n_{rd} denote the number obtained from listed numbers sampled from the telephone number frame, and n_{ru} denote those obtained from unlisted numbers in the RDD frame. There are several ways to combine the results from these three frames for estimation. One is to screen out RDD households that are also in the directory. Screening for these households may involve unreliable respondent reports about listed status, a potentially serious problem for determining the domain of RDD households. In addition, since a major portion of interviewing costs is incurred by the time a household is contacted, the marginal cost savings associated with not attempting interviews at listed households in the RDD sample is so small that the screening operation is not as efficient as other approaches. In particular, it is usually more efficient to attempt interviews at all households in the RDD sample and combine data from domains by the use of poststratified estimators for multiple frame surveys suggested by Hartley (1964, 1974).

Let $\bar{y}_d$, $\bar{y}_{rd}$, and $\bar{y}_{ru}$ denote the respective estimated means from each domain, and let p_d denote an estimate of the proportion of telephone households that are in the directory. The dual frame estimated mean is

---

[8] This latter approach has the advantage that advance letters can be mailed to all listed cases selected in the second phase sample.

$$\bar{y} = (1 - p_d)\,\bar{y}_{ru} + p_d\left(\theta\,\bar{y}_{rd} + (1 - \theta)\,\bar{y}_d\right),$$

where θ is an arbitrary mixing parameter which can be chosen to minimize the variance (or mean square error) of $\bar{y}$. The dual frame estimate combines the two domain estimates for the directory population, $\bar{y}_d$ and $\bar{y}_{rd}$, through the mixing parameter in the expression in large parentheses. This combined directory frame estimate is then combined with the unlisted domain estimate $\bar{y}_{ru}$ by the weight p_d. Appropriate poststratified variance estimators are described in Groves and Lepkowski (1985).

Groves and Lepkowski (1986) describe the poststratified estimation approach for a dual frame telephone survey. They observe that the poststratified approach can have an advantage over the screening approach whenever there are differences in sample estimates among the poststrata. Since listed and unlisted households are different, a poststratified estimate is likely to have smaller variances than a screening type estimate obtained at the same cost.

The sample of listed numbers affords the opportunity to send advance letters to improve response rates (Traugott, Groves, and Lepkowski, 1987). The extent of bias reduction attributable to lower nonresponse rates depends partly on the allocation of sample to the listed number frame. For moderate and large size samples, even small reduction in bias leads to a much larger allocation to the listed number frame and an important reduction in mean squared error (Groves and Lepkowski, 1986).

In order to implement the dual frame methodology, the directory status of each RDD interview must be known. To avoid using unreliable respondent reports about the directory status of their number, numbers from the RDD frame can be matched to the directory frame at the time the directory sample is drawn. Since with many RDD methods, numbers may be generated during the course of interviewing, the matching should be done for a larger set of telephone numbers than may actually be dialed. For example, if a two-stage RDD method is used, all numbers in the selected prefix areas can be matched to the directory prior to the start of interviewing.

On the surface, the dual frame design requires a more complicated sampling administrative operation. Since many survey designs utilize multiple frames already, the dual frame approach does not necessarily introduce a substantial increase in the complexity of sampling operations. Costs are increased by the need to match RDD frame cases to the directory, by the mailing of advance letters, and by the use of the more complicated poststratified estimator. Perhaps the largest drawback to the dual frame approach is the need to use a special estimator. Investigations of alternative estimators are needed to determine whether standard estimation procedures can provide reasonably accurate results for dual frame samples.

The dual frame approach offers cost reductions compared to RDD methods because of the higher proportion of productive numbers in the directory sample. Further, poststratified dual frame estimators have smaller variances than RDD sample estimators since listed and unlisted households have different characteristics that are separated in the poststratified estimator. The advance letter can also decrease nonresponse rates for a substantial portion of the sample. However, the gains in precision and nonsampling error appear at present to be modest, and further investigation is required to establish the value of dual frame approaches.

6. CONCLUDING REMARKS

It seems clear from this review that there are numerous opportunities for the development of improved telephone household sample designs in the United States. A more thorough and careful exploration of sample designs for directory based master address lists may have great benefits for improving the efficiency and accuracy of telephone surveys. There are numerous features of the two-stage RDD sampling methods that invite innovation, refinement, investigation, and relatively inexpensive experimentation. Several chapters in this volume offer design variations for improving telephone sampling methods. The list assisted and dual frame designs may also provide improvement over existing methods. Further elaboration and exploration of these designs and their various modifications are needed.

Development of new telephone sample designs is only one area in which additional research is needed. Analytic and empirical comparisons of the properties of competing designs are also needed. For example, although the procedures recently proposed by Potthoff are generalizations of the Mitofsky-Waksberg procedure, it is not clear whether these procedures lead to cost reductions in survey operations. The dual frame and list assisted two phase sample designs may provide advantages over RDD methods because of reductions in nonresponse rates, but it is not clear that these gains are justified when cost and other error considerations are taken into account.

CHAPTER 6

IMPLEMENTING THE MITOFSKY-WAKSBERG SAMPLING DESIGN WITH ACCELERATED SEQUENTIAL REPLACEMENT

G. J. Burkheimer and J. R. Levinsohn
Research Triangle Institute

The Mitofsky-Waksberg sampling design (Waksberg, 1978) provides a means of more efficiently identifying households in random digit dialing (RDD) telephone surveys through the use of a two stage sampling approach, while maintaining basically an equal probability of residential selection (see also this volume, Chapter 5). The identification of households in the first stage (primary units) remains a stochastic process, related to the (presumably known or estimable) probability of household identification [which in some applications may vary from stratum to stratum (see this volume, Chapter 8)]. For a fixed number of primary telephone numbers, the primary unit yield is a random variable, and strictly speaking, all first stage households so identified (together with associated second stage households) should be considered in the sample.

Typically telephone survey designs are driven by a targeted number of households to be identified (to meet precision requirements with minimum sample size). The use of a fixed number of initial random telephone numbers will quite likely yield either more or less than the targeted number of first stage households, and this "overidentification" or "underidentification" is magnified for total households by the cluster size factor *b*. Either over- or underidentification can be problematic, introducing a potential for sacrificing sample integrity or for additional costs and/or time required for completion.

Consequently, iterative approaches to implementing the Mitofsky-Waksberg procedure for sampling/identifying primary households, through planned replacement sampling from an ordered list of previously selected (purposefully oversampled) random telephone numbers, are frequently used for designs involving fixed numbers of primary households. Replacement of verified nonhouseholds is generally accomplished at fixed operational "stages" within the sampling/identification process (e.g., once per day or between shifts). Such sequential replacement procedures, while ameliorating the over- and underidentification problems, can introduce nontrivial operational complexities. These problems are magnified when design or time constraints require that first and second stage calling overlap, since identification of the primary unit is essential before associated secondary units can be identified.

A very conservative (in avoiding overidentification) sequential replacement approach involves one for one replacement of "nonhousehold" primary numbers, starting with a set of n telephone numbers, where n is the required primary household sample size. This approach, requiring confirmed "nonhousehold" status of a number before that number is replaced, ensures that overidentification will not occur; however, implementation introduces logistical problems. Specifically, delays are introduced by working, during any operational stage, only a limited set of the total telephone numbers that will ultimately have to be resolved.

For a national sample, about 80 percent of the first stage numbers to be worked will be discarded. Further, well designed calling plans are such that determining the status of a telephone number can take seven or more days in some cases. Consequently, the waiting period for a series of replacement telephone numbers can be very long, particularly in low yield areas. Not only does this introduce a significant time delay in operations but it also poses a serious problem of scheduling work flow and staff assignments within the established calling capacity for the study, since the identification process is essentially random, with random time delays.

In the remainder of this chapter we consider an alternative approach to the sequential first stage sampling/identification process, to resolve some problems of the one to one replacement scheme. The purpose of this approach is to reduce the number of separate stages during which additional primary telephone numbers are assigned for determining household status, while still meeting the targeted number of households. The approach accelerates, within the basic Mitofsky-Waksberg design, the sequential sampling/identification process, reducing time delays and balancing workload yet maintaining, within a specified level of risk for overidentification, the sample integrity. Other approaches have been advanced previously to address operational efficiency problems; however, they involve the selection of more than one primary number within a primary sampling unit. An earlier approach (Swint and Powell, 1983) was subsequently criticized by Waksberg (1984) on its mathematical properties; however, a more recent approach (Potthoff, 1987) appears more rigorous. This latter approach also provides for a more liberal definition of an "acceptable" first stage identification (see also this volume, Chapter 5).

In the following section, we provide, briefly, a mathematical model for the sequential identification and replacement process and introduce some necessary definitions. In subsequent sections, respectively, we (a) present a heuristic solution to accelerate the replacement process and discuss its acceleration properties, (b) discuss some operational constraints, and (c) provide a comparison of the accelerated procedure with the one to one replacement procedure.

1. A MODEL FOR IDENTIFICATION AND SEQUENTIAL REPLACEMENT

The replacement sampling/identification process is sequential over M stages (stages may be considered work shifts, days, or some other time units); the goal is identifying (at least) n total primary households. Procedures are initiated by placing N_0^0 random telephone numbers into the system to be worked; these numbers are obtained from an ordered list of B_0 previously selected random telephone numbers. If several strata are involved, the procedure is simply replicated within each stratum.

During a given operational stage, i ($i = 1, ..., M$), two separate operations occur. In the *identification operation* any unresolved telephone numbers in the system are called (to the extent of stage calling capacity), resulting in: (a) identifying the number as a household; (b) identifying the number as a nonhousehold; or (c) not resolving the number (e.g., not worked, unanswered, busy). Before beginning the next stage $(i+1)$, a *resolution and replacement operation* identifies nonhouseholds to be removed from the system and replaced with N_i^0 new telephone numbers from the B_i numbers remaining. For current purposes, we assume that numbers remaining unresolved after Q calls (Q fixed in advance) are then considered to be "nonhouseholds"; working a given number only once per operational stage (however defined) is also assumed in the discussion. This sequential process repeats until stage M is reached (i.e., until all telephone numbers introduced into the system have been resolved and (at least) n households have been identified).

The state of the system at the start of stage i can be represented as the vector $\underset{\sim}{S}_i$, where

$$\underset{\sim}{S}_i = [\, B_i, U_i^0, U_i^1, \ldots, U_i^Q, H_i, NH_i \,]$$

where B_i = the number of previously selected random telephone numbers unused at the start of stage i

U_i^q = the number of unresolved telephone numbers with q prior telephone calls at the start of stage i $(q = 1, 2, \ldots, Q)$

H_i = the number of identified households at the start of stage i

NH_i = the number of identified (and default) nonhouseholds at the start of stage i

The identification operation at stage i allows transition only from the unresolved cases (those in states U_i^q); H_i and NH_i are absorbing states and the B_i cases are not affected. If a telephone number in state U_i^q is not worked during the stage, it will obviously remain in the same state. If that number is worked, then the state change will be to either U_i^{q+1}, H_i , or NH_i . As indicated above, U_i^Q is empty at the start of each stage.

The system state after the identification operation at stage i can be represented by the vector $\underset{\sim}{S}_i'$,

$$\underset{\sim}{S}_i' = [B_i, U_i^{0'}, U_i^{1'}, \ldots, U_i^{Q'}, H_i', NH_i'] \,.$$

At this point, the resolution and replacement operation assigns those numbers in state $U_i^{Q'}$ to state NH_i' and adds N_i^0 additional numbers from B_i to state U_i^0 . The rate of adding numbers from B_i will be defined as ψ_i $(1 \geq \psi_i \geq 0)$, where

$$\psi_i = N_i^0 / B_i.$$

This operation on $\underset{\sim}{S}^1$ produces $\underset{\sim}{S}_{i+1}$, the state of the system at the start of the next stage (stage $i+1$). The transition process continues through M stages, at the end of which no cases remain in the U states and the number of identified households (cases in state H) is (at least) n.

The general nature of an optimization problem within this model is fairly clear. Costs generally increase with the number of stages M required to complete the process. Also, costs are incurred to the extent that the number of cases in the H state after stage M exceed the desired sample n. While transition equations can be written for the M stage process, the number of stage-specific parameters (particularly the transition probabilities from a given U state to the next U state or to one of the absorbing states H or NH which depend rather heavily on other operational considerations) would tend to overspecify the model and, thus, complicate optimization approaches.

Nonetheless, the nature of the solution is to minimize the number of stages in which the n targeted number of households can be identified, subject to acceptable departures of H from n at the end of stage M. Assuming that all feasible operational steps have been taken to optimize the values of the identification transition probabilities, the only free parameters remaining in the model are the ψ_i. The conservative one to one replacement procedure uses $\psi_0 = n/B_0$ and, subsequently, $\psi_i = \eta_i/B_i$, where η_i is the number of nonhousehold classifications made during stage i (i.e., $\eta_i = NH_i' + U_i^Q - NH_i$). This implies that $N_0^0 = n$, and $N_i^0 = \eta_i$.

2. AN ACCELERATION PROCEDURE

For some stage i of the identification/replacement process, define

$$U_i^+ = \sum_{j=0}^{Q-1} U_i^{j'} ,$$

as the total number of unresolved telephone numbers in the system after the identification operation (exclusive of those in state U_i^Q, which will default to nonhouseholds). We wish to find some number of additional random telephone numbers N_i^0, which, when added to the existing U_i^+ unresolved numbers, will produce a total of N_i^* unresolved numbers in the system so that the probability of identifying more than $n - H_i'$ additional households from these N_i^* numbers is equal to or less than some specified risk level α. (Note that setting $\alpha = 0$ results in the one for one replacement approach.)

Assume that the identification of a randomly selected telephone number as a primary household within the Q allotted calls follows a Bernoulli process, with known (or estimable) parameter π. The number of successes (i.e., identified primary households) x in N independent Bernoulli trials is distributed binomially with

$$\varepsilon(x) = N\pi$$

$$\text{var}(x) = N\pi(1 - \pi) .$$

The probability that x equals or exceeds some value k $(0 \leq k \leq N)$ can be computed directly or estimated from the (limiting) normal distribution [the approximation is considered reasonably accurate in the case of $N\pi(1 - \pi) \geq 3$]. Given π and some specified number of successes (say k^*), one can then determine the value of N (say N^*, $N^* \geq k^*$) such that the probability of more than k^* successes within those N^* trials is less than the specified risk level α ($0 \leq \alpha < .5$).

Let c ($c \geq 0$ by definition) be the unit normal value, such that $\Pr(z_x \geq c) = \alpha$. Under the normality assumption, k^* is given by

$$k^* = N^*\pi + c\sqrt{N^*\pi(1-\pi)}\,. \tag{6.1}$$

Isolating the term involving the radical, squaring, and combining terms yields a quadratic equation in N^*, the solution for which is

$$N^* = \frac{2k^* + c^2(1-\pi) - \sqrt{c^2(1-\pi)[4k^* + c^2(1-\pi)]}}{2\pi}\,. \tag{6.2}$$

It should be noted that both solutions to the quadratic equation are positive and real; however, the solution given in Eq. (6.2) is the only one satisfying the constraint of $N^* \leq k^*/\pi$ [from Eq. (6.1) when $c \geq 0$, as required].

Define N^*/k^* (which always equals 1 in the one to one replacement approach) as the index of acceleration. From Eq. (6.2) we can obtain

$$\frac{N^*}{k^*} = \frac{1}{\pi} - \left(\frac{\sqrt{a^2 + 4ak^*} - a}{2\pi k^*}\right), \tag{6.3}$$

where

$$a = c^2(1-\pi).$$

The term in parentheses in Eq. (6.3) can obviously not be negative, and tends to zero with increasing k^*; thus, the maximum possible acceleration index is $1/\pi$. [For $k^*=0$, N^* will also be zero, and we will define the acceleration index as 1 for this degenerate case, as well as for other cases for which Eq. (6.3) yields a value less than 1.]

The nature of the acceleration index can be seen in Figures 6.1 and 6.2, plots of N^*/k^* against k* for combinations of α and π. Acceleration increases with increases in k^* and α and with decreases in π. The parameter π places the upper bound on the acceleration index and α (as reflected in c) in combination with π determines the rate at which the acceleration index approaches the limiting value under increasing k^*. For minimum allowable α ($c = 0$), the convergence of the index to $1/\pi$ occurs at $k^* = 1$. Figure 6.2 also shows clearly the extremely conservative nature of the one for one replacement algorithm, since using a risk level as stringent as 0.001 still yields a better than 7 to 1 ratio for $k^* \geq 70$ and $\pi = 0.1$. (While not shown, the ratio exceeds 9 for k^* of 800 or greater, the maximum possible value is 10 for this π.)

In practice, one should integerize the value of N^* given in Eq. (6.2). Also, since the solution for N^* shown in Eq. (6.2) is an approximation, a more precise solution may be desirable [particularly in cases for which $N^*\pi(1 - \pi)$ is relatively small]. Exact solutions are straightforward (albeit computationally burdensome), using exact binomial probabilities and iterating over successive approximations of N^*, using an integerized estimate of N^* provided by Eq. (6.2) as a starting value.

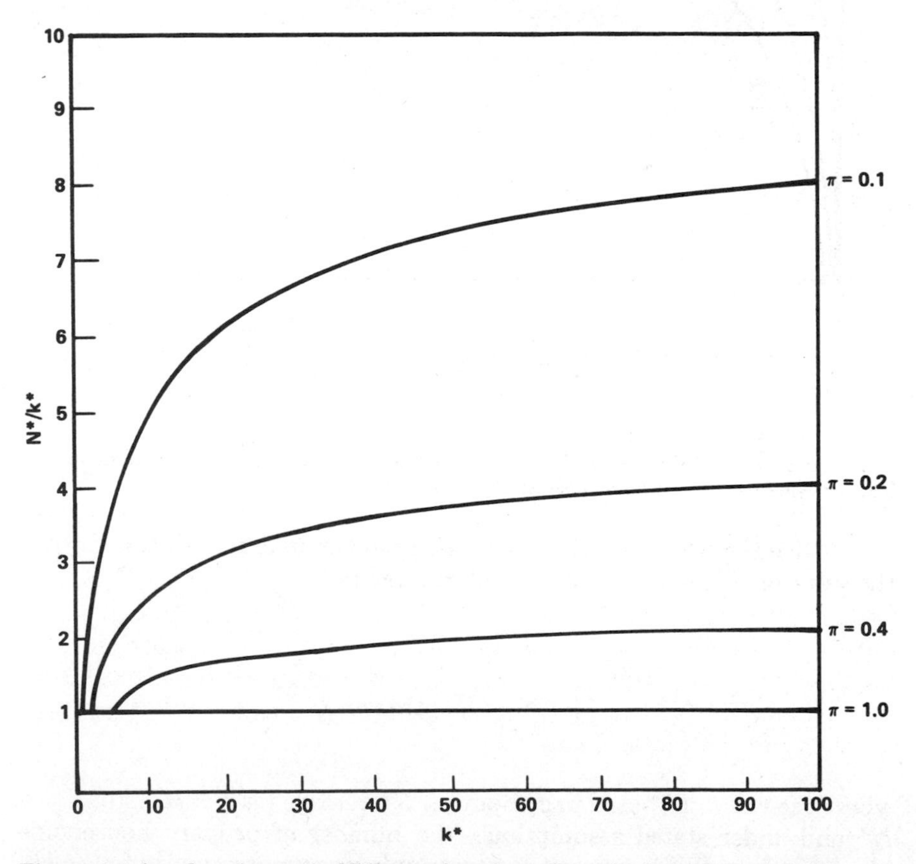

Figure 6.1 Acceleration ratio N^*/k as a function of required number of additional households k^* for $\alpha = 0.01$ and selected values of π.

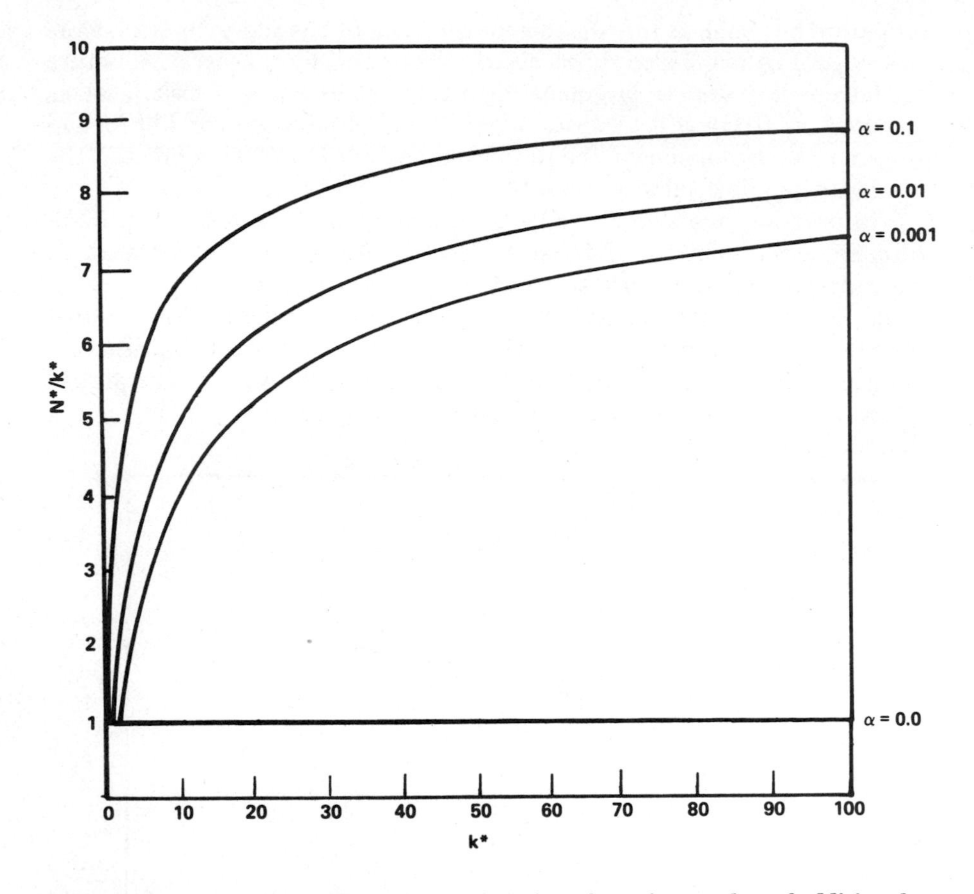

Figure 6.2 Acceleration ratio N^*/k^* as a function of require number of additional households k^* for $\pi = 0.1$ and selected values of α.

With either the integerized approximation or an exact solution for N^*, the value of N_i^0 can now be easily determined as

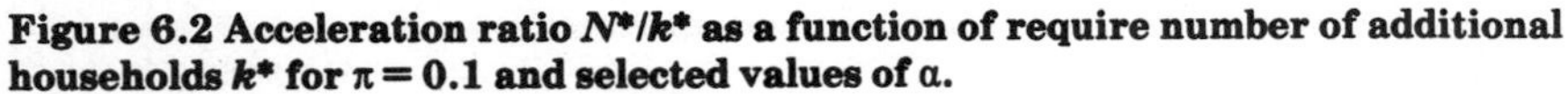

$$N_i^0 = \begin{cases} 0, \text{ if } U_i^+ \geq N^*, \\ N^* - U_i^+, \text{ otherwise,} \end{cases} \tag{6.4}$$

where the value of k^* used in computing N^* at stage i is $n - H_i'$. For $U_i^+ \leq N^*$ and under stated assumptions, the number of primary households identified from this increment of new telephone numbers, added to the U_i^+ unidentified numbers still in the system will exceed $n - H_i'$ with probability equal to or less than α. The case of $U_i^+ > N^*$ represents a situation for which past acceleration may already have been too great and probability of overidentification already exceeds α, using only the unidentified telephone numbers currently in the system. To maintain sample integrity, Eq. (6.4) does not allow for removal of numbers from the system once they have been added.

3. SOME OPERATIONAL CONSIDERATIONS

Some assumptions made in developing Eq. (6.4) need to be explored further in light of operational realities. First is the assumption that π is reasonably well known (or accurately estimable). Since the solution for N^* is quite sensitive to changes in π (as shown in Figure 6.1), it is important that the best possible estimate of π be used in computations. Estimates obtained from prior studies can be used (this volume, Chapter 8), provided, of course, that the household identification parameter has not changed. Also, revision of the estimate can be made at each stage, prior to computing N^*, using Bayesian approaches, in order to incorporate information from the current study as well.

Even the most rapidly converging revision algorithm will be of little value, however, if the initial value of N^* at stage 0 is too large as a result of an inappropriately low estimate of π, since this creates an excess of total unresolved numbers for subsequent stages that cannot be corrected by Eq. (6.4) even with subsequent revisions to estimates of π. This danger can be reduced (with an obvious loss in initial acceleration) by inflating the estimate of π used to start the process and using a rapid convergence probability revision algorithm based on experience during the current study.

A second assumption in the development is that the primary household identification rate π is equally applicable to "fresh" telephone numbers as to unresolved numbers in the system with some history of having been worked, regardless of the number of prior attempts q. This assumption is certainly untrue, since the conditional probability of household identification is known to vary with calling history. A multinomial model could be used to account for these differences (with considerable increase in conceptual and computational complexity), but experience in actual use of the acceleration approach suggests that the binomial model with a single household identification parameter works adequately, given good adherence to the study calling plan.

Finally, the risk level α for overidentification is established for a single application of the accelerated replacement approach. Due to the sequential nature of the process over the M operational stages, the replacement approach can be replicated a number of times in practice [specifically at every stage for which Eq. (6.4) yields a value of $N_i^0 > 0$]. Even though exact solutions for N^* will produce a single application risk level less than α in virtually all cases (due to the discrete probability points of the binomial distribution), repeated applications of the accelerated replacement procedure will generally yield an overapplications risk level greater than α.

Letting φ_r represent the actual risk level for the rth replication of accelerated replacement, the risk for overidentification Φ_R over all R replications is given by

$$\Phi_R = \sum_{r=1}^{R} \varphi_r \left(\prod_{s=0}^{r-1} (1-\varphi_s) \right),$$

where φ_s when $s = 0$ is defined as 0. For constant φ_r, the terms being summed are recognized as terms from the geometric distribution, in which case the value of Φ_R would be given by

$$\Phi_R = 1 - (1-\varphi_r)^R,$$

which approaches 1 as a limit. Fortunately successive applications of the accelerated replacement typically will involve smaller and smaller values of k^*, and will eventually converge to a one for one replacement approach (see Figure 6.1) as the value of $k^* (n - H_i')$ becomes sufficiently small (i.e., φ_r goes to 0 after a small number of replications). For example, with $\alpha = .01$ and $\pi \geq .3165$ replacement is restricted to one for one when $n - H_i \leq 3$. Simulations have suggested that, as a rule of thumb, expected Φ_R will not exceed 3α, for $.005 \leq \alpha \leq .05$. This relationship can be used in establishing an acceptable overall risk level.

The considerations discussed relate to the potential use of the accelerated replacement procedure for secondary telephone numbers. With reasonably small values of α and relatively small cluster sizes (and the relatively high average values of π expected in identified clusters), it is clear that the benefits of applying the accelerated approach will not be large (effective risk level will be 0 in many cases, thus resulting in no acceleration). On the other hand, overidentification costs of using acceleration with secondary numbers can be high. Considerable variability of π exists between clusters (even those within fairly tight geographic bounds); thus an estimate of π for a specific cluster will be relatively imprecise. As indicated previously, an inappropriately low estimate of π tends to create an uncorrectable overidentification.

4. SOME COMPARISONS

The accelerated replacement approach was implemented for the 1986 Youth Attitude Tracking Study (YATS-86), an annual RDD study by the Defense Manpower Data Center. That study required identification of 12,448 primary households and 70,786 secondary households, in addition to the subsequent household rostering and interviewing of approximately 10,000 identified target group members, all in a period of 14 weeks. The stratified design for this study is described elsewhere (Immerman and Mason, 1986); the sample was equally divided (within each stratum) between the Research Triangle Institute (RTI) and a subcontractor.

Direct comparisons with results from prior years of YATS are not possible, since a large number of other operational changes were also introduced in YATS-86. In an effort to produce meaningful comparisons of different approaches, simulations of operations using different replacement algorithms were conducted. These simulations used over 250 operational parameters (including those in the acceleration model and taking into account differences between strata), set to values actually realized in YATS-86. With other simulation parameters equal, the risk level α of the process was varied over two different values, 0 and 0.01 (the latter value was used during YATS-86). It was also assumed that, at each stage of the identification/replacement process, all remaining unresolved primary numbers (up to calling capacity) would be worked prior to working any secondary numbers in the system (a practice that was attempted, but not always realized, during YATS-86).

Given the parameters, it was possible to simulate the work flow and outcomes for the YATS-86 study under accelerated and nonaccelerated replacement approaches. For each value of α, 100 replications of the complete YATS-86 half-sample were simulated. The actual results of YATS-86 were not fully reproduced in any simulation (even with 250 simulation parameters, one could not expect to capture the large number of variables affecting actual operations). However, the simulations using $\alpha = .01$ sufficiently approximated the general form of workflow during actual YATS-86 operations to serve current purposes. More importantly, identical parameters were used for both values of α, providing the necessary comparability between the two conditions.

The simulations were conducted mainly to evaluate the effectiveness of the acceleration methods in supporting complex CATI logistics, but also allowed examination of the extent to which repeated applications of acceleration overidentify the primary households (i.e., the nature of Φ_R). Effectiveness was desired in (1) fewer stages for identification of primary households, and (2) greater flexibility for call scheduling. Such flexibility is necessary to implement scheduling optimization, for reducing costs while maximizing the likelihood of finding household members at home (see this volume, Chapter 25). To use optimal approaches (or for that matter any nontrivial calling plan) within a fixed calling capacity, a reasonable backlog of cases must be available in the system so that it is possible to set cases aside for optimal (specified) times of day and day of week.

Results of the simulations are presented in Tables 1 and 2. The measures of speed of identification and of overidentification are shown in Table 1. The results demonstrate the "tailing off" phenomenon of replacement approaches associated with closing out the identification of all residual primary households (where caseload is small and luck of the draw is most telling). Results also demonstrate that an accelerated replacement approach provides significant and meaningful gains in completion of primary household identification (approximately halving the time to reach the 95 percent completion point). In terms of CATI logistics, the 17 stage difference at the 95 percent completion point translates to an 8 to 9 day savings (assuming two shifts per day). Under tight time lines this savings may be quite important, but it also means that substantial work on subsequent CATI phases (e.g., rostering, interviewing) can be initiated that much sooner.

The cost of acceleration is also shown in Table 1. The simulations estimate indicates, for this example, that the targeted number of primary households was exceeded, on average, in 4.29 strata and that the total number of overidentified primary households averaged 5.06, less than 0.1 percent of target. Since 142 nonempty strata were defined, the estimated value of Φ_R for the multiple applications of accelerated replacement is approximately 0.03. The logistical implication of this overidentification is that these additional primary numbers, plus associated secondary cases, must be worked (even if they are subsequently deleted through subsampling). In most applications this would be a very reasonable trade-off for the improved speed of the process.

Table 2 presents the simulation results using a stagewise measure of case backlog, C, defined as the excess number of unresolved primary and secondary telephone numbers in ratio to the stage calling capacity. The comparison statistic used is the percent of trials in which the value of C falls within each of 5 categories: high flexibility ($C > 2.0$); moderate flexibility ($2.0 \geq C > 1.0$), low flexibility ($1.0 \geq C > 0.5$); some flexibility ($0.5 \geq C > 0.0$); and no flexibility ($C \leq 0$). The table shows that, in addition to faster identification, the accelerated replacement procedure maintains a steady backlog of cases exceeding twice the calling capacity through the 95 percent completion point. Acceleration also allows a higher degree of scheduling flexibility at any level for a much longer portion of the identification process.

Table 1. Production Statistics for 100 Simulations under Two Stagewise Replacement Risk Levels

Production Statistic	Mitofsky-Waksberg Stagewise Replacement with Risk Level 0.00		Mitofsky-Waksberg Stagewise Replacement with Risk Level 0.01		Difference	
	Average	Standard Deviation	Average	Standard Deviation	Average Diff.	Est. Standard Error[a]
Number of stages to 75% completion	17.38	0.49	11.51	0.50	5.87	0.07
Number of stages to 90% completion	28.98	0.55	15.97	0.17	13.01	0.06
Number of stages to 95% completion	38.19	0.78	20.50	0.54	17.69	0.09
Number of stages to completion	177.90	44.79	115.10	27.92	62.80	5.28
Number of strata with overage (Base of 142 Strata)	0.00	0.00	4.29	2.04	−4.29	0.20
Total cluster overage (Target of 6,224 Clusters)	0.00	0.00	5.06	2.55	−5.06	0.25

Note: Other parameters used in the simulation were chosen to approximate the conditions of the 1986 YATS CATI survey.
[a]Standard errors of differences estimated assuming independence.

Table 2. Scheduling Flexibility Statistics for 100 Simulations under Two Stagewise Replacement Risk Levels

Flexibility Statistic	Mitofsky-Waksberg Stagewise Replacement at Risk Level 0.00: Number of Stages with					Mitofsky-Waksberg Stagewise Replacement at Risk Level 0.01: Number of Stages with				
	High Flex. $C \geq 2$[a]	Mod. Flex. $2 > C \geq 1$[a]	Low Flex. $1 \geq C > .5$[a]	Some Flex. $.5 \geq C > 0$[a]	No Flex. $C \leq 0$[a]	High Flex. $C \geq 2$[a]	Mod. Flex. $2 > C \geq 1$[a]	Low Flex. $1 > C \geq .5$[a]	Some Flex. $.5 > C \geq 0$[a]	No Flex. $C \leq 0$[a]
Through 75% Completion										
Number of Stages	8.84	5.54	1.10	1.90	0.00	11.51	0.00	0.00	0.00	0.00
Standard Deviation	0.60	0.50	0.30	0.30	0.00	0.50	0.00	0.00	0.00	0.00
Percent of Total[b]	50.90%	31.90%	6.30%	10.90%	0.00%	100.00%	0.00%	0.00%	0.00%	0.00%
Through 90% Completion										
Number of Stages	18.44	7.54	1.10	1.90	0.00	15.97	0.00	0.00	0.00	0.00
Standard Deviation	0.74	0.50	0.30	0.30	0.00	0.17	0.00	0.00	0.00	0.00
Percent of Total[b]	63.60%	26.00%	3.80%	6.60%	0.00%	100.00%	0.00%	0.00%	0.00%	0.00%
Through 95% Completion										
Number of Stages	21.50	11.68	3.11	1.90	0.00	20.50	0.00	0.00	0.00	0.00
Standard Deviation	0.52	0.97	0.31	0.30	0.00	0.54	0.00	0.00	0.00	0.00
Percent of Total[b]	56.30%	30.60%	8.10%	5.00%	0.00%	100.00%	0.00%	0.00%	0.00%	0.00%
Through 100% Completion										
Number of Stages	21.50	13.41	7.05	4.04	131.90	34.00	8.16	3.80	0.60	68.48
Standard Deviation	0.52	0.58	0.36	0.45	44.79	0.00	0.37	0.40	0.57	27.98
Percent of Total[b]	12.10%	7.50%	4.00%	2.30%	74.10%	29.50%	7.10%	3.30%	0.50%	59.50%

Note: Other parameters used in the simulation were chosen to approximate the conditions of the 1986 YATS CATI survey.
[a]The quantity C is the ratio of the excess unresolved telephone numbers in the system at a stage (i.e., total number of unresolved telephone numbers minus calling capacity) to the stage calling capacity. Secondary telephone numbers were included in the total.
[b]Base used is average number of total stages (from Table 1) to reach the specified percent completion.

CHAPTER 7

CUTOFF RULES FOR SECONDARY CALLING IN A RANDOM DIGIT DIALING SURVEY

Charles H. Alexander
U. S. Bureau of the Census

Using the Mitofsky — Waksberg method, a relatively small number of "sparse" primary sampling units (PSU)s may require exorbitant secondary screening to locate the required $k+1$ residences; some may not even have $k+1$ residences. These situations are an irritant to interviewers, especially at the end of the survey period when a few "bad" PSUs may contribute an appreciable portion of the unfinished work. In practice, secondary sampling in these PSUs often ends with fewer than $k+1$ residences when time runs out at the end of the survey period. This suggests a need for a formal rule for cutting off secondary screening with fewer than $k+1$ residences, after reaching some limit on the total number of attempts in the PSU. Such a rule could reduce scheduling problems and assist in maintaining interviewer morale.

Looking beyond the very sparse PSUs, if the cost of unproductive secondary screening is appreciable compared to interviewing and processing costs, a cutoff rule conceivably could reduce the allocation of a sample to PSUs where the total cost is greatest. It has been speculated that this could reduce the variance for a given total cost, even if differential weights are used to eliminate the potential bias from the unequal selection probabilities. However, in the examples considered in this chapter the variance reduction is trivial. Thus, the main justification for using a cutoff rule is the resulting operational convenience. The results of the chapter do suggest that an appropriate rule can achieve this convenience with little or no increase in variance for constant cost.

A *cutoff rule* will correspond to a vector $c = (c_1, \ldots, c_k)$, where $1 < c_i \leq 100$. Two kinds of rules will be considered:

a. *Increasing rules*, with $1 < c_1 \leq \ldots \leq c_k$; the rule is to stop as soon as c_i numbers have been called, if i or fewer residences have been found.

b. *Decreasing rules*, with $100 \geq c_1 \geq \ldots \geq c_k \geq k + 1$; the rule is to stop calling when i residences have been found, if at least c_i total calls have been made.

The fixed rule $c_1 = \ldots = c_k$ may be viewed as a special case of either increasing or decreasing rules.

An increasing rule assumes that the worse you are doing in a PSU, the more you should think of giving up. A decreasing rule assumes that the worse you are doing, the harder you should try. Section 5 discusses why either could be appropriate, depending on circumstances.

Under either kind of rule, calling ceases as soon as $k + 1$ residential numbers have been found. A weight $W = (k + 1)/r$ will be applied to each case in the PSU, where r is the final number of residences found in that PSU. This weighting procedure is essentially the same as one described in Waksberg (1984). Under mild assumptions, it can be shown that these weights lead to an unbiased estimator of the population mean.

In this chapter, both increasing and decreasing rules will be evaluated according to their expected cost under a simple model, assuming a constant target variance. The mathematical derivations are based on an idealized model for RDD sampling, ignoring refusals, problems in determining residential status, and so forth. Some of these practical issues will be discussed in the final section.

1. NOTATION

Each PSU in the population will be a bank of 100 telephone numbers defined by the three-digit area code, the three-digit prefix code, and the first two digits of the suffix. A "residential" PSU has one or more residences.

The following notation will be used:

M = number of residential PSUs in the population
m = number of sample PSUs
N_j = number of residences in jth sample PSU
$p(r) = P(N_j = r)$, for $r = 1, \ldots, 100$
C_u = cost of screening a nonresidential number
C_p = total cost for screening, interviewing, and data processing for a residential number

X_{ji} = survey statistic for ith sample unit in jth sample PSU
ρ = intraclass correlation within PSU
$\sigma^2 = \text{Var}(X_{ji})$
Y = Expected number of unproductive calls to obtain one sample PSU during primary screening

Note that C_u, C_p, ρ, and σ are assumed to be constant. The following random variables depend on the cutoff rule c.

$R_j(c)$ = number of sample residences from jth sample PSU
$U_j(c)$ = number of nonresidential numbers called in jth sample PSU
$W_j(c) = (k+1)/R_j(c)$

Also, let

$$a_r(c) = E(R_j(c) \mid N_j = r)$$

$$b_r(c) = E(R_j(c) + U_j(c) \mid N_j = r)$$

$$d_r(c) = E(1/R_j(c) \mid N_j = r)$$

$$A(c) = E(R_j(c)) = \sum_{r=1}^{100} p(r) a_r(c)$$

Similar expressions relate $B(c)$ to $b_r(c)$ and $D(c)$ to $d_r(c)$.

2. COST FUNCTIONS

The expected cost for screening and interviewing m sample PSUs using cutoff rule c is

$$mC_u Y + E\left[\sum_{j=1}^{m} (C_p R_j(c) + C_u U_j(c))\right] \tag{7.1}$$

$$m\left[C_u Y + C_p A(c) + C_u\,(B(c) - A(c))\right]$$

2.1 Expected Cost for Constant Expected Variance

Suppose that the constraint is that the expected variance of the estimated mean should be approximately V_0. Conditional on $R_1(c)$, ..., $R_m(c)$, the variance of the weighted mean is

$$\text{Var}\left[\left(\sum_{j=1}^{m}\sum_{i=1}^{R_j(c)} W_j X_{ji}\right) / \left(\sum_{j=1}^{m}\sum_{i=1}^{R_j(c)} W_j\right)\right]$$

$$= \frac{\sigma^2}{m^2}\left[m\rho + \left(\sum_{j=1}^{m} 1/R_j(c)\right)(1-\rho)\right]. \tag{7.2}$$

The expected value of Eq. (7.2) with respect to $R_j(c)$ is

$$\frac{\sigma^2}{m}\left[\rho + (1-\rho)\, D(c)\right]. \tag{7.3}$$

For the expected variance, Eq. (7.3), to equal V_0, we need

$$m = \sigma^2\left[\rho + (1-\rho)\, D(c)\right] / V_0 . \tag{7.4}$$

Combining Eqs. (7.1) and (7.4), the expected cost necessary to obtain the desired expected variance is

$$E(\text{cost}) = (C_u \sigma^2 / V_0)\, K(c),$$

where

$$K(c) = \left[Y + A(c) \cdot C_p/C_u + B(c) - A(c)\right] \cdot \left(D(c) + \rho\left[1 - D(c)\right]\right). \tag{7.5}$$

This criterion is defined in terms of the expected variance. For a given sample of m PSUs, the actual sample size and variance may be higher or lower than the expected value. By contrast, if no cutoff rule is used, the sample size and variance for m PSUs is known in advance. For large samples, this is not a severe problem. It would be possible to select sample PSUs sequentially until the actual variance, Eq. (7.2), is equal to the target. However, such a sequential selection of PSUs is less convenient than having a preselected sample size m.

3. EVALUATION OF THE COST

In Eq. (7.5) the factor $(C_u \sigma^2 / V_0)$ does not depend on the cutoff rule. Thus, the minimum-cost cutoff rule will be the one which minimizes $K(c)$. To calculate $K(c)$ requires knowledge of Y, the ratio C_u/C_p, and the within-PSU correlation ρ. Also, to calculate $A(c)$, $B(c)$, and $D(c)$, we need to know the distribution $p(r)$. For fixed, increasing, or decreasing rules, the conditional expectations $a_r(c)$, $b_r(c)$, and $d_r(c)$ can be calculated from elementary combinatorial identities, as described in Sections 5, 6, and 7.

Note that other cost criteria, such as minimizing the expected cost for a given target expected sample size, could be considered. The requirement of constant variance seems reasonable, since it takes into account the loss of precision done to the unequal weights W_j.

Using data from Hogue and Chapman (1984), we assume $Y \cong 4$ and

$$p(r) \cong \begin{cases} 0; r = 1, \ldots, 3 \\ (r+2)/4800; r = 4, \ldots, 52 \\ (25r - 46)/100{,}000; r = 53, \ldots, 77 \\ (115 - r)/2000; r = 78, \ldots, 100 \end{cases}$$

4. INCREASING OR DECREASING RULES

Only increasing and decreasing rules will be considered. Mixtures of these rules are possible, but the combinatorial problem becomes very complex.

Intuitively, whether an increasing or a decreasing rule is preferable in a given situation depends on the distribution of residential numbers in the population PSUs, $p(r)$, and the relative importance of three effects:

1. The fewer residential numbers which have been found after T numbers have been called, the greater the posterior probability that N_j is small. See Hogue and Chapman (1984).

2. Conditional on N_j, the fewer residential numbers which have been found after T numbers have been called, the more residential numbers are left in the $100 - T$ numbers which have not been called.

3. The fewer residential numbers which are found, the higher the weight W_j applied to those numbers.

Effects 2 and 3 tend to imply that the fewer residential numbers one has found, the greater the incentive to continue. This would tend to favor a decreasing rule. Effect 1 leads to the opposite conclusion, favoring an increasing rule. Note that for an increasing rule, if cutoff occurs with $r < k+1$ residences, exactly c_r total calls will have been made.

For a decreasing rule, cutoff with r residences may occur after c_r, $c_r + 1$, ..., c_{r-1} total calls. The condition $c_k \geq k+1$ is imposed for decreasing rules because otherwise there would be no possibility of selecting $k+1$ units from a PSU using the cutoff rule.

5. COMBINATORIAL IDENTITIES

The following definitions pertain to sequential sampling without replacement from a set of l numbers, r of which are residential. In this section, it is not assumed that the first number selected is residential.

$f_1(l, r, s, t)$ = probability that exactly s residential units are found when t units are selected; $0 \leq s \leq t \leq l, s \leq r, t - s \leq l - r$.

$f_2(l, r, s, t)$ = probability that exactly s residential units are found when t units are selected, with the sth residential unit being the last of the t; $1 \leq s \leq t \leq l, s \leq r, t - s \leq l - r$.

$e(l, r, s)$ = expected number of units which must be selected until exactly s residential units have been located; $1 \leq s \leq r \leq l$.

$C(t, s) = t!/(s!(t-s)!)$, for $0 \leq s \leq t$.

The quantities f_1, f_2, and e are found as follows:

$$f_1(l, r, s, t) = C(r, s)\, C(l-r, t-s)/C(l, t)$$

$$f_2(l, r, s, t) = P(t\text{th unit residential}) \cdot P(s{-}1 \text{ residences in first } t-1 \mid t\text{th unit is a residence})$$

$$= (r/l) f_1(l{-}1, r{-}1, s{-}1, t{-}1)$$

$$e(l, r, s) = \sum_{t=s}^{l-r+s} t f_2(l, r, s, t)$$

$$= \sum_{t=s}^{l-r+s} t(r/l) C(r{-}1, s{-}1) C(l-r, t-s)/C(l{-}1, t{-}1)$$

The upper limit of summation is $l-r+s$, because once $l-r+s$ units have been selected, there are only $r-s$ units left. Even if all these were residences, s residences would already have been found.

6. CONDITIONAL EXPECTATIONS FOR INCREASING RULES

In sample PSU of 100 numbers with r residences, the first number selected is a residence. The remaining $r{-}1$ residences and $100{-}r$ nonresidences are selected one at a time in random order, until cutoff occurs according to the cutoff rule c. For an increasing rule, cutoff at c_t occurs only if exactly t residences have been found after c_t calls.

The case $r \geq k+1$. Let $g_c(t, s)$ be the probability of the event that (1) there was no cutoff at $c_1, \ldots, c_{t-1}$, and (2) exactly s residences were located on the first c_t calls, where $1 \leq t \leq s \leq k$. For $t = 1$, the first condition is automatically met.

Note that $g_c(t, t)$ is the probability that cutoff occurs at c_t with t residential units located, for $1 \leq t \leq k$.

The values $g_c(t, s)$ may be found from the following recurrence relation:

$$g_c(t, s) = \sum_{i=t}^{s} g_c(t{-}1, i)\, f_1(100{-}c_{t-1}, r-i, s-\mathrm{i}, c_t - c_{t-1})\,. \tag{7.6}$$

Indeed, for the event defining $g_c(t, s)$ to occur, there must have been at least t (but no more than s) residences located by the cutoff point c_{t-1}, with no prior cutoff, and there must be exactly enough residences located between c_{t-1} and c_t to bring this number up to s.

The initial conditions are $c_0 = 1$, $g_c(0, 1) = 1$, and $g_c(0, s) = 0$ for $s > 1$. The initial conditions and Eq. (7.6) correspond to the relation

$$g_c(1, s) = f_1(99, r-1, s-1, c_1-1), \text{ for } 1 \leq s \leq k .$$

$1 - \sum_{t=1}^{k} g_c(t, t)$ gives the probability that there is no cutoff due to the cutoff rule, but instead interviewing stops when $k+1$ residences are found. Thus,

$$a_r(c) = \sum_{t=1}^{k} t g_c(t, t) + (k+1)\left(1 - \sum_{t=1}^{k} g_c(t, t)\right)$$

$$d_r(c) = \sum_{t=1}^{k} t^{-1} g_c(t, t) + (k+1)^{-1}\left(1 - \sum_{t=1}^{k} g_c(t, t)\right)$$

Finally, $b_r(c)$ may be found by subtracting, from the "no cutoff" expected value $1 + e(99, r-1, k)$, the expected number of numbers called after a cutoff would have taken place, i.e.,

$$b_r(c) = 1 + e(99, r-1, k) - \sum_{t=1}^{k} g_c(t, t)\, e(100-c_t, r-t, k+1-t) .$$

The case $r < k+1$. If $r < k+1$, then $g_c(t, t)$ is found as before, except that Eq. (7.6) only applies to $1 \leq t \leq s \leq r$.

In this case:

$$a_r(c) = \sum_{t=1}^{r} t g_c(t, t) + r\left(1 - \sum_{t=1}^{r} g_c(t, t)\right)$$

$$d_r(c) = \sum_{t=1}^{r} t^{-1} g_c(t, t) + r^{-1}\left(1 - \sum_{t=1}^{r} g_c(t, t)\right)$$

$$b_r(c) = 100 - \sum_{t=1}^{r} g_c(t, t)(100 - c_t) .$$

These three conditional expectations given r may be used to find the expected cost for the cutoff rule c, as described in Section 3. Note that the conditional expectation formulas still would apply regardless of the cost model.

7. CONDITIONAL EXPECTATIONS FOR DECREASING RULES

Let $h_c(t, s)$ be the probability that interviewing is cut off with s residences after exactly t total calls in the PSU, for $k+1 \leq t \leq 100$ and $1 \leq s \leq k+1$. Assume that the PSU contains r residences and that the first number was residential.

For a decreasing rule, cutoff is not restricted to the cutoff points c_t. Interviewing will stop after i residences with a total number of calls anywhere between c_i and c_{i-1} inclusive. There would never be more than c_1 numbers called. To get the full $k+1$ residences, they would have to be located on or before a total of c_k numbers had been called.

Three cases need to be considered.

The case s = 1.

$$h_c(c_1, 1) = f_1(99, r-1, 0, c_a-1)$$

The case s > 1 and t = c_s. For calling to stop at $t = c_s$ calls with s residences, where $2 \leq s \leq k+1$ and $s \leq t$, the sth residence must have been found on the jth call, for $s \leq j \leq c_s$, and then no residences were found between the $(j+1)$th call and call c_s. Therefore,

$$h_c(c_s, s) = \sum_{j=s}^{c_s} f_2(99, r-1, s-1, j-1)\, f_1(100-j, r-s, 0, c_s-j)\,.$$

This includes the case $t = s = k+1$, where $c_{k+1} = k+1$.

The case s > 1 and t > c_s. In this case, for calling to stop at t calls with s residences, where $2 \leq s \leq k+1$ and $s \leq t$, it is only necessary for the sth residence to be found on exactly the tth call. If this occurs, then cutoff would not have occurred on any earlier call. In this case,

$$h_c(t, s) = f_2(99, r-1, s-1, t-1).$$

The three cases listed include all possibilities for stopping. Before c_s total calls have been made, one would not stop with s residences. (This includes the case $s = k+1$ where $c_{k+1} = k+1$, since clearly $k+1$ residences could not be obtained before c_{k+1} calls.) For $t > c_{s-1}$ it would be impossible to stop with s residences after t calls. Indeed, to have s residences by the tth call, there must have been at least $s-1$ residences after call $t-1$. Since $t-1 \geq c_{s-1}$, calling would have stopped at call $t-1$, if not earlier, so that the tth call would not be made.

If $r < k+1$, then $h_c(t, i) = 0$ for $r < i$. The conditional expectations given r may be found from the following formulas, using the probabilities calculated above:

$$a_r(c) = \sum_{i=2}^{k+1} \sum_{t=c_i}^{c_{i-1}} i h_c(t, i) + h_c(c_1, 1)$$

$$d_r(c) = \sum_{i=2}^{k+1} \sum_{t=c_i}^{c_{i-1}} i^{-1} h_c(t, i) + h_c(c_1, 1)$$

$$b_r(c) = \sum_{i=2}^{k+1} \sum_{t=c_i}^{c_{i-1}} t h_c(t, i) + c_1 h_c(c_1, 1) .$$

In these expressions, let $c_0 = 100$.

8. COMPUTATIONAL CONSIDERATIONS

In evaluating the expressions for $a_r(c)$, $b_r(c)$, and $d_r(c)$ in Sections 6 and 7, some care must be taken in planning the calculations to avoid exceeding the available computer memory. For decreasing rules, one approach is to store the binomial coefficients $C(r, s)$ for $0 \leq r, s \leq 100$ and calculate all other values as needed. For increasing rules, it is also necessary to store the values $g_c(t, s)$ for $s = 1, ..., t$ as a vector. A second vector would be used to save the "lagged values" $g_c(t-1, s)$ for $s = 1, ..., t-1$; after the calculations for t are completed, the results are transferred to the "lagged values" vector.

9. EXAMPLES OF THE EFFECT OF CUTOFF RULES

In illustrating the effect of cutoff rules, only optimal values of k will be considered. The optimal k with no cutoff is given by Waksberg (1978). (The formula assumes that there are no residential PSUs with fewer than $k+1$ residences. This is not true for our example, so the optimal value will only be approximate.) Using values of k greater than the optimum would produce spurious advantages for the cutoff rule. For example, if the selected value k is larger than the optimal value k_{opt}, then the decreasing cutoff rule given by

$$c_i = 100; \quad i = 1, ..., k_{opt}$$

$$c_i = k_{opt} + 1; \quad i = k_{opt} + 1, ..., k$$

is the same as using k_{opt} with no cutoff. This does not really demonstrate any advantage to using a cutoff rule, since the same effect could be achieved more easily by using k_{opt} with no cutoff.

The real question is whether a cutoff rule with $k = k_{opt}$ can make a substantial improvement over using k_{opt} with no cutoff. Calculations based on the $p(r)$ distribution given by Hogue and Chapman (1984) show that the improvement is trivial except when the cost ratio C_p/C_u is unrealistically small.

For example, Table 1 gives illustrative results for fixed rules, assuming a moderate cost ratio $C_p/C_u = 5$, using $\rho = 0.017$ and $\rho = 0.0053$ for which $k_{opt} = 4$ and 8, respectively are optimal. The cost ratio of 5 is lower than what is usually found in practice; for higher cost ratios, the savings would be even less.

One reason for the trivial cost reduction is that, in many cases, the probability of cutoff is small. For example, for cutoff at 70 with $k = 8$, the probability of a cutoff is 0.0175.

Table 1. Cost K_2(c) for Fixed Cutoff Rules, Relative to No Cutoff

Cutoff Point	Cost for $k = 4 (\rho = 0.017)$	Cost for $k = 8$ $(\rho = 0.0053)$
10	1.0591	1.1910
20	1.0075	1.0611
30	0.9995	1.0207
40	0.9980	1.0071
50	0.9981	1.0019
60	0.9985	0.9998
70	0.9989	0.9992
80	0.9994	0.9992
90	0.9997	0.9995
100	1.0000	1.0000

Assumes cost ratio $C_p/C_u = 5$.

Costs for some increasing and decreasing rules were also calculated for this example. The increasing rules were generated by the following algorithm. To save computer time, only values of c_i which are multiples of 5 were used.

Step 0: start with $c_1 = ... = c_k = 100$.

Step i $(i = 1, ..., k)$: choose c_i which minimizes $K(c)$ given the other cutoff points.

Steps 1, ..., k are repeated until the process converges.

A similar algorithm was used for decreasing rules.

These algorithms seem to be plausible ways to look for a good rule, although they may not necessarily lead to the optimal cutoff rule. The results given by these algorithms are sufficiently discouraging that further research to demonstrate an optimal rule would not seem to be worthwhile. Table 2 shows the best decreasing rules generated by the algorithm, showing their $K(c)$ value relative to that for k_{opt} with no cutoff. As before $C_p/C_u = 5$. These decreasing rules make a trivial improvement in variance. No increasing rules are shown, since no increasing rule was found which did better than k_{opt} with no cutoff. Thus in the example, decreasing rules seem to work better than increasing rules, but neither offers much variance improvement.

Table 2. Cost of "Best" Decreasing Rule, Relative to No Cutoff

		"Best" Cutoff Rule								
K	ρ	c_i	c_2	c_3	c_4	c_5	c_6	c_7	c_8	Cost Relative to No Cutoff
4	0.017	80	60	30	15	—	—	—	—	0.9864
8	0.0053	100	100	100	85	65	45	30	20	0.9922

Assumes cost ratio $C_p/C_u = 5$.

The results given in Tables 1 and 2 admittedly depend on the assumed distribution $p(r)$. For artificial distributions, very different results may be obtained. For example, let $p(4) = p(100) = 0.5$, so that there are only 100 percent residential PSUs and very sparse PSUs with $k = 4$. In this case, cutoff rules can be very beneficial, and increasing rules seem to perform better than decreasing rules in the cases the author has considered.

However, the distribution which was assumed in producing Tables 1 and 2 seems to be generally realistic. Somewhat similar, though by no means identical, actual distributions are given in Waksberg (1984). The basic results do not seem to be extremely sensitive to the assumed distribution. Even altering the Hogue-Chapman (1984) distribution by shifting 10 percent of the probability from $r = 81, ..., 100$ to $r = 1, ..., 20$ makes only a 2 percent improvement in variance for the best decreasing cutoff rule. Although these limited simulations cannot be regarded as definitive, they contribute to a pessimistic view of the potential for reducing variance through the use of a cutoff rule.

On the positive side, the simulations suggest that it is not necessary to be very close to the optimal cutoff to have a slight improvement in variance, or at worst a trivial increase, compared to no cutoff. Based on Table 1, a fixed cutoff at about 40 would appear to be reasonable for $k = 4$, and a cutoff at about 70 for $k = 8$.

10. SOME PRACTICAL CONSIDERATIONS

Cutoff rules can be applied with varying degrees of formality. One possibility would be to have a computerized call scheduler automatically stop selecting numbers in a PSU once the cutoff had been reached. Another alternative would be to give the operational supervisors the option to stop the sampling in a PSU after the cutoff had been reached. The second alternative may be more practical. However, it theoretically could lead to selection bias if the supervisors applied the rule selectively based on the characteristics of the residences which had already been interviewed in the PSU.

The model presented in this chapter makes several simplifying assumptions. These will inevitably be violated in practice, at least to some extent. This is another reason not to try to be too precise about choosing an optimal cutoff rule. One problem is that surveys are usually conducted to measure several variables, which may have different values of ρ. Thus, the cutoff rule, along with the value of k, must be a compromise solution. Even for a single variable, the value of ρ will not be known precisely.

The assumption that screening costs are exactly proportional to the number of cases screened may not be realistic, especially when small differences are involved. If interviewers are hired to work shifts of a fixed number of hours, then eliminating a few calls may not reduce salary costs at all, unless the reduction is great enough to have an interviewer not come in for a shift. On the other hand, if an interviewer has to come in just to make secondary screening calls in one remaining "sparse" PSU, those calls could be much more expensive than the average screening cost per case.

The model ignores the possibility of refusals or other failures to interview residential numbers. The extension of the model to handle refusals would be simplest when other residential numbers within the PSU are substituted for the noninterviews. If only interviewed residential numbers are included as "residential" at both primary and secondary selection, then the problem is essentially unchanged. However, if the PSU is selected even when the primary number is a noninterview, the problem becomes more complicated. A different cutoff rule may be needed for such PSUs than for PSUs in which the primary number was interviewed. One possibility would be to apply the cutoff rule only when the primary number is interviewed.

A similar problem arises for surveys which do not include the primary screening number in the sample at the secondary stage. For such surveys, a fixed cutoff rule may create bias problems which the weights $W_j(c)$ cannot eliminate. Our model does not easily extend to this case.

When weighting, rather than substitution, is used to adjust for nonresponse, the problem is still more complex. Here the basic quota is $k+1$ residences, whether or not they are interviewed. In this case, it is not clear whether the cutoff should be stated in terms of the number of residences found, or the number of interviewed residences. (Stating the cutoff in terms of interviewed residences does give more control over the weights in the PSUs.) Either way, our model is not easily extended to this case.

Although decreasing rules may reduce variance more than fixed rules, they tend to require more calling in the sparsest PSUs, which are the most irritating to interviewers. The variance reduction seems to be trivial even with the decreasing rules, so fixed rules would seem to be preferable based on considerations of interviewer morale. Since the results for fixed rules in Section 10 did not seem to be very sensitive to the exact cutoff point, those results may also be reasonable guidelines for when to permit a cutoff even in situations when the model does not apply exactly.

CHAPTER 8

MINIMUM COST SAMPLE ALLOCATION FOR MITOFSKY-WAKSBERG RANDOM DIGIT DIALING

R. E. Mason and F. W. Immerman
Research Triangle Institute

Many surveys, perhaps most, are faced with the problem of determining a sample allocation that will simultaneously satisfy multiple variance requirements. Kuhn-Tucker theory (see, for example, Simmons, 1975) provides a basis for determining the sample allocation that will satisfy a set of variance constraints for the least cost. The first application of Kuhn-Tucker theory in this context is described by Folsom et al. (1979). A general exposition of the procedure is presented by Chromy (1987).

In the allocation procedure, equations are developed to describe the variances of an arbitrary set of parameter estimates and the variable survey costs. The cost and variance equations are then combined with the specified variance constraints in an objective function and solved simultaneously to obtain the least cost allocation. The procedure also identifies the relative importance of each constraint in the determination of the allocation solutions and hence their relative contributions to the total cost of the survey. Unduly costly constraints can be identified and consideration given to their relaxation as a cost reduction measure.

As applied to Mitofsky-Waksberg random digit dialing (RDD) (Waksberg, 1978), the procedure yields the number of clusters and cluster sizes to be selected from within each of the design strata. The cost model makes use of the fact that the expected frequency with which Mitofsky-Waksberg clusters are identified varies considerably across the country, as does the conditional frequency of identifying a residence given a cluster. Both frequencies are important determinants of survey cost. Most surveys with broad geographic coverage benefit if information about the spatial distribution of the frequencies is used to allocate the sample.

Wyoming might be contrasted with California by way of example. First, the populations in each state might, given specific survey objectives, contribute differentially to different subpopulations (domains) of interest. Wyoming might assume more importance in estimating parameters describing intensely rural populations, whereas California would contribute heavily to parameter estimates for heavily urbanized populations. The procedure described in this chapter allows the investigator to simultaneously impose different variance constraints on estimates from both domains. Second, such differences as exist between the two states in population densities in relation to telephone exchange densities generate different costs associated with initially identifying clusters and subsequently identifying residences within clusters. The number and size of clusters to be used in each state are those that will satisfy the variance constraints for the least cost.

1. KUHN-TUCKER CONDITIONS

The variances of the parameter estimates can be expressed as a function $V_d\{n\}$ of the sample sizes $\{\underline{n}\}$ selected from within the strata at the various stages or phases of sampling specified by the sampling design. The subscript d identifies the domains chosen by the investigator for the purpose of determining the sample allocation. The variable cost of the survey can be expressed as a function $C\{\underline{n}\}$, of the same sample sizes.

The sample allocation problem can be stated in terms of minimizing the cost function $C\{\underline{n}\}$ subject to the inequality variance constraints

$$V_d\{\underline{n}\} \leq K_d \,.$$

The constraint values K_d are chosen by the investigator along with the domain parameters to which they apply. The solutions sought, denoted by $\underline{n}^*$, are the sample sizes that minimize the objective function,

$$0\,\{\underline{n}, \underline{\lambda}\} = C\{\underline{n}\} + \sum_d \lambda_d \,[V_d\{\underline{n}\} - K_d] \tag{8.1}$$

where λ_d are Lagrange multipliers (one for each of the variance constraints imposed).

Taking derivatives of the objective function with respect to the vector of sample sizes and equating to zero yields equations of the form

$$\frac{\delta\{C\{\underline{n}\}\}}{\delta\{\underline{n}\}} = -\sum_d \lambda_d \frac{\delta\{V_d\{\underline{n}\}\}}{\delta\{\underline{n}\}} \tag{8.2}$$

for all values of *d*.

If the constraints hold, then at $\underline{n}^*$ there must exist values of the Lagrange multipliers $\lambda_d{}^*$ such that Eq. (8.2) evaluated at $\underline{n}^*$ is true and additionally

$$V_d\{\underline{n}^*\} \le K_d\,, \tag{8.3}$$

$$\lambda_d^* \ge 0\,, \tag{8.4}$$

$$\lambda_d^*[V_d\{\underline{n}^*\} - K_d] = 0\,. \tag{8.5}$$

Equations 8.2 through 8.5 are the Kuhn-Tucker necessary conditions [see, for example, Simmons (1975), pages 186 and 192]. For most problems, given the equations $C\{\underline{n}\}$ and $V_d\{\underline{n}\}$, the values $\underline{n}^*$ are found by using an iterative numerical procedure.

2. SPECIFICATION OF THE VARIANCE CONSTRAINTS

The choice of parameters to use as the basis for determining the sample allocation is limited in practice by the necessity of knowing (for design purposes) something about population variances. Such information is often lacking, at least prior to the survey.

A convenient way around the problem is to specify the design requirements in terms of the variances to be associated with estimates of the relative sizes of specified domains. The (binomial) population variances are then known. Specifying the problem this way has some generality and provides a useful surrogate for other parameters. Certainly parameters describing other domain characteristics are unlikely to be reliably estimated if the relative domain sizes themselves cannot be. This choice of parameters is not restrictive if the requisite variance information is known. However, domain size parameters are used in the remainder of the paper.

The relative domain sizes used for allocating the sample can be chosen based on what is known about their distribution in the inferential population or on the basis of those values that have policy or program implications. For example, an abatement program might be required if the frequency of positive cases exceeds some threshold. A reasonable design would seek to provide specified reliability for estimating the threshold level even if the existing number of positive cases was thought to be different.

Notationally, the relative domain sizes P_d to be used to allocate the sample are specified together with the variance constraints K_d that are required of the sample estimates $\hat{P}_d$. The domain size specifications can be different in different design strata. The variance constraints are set to provide the level of precision required by data analysts, policy makers, and other users of the survey data. The constraints often represent a compromise between the level of precision ideally desired and that which is affordable given the resources available for the survey.

3. VARIANCE MODEL

If the procedure described by Waksberg (1978) is followed in selecting telephone numbers, the sampling design can be described as a two stage, equal probability, with replacement selection of residences. If strata, geographic or otherwise, are constructed to classify whole telephone exchanges, then the self-weighting feature of the design applies at least within strata. To describe the variance of a sample estimate $\hat{P}_d$, the number of residences in each of the design strata is assumed known. In practice, housing unit counts that are reasonably current and have acceptable accuracy and geographic resolution can be obtained commercially. The geographic resolution provided, however, is likely to describe areas such as counties rather than telephone exchanges.

The number of housing units in the ath design stratum is denoted notationally by

$$N(a), \quad a = 1, 2, \ldots,$$

and the first stage and second stage sample sizes by $n_1(a)$ and $n_2(a)$, respectively. Following this notation, sample estimates of the parameters P_d, are given by

$$\hat{P}_d = \sum_a \frac{N(a)}{N} \hat{P}_d(a) ,$$

where $\hat{P}_d(a)$ is the per sampling unit estimated domain size in the ath stratum, and

$$N = \sum_a N(a) .$$

The associated sampling variances are given by

$$V_d\{\underline{n}\} = \sum_a \left(\frac{V_{d,1}(a)}{n_1(a)} + \frac{V_{d,2}(a)}{n_1(a)\, n_2(a)} \right) . \tag{8.6}$$

Expressing the variance in terms of the components associated with the two stage selection of the sample facilitates taking derivatives. The variance components are defined by

$$V_{d,1}(a) = \left(\frac{N(a)}{N}\right)^2 V_d(a)\, R_d(a) ,$$

$$V_{d,2}(a) = \left(\frac{N(a)}{N}\right)^2 V_d(a)\, (1 - R_d(a)) ,$$

where

$$V_d(a) = P_d(a)\, (1 - P_d(a)) ,$$

and the values $R_d(a)$ are the intracluster correlations arising from sampling residences within the same first five digits of a seven-digit telephone number.

With stratification, different variance constraints can be imposed on the same domain in different strata, thereby effectively oversampling some strata relative to others. Oversampling could result, for example, from a requirement to make stratum level comparisons. If different stratum level variance constraints are imposed, an additional separately specified overall constraint is advisable to keep unequal weighting effects on aggregate estimates at acceptable levels.

4. COST MODEL

A form of the cost equation that is compatible with Eq. (8.1) is simply

$$C\{\underline{n}\} = \sum_a \left(n_1(a)\, C_1(a) + n_1(a)\, n_2(a)\, C_2(a)\right) , \tag{8.7}$$

where

$C_1(a)$ = the per first stage unit (cluster) average cost in the ath design stratum

$C_2(a)$ = the per second stage unit (residence) average cost in the ath design stratum.

Equation (8.7) does not, however, provide a particularly useful description of the survey costs.

Waksberg (1978) describes the costs in terms of the average cost of an unproductive call C_u and the average cost of a productive call C_p. Waksberg's cost coefficient C_p can be usefully broken into two parts according to whether the productive call yields screening or substantive information. In many surveys, eligibility criteria may serve to exclude some number of identified residences from participation in the survey. Screening questionnaires developed to resolve eligibility issues are typically shorter and less costly to administer than are the substantive questionnaires. Notationally the average cost of a productive call can be divided into the average cost of administering and processing a screening questionnaire, denoted by C_s, and that for a substantive questionnaire, denoted by C_q, by

$$C_p(a) = C_s(a) + p_e(a)\, C_q(a)\,,$$

where $p_e(a)$ is the eligibility rate in the ath design stratum.

Waksberg's cost model (1978, equation 5.5) makes use of two other quantities, namely the proportion of residential numbers in the population π and the proportion of prefix areas with no residential numbers t. The first quantity is the expected frequency with which clusters are identified; that is, the probability of obtaining a residence using a four-digit random number subtended to an assigned area code/prefix (i.e., NPA/NXX) combination. Notationally the frequency is denoted by $p_c(a)$. Developing a cost model in the form of Eq. (8.7) is facilitated by replacing the second quantity t by the conditional expected frequency of identifying a household given an identified cluster; that is, using an independent selection of random digits in positions nine and ten subtended to the eight digit sequence that identifies a cluster. The conditional probability is denoted in what follows by $p_r(a)$.

With these changes the first and second stage cost coefficients in Eq. (8.7) can be written as

$$C_1(a) = \left(\frac{1-p_c(a)}{p_c(a)} - \frac{1-p_r(a)}{p_r(a)} \right) C_u(a)\,,$$

$$C_2(a) = \left(\frac{1-p_r(a)}{p_r(a)} \right) C_u(a) + C_s(a) + p_e(a)\ C_q(a)\,.$$

Note that the per cluster costs $C_1(a)$ exclude the cost of administering screening and substantive questionnaires to the residences identifying sample clusters; that is, those residences obtained using the initially selected ten-digit number. These costs are included in the second stage per residence cost coefficient $C_2(a)$.

5. ALLOCATION SOLUTIONS

The design-specific form of the objective function, Eq. (8.1), is provided by the variance and cost equations, Eqs. (8.6) and (8.7). Taking partial derivatives of the objective function with respect to the sample sizes $n_1(a)$ and $n_2(a)$ and equating to zero provides allocation solutions taking the form

$$n_1(a) = \left(\frac{\sum_d \lambda_d \, V_{(d),1}(a)}{C_1(a)} \right)^{1/2},$$

$$n_2(a) = \left(\frac{C_1(a) \sum_d \lambda_d \, V_{d,2}(a)}{C_2(a) \sum_d \lambda_d \, V_{(d),1}(a)} \right)^{1/2}.$$

In most cases, there is no direct solution to this set of equations, and values for the Lagrange multipliers are found iteratively. Initial values for the Lagrange multipliers must be supplied to begin the iterative process. Informative initial values for the Lagrange multipliers can be obtained by first computing the cluster sizes

$$\tilde{n}_2(a) = \left(\frac{C_1(a)\,[1 - R_d(a)]}{C_2(a) R_d(a)} \right)^{1/2}$$

for each stratum and then computing

$$(\tilde{\lambda}_d)^{1/2} = \frac{1}{K_d} \sum_a \left(\left(\frac{V_{d,2}(a)}{\tilde{n}_2(a)} + V_{d,1}(a) \right) (\tilde{n}_2(a)C_2(a) + C_1(a)) \right)^{1/2}.$$

The domain sample sizes computed using these values of the Lagrange multipliers will individually satisfy the variance constraints.

Following definition of the initial Lagrange multiplier values, iterative numerical procedures are used to adjust the values so as to yield sample sizes that simultaneously satisfy all variance constraints. If initial values are set equal to the values that satisfy the variance constraints individually, then those constraints that are important in determining the allocation solutions can be identified. They will have final Lagrange multiplier values closest to the initial values. Constraints that are in fact superfluous (i.e., are coincidentally satisfied with the imposition of other constraints) will have final Lagrange multiplier values of zero. In every case the final Lagrange multiplier values will be less than or equal to initial values computed as above.

A comparison of initial and final values is often useful in reconciling initially perceived reliability requirements with budget realities. By relaxing one or two of the variance constraints that essentially drive the allocation solutions, quite impressive cost reductions can be achieved.

However, note that the sample sizes obtained are those that simultaneously satisfy fixed variance constraints for the least cost. Unlike the single constraint case, the multiple constraint problem cannot be uniquely specified in terms of fixing the cost.

6. AN ITERATIVE PROCEDURE

One way to obtain the Lagrange multiplier values is to simply multiply their values at successive iterations by the ratio of the variances, given the sample sizes as determined at that point, to the imposed constraints. The ratios increase the Lagrange multiplier values when the variances exceed the constraints (i.e., sample sizes are still too small) and decrease them when the variances are smaller than the constraints.

The iterative process can be speeded up by identifying those values of the Lagrange multipliers that approach zero and setting them to exactly zero. Some care is needed to ensure that the multiplier is actually on its way to zero since values set to zero are subsequently fixed at this value. Although the reason is not clear, using the square roots of the Lagrange multipliers in the calculations reduces the number of iterations needed for convergence. Sufficient accuracy is ensured by continuing the iterative process until Eq. (8.5), squared and summed over the set of imposed constraints, is less than some arbitrary, very small value, such as 10^{-6}.

The above iterative procedure is capable of solving large allocation problems. The largest allocation problem so far undertaken by the authors was for the Youth Attitude Tracking Study (YATS), conducted annually by the Department of Defense. The YATS allocation currently involves first and second stage sample sizes within 66 design strata subject to 173 inequality variance constraints. The large number of constraints results from the different geographic reporting domains of separate interest to the Army, Navy, Marine Corps, and Air Force.

7. A NUMERICAL EXAMPLE

Quite acceptable assessments of the frequencies $p_c(a)$ can be made by simply dividing current housing unit counts by 10,000 times the number of NPA/NXX codes in specified geographic areas. Current housing unit information in computer accessible form can be commercially obtained on a county basis. Rate Center States and Cities on current NPA/NXX and V+H Coordinates Tapes (available from Bell Communications Research, Inc.) can be used to approximate the county distribution of NPA/NXX codes. Assessments computed in this way assume that complete and accurate response is obtained during cluster identification, and that all housing units have telephones. To the extent that these assumptions are not met, observed values for $p_c(a)$ will be lower than these "optimal" assessments (see this volume, Chapter 3). However, for design purposes, the above calculation provides the information needed to quantitate the relative geographic variability expected in $p_c(a)$.

Experience with the YATS at the level of geographic definition provided by the 66 design strata indicates that values of the frequencies $p_c(a)$, computed as described above, can range from 0.077 (North Dakota) to 0.369 (Miami and Tampa areas of Florida), averaging 0.263 nationally. This national average is, as expected, higher than the frequency typically experienced in many surveys. Observed values of the conditional frequencies $p_r(a)$ range from 0.242 (also North Dakota) to 0.664 (Mississippi), averaging 0.490 nationally, which is lower than the average reported in the literature. Table 1 profiles the variability observed in values of $p_c(a)$ and $p_r(a)$. The information is taken from the 1986 YATS (Immerman and Mason, 1986).

To numerically illustrate the allocation procedure, a simplified version of the YATS sampling design problem is presented below. First, some background on YATS is necessary. The primary objective of YATS is to estimate the proportion $\hat{P}_d^*$ of a specific subpopulation of young males

Table 1. Assessments of the Cost Model Frequencies $p_c(a)$ and $p_r(a)$

Census Division/State	Urban/Rural	Frequency of Identifying Clusters $p_c(a)$	Frequency of Identifying Residences Within Clusters $p_r(a)$
New England			
Connecticut	–	0.312	0.386
Maine	–	0.155	0.611
Massachusetts	U	0.336	0.512
	R	0.295	0.559
New Hampshire	–	0.144	0.540
Rhode Island	–	–	–
Vermont	–	–	–
Middle Atlantic			
New Jersey	–	0.343	0.435
New York	U	0.332	0.495
	R	0.237	0.548
Pennsylvania	U	0.341	0.504
	R	0.256	0.499
South Atlantic			
Delaware	–	–	–
Florida	U	0.368	0.465
	R	0.301	0.485
Georgia	–	0.287	0.507
Maryland	–	0.196	0.476
North Carolina	–	0.288	0.505
South Carolina	–	0.290	0.618
Virginia	–	0.247	0.534
West Virginia	–	0.253	0.522
East North Central			
Illinois	–	0.293	0.491
Indiana	–	0.210	0.536
Michigan	–	0.281	0.422
Ohio	U	0.267	0.479
	R	0.232	0.466
Wisconsin	–	0.222	0.505
East South Central			
Alabama	–	0.270	0.600
Kentucky	–	0.242	0.547
Mississippi	–	0.235	0.664
Tennessee	–	0.291	0.499

Table 1 (Continued)

Census Division/State	Urban/Rural	Frequency of Identifying Clusters $p_c(a)$	Frequency of Identifying Residences Within Clusters $p_r(a)$
West North Central			
Iowa	–	0.129	0.381
Kansas	–	0.139	0.535
Minnesota	–	0.175	0.582
Missouri	–	0.179	0.443
Nebraska	–	0.106	0.453
North Dakota	–	0.077	0.242
South Dakota	–	0.084	0.379
West South Central			
Arkansas	–	0.182	0.485
Louisiana	U	0.307	0.532
	R	0.234	0.480
Oklahoma	–	0.183	0.500
Texas	U	0.229	0.475
	R	0.119	0.577
Mountain			
Arizona	–	0.267	0.434
Colorado	–	0.197	0.455
Idaho	–	0.136	0.640
Montana	–	0.102	0.225
Nevada	–	–	–
New Mexico	–	–	–
Utah	–	0.176	0.484
Wyoming	–	–	–
Pacific			
California	U	0.326	0.488
	R	0.272	0.487
Oregon	–	0.251	0.433
Washington	U	0.292	0.483
	R	0.181	0.500

With modification from: Immerman, F. W., and R. E. Mason, 1986, "Sampling Design and Sample Selection Procedures," Youth Attitude Tracking Study 1986, Research Triangle Institute, Report Number RTI/3624/01-01W, 66 pp.

that express positive propensity toward military service. Because YATS is conducted annually, relatively recent propensity estimates are available for design purposes. Also available are estimates of the per household incidence rate of target population members (i.e., $p_e(a)$). Note that

$$\hat{P}_d(a) = \hat{P}_d^*(a) p_e(a)\ .$$

For this example, nine geographic design strata, defined by the nine U.S. Census Divisions, will be considered. Table 2 presents the division level values used as inputs to the procedure. The first and second stage stratum level variance components are also shown. Variance computations for all allocations incorporate $R_d(a)$ computed using an average, individual level intracluster correlation, say R^*, of 0.24. The individual and household level intracluster correlations are related by

$$R_d(a) = R^* \frac{p_e^2(a) \hat{P}_d^*(a) \left(1 - \hat{P}_d(a)\right)}{V_d(a)}\ .$$

Stratum level values of $p_c(a)$ and $p_r(a)$ used in the allocation are listed in Table 3. The values were computed as simple means of the values shown in Table 1. The cost models for all three allocations are based on national level cost coefficients of $C_u = \$2.80$, $C_s = \$5.32$, and $C_q = \$24.74$.

Suppose that the sample must satisfy variance constraints for three reporting domains — the first consisting of the entire United States, the second of the five eastern Census Divisions, and the third of the four western Census Divisions. If these three constraints are arbitrarily defined to be 0.0001, 0.00015, and 0.00015, respectively, the procedure yields the allocation presented in Table 4. Convergence required 62 iterations. All three constraints were binding (that is, none of the Lagrange multipliers approached zero).

The cluster sizes and numbers of clusters presented in Table 4 are the deterministically rounded integer values of the optimal cluster sizes and numbers of clusters produced by the procedure. For surveys in which sample sizes are small and/or cluster sizes are very small (e.g., 1 or 2 households per cluster), rounding can yield an allocation that is relatively far from the optimum computed by the procedure. In such instances, alternatives to deterministic integer rounding, such as random rounding to achieve in expectation the noninteger allocation, might be considered.

Table 2. Input Values for Variance Model Variables

Census Division	Number of Households[a] ($N(a)$)	Per Household Eligibility Rate ($p_e(a)$)	Positive Propensity Rate[b] ($P^*(a)$)	Variance Components	
				First Stage $V_{d,1}(a)$ ($\times 10^{-3}$)	Second Stage $V_{d,2}(a)$ ($\times 10^{-3}$)
New England	4,674,484	0.094	0.287	0.434	0.258
Middle Atlantic	14,503,499	0.092	0.288	0.417	0.254
South Atlantic	14,364,551	0.089	0.332	0.422	0.283
East North Central	14,044,025	0.096	0.283	0.449	0.260
East South Central	5,997,176	0.089	0.317	0.412	0.270
West North Central	7,666,568	0.087	0.287	0.372	0.240
West South Central	9,868,176	0.091	0.307	0.423	0.267
Mountain	4,593,717	0.088	0.287	0.380	0.242
Pacific	12,772,074	0.092	0.251	0.382	0.222
United States	88,484,270	0.091	0.293		

[a]Source: Market Statistics, Inc., projections.
[b]Person level rates.

Table 3. Census Division Cost Model Inputs

Census Division	Frequency of Identifying Clusters $p_c(a)$	Frequency of Identifying Residences Within Clusters $p_r(a)$
New England	0.284	0.530
Middle Atlantic	0.302	0.496
South Atlantic	0.279	0.514
East North Central	0.251	0.483
East South Central	0.260	0.578
West North Central	0.127	0.431
West South Central	0.209	0.508
Mountain	0.176	0.448
Pacific	0.264	0.478
United States	0.249	0.495

Table 4. Sample Allocation — Three Variance Constraints, Division Level p_c and p_r

Census Division	Cost Components: First Stage $C_1(a)$	Cost Components: Second Stage $C_2(a)$	Number of Clusters $n_1(a)$	Cluster Size $n_2(a)$	Total Households	Expected Total Numbers Dialed	Expected Interviews	Estimated Cost ($000)
New England	4.58	10.13	379	5	1,895	4,195	178	20.9
Middle Atlantic	3.63	10.44	1,293	5	6,465	14,709	595	72.2
South Atlantic	4.59	10.17	1,146	5	5,730	13,026	510	63.5
East North Central	5.36	10.69	1,069	5	5,345	13,112	513	62.9
East South Central	5.92	9.57	416	6	2,496	5,199	222	26.3
West North Central	15.55	11.17	497	9	4,473	13,138	389	57.7
West South Central	7.89	10.28	959	7	6,713	15,915	611	76.6
Mountain	9.66	10.95	383	7	2,681	7,306	236	33.0
Pacific	4.75	10.65	1,520	5	7,600	18,477	699	88.2
United States			7,662	5.7	43,398	105,076	3,953	501.4

CHAPTER 9

WEIGHTING ADJUSTMENTS FOR RANDOM DIGIT DIALED SURVEYS

James T. Massey and Steven L. Botman
National Center for Health Statistics

There are a number of potential biases associated with telephone surveys that affect the accuracy of the telephone survey results. This chapter describes a number of post survey weighting adjustments employed to eliminate or reduce potential biases from such factors as multiple telephone numbers in the same household, noncoverage of households without telephones, and nonresponse. A national random digit dialing telephone survey on smoking carried out in 1980 is used to illustrate their effects. Four separate estimators, derived using different combinations of the weighting adjustments, were compared to the unweighted telephone survey estimator and the impact of the adjustments on the smoking survey statistics was examined.

Post survey weighting adjustments are frequently used in national household surveys to compensate for undercoverage and nonresponse, and are an accepted practice for many national surveys using face to face household interviews. For example, detailed weighting adjustments are described in Bailar et al. (1978) and Bean (1974) for two of the largest continuous national surveys, the Current Population Survey (CPS) and the National Health Interview Survey (NHIS), respectively. For these two surveys weighting adjustments are used in estimation to account for unequal probabilities of selection and nonresponse, and to adjust for undercoverage of the U.S. population. For telephone surveys, where noncoverage, undercoverage, and nonresponse are generally higher, the impact of post survey adjustments on survey estimates may be greater than the impact of post survey adjustments in face to face surveys. This chapter examines the impact of post survey adjustments in a national telephone survey.

1. DATA SOURCES

An RDD telephone survey was used by the National Center for Health Statistics (NCHS) to obtain information on smoking and health related characteristics of the civilian noninstitutionalized population in the conterminous United States. Although two interviews were scheduled six months apart for each sample household, this chapter presents only the results from the first household interview conducted during the last six months of 1980. The smoking questionnaire required an average of 15 minutes to complete in each household. All adults 17 years old or older were eligible for the interview and self-response was required for all adults.

The sample design was patterned after the Mitofsky-Waksberg approach (Waksberg, 1978) using a cluster size of 9 households per primary sampling unit (PSU). The sampling method is described by Lepkowski (this volume, Chapter 5). The RDD sampling procedure produced a total of 5,114 households. All adults 17 years of age or older were interviewed separately with one or more interviews completed in 4,301 households. Within those households a total of 7,736 interviews were completed out of a possible 8,999. Based on these numbers the household response rate was estimated to be 84 percent (households with at least one completed interview) and the person response rate was estimated to be 73 percent.

During the data collection period for the RDD smoking survey, a smoking supplement was incorporated concurrently into the face to face NHIS. The questions on the NHIS smoking supplement were a small subset of the questions on the RDD study. Interviews were conducted in approximately 40,000 households as part of the 1980 NHIS. Within these households, 11,333 individuals were selected to complete the smoking supplement and a final response rate of 94 percent was obtained. Selected results from the NHIS smoking supplement were compared to the RDD telephone survey results.

2. WEIGHTING ADJUSTMENTS

2.1 Basic Probability of Selection

The basic probability of selection refers to the initial probability that a residential telephone number is selected in a telephone survey. The Mitofsky-Waksberg two-stage sampling design used for the RDD telephone survey on smoking is approximately self-weighting. That is, each household selected for the survey has approximately the same chance of being selected. This assumes that all households with telephones have only one telephone number and that all blocks of telephone numbers sampled at the second stage of selection contain a minimum prespecified number of residential telephone numbers. With these assumptions the basic probability of selection for a household with a telephone should be equal to the proportion of telephone households sampled.

For the telephone survey on smoking the probability of selection for any household is given by

$$p_{ij} = \frac{n_i m}{100\, M}$$

where p_{ij} = probability of selecting the jth sampled household in the ith PSU

n_i = cluster size in the ith PSU or the number of residential numbers sampled in the ith PSU (usually 9)

m = number of primary sample numbers dialed at the first stage of selection

M = total number of PSUs in the population.

A more detailed explanation of the basic probability of sample selection is given by Waksberg (1978). For the telephone survey on smoking

$$p_{ij} = \frac{9\ (2{,}441)}{100\ (3{,}201{,}700)} = 0.0000686$$

The reciprocal of the probability of sample selection determines the basic household weight. For the telephone survey on smoking the basic household weight was equal to 14,574.

2.2 Multiple Telephones in Household

Households with more than one telephone number have a higher probability of selection for the sample than households with only one telephone number based on using a frame of all telephone numbers. The probability of selecting a household in an RDD telephone survey is proportional to the number of distinct telephone numbers in the household. The probability of selection, therefore, must take into account the number of telephone numbers in the household. For the RDD telephone survey on smoking 6.3 percent of all sampled persons resided in a household with more than one nonbusiness telephone number.

The adjustment A_{ij} for the number of nonbusiness telephone numbers in the jth household in the ith PSU is given by

$$A_{ij} = 1/[\text{minimum } (T_{ij}, 2)]$$

where T_{ij} is the number of nonbusiness telephone numbers in the jth household in the ith PSU.

The multiple telephone adjustment was truncated at one-half because it was felt that any further reduction in bias would be offset by an increase in variance due to the differential weighting. Moreover it was unclear in some cases where a large number of telephone numbers was reported whether the number of extensions was being mistakenly reported as the number of telephone numbers.

Combining the first two adjustments one calculates a household weight W_{ij} for the jth sample household in the ith PSU by

$$W_{ij} = A_{ij}/P_{ij}\,.$$

It should be noted that each adult in the household is assigned the household weight because all adults in the sample households were selected for the telephone survey on smoking.

2.3 Household Nonresponse

In most household surveys of the population some sample households and sample persons are not interviewed for a variety of reasons. There is some evidence (Herzog et al., 1983; Weaver et al., 1975; Massey et al., 1981; and DeMaio, 1984) that while the telephone survey response does not differ greatly by geographic region, there do appear to be significant differences in response rates by age, race, education, and income. Although these are primarily person characteristics, many of the nonresponse differences will show up as neighborhood differences.

For the telephone survey household nonresponse adjustments were made at the PSU level. The nonresponse household adjustment B_i for the ith sampled PSU is given by

$$B_i = \text{minimum } (n_i/n_i', 2)$$

where n_i' is the number of sample households in the ith PSU with one or more completed interviews.

2.4 Households Without Telephones

A number of studies have shown that the telephone and nontelephone populations are quite different with respect to demographic, economic, and health characteristics (Groves and Kahn, 1979; Banks, 1983; Thornberry and Massey, this volume, Chapter 3). While the estimates of characteristics for the total population are not affected very much by the omission of the nontelephone households, some of the subdomain estimates could have large biases due to the noncoverage of households without telephones. Telephone coverage is particularly low for such subdomains of the population as blacks in the South, persons with low incomes, persons in rural areas, persons with less than 12 years of education, persons in poor health, and heads of households under 25 years of age. In some cases the noncoverage exceeds 50 percent.

In deciding which variables should be used to adjust national RDD telephone estimates for noncoverage, several factors need to be considered. The variables selected should: be correlated to telephone noncoverage; be important domain variables for national statistics; be reasonably stable over time; be routinely estimated by the Bureau of the Census; not be subject to a large response error; and have a low item nonresponse rate. It is also critical that the variable definitions used in an RDD telephone survey be consistent with the definitions used to define the benchmark totals obtained from the Bureau of the Census.

For the telephone smoking study the two variables that satisfy these criteria are geographic region and race. Although family income and poverty status are other important variables in explaining noncoverage differences, both variables fluctuate over time and have a high item nonresponse rate. The adjustments by region and race were calculated by using the ratio of the estimated total population to the estimated telephone population. The ratios were obtained from the 1980 NHIS. Table 1 presents the ratio factors used to adjust for telephone noncoverage. The adjustment factors range from 1.04 to 1.26. The noncoverage adjustments are applied by multiplying the basic weight of each person falling into a region and race stratum by the respective factor for that stratum.

2.5 Poststratification

The last telephone survey adjustment considered is a poststratified ratio adjustment. This type of adjustment is commonly used in face to face national surveys to bring the survey estimates of subdomain totals in agreement with independent population figures for the subdomains. Bean

Table 1. Noncoverage Ratio Adjustment for a National RDD Telephone Survey on Smoking by Region and Race, 1980

Region and Race	Adjustment Factor
Northeast	
Black	1.1283
Races other than black	1.0395
North Central	
Black	1.1080
Races other than black	1.0413
South	
Black	1.2555
Races other than black	1.0880
West	
Black	1.0748
Races other than black	1.0540

(1974) describes the poststratification procedure used for the NHIS and Census Bureau Technical Report 40 (1978) describes the poststratification procedures used for the CPS. For the RDD telephone survey on smoking, the poststratification ratio adjustment attempted to achieve a number of objectives: to reduce the potential bias due to noncoverage, undercoverage, and nonresponse; to improve the precision of the survey estimates; and to provide consistency with other statistics.

Undercoverage refers to nonenumerated persons within households and is discussed by Maklan and Waksberg (this volume, Chapter 4). Thornberry and Massey (1983) have shown that nonresponse, including both household nonresponse and that due to some adults in a household being interviewed with others not being interviewed, varies significantly across domains. For example, while almost all persons 65 years of age and over reside in households with telephones, their response rate is often 10 to 15 percentage points below the response rates for other age groups in a telephone survey. The situation can even be worse for the low income population and those with less than a high school education for whom both coverage and response rates are low for telephone surveys.

In the RDD telephone survey on smoking a single set of poststratification variables was used to adjust for noncoverage, undercoverage, and residual nonresponse not taken care of by the household nonresponse adjustment. The variables selected were age, sex, and educational attainment, which were used to subdivide the population into 24 strata.

The NHIS estimates of the total population in each stratum (previously adjusted to independent Bureau of the Census estimates),

Table 2. Poststratification Adjustment Factors by Sex, Educational Attainment, and Age for the Fully Weighted RDD Survey

Sex and Age	Educational Attainment (in Years)		
	<12	12	>12
Males			
17–24 years	1.4221	1.2736	0.8659
25–44 years	1.7867	1.1206	1.0213
45–64 years	1.3440	1.2917	1.0772
65 years and over	1.5278	1.0599	0.9852
Females			
17–24 years	1.4550	1.0679	0.9329
25–44 years	1.5668	1.0144	0.8749
45–64 years	1.2064	1.0017	0.8778
65 years and over	1.5103	1.1148	1.0090

were divided by the sum of the telephone survey weights for each of the strata. Table 2 presents the ratio adjustments for each of the 24 poststratification strata after all other adjustments had been applied. The poststratified ratio adjustments ranged from 0.88 to 1.79. These adjustments indicate just what a potential problem nonresponse and noncoverage may be in telephone surveys.

The poststratified ratio adjustments are applied within each of the 24 strata by multiplying each person's previously adjusted weight by the poststratification factor.

3. EVALUATING ESTIMATORS FOR THE TELEPHONE SURVEY ON SMOKING

Four separate estimators were derived using combinations of the weighting adjustments described above. Several criteria were used to evaluate the impact of the combinations of weighting adjustments on the telephone survey statistics. The overall effect of the weighting adjustments was evaluated by comparing the weighted subdomain distributions with the unweighted subdomain distributions. The magnitude of the change in the telephone estimates was further compared to the effect of similar adjustments on the NHIS estimates. The direction of the change in the telephone estimates was then compared to the direction of the NHIS estimates and the direction of the difference between NHIS estimates for telephone and nontelephone households. Finally the precision of the weighted estimates was examined.

3.1 Description of Estimators Examined in This Study

A number of estimators for the telephone survey on smoking could be constructed depending on whether all or only some of the post survey weighting adjustments described in Section 2 are used. It was decided, however, that two restrictions should be placed on the estimators to be examined. First, the estimators should include an adjustment for the probability of household sample selection. Second, the estimated population totals for important domains of study should coincide with the population controls for these subdomains. This objective was accomplished by the inclusion of a poststratified ratio adjustment for all estimators. The need for a poststratified adjustment in any weighted estimator is supported by the size of these adjustments shown in Table 2.

With these restrictions, the combinations of post survey weighting adjustments yielded four estimators for the telephone survey on smoking. The composition of these estimators in terms of the proposed weighting adjustments is presented in Table 3.

Estimator 4 is considered the fully adjusted estimator since it contains each of the proposed post survey weighting adjustments. Analytic expressions for the four estimators are not shown, but expressions for similar estimators are shown in Bean (1974) and in a recent NCHS report (1987). The latter report describes a study conducted by the Survey Research Center (SRC) of the University of Michigan for NCHS. The RDD telephone survey on smoking described in this chapter and the SRC study were conducted about the same time, and the research conducted on the weighting of telephone estimates between the studies is similar in many respects. The parallel research was intentional and discussions were held between NCHS and SRC staff to coordinate the research.

Table 3. Post Survey Weighting Adjustments Used to Form Estimators for a Telephone Survey on Smoking

	Weighted Estimator			
Post Survey Adjustment	1	2	3	4
Probability of selection	x	x	x	x
Household nonresponse		x		x
Telephone noncoverage			x	x
Poststratification	x	x	x	x

For each estimator, the sampling variances for the telephone survey on smoking were approximated using a balance repeated replication procedure (McCarthy, 1966, 1969).

3.2 Weighted Versus Unweighted Estimates from the Telephone Survey

The four weighted estimators were first compared to the unweighted telephone survey estimator using the results of the telephone survey on smoking. The effects of the weighting adjustments on the percentage distributions for socio-demographic subdomains are shown in Table 4. Since all of the weighted estimators were poststratified by sex, age, and education, the percent distribution for these characteristics is the same for all of the weighted estimators. However, the differences between the weighted and unweighted distributions are dramatic. The weighting adjustments increase the estimated percent of persons over 65 years of age by 17.8 percent, the estimated percent of blacks by 17.2 percent, and the estimated percent of persons without a high school education by 32.1 percent. The weighted adjustments further adjust the telephone sample for the underrepresentation of males and the overrepresentation of high income families. To the extent that the main survey variables are related to age, sex, and education, the weighting adjustments would have a substantial impact on the survey results.

The effect of the weighting adjustments on the estimate of the percent of blacks in the population depends to a large extent on whether the variable "race" was used for the adjustments. The weighted estimates (1 and 2) that exclude the household noncoverage adjustment do increase the estimated percentage of blacks in the population, but not nearly as much as the weighted estimates that include the noncoverage race adjustment. The effect of the weighting adjustments on the distribution of family income is relatively small. The adjustments increase the estimate of the number of persons reporting family incomes under $10,000 from 18 percent to 20 percent. This percentage is still significantly smaller than the estimated percentage of persons with family incomes under $10,000 in the NHIS. A critical problem, however, with the reporting of income for the telephone survey is the high item nonresponse. In the smoking survey, 14 percent of the respondents did not report family income. With the exception of race the four weighted estimators are about equal for all of the socio-demographic subdomains shown in Table 4, but many of these items show fairly large differences between weighted and unweighted estimates.

Table 4. Comparison of Weighted and Unweighted Telephone Survey Estimates of Socio-Demographic Percentage Distributions

Subdomain	Unweighted Estimate	Alternative Weighted Estimates 1	2	3	4	Standard Error for Estimator 4
Sex						
Males	44.4	47.2	47.2	47.2	47.2	0.0
Females	55.6	52.8	52.8	52.8	52.8	0.0
Age						
17–24	20.1	20.1	20.1	20.1	20.1	0.0
25–44	39.9	37.8	37.8	37.8	37.8	0.0
45–64	27.2	27.2	27.2	27.2	27.2	0.0
65+	12.8	15.0	15.0	15.0	15.0	0.0
Race						
Black	8.7	9.1	9.2	10.1	10.2	0.3
Other	91.3	90.9	90.8	89.9	89.8	0.3
Education						
<12 years	23.7	31.3	31.3	31.3	31.3	0.0
12 years	38.9	37.6	37.6	37.6	37.6	0.0
13+ years	37.4	31.1	31.1	31.1	31.1	0.0
Income (1,000's)						
<$5	6.0	6.6	6.6	6.7	6.7	0.3
$5–9	12.0	13.5	13.6	13.5	13.6	0.4
$10–14	16.3	16.7	16.8	16.8	16.9	0.4
$15–24	26.8	26.3	26.5	26.3	26.5	0.5
$25+	24.5	22.5	22.3	22.4	22.2	0.4
Unknown	14.4	14.4	14.2	14.3	14.1	0.4

Note: The domain estimates used for poststratification are not considered subject to sampling error.

In Table 5 the unweighted and weighted telephone estimates are compared when estimating the percent of the population who never smoked. The smoking status variable "never smoked" was considered to be better defined and less subject to response error than any of the other smoking status variables. All of the weighted estimates reduced the estimated percentage of persons 17 years of age or older who never smoked by approximately 1 percent. The reduction in percent of persons who never smoked was almost 3 percent for blacks between the ages of 25 and 65. The direction of the overall change is consistent with the fact that a much higher percentage of females have never smoked, and that females are overrepresented in the unadjusted telephone sample. In general, the fully adjusted estimate had a greater impact on the unweighted estimate than any of the partially adjusted estimators, although there were smaller differences between the weighted estimates than between the unweighted and weighted estimates.

Other selected smoking and health statistics are shown in Table 6. In this table the unweighted telephone estimate is only compared to the fully adjusted telephone survey estimate. The impact of the weighting adjustments on the proportions shown in Table 6 varied from a very small percentage point change (0.2) to a change of 2.2 percentage points for persons in fair or poor health. It is difficult to access the significance of these relatively small percentage point changes. The changes appear larger, however, when comparing the impact of similar weighting adjustments to the face to face NHIS estimates. In the NHIS the post survey weighting adjustments very rarely change the unweighted estimate by more than one percentage point. The impact of the telephone weighting adjustments also appears more significant when compared to changes in smoking behavior over time. For example, a change of one or two percentage points of the proportion of persons who smoke is viewed as a very important change with respect to the health of the nation.

Table 5. Comparison of Estimates of Percentage of Population Who Never Smoked for Unweighted and Weighted Telephone Survey Estimators

Age and Race Subdomain	Unweighted Estimate	Alternative Weighted Estimators				Standard Error for Estimator 4
		1	2	3	4	
All 17+ years	45.6	44.5	44.6	44.6	44.7	0.5
White	44.9	43.7	43.8	43.7	43.8	0.5
Black	50.0	49.7	48.9	50.0	49.2	1.6
17 − 24 years	58.5	57.5	57.3	57.6	57.4	1.7
White	57.4	56.1	55.9	56.1	55.9	1.3
Black	63.7	63.9	64.0	64.4	64.4	1.6
25 − 44 years	44.0	41.9	41.9	41.9	41.9	1.0
White	43.5	41.3	41.4	41.4	41.4	0.9
Black	43.3	41.8	40.5	41.9	40.6	2.5
45 − 64 years	36.1	34.7	34.8	34.7	34.9	1.1
White	35.8	34.4	34.6	34.4	34.6	1.0
Black	40.1	38.2	37.3	38.3	37.4	3.3
65+ years	50.8	51.7	52.2	51.9	52.3	1.9
White	49.7	50.5	51.0	50.5	51.0	1.4
Black	65.7	66.5	66.8	67.1	67.3	4.2

Table 6. Comparison of Weighted and Unweighted Estimates for an RDD Telephone Survey on Smoking with Estimates from the Face to Face NHIS[a]

	Telephone Survey			Face to Face NHIS		
Survey Variable	Unweighted	Fully Adjusted	Percent Change	Telephone Households	All Households	Percent Change
Smoking status						
Present smoker	31.2	32.0(0.5)	2.6	31.9(0.4)	32.8(0.5)	2.8
Former smoker	23.1	23.3(0.4)	0.9	21.3(0.4)	20.7(0.4)	−2.8
Never smoked	45.7	44.7(0.5)	−2.2	46.7(0.4)	46.4(0.5)	−0.6
Current smoker with prior attempt to quit						
No quit attempts (< 1 year)	54.6	53.3(1.2)	−2.4	58.4(1.0)	57.1(1.0)	−2.2
Health Status						
Excellent health	35.3	33.1(0.5)	−6.2	46.1(0.4)	45.2(0.5)	−2.0
Poor or fair health	16.2	18.2(0.4)	12.3	14.8(0.3)	15.2(0.3)	2.7

[a]Standard error shown in parentheses.

3.3 Effect of the Post Survey Adjustments on the Agreement Between Telephone and Face to Face Survey Results

The same set of questions on cigarette smoking asked in the telephone survey was also added as a supplement to the face to face NHIS in 1980. In this NHIS weighted estimates were calculated for the total population and for the households with telephones. The NHIS results are adjusted for nonresponse and undercoverage. The effect of these adjustments on the unweighted proportions was very small, however, since household nonresponse and undercoverage are both less than 10 percent.

Since the NHIS results are not subject to a telephone noncoverage bias and are less likely to have a serious nonresponse bias, it was hypothesized that the telephone adjustments should bring the telephone results into closer agreement with the NHIS results. Table 6 presents the telephone survey and NHIS estimates for six of the major survey variables. In general, the weighted adjustments for the telephone survey did not bring the unweighted estimates into closer agreement with the NHIS results. In fact, the telephone adjusted estimates were further from the NHIS estimates for five of the six variables shown in Table 6. The one variable for which the post survey adjustments did bring the telephone survey results into closer alignment with the NHIS results was percent of present smokers. When this variable was further analyzed within age and sex subdomains, the weighting adjustments consistently brought the telephone results into closer alignment with the NHIS results.

A closer inspection of the results shown in Table 6, however, did show an interesting effect produced by the telephone survey adjustments. For the NHIS the difference between the estimates for the telephone population and the total population corresponded to the direction of the difference between the unadjusted and adjusted telephone survey estimates for all but one variable. From this analysis it appears that the telephone post survey adjustments have partially accounted for the telephone coverage bias. The fact that the post survey adjustments did not account for the difference between the telephone survey and NHIS results indicates that either the differences are caused by other factors, such as mode differences, or the noninterviewed persons in the post survey adjustment strata are not similar to the interviewed persons in the strata.

The rather inconclusive evaluation of the impact of post telephone survey adjustments is further supported by the research results from the SRC study. Table 7 contrasts the unadjusted and the fully adjusted SRC telephone survey estimates with the NHIS estimates for all households and households with telephones for selected health statistics. The directions of the telephone survey adjustments are not consistently in the

Table 7. Comparison of SRC Telephone Survey Unadjusted and Post Survey Adjusted Estimates with NHIS Estimates for All Households and Households with Telephones for Selected Health Characteristics

	SRC[a] Telephone Estimate			
Health Variable	Unadjusted	Adjusted[b]	Households with Telephone	Total Households
2 week bed days	.290	.250	.328	.333
2 week work loss	.300	.282	.171	.171
12 month hospital episodes	.161	.155	.158	.160
Acute conditions	.183	.194	.116	.116
Chronic conditions	.473	.412	.415	.417

[a]Study conducted by the Survey Research Center of the University of Michigan for the U.S. National Center for Health Statistics (NCHS, 1987).
[b]Estimate adjusted for probability of selection, nonresponse, telephone coverage, and age, sex, and race.

direction of the NHIS estimates or in the direction of the noncoverage bias. Other confounding factors are clearly present. It should be noted that the face to face interview results do not represent a standard of comparison, just a different measure that can be used to contrast the characteristics of persons in households with and without telephones.

There have been a number of recent papers comparing telephone survey results with face to face survey results, and several explanations have been given to explain response differences between the two modes of data collection. Jordan et al. (1980) suggest that more extreme values are reported in telephone surveys, while Locander et al. (1976) report that socially desirable acts may be overreported in telephone surveys. Groves and Kahn (1979) indicate that telephone respondents are less candid about sensitive topics. Whether questions on cigarette smoking are sensitive or threatening to persons 17 years old or older is not known. Under the assumption that the questions on cigarette smoking are sensitive to some respondents, none of the explanations above explains all of the differences shown in Table 6. It does appear that socially desirable acts, such as being a former smoker or having made an attempt to quit smoking in the past year, are reported at higher levels in the telephone survey. The proportion of respondents reporting being in excellent health, however, is substantially lower in the telephone survey. One possible explanation for this large difference in the reporting of health status is the fact that not all of the respondents in the NHIS reported for themselves. About one-third of the adult respondents in the NHIS had their health status reported by another adult in the same household (self-reporting was required for the questions on cigarette smoking). Another explanation of the differences between the two modes of data collection is the questionnaire content and format. Although all of the questions being compared were identical in wording for the two surveys, the NHIS survey contained a number of additional questions about various health topics and the questions being compared were not in the same locations on the two survey questionnaires. When SRC (U.S. NCHS, 1987) examined the effect of similar post survey telephone adjustments on health variables, the direction of the changes was not significantly related to the direction of the difference between telephone surveyed households and all households.

The telephone weighting adjustments changed the unweighted estimated percentages shown in Table 6 by as much as 2 percentage points. For the health status variables this represents percent changes of 7 and 12 percent (actual change/unweighted percent). This indicates that there are major differences in health status among the domains chosen as the post survey adjustment cells. Whether the bias associated with the telephone health status estimates is reduced is still open to conjecture. One way to partially address this issue is to apply the weighting adjustments to the households with telephones interviewed in the face to face survey and observe whether the adjusted estimates for households with telephones better approximates the estimates for all households. Thornberry and Massey (1978), Banks (1983), and Banks and Anderson (1984) did this type of analysis and found that many of the adjusted estimates for households with telephones did move toward the estimates for all households, but that differences between the estimates for the total and the telephone populations still existed. More recently, three contributed papers at the 1987 International Conference on Telephone Survey Methodology (Smith; Steel and Boal; and Robinson and Triplett) all showed that post survey adjustments for telephone households from face to face conducted surveys did account for much of the differences in the characteristics of interest between telephone households and all households.

Table 8 presents some of the results from the Thornberry and Massey (1978) paper. Two poststratification adjustments were used to reweight the data for households with telephones for a number of health variables. The effect of the adjustments was generally consistent with the direction of the noncoverage bias, although the adjustments often produced very little change.

Table 8. Unadjusted and Ratio Adjusted NHIS Rates for Selected Health Characteristics: United States, 1976

Characteristics	Nontelephone Population	Telephone Population	Total Population[c]	Telephone Adjusted by IRC[a]	Telephone Adjusted by ASC[b]
Number of physician visits per person per year	4.2	5.0	4.9(.05)	5.0	5.0
Number of dental visits per person per year	0.7	1.7	1.6(.03)	1.6	1.7
Percent seeing a physician within last six months	53.5	59.9	59.3(1.7)	59.9	59.9
Percent seeing a dentist within past six months	18.0	36.5	34.9(1.2)	35.7	36.3
Number of hospital discharges per 100 persons per year	14.6	14.1	14.1(.20)	14.3	14.0
Percent with limitation of major activity due to chronic conditions	12.6	10.6	10.8(.15)	11.2	10.4
Days of restricted activity associated with acute conditions per 100 persons per year	1,175	934	957(17)	947	934
Days of bed disability associated with acute conditions per 100 persons per year	579	429	443(11)	435	429
Number of acute conditions per 100 persons per year	241	217	219(3.7)	216	218
Population (millions)	19.9	191	211(5)	211	211

[a]Income-region-color.
[b]Age-sex-color.
[c]Standard errors shown in parentheses.

3.4 Effect of Post Survey Weighting Adjustments on the Sampling Variance

The use of post survey weighting adjustments in an estimator for a face to face household survey has been found to reduce the sampling variance. In particular, Simmons and Bean (1969) and Jones and Chromy (1982) reported that the use of a poststratified ratio adjustment in an estimator reduced the sampling variance.

No differences were detected in the sampling variances among the four poststratified estimators. Because of the similarity in the sampling variances for these estimators, the sampling error for estimator 4 presented in this chapter may also be used to approximate the sampling error for the same characteristic of interest obtained for estimators 1–3.

4. CONCLUSIONS

This chapter discusses alternative procedures for making post survey weighting adjustments for RDD national telephone surveys in order to reduce the potential biases due to unequal probabilities of selection, noncoverage, and undercoverage of the population, and nonresponse. Although national telephone surveys have become more widely accepted, many telephone surveys still have combined nonresponse and noncoverage rates of more than 30 percent. The basic method used to adjust the telephone survey estimates for undercoverage and noncoverage of the population and nonresponse realigns the telephone sample distribution of persons to either known or estimated population distributions, as reported by the Bureau of the Census. The variables used in this research to define the domains for weighting cell adjustments are region, race, age, sex, and educational attainment. Large differences in telephone coverage and response exist for different levels of each of these variables. The post survey adjustments will reduce the biases due to undercoverage and nonresponse, with the amount of the reduction depending on the extent of similarity between respondents and nonrespondents.

For a national RDD telephone survey on smoking conducted by the NCHS, four different weighting procedures were compared. The four differed in the number and the type of adjustments used prior to a final poststratification. In general, the four weighted estimates all differed from the unweighted estimate, but did not differ much from each other when estimating the smoking characteristics of the population.

For two of the smoking and health characteristics observed, the adjusted estimated proportion for the total population was changed by as much as 2 percent. Since the estimated variances for the four weighting procedures were almost identical, the sampling variance for these estimators could not be used to differentiate between the survey adjustment procedures tested. These results, however, may have been different for estimates of other characteristics of interest. By looking at the estimated distribution among domains of the population it becomes apparent that a poststratification realignment by age, sex, and education will not align the telephone population very well by race and region. It is recommended, therefore, that all of the key domain variables be used in the post survey adjustments, including race. A raking post survey adjustment procedure would seem to be a logical choice to simultaneously control marginal totals. Our investigation showed that the household nonresponse adjustment could be omitted without any significant effect on the survey estimates. The fully adjusted estimator, however, was selected for further comparisons.

When the effect of the fully adjusted estimator was compared to the NHIS face to face survey results, we found that the post survey weighting adjustments had about the same size effect as the difference between the estimates for households with telephones and all households for the NHIS. This would seem to indicate that the undercoverage, noncoverage, and nonresponse biases are being reduced. The fact that there still remained considerable differences between the telephone survey and NHIS results for some of the variables indicate that other factors may be responsible for the differences. Factors such as respondent rules, questionnaire format, and response error may cause more of an impact on the survey results than either the undercoverage or the nonresponse bias.

In order to separate and accurately measure all of the factors that can affect the telephone survey results, more controlled studies are needed. Record check studies would be invaluable in estimating response errors associated with telephone surveys. Comparisons with face to face survey results are almost always confounded with extraneous effects. Because of such advantages as cost, smaller design effects, and smaller interviewer effects, telephone surveys are here to stay. To produce reliable national estimates from telephone surveys it is critical that the nonsampling errors be understood and, when possible, the survey results adjusted to reduce the magnitude of the nonsampling errors.

CHAPTER 10

STRATIFICATION OF PREFIX AREAS FOR SAMPLING RARE POPULATIONS

Leyla Mohadjer[1]
Westat, Inc.

Survey researchers are frequently interested in specific subgroups of the population that comprise rather small percentages of the total population, e.g., of blacks, Hispanics, low income families, etc. With any method of sample selection, surveys of rare populations almost always require a considerable amount of screening. The frame generally used for random digit dialing (RDD) (Bell Communications Research (BCR) tape; see this volume, Chapter 5) comprises all telephone households, and the subsets cannot be determined except as part of a screening procedure. Extensive screening is necessary to locate members of the rare population, and as a result, it is usually very costly to sample rare populations through RDD. For example, calculations based on the 1980 Census data indicate that it takes an average of 35 screened households to locate a black male between 20 and 29 years of age, and an average of 64 screened households to locate a Hispanic male in this age range. A common strategy is to use geographic stratification, classifying areas by the proportion of their population in the specified subgroup, with higher sampling rates in the strata in which the subgroups are concentrated. The information that is available about prefix areas on the BCR tape used as the sampling frame is inadequate for stratification purposes. The only information that could be used is a rather crude geographic indicator.

This chapter discusses the use of a commercially available tape that contains census characteristics for prefix areas, and evaluates the quality of the information for some key items. Possible improvements in

---

[1]The author wishes to thank Joseph Waksberg for his guidance and suggestions, the reviewers David Chapman and Graham Burkheimer, for their helpful comments, and Elizabeth Gaughan for her expert typing support.

efficiency from using the tape to sample rare populations are presented. The tape developed by Donnelley Marketing Information Services contains the 1980 Census and updates of census characteristics for prefix areas, and it can be used for stratification useful in the selection of respondents from rare populations.

Section 1 provides a review of related studies done in the area of sampling rare populations. Section 2 presents a description of the Donnelley tape, how it is compiled, and some tabulated data on rare populations, extracted from the tape. A comparison of the information given by Donnelley with comparable findings from an RDD survey is given in Section 3. Section 4 provides a discussion of stratification by concentration of rare groups when (1) the rare group is sampled as part of the general population, and (2) the rare group is the only population of interest. Finally, a brief summary and conclusion are given in Section 5.

1. HISTORICAL REVIEW OF OTHER WORK

There have been several recent articles on possible methods of increasing the yield of a rare population with RDD. The Mitofsky-Waksberg sampling method (Waksberg, 1978) can be adapted to subsets of the population. Articles by Blair and Czaja (1982) and Waksberg (1983) point out that the same theory applies if the acceptance-rejection scheme is defined in terms of the rare population, rather than for the total households. When the rare population is disproportionately distributed among prefix areas, and if many of these locations have no subgroup members, the efficiency of the screening design can be improved. The reduction in screening workloads are possible in the same way that the procedure reduces the screening of telephone numbers needed for a fixed sample size of households. However, the gain cannot always be achieved. In order for this approach to be practical, most blocks of telephone numbers that are retained need to have at least the number of members of the rare population that one attempts to achieve per block. This will frequently not occur for rare populations, and differential weighting is necessary for unbiased results. The increase in variance due to differential weighting may more than offset the gains from the reduced screening. An article by Inglis et al. (1987) shows that increasing the block size from 100 consecutive numbers to 200 and 400 will increase the efficiency of this approach.

A different approach was taken in a research study carried out at the University of Michigan (Groves, Lepkowski, and Landenberger, 1984). The Michigan study investigated the feasibility of matching census small area data to telephone sampling units for the purposes of stratification.

The prefix area is the smallest geographic area that can be determined by the BCR frame. Since prefix area information is not available on census individual records, the Michigan study compared alternative techniques of associating the telephone prefix with census areas. The authors reported that the matching procedures which seemed to work best were also the most labor intensive and costly. From this study, it is not clear whether conditions for important gains exist. They also report that, there are some commercial firms that have assembled data bases from computer matching of prefix areas and census information, and considerable savings in cost would result from using these tapes when compared to the approach of this study, although no information was available on the accuracy of these data bases.

This chapter reports on an evaluation of one of these commercial data bases, and describes how it can be used in an RDD survey.

2. DESCRIPTION OF THE DONNELLEY TAPE

Donnelley Marketing Information Services (DMIS), located in Stamford, Connecticut, provides services to help companies with their marketing needs. One of the services that can be provided by Donnelley is a computer tape of telephone prefix areas with their associated demographics.

The tape contains 1980 Census and updates of 1980 Census characteristics for prefix areas. The information for this tape is compiled from two sources. The first source is telephone directories. The "telephone list" is compiled from over 4,700 directories. Donnelley has informed us that this list, which yields almost national coverage, is updated daily. The second source of information is official records of state Motor Vehicle Registration departments. This "auto list" is currently available in 36 states plus the District of Columbia. Where available, the auto list is updated annually (twice a year for certain states). The addresses and telephone numbers on these lists have been used to associate prefix areas to ZIP codes. The 1980 Census ZIP code tabulations provided the basis for the estimates of 1980 Census characteristics for prefix areas.

Estimates and projections are produced annually for the following items: total population, total households, population by age and sex, population by race and ethnicity, households by household income, householder by age, and household income by age of householder. The Donnelley estimates and projections are produced once a year, and are dated January 1 of the relevant estimate or projection year.

The tape used for this study contains 1980 Census and 1985 updates of 1980 Census characteristics for prefix areas, and can be used for stratification and for the selection of respondents from rare populations. The tape contains all ZIP codes in the United States with their associated area codes and prefix areas. Appended to each record are ZIP level demographics such as Federal Information Processing Standards (FIPS) state code and metropolitan statistical area (MSA) code. 1980 Census data and 1985 updates of the 1980 Census are provided for population by age, sex, race and ethnicity, income, and household data. There is a record for each ZIP/area code prefix combination with the following census information: ZIP code, area code, and prefix, FIPS state code, county code number, MSA code, post office name, 1980 group quarters population, 1980 population by race (total, white, black, Indian and other Asian, Spanish), 1980 housing units by occupancy status, 1980 housing units by household type, 1980 households with children, 1980 households with persons 65 and over, 1980 householders 65 and over, 1980 housing units by tenure, 1980 housing units by race of householder, and housing units by household income in 1979. The 1985 updates of the associated 1980 Census data are also provided. The following three fields were added to the tape at Westat's request:

1. Number of Donnelley households in the ZIP/area code and prefix combination

2. Number of Donnelley households in the ZIP

3. Percent of households for the ZIP within the prefix area.

Tables 1, 2, and 3 provide the distribution of blacks, Hispanics, and Asians as reported in Donnelley, with prefix areas divided into classes based on the percent of the rare population they contain, using the three new fields added. We used the 1980 Census for the purpose of this chapter to avoid uncertainty about the accuracy of the 1985 updates.

3. EVALUATION OF DONNELLEY TAPE QUALITY

The empirical results are based on comparisons of the Donnelley tape with findings from a survey on smoking carried out in 1986 (September 1986 through January 1987) by Westat for the Department of Health and Human Services (DHHS). The purpose of the smoking survey was to ascertain the present smoking behavior of adults in the United States, and their knowledge and attitudes relating to health risks of smoking.

Table 1. Estimated Percent of the U.S. Black Population, by Percent of Blacks in Prefix Areas as Reported by Donnelley

Percent Black in Exchange	Total Population in Category[a]	Black Population in Category[a]	Number of Exchanges in Category	Percent of Blacks in Category	Percent of Exchanges in Category
<10%	155,288,772	3,784,815	24,678	14.33	76.33
10-19%	27,021,517	3,892,639	2,935	14.74	9.08
20-29%	14,580,247	3,594,217	1,594	13.61	4.93
30-39%	10,165,232	3,560,519	1,055	13.48	3.26
40-49%	5,665,619	2,521,695	701	9.55	2.17
50-59%	4,167,272	2,285,326	484	8.65	1.50
60-69%	2,668,337	1,734,612	305	6.57	.94
70-79%	3,052,879	2,277,627	318	8.62	.98
80-89%	2,415,071	2,051,615	201	7.77	.62
>90%	754,656	711,803	58	2.69	.18
Total	225,779,602	26,414,868	32,329[b]	100.00	100.00

[a]1980 Census population.
[b]This number excludes about 5,000 prefix areas for which Donnelley had no record of having any households.

Table 2. Estimated Percent of the U.S. Hispanic Population, by Percent of Hispanics in Prefix Areas as Reported by Donnelley

Percent Hispanic in Exchange	Total Population in Category[a]	Hispanic Population in Category[a]	Number of Exchanges in Category	Percent of Hispanics in Category	Percent of Exchanges in Category
<10%	187,906,649	3,922,134	28,058	26.93	86.79
10-19%	17,603,022	2,538,945	2,023	17.43	6.26
20-29%	8,372,164	2,085,136	949	14.32	2.94
30-39%	4,208,710	1,468,105	503	10.08	1.56
40-49%	2,897,627	1,288,914	285	8.85	.88
50-59%	1,601,846	884,463	168	6.07	.52
60-69%	1,416,155	923,264	151	6.34	.47
70-79%	655,469	487,896	75	3.35	.23
80-89%	867,131	733,378	89	5.04	.28
>90%	250,830	231,781	28	1.59	.09
Total	225,779,603	14,564,016	32,329[b]	100.00	100.00

[a]1980 Census population.
[b]This number excludes about 5,000 prefix areas for which Donnelley had no record of having any households.

The conceptual universe for this study consisted of the noninstitutional, civilian population of the United States 17 years and over. The actual interviewing was restricted to that part of the population residing in housing units with telephones (about 92 percent). The results were weighted in an attempt to produce statistics that reflect characteristics of the desired population. The RDD selection of

Table 3. Estimated Percent of the U.S. Asian Population, by Percent of Asians in Prefix Areas as Reported by Donnelley

Percent Asian in Exchange	Total Population in Category[a]	Asian Population in Category[a]	Number of Exchanges in Category	Percent of Asians in Category	Percent of Exchanges in Category
<10%	220,956,629	2,298,965	31,738	65.94	98.17
10–19%	3,174,665	415,944	334	11.93	1.03
20–29%	418,898	96,036	66	2.75	.20
30–39%	362,851	118,916	55	3.41	.17
40–49%	111,666	50,902	22	1.46	.07
50–59%	157,197	83,848	35	2.41	.11
60–69%	346,205	229,049	43	6.57	.13
70–79%	174,901	130,687	27	3.75	.08
80–89%	76,590	61,889	9	1.78	.03
>90%	0	0	0	0	0
Total	225,779,602	3,486,236	32,329[b]	100.00	100.00

[a]1980 Census population.
[b]This number excludes about 5,000 prefix areas for which Donnelley had no record of having any households.

households was carried out through a variant of what is generally referred to as the Mitofsky-Waksberg method (Waksberg, 1978). The variant designates an equal number of telephone numbers within each block instead of an equal number of households. Unbiased estimates can be prepared by the application of appropriate weights. The variant permits the sample selection to be completed more quickly.

The RDD sample consisted of 2,297 clusters (blocks), with 15 telephone numbers per block. The average number of screened households per block was about 8. A two stage weighting procedure was used. The first stage used the reciprocals of the probabilities of selection. The second stage was the poststratification weights, adjusted for anomalies in the sample, nonresponse, and for the fact that telephone surveys exclude nontelephone households. The variables used in poststratification were geography, education, race, age, and sex. A total of 13,031 respondents completed the questionnaire with 1,096 blacks, 584 Hispanics, and 185 Asians. The response rate for the screener was about 86 percent and for the questionnaire about 87 percent.

Tables 4, 5, and 6 provide comparisons of percents of black, Hispanic, and Asian populations, respectively, with the information provided on the Donnelley tape for the 2,297 blocks in the smoking survey. It should be noted that the smoking survey included persons 17 years or older, whereas Donnelley reports the total population of the United States. Another difference that should be noted is that Donnelley figures are 1980 census data and the smoking survey was mainly accomplished in 1986. Furthermore, the classification of each prefix area to categories for the

Table 4. Estimated Percent of Blacks in Prefix Areas, Classified by Concentration of Blacks in the Exchanges as Reported by Donnelley and Those Observed in the Smoking Survey for the 2,297 Exchanges Selected for the Smoking Survey

Percent Black in Exchange Reported by Donnelley	Number of Exchanges in Category in Smoking Survey	Percent of Exchanges in Category in Smoking Survey	Percent Black in Category Reported by Donnelley[a]	Percent Black in Category in Smoking Survey
<10%	1,601	69.70	14.91	13.79
10–19%	281	12.23	14.03	12.58
20–29%	140	6.10	13.05	13.62
30–39%	99	4.31	13.47	12.18
40–49%	49	2.13	8.11	8.70
50–59%	32	1.39	7.09	5.67
60–69%	24	1.05	6.45	7.74
70–79%	39	1.70	10.33	13.65
80–89%	25	1.09	9.38	8.91
>90%	7	.30	3.18	3.16
Total	2,297	100.00	100.00	100.00

[a]1980 Census population.

Table 5. Estimated Percent of Hispanics in Prefix Areas, Classified by Concentration of Hispanics in the Exchanges as Reported by Donnelley and Those Observed in the Smoking Survey for the 2,297 Exchanges Selected for the Smoking Survey

Percent Hispanic in Exchange Reported by Donnelley	Number of Exchanges in Category in Smoking Survey	Percent of Exchanges in Category in Smoking Survey	Percent Hispanic in Category Reported by Donnelley[a]	Percent Hispanic in Category in Smoking Survey
<10%	1,892	82.37	24.93	37.19
10–19%	203	8.84	18.91	16.41
20–29%	93	4.05	16.18	14.03
30–39%	46	2.00	10.77	5.19
40–49%	25	1.09	8.33	5.83
50–59%	12	.52	4.56	5.02
60–69%	11	.48	4.91	6.36
70–79%	6	.26	5.17	1.66
80–89%	7	.30	4.44	6.98
>90%	2	.09	1.79	1.33
Total	2,297	100.00	100.00	100.00

[a]1980 Census population.

smoking survey are based on an average of eight households per prefix area.

Table 4 compares the black population in the different categories as estimated in the smoking survey and reported by Donnelley. It can be noted that the percentage of blacks reported by Donnelley is very close to the percentages found in the smoking survey. Although the population

Table 6. Estimated Percent of Asians in Prefix Areas, Classified by Concentration of Asians in the Exchanges as Reported by Donnelley and Those Observed in the Smoking Survey for the 2,297 Exchanges Selected for the Smoking Survey

Percent Asian in Exchange Reported by Donnelley	Number of Exchanges in Category in Smoking Survey	Percent of Exchanges in Category in Smoking Survey	Percent Asian in Category Reported by Donnelley[a]	Percent Asian in Category in Smoking Survey
<10%	2,244	97.69	69.20	84.48
10–19%	38	1.65	14.72	6.27
20–29%	6	.26	4.38	2.03
30–39%	4	.17	2.19	1.82
40–49%	0	00.	—	.00
50–59%	2	.09	.79	.32
60–69%	2	.09	5.35	4.21
70–79%	1	.04	3.37	.87
80–89%	0	.00	—	.00
>90%	0	.00	—	.00
Total	2,297	100.00	100.00	100.00

[a]1980 Census population.

figures given by Donnelley are about seven years old, they are still adequate for stratifying blacks by their concentration.

Table 5 provides the same comparison for the Hispanic population. The percentages of concentration reported by Donnelley are not as close to the findings of the survey as they were for the blacks. One reason may be the difference in the distribution of Hispanics by age. The smoking survey only includes persons who are 17 years or older. Another reason may be the fact that the Donnelley data are seven years old and the distribution of Hispanics may have changed since 1980. A third possibility is that the estimating technique used by Donnelley for associating prefix areas to census areas does not work as well for Hispanics as for blacks. Furthermore, the sample size is smaller for Hispanics than for blacks. Finally, Table 6 shows the comparison between the Donnelley data and the survey for Asians. Considering the small sample size for Asians, the percentages given by Donnelley and the survey are moderately close. Both sources indicate that geographic stratification is not effective for the Asian-American population.

4. APPLICABLE THEORY ON STRATIFICATION BY CONCENTRATION OF RARE GROUPS

The information available on the Donnelley tape can be used for stratification purposes. There are basically two types of surveys that involve sampling rare populations: studies that are restricted to the rare population, and studies of the total population that require some particular subsets (rare populations) to be oversampled. This section discusses the optimum allocation of the sample for the two cases using the Donnelley tape to construct the strata.

4.1 Rare Population Is the Only One Surveyed

When a study involves sampling only the rare population, as shown by Waksberg (1973) and Kalton et al. (1986), stratification will produce a significant reduction in the level of screening when: (a) a high percentage of the rare population is in the strata established for oversampling, and (b) these strata contain a small part of the total population (or contain a substantial proportion of the rare population). The approach to be taken is, thus, to stratify the prefix areas by concentration of the rare population, and oversample the strata with high concentrations.

Starting with the simplest case, suppose that we establish two strata, one with a high concentration of the population of interest (stratum 1), and the other with a lower concentration (stratum 2).

Assume a single cost function in which the total cost of interviewing n_i units within stratum i, $i = 1, 2$ is given by

$$C = (r_1 n_1 + r_2 n_2) c_1 + (n_1 + n_2) c_2$$

where n_i is the sample size in stratum i, r_i is the average amount of screening required to locate one member of the rare group in stratum i, c_1 is the average cost of a screening call, c_2 is the average cost of interviewing one member, and C is the total cost. If we minimize the sampling variance subject to a fixed cost, we obtain the optimum allocation sample sizes. The ratio of sample sizes is given by

$$\frac{n_1}{n_2} = \frac{\sigma_1 N_1}{\sigma_2 N_2} \sqrt{\frac{r_2 + \dfrac{c_2}{c_1}}{r_1 + \dfrac{c_2}{c_1}}} = \frac{\sigma_1 N_1}{\sigma_2 N_2} \sqrt{\frac{r_2 + z}{r_1 + z}} \tag{10.1}$$

where N_i is the population of the rare group in stratum i, and z is the ratio of c_2 to c_1.

Table 7 provides the value of the ratio of sample sizes for different values of N_1/N_2, z, r_1 and r_2, when $\sigma_1 = \sigma_2$. Several conclusions can be derived from Table 7.

- If the amount of screening required in stratum 1 is the same as in stratum 2, $r_1 = r_2$, then the optimum allocation reduces to the proportional allocation, $n_1/n_2 = N_1/N_2$. The terms r_1 and r_2 are equal when the proportion of the rare population compared to the general population is the same in both strata.

- As the cost of interviewing over the cost of screening one member of the rare group (c_2/c_1) increases, the ratio of sample sizes (n_1/n_2) decreases.

- As the ratio of the amount of screening required to locate one member of the rare group in stratum 2 over stratum 1 (r_2/r_1) increases, the ratio of sample sizes (n_1/n_2) increases.

Table 7 confirms the two conditions stated by Waksberg (1973) and Kalton et al. (1986) on the efficiency of stratification. That is, stratification can produce a significant reduction in the level of screening when N_1/N_2, and r_2/r_1 are high. The table shows that the ratio of n_1/n_2 is the highest when the two conditions are satisfied.

Having the population distributions for blacks and Hispanics, as given in Tables 1 and 2, we can use Table 7 to derive the ratio of the optimum samples for various two-stratum designs. For example, let stratum 1 include all prefix areas in Table 1 for which the percent of blacks is more than 30, and stratum 2 include the remainder of the prefix areas. The ratio of the black population sizes in the two strata, N_1/N_2, is about 60/40. The screening ratio in stratum 1, r_1, is about 2, and in stratum 2, r_2, about 18. For the purpose of illustration, assume that $c_2/c_1 = 5$, $N_1/N_2 = 60/40$, $r_2/r_1 = 10$, and $r_1 = 1$. Then the ratio of the optimum sample sizes for the two strata, as given in Table 7, is equal to 2.4. Equation (10.1) can be used for the derivation of the exact values of the ratios.

For a more general case, we divide the population into L strata based on the concentration of the rare population within clusters of area code prefix numbers. Let n_i be the sample size (number of interviews) in stratum i, $i = 1, 2, \ldots, L$.

Also, define r_i, c_1, c_2, C, and N_i the same way as in the two-stratum case.

The total cost of interviewing n_i eligibles in stratum i, $i = 1, 2, \ldots, L$, is given by

Table 7. Variations of the Sample Size Ratio n_1/n_2 for Various Values of Population Ratios N_1/N_2; Ratio of the Cost of Interviewing Over Screening c_2/c_1; Amount of Screening in Stratum 1 (r_1) and Stratum 2 (r_2) for Two-Stratum Design

		C_2/C_1 (N_1/N_2 = 1)				C_2/C_1 (N_1/N_2 = 60/40)				C_2/C_1 (N_1/N_2 = 70/30)				C_2/C_1 (N_1/N_2 = 80/20)				C_2/C_1 (N_1/N_2 = 90/10)			
r_1	r_2/r_1	1	5	10	20	1	5	10	20	1	5	10	20	1	5	10	20	1	5	10	20
1	1	1.0	1.0	1.0	1.0	1.5	1.5	1.5	1.5	2.3	2.3	2.3	2.3	4.0	4.0	4.0	4.0	9.0	9.0	9.0	9.0
1	5	1.7	1.3	1.2	1.1	2.6	1.9	1.8	1.6	4.0	3.0	2.7	2.6	6.9	5.2	4.7	4.4	15.6	11.6	10.5	9.8
1	10	2.4	1.6	1.4	1.2	3.5	2.4	2.0	1.8	5.5	3.7	3.2	2.8	9.4	6.3	5.4	4.8	21.1	14.2	12.1	10.8
1	20	3.2	2.0	1.7	1.4	4.9	3.1	2.5	2.1	7.7	4.8	3.9	3.2	13.0	8.2	6.6	5.5	29.2	18.4	14.9	12.4
5	1	1.0	1.0	1.0	1.0	1.5	1.5	1.5	1.5	2.3	2.3	2.3	2.3	4.0	4.0	4.0	4.0	9.0	9.0	9.0	9.0
5	5	2.1	1.7	1.5	1.3	3.1	2.6	2.3	2.0	4.9	4.0	3.6	3.1	8.3	6.9	6.1	5.4	18.7	15.6	13.8	12.1
5	10	2.9	2.4	2.0	1.7	4.4	3.5	3.0	2.5	6.8	5.5	4.7	3.9	11.7	9.4	8.0	6.7	26.2	21.1	18.0	15.1
5	20	4.1	3.2	2.7	2.2	6.2	4.9	4.1	3.3	9.6	7.6	6.3	5.1	16.4	13.0	10.8	8.8	36.9	29.2	24.4	19.7
10	1	1.0	1.0	1.0	1.0	1.5	1.5	1.5	1.5	2.3	2.3	2.3	2.3	4.0	4.0	4.0	4.0	9.0	9.0	9.0	9.0
10	5	2.2	1.9	1.7	1.5	3.2	2.9	2.6	2.3	5.0	4.5	4.0	3.6	8.6	7.7	6.9	6.1	19.4	17.2	15.6	13.8
10	10	3.0	2.7	2.4	2.0	4.6	4.0	3.5	3.0	7.1	6.2	5.5	4.7	12.1	10.6	9.4	8.0	27.3	23.8	21.1	18.0
10	20	4.3	3.7	3.2	2.7	6.4	5.6	4.9	4.1	10.0	8.6	7.6	6.3	17.1	14.8	13.0	10.8	38.5	33.3	29.2	24.4
20	1	1.0	1.0	1.0	1.0	1.5	1.5	1.5	1.5	2.3	2.3	2.3	2.3	4.0	4.0	4.0	4.0	9.0	9.0	9.0	9.0
20	5	2.2	2.1	1.9	1.7	3.3	3.1	2.9	2.6	5.1	4.8	4.5	4.0	8.8	8.2	7.7	6.9	19.7	18.4	17.2	15.6
20	10	3.1	2.9	2.7	2.4	4.6	4.3	4.0	3.5	7.2	6.7	6.2	5.5	12.4	11.5	10.6	9.4	27.8	25.8	23.8	21.1
20	20	4.4	4.0	3.7	3.2	6.6	6.0	5.6	4.9	10.2	9.4	8.6	7.6	17.5	16.1	14.8	13.0	39.3	36.2	33.3	29.2

$$C = \sum_{i=1}^{L} r_i n_i c_1 + n_i c_2 .$$

Minimizing the variance subject to a fixed cost C, we obtain

$$\frac{n_1}{n_i} = \frac{\sigma_1 N_1}{\sigma_i N_i} \sqrt{\frac{r_i + z}{r_1 + z}} . \tag{10.2}$$

Equation (10.2) can be used to derive ratios of optimum sample sizes for multiple-stratum designs for the black and Hispanic population distributions, as given in Tables 1 and 2. The application is similar to the one described earlier for the two-stratum design.

4.2 Rare Population is Oversampled as Part of the General Population

Most of the discussion given in the previous section also applies to studies of the total population which require some particular subsets to be oversampled. Overall, the same conclusions hold when oversampling the subgroup populations as when the study is restricted to the subgroup population. However, the effect of the variable sampling rates resulting from optimum allocation on other domains to be analyzed needs to be taken into consideration.

A simple, but direct, approach is to divide the prefix areas in the United States into two geographic strata — areas with high concentrations of the rare population and the remaining prefix areas. The Donnelley tape is used to establish the "rare" stratum. The sampling rate needed for the nonrare population is used throughout the entire United States, using the prefix areas in the BCR tape, and a sample of both the rare and nonrare populations selected at this rate. A supplementary sample is selected from the rare stratum, which is screened and only members of the rare population are retained. Note that under this design both parts of the sample will contain members of the rare population. Table 7 can be used to estimate sample sizes for the two strata for various screening costs and population proportions, in order to determine whether the sample size in the nonrare stratum should be increased above the level required for the general population.

5. SUMMARY

Surveys of rare populations almost always require a considerable amount of screening. A common strategy is to stratify areas by the proportion of their population in the specified subgroup, with higher sampling rates in the strata in which the subgroups are concentrated. The frame generally used for RDD (BCR tape) is inadequate for stratification purposes. This chapter discussed the use of the Donnelley data base, a commercially available tape that contains census characteristics for prefix areas for stratification purposes. The chapter evaluated the quality of the information for sampling black, Hispanic, and Asian-American populations. The comparisons between the Donnelley information and findings from a nationwide survey showed that the Donnelley data base is adequate in quality and considerable cost savings can result from using it for stratification purposes for blacks and Hispanics. The Asian-Americans are not heavily concentrated in any area and, thus, geographic stratification is not effective for this subgroup. We have not examined potential gains for oversampling other social or economic domains.

CHAPTER 11

SAMPLING VARIANCE AND NONRESPONSE RATES IN DUAL FRAME, MIXED MODE SURVEYS

Monroe G. Sirken
National Center for Health Statistics

Robert J. Casady
U. S. Bureau of Labor Statistics

When the available sampling frame is complete but requires a costly data collection mode, efficiency considerations suggest the joint use of the complete frame and an incomplete frame, provided that data collection in the latter is less costly than in the former. The National Health Interview Survey (NHIS), for example, is currently noninstitutionalized population of the United States, but data collection is expensive because it requires a face to face interview mode. On the other hand, the telephone interview mode of data collection is less expensive but the exclusive use of a telephone frame would limit coverage to the approximately 93 percent of U.S. households with telephones. Current available evidence indicates that for most subjects the quality of reporting is not importantly different in telephone and face to face interviewing, therefore, efficiency considerations suggest augmenting the area/list household frame with a frame of telephone numbers.

This strategy for improving design efficiency is not new. Hansen, Hurwitz, and Madow (1953) suggest the use of a partial list together with an area frame for a proposed study of a Jewish population. Building on the work of Hartley (1962) more than two decades ago, substantial progress has been made in designing sampling efficient dual frame surveys. Closed formulas, reflecting frame differentials in data collection costs, sampling variability, and coverage completeness, exist for optimizing the sample allocation to the two frames. Using a dual frame estimator and optimum allocation procedure suggested by Lund (1968),

Casady, Snowden, and Sirken (1981) investigated the possible savings of redesigning the NHIS as a dual frame, mixed mode survey that would be based on an area/list frame and a frame of randomly dialed telephone numbers. They demonstrated that for fixed reliability the costs would be consistently smaller for the dual frame design than for the existing single frame NHIS design if approximately 60 to 80 percent of the sample was allocated to the telephone frame. More recently dual frame, mixed mode surveys have received considerable attention in the literature. Groves and Lepkowski (1985) developed a very detailed cost model, Biemer (1983) extended the error model to include both response errors and bias in the telephone frame, and Lepkowski and Groves (1986) further extended the error model to include bias in the area frame. Lepkowski (this volume, Chapter 5) explores applications of dual frame methodology in the design of telephone surveys under the assumption that the telephone population is the target population. The dual frame, single mode survey designs discussed by Lepkowski may prove to be very cost efficient relative to the single frame telephone survey designs commonly in use today.

The sampling variability and nonresponse effects of changing the NHIS design from a single frame of household addresses to a dual frame of household addresses and randomly dialed telephone numbers are investigated in this chapter. For our purposes, the response rate will be used to measure the nonresponse effect of survey design. Obviously a better measure of nonresponse effect would be the nonresponse bias of key variates. However, as these biases are rarely known, design specifications are usually expressed in terms of minimum response rate rather than maximum nonresponse bias. The decision to express nonresponse effects in terms of response rate was motivated by this practical convention.

Both the sampling variance and the nonresponse rate in a dual frame survey depend on the fraction of the sample allocated to each frame. The dual frame sample allocation that would be optimal for minimizing sampling variance would not necessarily be optimal for assuring acceptable response rates unless the response rates were about the same for both frames. Clearly, if the response rates were unequal, allocating 100 percent of the sample to the frame with the higher response rate would be the optimal strategy for minimizing the dual frame nonresponse rate, though a smaller percentage might be optimal for minimizing sampling variance. In fact current experience indicates that response rates in telephone interviews are likely to be substantially smaller than the face to face interview nonresponse rates which are currently about 95 percent in the NHIS. For this reason, Sirken and Casady (1982) concluded that it would be premature to plan to redesign the NHIS as a dual frame survey until the nonresponse effects were more fully understood.

In the next section a dual frame response rate is defined and a simple dual frame cost model is specified. Also, the sampling variance of a proposed dual frame estimator is derived under a simple error model. The more detailed cost and error models mentioned previously were not utilized as the requisite data for estimating the model parameters were not routinely available. It is encouraging to note that there is a steadily increasing body of information regarding cost and error factors in complex sample surveys but our knowledge is still fragmentary and in many cases conjectural. The need for such information is especially critical at the planning and design stage, so it is essential that survey organizations make every effort to routinely document and disseminate information on the cost and design characteristics of their current surveys.

In Section 2, results derived under the proposed cost/error model are used to investigate the sensitivity of the sampling variance and response rate in a dual frame NHIS to changes in the telephone response rate and the fraction of the dual frame sample allocated to the telephone frame. These findings are used in Section 3 to analyze the sampling variance effects of redesigning NHIS as a dual frame survey subject to the condition that the dual frame nonresponse rate would not exceed a specified minimum level. The sensitivity of dual frame sampling variance and response rate to changes in the telephone and face to face response rates and changes in the allocation of the sample to the two frames are briefly summarized in Section 4.

1. THE STATISTICAL MODEL

The problem that we wish to address can be simply stated. Subject to a fixed expected survey cost, say C^*, we want to determine the proportions of the dual frame sample that should be allocated to the frames so that the sampling variance is minimized subject to the condition that the dual frame response rate meets or exceeds a specified minimal level.

To solve this allocation problem we must:

a. Develop an expression for the dual frame response rate

b. Develop a dual frame cost model which incorporate the major sources of survey cost

c. Propose a dual frame estimator and derive its variance.

The remainder of this section will be devoted to these three tasks.

1.1 Response Rate

Suppose that the total expected sample size for the dual frame survey is n persons. (Since the sampling unit in both frames is a household or group of households, expected values apply when discussing sample size in terms of persons.) If we let θ represent the proportion of the dual frame sample from the telephone frame F_2 then the expected number of sample persons from F_2 is given by $n_2 = \theta n$ and the expected number of sample persons from the area frame F_1 is given by $n_1 = (1-\theta)\,n$. Assuming the response rates for the two frames are π_1 and π_2 respectively, then the expected number of respondents from F_1 is $\bar{n}_1 = \pi_1 n_1$, the expected number of respondents from F_2 is $\bar{n}_2 = \pi_2 n_2$, and the expected dual frame response rate is given by

$$R(\pi_1, \pi_2, \theta) = (1-\theta)\,\pi_1 + \theta\pi_2\,. \tag{11.1}$$

Of course the response rate could have been defined in many other ways. The authors feel that the "unweighted" response rate given in Equation (11.1) is the most appropriate from the point of view of the National Center for Health Statistics, which sponsors the NHIS. Other definitions may be more appropriate when viewed from other perspectives.

1.2 Cost Model

The cost model adopted for this chapter is composed of two components:

a. Overhead costs which primarily include the central office administration and survey maintenance costs

b. Variable interviewing costs which primarily include the direct and indirect interviewing costs.

The overhead costs can be represented as:

$$C\,(\theta) = \begin{cases} K_1 \text{ if } \theta = 0 \\ K_2 \text{ if } 0 < \theta < 1 \end{cases} \tag{11.2}$$

where K_1 represents the overhead costs for an area frame survey while K_2 represents the overhead costs for a dual frame survey. As K_2 represents overhead costs for maintaining both a face to face and telephone interviewing capability it should be assumed that K_2 will always be considerably larger than K_1. Interviewing costs are given by:

C_{11} = cost of a completed face to face interview (i.e., a frame F_1 interview)
C_{12} = cost of a nonresponse face to face interview
C_{21} = cost of a completed telephone interview (i.e., a frame F_2 interview)
C_{22} = cost of a nonresponse telephone interview.

These costs are assumed to include general survey related items such as direct costs, indirect costs, training, and supervision as well as interview mode specific items such as travel and revisit for the face to face mode and nonhousehold contact and call backs for the telephone mode. We can express the total expected survey cost as:

$$C^* = C(\theta) + \overline{C}_1 n_1 + \overline{C}_2 n_2 \tag{11.3}$$

where

$$\overline{C}_1 = C_{11}\pi_2 + C_{12}(1-\pi_2)$$
= average variable cost for sample persons from F_1

$$\overline{C}_2 = C_{21}\pi_2 + C_{22}(1-\pi_2)$$
= average variable cost for sample persons from F_2.

It should be noted that $\overline{C}_1$ and $\overline{C}_2$ reflect variable costs for both respondents and nonrespondents.

1.3 A Dual Frame Estimator and Its Variance

All persons in the universe are included in one of two mutually exclusive domains. The first, denoted D_1, consists of all persons in households *without* telephones and the second, denoted by D_2, consists of persons who reside in households *with* telephones. The parameter α will be used to denote the proportion of the population in D_1 and we let

μ_1 = mean level of characteristic Y in D_1

μ_2 = mean level of Y in D_2

σ_1^2 = variance of Y in D_1

σ_2^2 = variance of Y in D_2.

It should be noted that the population mean μ can be expressed as

$$\mu = \alpha\,\mu_1 + (1-\alpha)\,\mu_2\,.$$

Based on the work of Lund (1968), Casady, Snowden, and Sirken (1981) suggest the following dual frame estimator for μ:

$$\overline{Y} = \hat{\alpha}\,\overline{Y}_{11} + (1-\hat{\alpha})\,[\hat{\lambda}\,\overline{Y}_{12} + (1-\hat{\lambda})\overline{Y}_{22}]$$

where $\hat{\alpha}$ = estimator of α based on respondents from F_1, or known from independent sources

$\overline{Y}_{11}$ = estimator of μ_1 based on respondents from $F_1 \cap D_1$,

$\overline{Y}_{12}$ = estimator of μ_2 based on respondents from $F_1 \cap D_2$

$\overline{Y}_{22}$ = estimator of μ_2 based on respondents from F_2.

The weighting factor $\hat{\lambda}$ is given by

$$\hat{\lambda} = (\delta_{22}/n_2') \;/\; (\delta_{12}/n_{12}' + \delta_{22}/n_2')$$

where n_{12}' = observed number of respondents from $F_1 \cap D_2$
n_2' = observed number of respondents from F_2

and δ_{12}, δ_{22} are the design effects (deffs) of $\overline{Y}_{12}$, $\overline{Y}_{22}$, respectively.

If we assume that:

a. The samples from the two frames are selected independently

b. $\hat{\alpha}$, $\overline{Y}_{11}$, $\overline{Y}_{12}$, and $\overline{Y}_{22}$ are asymptotically unbiased for their respective parameters

c. Cov $(\hat{\alpha}, \overline{Y}_{11})$ and Cov $(\hat{\alpha}, \overline{Y}_{22})$ vanish asymptotically

then it can be verified that the asymptotic expectation and variance of $\overline{Y}$ are

$$E(\overline{Y}) = \mu \tag{11.4}$$

$$\mathrm{Var}(\overline{Y}) = \left(\frac{1}{n}\right) \left[\begin{array}{c} \dfrac{\alpha\delta_{11}\sigma_1^2}{\pi_1\,(1-\theta)} \\ + \dfrac{(1-\alpha)^2\,\sigma_2^2\,\delta_{12}\,\delta_{22}}{(1-\alpha)\pi_1\,(1-\theta)\delta_{22} + \pi_2\,\theta\,\delta_{12}} \\ + \dfrac{2\rho^*(1-\alpha)^2\,\sigma_1\sigma_2\delta_{22}\,\left((\alpha/(1-\alpha))\delta_{11}\delta_{12}\right)^{1/2}}{(1-\alpha)\pi_1\,(1-\theta)\delta_{22} + \pi_2\,\theta\delta_{12}} \\ + \dfrac{\alpha\,(1-\alpha)\,\delta_\alpha\,(\mu_1-\mu_2)^2}{(1-\theta)\,\pi_1} \end{array} \right] \tag{11.5}$$

where δ_α, δ_{11} are the deffs of $\hat{\alpha}$, $\overline{Y}_{11}$, respectively, and ρ^* is the correlation between $\overline{Y}_{11}$ and $\overline{Y}_{12}$. Note that assumption (b) asserts the absence of nonresponse bias for both interview modes so that the impact of nonresponse is limited to increasing the sampling variance. The variance of $\overline{Y}$ can also be expressed in terms of the cost parameters instead of the total sample size n, by solving Equation (11.3) for n and substituting the result into Equation (11.5).

Assuming that reasonable estimates of the population and design parameters are available and the cost parameters are known then Equation (11.5) can be used to obtain a value of the allocation parameter θ, say $\tilde{\theta}$, that is optimal in the sense that Var($\overline{Y}$) is minimized. In the case of NHIS, the population and design parameters have been estimated for a wide range of health variables and these estimates are presented in Table 1. Unfortunately even this amount of information regarding population and design parameters is rarely available at the survey planning and design stage. Futhermore the use of Equation (11.5) leads to a variable specific optimal allocation whereas for a multipurpose survey, such as NHIS, a single allocation that is near optimal for nearly all important variables is required.

Thus for the purpose of survey design the dependence of Equation (11.5) on variable specific population parameters will be eliminated through the following assumptions.

Assumption 1: $\sigma_1^2 \cong \sigma_2^2 \cong \sigma^2$ *and* $\delta_{11} \cong \delta_{12} \cong \delta_1$. The first part of this assumption asserts that the within − domain population variances are approximately equal and the second part of the assumption asserts that

Table 1. Estimated Design and Population Parameters for Health Variables in the National Health Interview Survey (NHIS) (All Estimates Based on 1976 Data)

	Population Parameters				Design Parameters for 1970 NHIS Design					Design Parameters for RDD Telephone Surveys
Health Variables	μ_1	μ_2	σ_1^2	σ_2^2	α	δ_{11}	δ_{12}	ρ^*	δ_α	δ_{22}
Health status	1.9	1.6	.65	.57	09	2.11	2.69	.04	5.9	2.10
Number of dental visits	.7	1.7	25.36	60.44	.09	1.25	1.45	.01	5.9	1.30
Number of doctor visits	3.5	3.9	83.45	90.88	.09	1.20	1.23	.03	5.9	1.26
Total number of chronic conditions	.6	.7	1.39	1.35	.09	1.52	1.88	.05	5.9	1.65
Proportion with one or more chronic conditions	.3	.4	.23	.24	.09	1.89	1.53	.04	5.9	1.54
Limited activity days for persons with chronic condition	3.0	3.3	1.36	1.08	.08	1.21	1.41	.02	5.9	1.09
Workdays lost (adults)	4.7	4.7	852.07	758.61	.08	.98	1.02	.00	5.9	1.01
School days lost (children)	5.6	5.1	693.82	509.45	.09	1.67	1.38	.02	5.9	1.29
Bed disability days	9.1	6.9	1828.74	1366.10	.09	1.18	1.19	.04	5.9	1.14
Health status for school children	1.7	1.4	.38	.34	.09	2.26	2.16	.01	5.9	1.88
Health status for workers (adults)	2.0	1.7	.76	.64	.08	1.55	2.00	.05	5.9	1.26
Hospital days	1.0	1.0	40.6	32.02	.09	1.06	1.05	.00	5.9	1.05

the F_1 sample elements of the two domains are subject to the same design effect. Although such assumptions are commonly utilized in survey design work, they should not be used without some degree of verification. The data in Table 1 indicate that Assumption 1 is reasonable for many NHIS variables.

Assumption 2: $\rho^* \cong 0$. This assumption asserts that the correlation between $\overline{Y}_{11}$ and $\overline{Y}_{12}$ is approximately zero. If the sample design for F_1 was simple random then this assumption would be true. Even for more complex designs ρ^* is usually very small. As can be seen in Table 1 the estimated value of ρ^* never exceeds .05. It is easily verified that under Assumptions 1 and 2 the third term of Equation (11.5) is approximately zero.

Assumption 3: Either α *is known* or $(\mu_1 - \mu_2)^2$ *is small relative to* σ^2. In many cases α is reasonably well known for the population of interest (see this volume, Chapter 3). If this is the case, it is not necessary to estimate α from the F_1 sample and it can be verified that the last term in Equation (11.5) vanishes. Even if α must be estimated, it may be reasonable to assume that the within-domain population variance is large relative to the squared difference between the mean levels for the telephone and nontelephone domains in which case the fourth term in Equation (11.5) is small relative to the first two terms. The data in Table 1 indicate that with the exception of the "health status" variables, the ratio $(\mu_1 - \mu_2)^2/\sigma^2$ is indeed small, and even for these variables the ratio (which ranges from .14 to .26) is not unreasonably large.

Thus under Assumptions 1, 2, and 3 we can write

$$\operatorname{Var}(\overline{Y}) = \frac{\delta_{1\cdot}\,\sigma^2}{n}\left[\frac{\alpha}{\pi_1(1-\theta)} + \frac{(1-\alpha)^2\,\delta_{22}}{(1-\alpha)\,\pi_1(1-\theta)\,\delta_{22} + \pi_2\theta\,\delta_{1\cdot}}\right] \tag{11.6}$$

or

$$\operatorname{Var}(\overline{Y}) = \frac{\delta_{1\cdot}\,\sigma^2}{(C^* - C(\theta))}\left[\frac{\alpha}{\pi_1(1-\theta)} + \frac{(1-\alpha)^2\,\delta_{22}}{(1-\alpha)\pi_1(1-\theta)\delta_{22} + \pi_2\theta\,\delta_{1\cdot}}\right]$$

$$\times\left[\overline{C}_1(1-\theta) + \overline{C}_2\theta\right]. \tag{11.7}$$

In what follows all parameters except π_2 and θ will be considered fixed, thus we can express the sampling variance effect of the dual frame design, relative to the area/list frame design, as the ratio of variances

$$\psi(\pi_2, \theta) = \frac{\operatorname{Var}(\overline{Y}; \pi_2, \theta)}{\operatorname{Var}(\overline{Y}; \pi_2, 0)} \tag{11.8}$$

Using Equation (11.7) we have

$$\operatorname{Var}(\overline{Y}; \pi_2, 0) = \frac{\delta_{1.}\, \sigma^2\, \overline{C}_1}{\pi_1(C^* - K_1)} \tag{11.9}$$

so that

$$\psi(\pi_2, \theta) = \begin{cases} 1 & \text{if } \theta = 0 \\ \dfrac{(C^* - K_1)}{(C^* - K_2)}\left(1 + \dfrac{\overline{C}_2}{\overline{C}_1}\left(\dfrac{\theta}{1-\theta}\right)\right) & \\ \times\left(\alpha + \dfrac{(1-\alpha)}{1 + \pi_2\theta\delta_{1.}/\pi_1(1-\theta)(1-\alpha)\delta_{22}}\right) & \text{if } 0 < \theta < 1 \end{cases} \tag{11.10}$$

It is interesting to note that for $0 < \theta < 1$ the algebraic expression for $\psi(\pi_2, \theta)$ factors into three terms which isolate the impact of overhead costs, variable cost, and design characteristics on sampling variance. This formulation should be very useful in evaluating alternative design strategies. Also, for π_2 fixed but arbitrary $\psi(\pi_2, \theta)$ is monotonically decreasing on the interval $(0, \tilde{\theta})$, and monotonically increasing on the interval $(\tilde{\theta}, 1)$ where $\tilde{\theta}$, the optimal allocation, is given by

$$\tilde{\theta} = \left(1 + \frac{D_2}{(1-\alpha)\,(D_3-1)}\right)^{-1} \tag{11.11}$$

and $D_1 = \overline{C}_2 / \overline{C}_1$
$D_2 = \pi_2\delta_{1.} / \pi_1\delta_{22}$
$D_3 = ((D_2 - (1-\alpha)D_1)/\alpha D_1)^{1/2}$

2. FINDINGS

In this section, we investigate the sampling variance and response rate effects of converting NHIS from a single frame to a dual frame survey. This involves studying the sensitivity of the dual frame sampling variance and response rate to changes in θ, the proportion of the dual frame sample allocated to the telephone frame, and π_2, the response rate in the telephone survey. We assume that the other parameters required by our model are fixed, including the cost components and the population variances and their design effects. Also, we assume that $\pi_1 = .95$, the current NHIS face to face response rate.

The sensitivity of $\psi(\pi_2, \theta)$, the dual frame sampling error effect, and $R(\pi_2, \theta)$, the dual frame response rate, to changes in π_2 and θ are examined in Table 2. We permit π_2 to vary between .70 and .85 as we expect the telephone response rate to fall in that range. We permit θ to vary between zero and .90. As $\theta = 0$ implies continuation of a single frame NHIS, it follows that $\psi(\pi_2, 0) = 1.00$ and $R(\pi_2, 0) = \pi_1 = .95$.

The overall effect on $\psi(\pi_2, \theta)$ of varying θ is essentially the same for $\pi_2 = .70, .75, .80,$ and .85. Table 2 indicates that an allocation of 70 percent to the telephone frame is nearly optimal for all π_2 [by Equation (11.11) the true optimal is approximately 75 percent for all values of π_2 considered in this chapter] and, as expected, $\psi(\pi_2, \theta)$ increases as θ deviates in either direction from the optimal. For $\theta < .20$, $\psi(\pi_2, \theta) > 1.00$ indicating an adverse sampling variance effect for a dual frame design. However, the sampling variance effect is uniformly favorable in the range $.30 < \theta < .90$.

On the other hand, $R(\pi_2, \theta)$ declines monotonically from its highest level at $R(\pi_2, 0) = .95$ as θ increases. For a fixed θ, increasing π_2 only modestly enhances the sampling variance effect but more substantially reduces the dual frame response rate. For example at near optimal allocation (i.e., $\theta = .70$) decreasing π_2 from .70 to .85 produces only modest declines in the sampling variance effect $\pi(.70, .70) = .841$ and $\psi(.85, .70) = .815$ but the dual frame response rate increases substantially from $R(.70, .70) = .775$ to $R(.85, .70) = .880$.

3. DISCUSSION

In establishing precision requirements for surveys, nonresponse rates are readily available as survey byproducts and often serve as surrogates for nonresponse biases which are rarely known. Although the relationship between the two measures is tenuous at best, the use of one for the other may be partially justified on the grounds that the larger the nonresponse rate the larger the possible nonresponse biases. In any event, survey

Table 2. Sampling Variance Effects and Response Rates in Dual Frame Surveys for Selected Telephone Response Rates (π_2) and Sample Allocations (π_1 Assumed Fixed at .95)

Percent of Sample Allocated to the Telephone Frame	Telephone Response Rate							
	$\pi_2 = .70$		$\pi_2 = .75$		$\pi_2 = .80$		$\pi_2 = .85$	
	Sampling Variance Effect	Dual Frame Response Rate	Sampling Variance Effect	Dual Frame Response Rate	Sampling Variance Effect	Dual Frame Response Rate	Sampling Variance Effect	Dual Frame Response Rate
θ	$\psi(\pi_2,\theta)$ in %	$R(\pi_2,\theta)$ in %	$\psi(\pi_2,\theta)$ in %	$R(\pi_2,\theta)$ in %	$\psi(\pi_2,\theta)$ in %	$R(\pi_2,\theta)$ in %	$\psi(\pi_2,\theta)$ in %	$R(\pi_2,\theta)$ in %
0	100.0	95.0	100.0	95.0	100.0	95.0	100.0	95.0
10	109.2	92.5	108.8	93.0	108.4	93.5	108.0	94.0
20	104.3	90.0	103.5	91.0	102.8	92.0	102.1	93.0
30	99.4	87.5	98.4	89.0	97.4	90.5	96.5	92.0
40	94.7	85.0	93.6	87.0	92.5	89.0	91.5	91.0
50	90.4	82.5	89.2	85.0	88.1	87.5	87.0	90.0
60	86.7	80.0	85.5	83.0	84.4	86.0	83.5	89.0
70	84.1	77.5	83.1	81.0	82.3	84.5	81.5	88.0
80	84.6	75.0	84.1	79.0	83.7	83.0	83.4	87.0
90	97.4	72.5	98.1	77.0	98.8	81.5	99.6	86.0

*R and ψ are calculated by Equations (11.1) and (11.10), respectively. The parametric values used to calculate ψ are $\alpha = .07$, $C^* = \$4{,}000{,}000$, $K_1 = \$800{,}000$, $K_2 = \$1{,}200{,}000$, $C_{11} = \$29.50$, $C_{12} = \$14.50$, $C_{21} = \$17.00$, $C_{22} = \$3.50$, $\delta_1 = 1.6$ and $\delta_{22} = 1.3$.

tolerance levels are usually expressed in terms of minimum response rates rather than maximum nonresponse biases. As this is the prevailing practice in designing single frame surveys, we suggest that it is an appropriate practice to opt for designing multiple frame surveys. In this section, we analyze the sampling variance effects of redesigning the NHIS as a dual frame survey subject to a minimal acceptable dual frame response rate, say $R(\pi_2, \theta) = R^*$, where we assume $\pi_2 \leq R^* \leq \pi_1$.

Substituting R^* for $R(\pi_1, \pi_2, \theta)$ in Equation (11.1) and solving for θ we obtain

$$\theta^* = (R^* - \pi_1) \,/\, (\pi_2 - \pi_1) \tag{11.12}$$

where θ^* is the maximally permissible allocation of the dual frame sample to the telephone frame compatible with R^*. Since the single frame NHIS response rate is $\pi_1 = .95$, it seems unlikely that a value of $R^* < .90$ would be acceptable. Hence, in the following discussion we fix $R^* = .90$ and $\pi_1 = .95$ and observe the direct effect of varying π_2 between .70 and .85 on θ^* and $\psi(\theta^*, \pi_2)$.

According to Equation (11.12), the constraints $R^* = .90$ and $\pi_1 = .95$ imply that $\theta^* = .20, .25, .33$, and $.50$ respectively when $\pi_2 = .70, .75, .80$, and $.85$. Referring to Table 2, we note that for the above values of θ^* the associated values of $\psi(\theta^*, \pi_2)$ are 1.04, 1.01, .97, and .87 respectively. Clearly the sampling variance effects associated with θ^* are substantially less favorable than they would have been at the near optimal allocation of $\theta = .70$. More specifically, $\psi(.70, \pi_2) = .841, .831, .823$, and $.815$ respectively for $\pi_2 = .70, .75, .80$, and $.85$.

When $R^* = .90$, the NHIS dual frame design is compatible with $\pi_2 = .85$, it would be minimally effective with $\pi_2 = .80$, and the single frame design would be preferred if $\pi_2 \leq .75$. Referring to Table 2, we note that when $R^* = .85$ and $.80$, the dual frame design would be effective for $\pi_2 \geq .75$, and $.70$ respectively.

For the full potential sampling variance gain of a dual frame design to be realized it is necessary that $\theta^* \geq \tilde{\theta}$ where $\tilde{\theta}$ is the optimal allocation. The response rate that satisfies the constraints that $R = R^*$ and $\theta^* = \tilde{\theta}$ is

$$\tilde{\pi}_2 = \frac{R^* - \pi_1(1 - \tilde{\theta})}{\tilde{\theta}} \tag{11.13}$$

In other words, $\tilde{\pi}_2$ is the minimum telephone response rate that allows the full potential sampling variance gain of dual frame sampling. For example, the minimum telephone response rate that would be required to assure the most favorable NHIS dual frame sampling variance effect when

$R^* = 90$ is $\tilde{\pi}_2 = .88$. Where $R^* = .85$ and .80, the corresponding values are $\tilde{\pi}_2 = .82$ and .75 respectively.

4. SUMMARY AND CONCLUSIONS

Substantial gains in survey design efficiency are sometimes realized when a complete but costly to use sampling frame is complemented by an incomplete and less expensive sampling frame. The cost/error models developed by Hartley and others for designing dual frame surveys are expanded in this chapter by taking into consideration the dual frame design effects on nonresponse rates as well as sampling variance. The dual frame estimator originally proposed by Lund is generalized to reflect nonresponse in either or both sampling frames and expressions are derived for the dual frame sampling variance and response rate which incorporate parameters for controlling nonresponse and sampling variance.

The cost/error model proposed in this chapter is used to investigate the sampling variance and response rate effects of changing NHIS from a single to a dual frame survey in which the existing area/list frame of households would be complemented by an RDD frame of telephone numbers. Data are presented to show how sensitive these effects would be to the telephone response rate and the proportion of the dual frame sample allocated to the telephone frame. Assuming that no more than a five percentage point reduction in the 95 percent single frame NHIS response rate would be an acceptable dual frame NHIS response rate implies a minimum acceptable telephone response rate of about 85 percent and approximately a 20 percent reduction in the current NHIS sampling variance.

It is concluded that the full potential sampling variance gain of an NHIS dual frame design would not be realized unless the telephone response rate can be maintained within about five percentage points of the minimally acceptable dual frame response rate. However, dual frame survey designs should not be dismissed solely on this criterion as significant reductions in sampling error are sometimes realized even when the differential exceeds five percentage points.

SECTION C

NONRESPONSE IN TELEPHONE SURVEYS

CHAPTER 12

AN OVERVIEW OF NONRESPONSE ISSUES IN TELEPHONE SURVEYS

Robert M. Groves
The University of Michigan

Lars E. Lyberg
Statistics Sweden

Survey nonresponse is the failure to obtain measurements on sampled units.[1] In telephone surveys nonresponse arises because (a) persons associated with sampled numbers are never contacted by an interviewer, (b) once contacted the sample person (or another household member) refuses to participate in the survey, or (c) the sample person has a physical, mental, or language disability and is unable to provide the survey information. All these types of nonresponse result in no survey information being obtained for some sampled units, and they are collectively labeled "unit" nonresponse. In addition to unit nonresponse, some sampled persons choose to provide only a subset of the data requested of them; hence, "item" nonresponse results.

Nonresponse is perhaps the most inscrutable of survey errors. Many other survey errors can, at least theoretically, be dealt with if the researcher has adequate financial and administrative resources. Nonresponse in voluntary surveys, however, is the result of behavior of persons who are outside the researcher's control. In the western world telephone survey nonresponse has become increasingly troublesome and threatens to eliminate the unique property of sample surveys among other

[1] Nonresponse to telephone surveys can be distinguished from another error of nonobservation, coverage error (reviewed in Chapter 1 of this book), by the fact that nonrespondent units are selected into the sample, but not measured, whereas noncovered units have no chance of being selected in any sample (i.e., they have no telephone number) and hence are never measured.

methods of social measurement — statistical inference to a known population.

The literature on nonresponse is extensive, and we do not intend to review it here. Excellent discussions of nonresponse are found in Madow et al. (1983), Bailar (1984), and Little and Rubin (1987). This chapter concentrates on nonresponse issues that are specific to telephone surveys, rather than general issues that are relatively independent of mode.

Nonresponse is a problem in telephone surveys for three reasons. First, to the extent that the nonrespondents are different from the respondents on the survey measures, statistics based on respondent data alone will be biased estimates of the full telephone population parameters of interest. This nonresponse error, potentially affecting both descriptive statistics (e.g., means and proportions) and analytic statistics (e.g., regression coefficients), is described below. Second, nonresponse reduces the size of the sample which forms the basis of the estimates. Hence, to the extent that sampling error is a function of the number of cases, the sampling variance of survey estimates is increased. Third, survey costs are increased by efforts to reduce nonresponse. These efforts include multiple dialings on sample telephone numbers which are not answered, attempts to persuade reluctant sample persons to cooperate, and extension of the survey period to permit repeated dialings on nonrespondent cases. Hence, a discussion of nonresponse problems in telephone surveys must also consider the cost implications of nonresponse.

This chapter focuses on nonresponse within cold household telephone surveys. It will largely omit discussion of telephone surveys of nonhousehold populations (e.g., businesses) and nonresponse in telephone reinterview surveys.

1. NONRESPONSE ERROR

For any linear survey statistic[2] nonresponse error produces its effects through two components, the nonresponse rate and the difference between nonrespondents and respondents to the survey, and the statistic can be expressed in the following way:

$$y_r = y_n + (nr)/n)(y_r - y_{nr})$$

---

[2] A linear statistic in this context refers to any computation based on the survey data which is a linear combination of attributes of the sampled units. In a clustered sample the sampled units are clusters of telephone numbers. In an element sample, the sampled units are individual telephone numbers themselves.

where y_r = the statistic estimated from the r respondents
y_n = the statistic estimated for all n sample cases
y_{nr} = the statistic estimated from the nr nonrespondents.

For an analytic statistic (e.g., a regression coefficient) a more complex expression is needed, a function of nonresponse rates, variances and covariances of variables involved, and the nature of the model estimated (see Heckman, 1979). Thus, the error introduced to the survey estimate is a function of the percentage of the sample not responding to the survey and the relative characteristics of respondents and nonrespondents. The nonresponse rate is one component of the error, but does not by itself indicate the magnitude of nonresponse error.

When response rate is used as a proxy indicator of nonresponse error, the researcher is essentially assuming that the relationship between the two components of nonresponse error is that of either line *A* or line *B* in Figure 12.1. Line *A* implies that the difference between respondents and nonrespondents is invariant for surveys with different nonresponse rates. Thus, nonresponse error declines linearly with nonresponse rate. Line *B* implies that if a low nonresponse rate is obtained, nonrespondents are more like respondents than when a high nonresponse rate is obtained. This means that nonresponse error increases more rapidly than a linear function of the rate. Line *C* is the case that destroys the utility of nonresponse rates as indicators of nonresponse error. When a low nonresponse rate is obtained in *C*, the remaining nonrespondents are quite different from the respondents. Theoretically nonresponse error can actually increase with lower nonresponse rates.

Which of the curves in Figure 12.1 apply in a specific survey? To answer that question an evaluation study is required, containing data on respondents and nonrespondents. An evaluation could assess the value of the $(y_r - y_{nr})$ term at different response rates. A weaker type of study, dependent solely on response distributions, examines the change in the respondent mean for different achieved response rates. For example, in a survey on the 1984 U.S. presidential election, Traugott (1987) found that increasing the response rate obtained more Republican respondents, and hence, produced higher estimated percentages planning to vote for Ronald Reagan.

The expression for nonresponse error given is deceptively simple. There are different reasons for nonresponse, each of which might be associated with the failure to measure different kinds of people. As noted above, in household telephone surveys sample persons are nonrespondents because they cannot be reached, because they are physically or mentally unable to respond, or because they refuse to

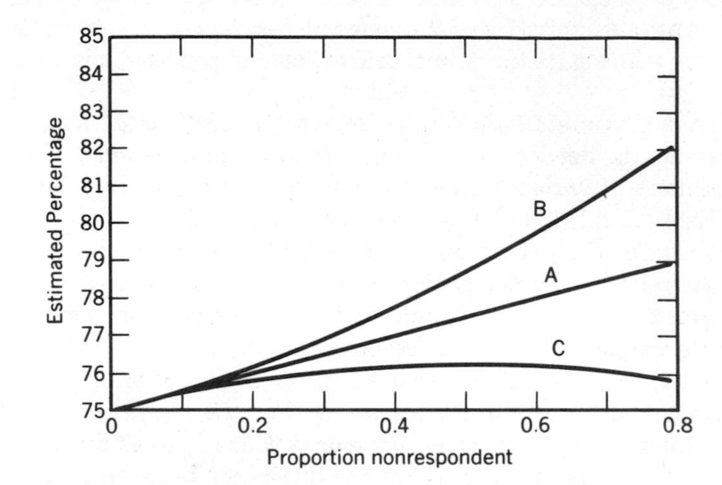

Figure 12.1 Estimated percentage under three nonresponse models (true % = 75).

cooperate with the request for the interview. Thus a more appropriate expression for a survey statistic might be

$$y_r = y_n + (nc/n)(y_r - y_{nc}) + (ni/n)(y_r - y_{ni})$$
$$+ (rf/n)(y_r - y_{rf}$$

where y_{nc} = the statistic for the *nc* noncontacted sample cases
y_{ni} = the statistic for the *ni* sample cases that are incompetent to provide the interview
y_{rf} = the statistic for the *rf* sample cases that refused the interview.

In this case, $nc + ni + rf = nr$ in the simpler expression. This more complex expression is more enlightening because there is little reason to believe that the characteristics of these different kinds of nonrespondents are similar.

Thus far we have illustrated how nonresponse in a particular administration of a survey design affects the achieved values of survey statistics. Nonresponse also affects the variability in estimates over possible replications of the survey design. For example, different telephone interviewers can affect the levels of nonresponse error in a statistic. Interviewers vary in their ability to pursue elusive respondents, in their patience in administering questionnaires to persons with limited language or cognitive abilities, and in their ability to persuade the reluctant that the survey interview is a valuable experience. For example, in one ongoing centralized telephone survey at The University of Michigan Survey Research Center, an average interviewer performance would result in about 84 percent of those contacted providing interviews. It is not uncommon, however, for the poorest 10 percent of the interviewers to have rates less than 65 percent and for the best 10 percent to have 100 percent of their contacted cases interviewed .[3] The coefficient of variation across interviewers in these rates would be about .14. Statistics Sweden shows higher variation in their labor force survey conducted by telephone, with an average coefficient of variation of interviewer level nonresponse rates of .65 on an average nonresponse rate of about 3.8 percent. Both Collins et al. (this volume, Chapter 13) and Oksenberg and Cannell (this volume, Chapter 16) present additional evidence on this point. Clearly, the choice of different interviewers can lead to different levels of overall nonresponse. That is, if the survey were replicated on the same sample with a different set of interviewers, lower or higher nonresponse rates and nonresponse error might result. Nonresponse errors thus can have variable and fixed components — both nonresponse *variance* and nonresponse *bias* may exist. The expressions above describe only the fixed components, under the assumption that the achieved rates and differences are fixed over replications. Researchers should choose to ignore the variability in response rates over interviewers only if they are unconcerned with the possibility that their survey results might vary if a different set of interviewers had done the work.

2. ISSUES IN NONRESPONSE RATES IN TELEPHONE SURVEYS

Response and nonresponse rates are often used as one way to evaluate surveys, but there are so many different ways of calculating response rates that comparisons across surveys are fraught with misinterpretations. Before comparing the many different calculations labeled "response rate," we examine the various design features affecting outcomes. There are four survey design aspects that affect the response rate calculations:

---

[3] Researchers in face to face surveys always suspected interviewer variability in gaining the cooperation of sample persons, but could not measure it because interviewer assignments were not randomized. Thus, the likelihood of responding was not constant over interviewer assignment groups. This can be eliminated with random assignment procedures in centralized facilities.

(a) Whether all units on the sampling frame are eligible for the survey (i.e., whether some units on the frame are not members of the target population). In almost all telephone samples, some numbers on the sampling frame are not household numbers and thus are not part of the population to be measured (e.g., business numbers, nonworking numbers).

(b) Whether each unit sampled contains one sample element or many. Some household telephone surveys attempt interviews with all members of the household (or all adult members) about their individual attitudes or behaviors; others use one informant to obtain information about attributes of the household. When no contact or cooperation can be obtained from the entire household, in the first design, several sample elements (persons) are lost; in the second, only one element (a household).

(c) Whether all sample persons have the same probability of selection. Some surveys oversample certain groups in the population to permit special analyses (e.g., those of certain age groups). In such designs, nonresponse error depends on the pattern of nonresponse across groups sampled with different probabilities.

(d) Whether substitution at the sampling stage is permitted by the design. Some designs permit the survey administrators to substitute a similar household for a sample one which cannot be contacted or which refuses the interview request. In such cases, a decision must be made about how the initial noninterview case contributes to the response rate.

2.1 Noninterview Outcomes Involving Answered Dialings

Some outcomes of dialing sample telephone numbers permit relatively unambiguous classification of eligibility for the number. This is true when a person answers the telephone and reports that the number is used only for business purposes (e.g., it is the office number of a retail store), that it is a public telephone on the street or in a business, or an automobile telephone. Other results of dialing unambiguously identify a nonworking number. These include an operator intercept (which tells the caller that the number dialed is not working), a nonworking number recording, a message that the number dialed has been changed to another number, and the myriad of other messages.[4] Other contacts after dialing yield more ambiguous information about the household status of a number. For example, what is the appropriate classification when a computer tone occurs after the number is answered, when an answering machine message follows the response, or when the telephone is answered but then immediately hung up?

Most survey organizations ask at least two eligibility questions of the person who answers the telephone aiming at: (a) verification that the sample number was the one reached, and (b) determination of the status of the unit to which the number is assigned (i.e., a residence, a business, a group quarters). For those cases described as businesses, some organizations also ask a question about whether any person uses the same number as his or her personal number. This might occur in a business operated from a residence or in a business with some staff members resident on the site. Since the person answering the telephone can make errors in responding to these questions, the classification of sample numbers that are answered is not completely under the control of the interviewer.

Some difficulties with classification of sample numbers arises from advances in the telephony technology. For example, the "call forwarding" feature now available in most areas of the United States permits a person to transfer calls from one instrument to another (which has a different telephone number). The feature, useful to someone who has two residences, will result in reaching a different number than the one dialed. Similarly, mobile cellular telephones can be physically located anywhere. This directly challenges the unambiguous linking of residences to telephone numbers, the very basis of the use of the telephone number frame to sample households. These developments and others may force future survey researchers to add questions to their introductory presentations to verify whether the number dialed is appropriately assigned to the household contacted.

The inability for the sample person to provide responses to the survey is often placed in a diverse category called "noninterview for other reasons." This category has varying definitions among and within survey organizations. The tradition at the National Opinion Research Center in many of its national surveys is to avoid interviews with persons who cannot speak English (i.e., they are removed from the denominator of the

---

[4] In most sample designs if the number sampled has been changed to another, the sample number is classified as nonworking, and thus ineligible. Continuing the calling on the new number (to which the sampled number was changed) in most circumstances gives the corresponding household two chances of selection, one on its own and another from the sampled number.

response rate calculations and from the target population).[5] Other organizations attempt to use translators or to hire interviewers with Spanish (or other language) skills in areas where such problems exist. A similar problem arises with the deaf and for some surveys requiring visual abilities, the blind.

Refusals can result from the reluctance of the entire household to answer any of the interviewer's questions, from the denial of the interview by proxy (e.g., a wife refusing access of the interview to the husband, who was chosen as the respondent), or from the direct denial given by the selected respondent. The first type of refusal produces more difficult problems than the latter two. When a household level refusal occurs, the interviewer typically is ignorant of how many eligible persons are household members. In a survey in which all persons in the household are to be measured, the interviewer is unaware of whether the household refusal corresponds to one or many nonrespondent cases, at the person level. In a survey in which one respondent is selected among those eligible, the interviewer is ignorant of the probability of selection of the resulting nonrespondent.

In telephone surveys there is a category of refusal that occurs only rarely in face to face interview surveys — the partial or broken off interview. Rarely are interviewers asked to leave the house after they begin a face to face interview, but people apparently feel freer to terminate in the middle of a telephone interview.

Finally, there are outcomes that are specific to individual surveys. If the sampling frame is a list of individuals, there may have been deaths since the frame was created. Deaths are most often treated as nonsample cases, although if the respondent is to report of past events, they might be treated as nonresponse cases. In some quota sampling schemes (with procedures specifying a certain number of interviews for different classes of persons) there are sample cases which are not interviewed because the quota class which they occupy has been filled.

2.2 Noninterview Outcomes Involving Unanswered Numbers

The number of noncontact cases encountered in a survey is a function of the number of repeated calls that interviewers make on sample cases. If a survey is conducted with a maximum of one call attempted for each

---

[5] The rationale for this stems, no doubt, from concerns both with nonresponse error and measurement error associated with this population. Even when this group might be willing to respond, the translation of survey questions into another language and of the respondent's answers to English might be productive of errors in the recorded data.

sample case, the vast majority of nonresponse cases are noncontacts. In household telephone samples of persons (either one or multiple respondents per household) "noncontact" is generally used to mean that the interviewer did not speak with anyone in the household, not merely that the designated sample person was not reached.

Whenever a sample is based on telephone numbers, the researcher is subjected to all possible results from dialing a telephone number. In RDD samples especially, many dialings yield an outcome other than an answered number. The most common is a number ringing repeatedly without answer. When a dialing results in no answer, because the interviewer cannot visually inspect the structure to which the number is connected, it is unclear whether the number is a working number. Families take extended vacations; temporary job assignments force people to leave their principal residences for short periods. An interviewer can repeatedly call without contact on a household with residents who are hard of hearing, or occupants choosing not to answer the telephone. On the other hand, some nonworking telephone numbers in the United States are connected to a ringing tone, so that whenever they are dialed the interviewer will hear the number ring interminably. Other sampled numbers may be pay telephone booths, where a ringing telephone may go unanswered. Early work reported by Groves and Kahn (1979) suggests that only about 5 percent of numbers dialed over 12 times were working household numbers. A more extensive survey effort in this area is reported by Sebold (this volume, Chapter 15).

One administrative rule that affects the proportion of dialings that are classified as unanswered is the number of rings that an interviewer is required to hear prior to terminating the call. Smead and Wilcox (1980) note that the vast majority of telephone numbers are answered before 5 rings. Among researchers using the telephone to interview elderly adults, however, it is common to use a rule with many more dialings, in order to compensate for the population's physical limitations.

Other dialings produce strange electronic switching noises, absolute silence, or clicking sounds, without any ringing tone. Still others result in what is called a "fast busy," a tone interrupted 120 times per minute instead of the 60 times a minute for a normal busy signal (reporting that the number is currently engaged). "Fast busy" signals are used to report busy circuits in the area being called, connection difficulties (as the result of storms, floods, fires, etc.), or nonworking status of a number. It is generally believed that numbers that receive the fast busy signal over repeated calls over many days are indeed nonworking numbers. Hence, most organizations set up calling rules for repeated dialings on numbers that yield fast busy signals prior to permit their final disposition.

A normal busy signal merely indicates that the number is in use, not that it is a working household number. Some organizations counsel interviewers to wait 5 to 15 minutes and then dial such a number again. This takes advantage of the higher probability of contact given that the number was engaged.

3. RESPONSE RATE CALCULATION

Response rates can be calculated in a myriad of ways, each implying to the naive reader different levels of success in measuring the full sample. The alternative forms of calculations have been the focus of attention of several investigators (Wiseman and McDonald, 1979; Kviz, 1977; Bailar and Lanphier, 1978; Swensson, 1983). There do exist recommendations for preferred estimates (CASRO, 1982), but there is no universal compliance with the guidelines. Rather than using a single preferred response rate calculation, survey researchers use several rates, each for a different purpose and each yielding a different measure of the completeness of the data collection on the sample.

In telephone surveys the following outcomes are relevant:

I = completed interviews
P = partial interviews
NC = noncontacted but known eligible numbers
NA = unanswered numbers
R = refused eligible numbers
NE = noneligible units
NI = other noninterviewed units .

One rate of interest to some survey personnel is the cooperation rate, defined by

$$I/(I+P+R);$$

that is, the ratio of completed interviews to all contacted cases capable of being interviewed. This rate might be used to characterize how successful interviewers were in persuading those able to do the interview to comply with the request.

Another rate of interest adds to the denominator those cases not contacted,

$$I/(I+P+NC+R),$$

still excluding those cases which cannot supply an interview because of mental or physical impairment. For example, the practice of excluding non-English speakers from the response rate calculations could be viewed as use of such an estimate. This estimate might be used to characterize how completely those sample cases which could provide survey measures did indeed produce them. Note, however, even for these purposes the estimate is flawed by underestimating the success of the field procedures. Not all of the NC noncontact cases would be physically or mentally able to provide the interview, but they are all included in the base.

A conservative response rate is

$$I/(I+P+NC+NA+R+NI),$$

where the denominator includes all sample cases in which an interview could have been completed plus all unanswered numbers. This rate more clearly estimates the proportion of all eligible persons measured by the survey procedures, but is an underestimate to the extent that some unanswered numbers are nonhousehold numbers.

Another response rate is

$$I/(I+P+NC+R+NI),$$

which counts all unanswered numbers as nonhousehold numbers.

As reviewed above, the major difficulty with computing the response rate for telephone surveys is the determination of eligibility for unanswered numbers. Given that difficulty, some researchers present the last two response rates, one including all unanswered numbers in the denominator, the other excluding all unanswered numbers. For example, Groves and Kahn (1979) give a response rate that ranges between 59 and 70 percent, using these two calculations. Other researchers prefer to estimate the proportion of eligible numbers among the unanswered numbers:

$$I/[I+P+NC+p(NA)+R+NI]$$

where p is the proportion of unanswered numbers that are working household numbers. The use of the p term in the expression above would not be necessary if, for the frame population sampled, the researcher knew what portion are ineligible numbers (e.g., nonresidential, nonworking numbers). Then the denominator of the response rate would merely be the number of eligible sample numbers. This information is not available in U.S. samples typically. There is some evidence that the likelihood that a number from an RDD sample is residential is a function of the number of repeated calls without answer (see this volume, Chapter 15).

It is important to note that these rates are *sample statistics*, subject to sampling variability. That is, a different response rate would have been obtained if a different set of persons had been drawn into the sample. The rates would vary depending on the variability in cooperation and ease of contact of the target population.

Perhaps the most important observation about response rates is that they are likely to have multiple purposes, and that calculation procedures should be fitted to the purpose at hand. One major purpose is the evaluation of the field activities. For this purpose several related rates are of interest:

a. The contact rate, (I+P+R+NI)/(I+P+R+NI+NC+NA), to assess how fully the sample was alerted to the survey

b. The cooperation rate, I/(I+P+R), to assess how well the field staff persuaded those contacted and able to respond

c. The refusal conversion rate, the proportion of cases which initially refused to provide an interview, but later agree to do so.

Clearly, the largest problem in calculating response rates for telephone surveys is the indeterminacy of eligibility, due to unanswered numbers, answering machines, immediate terminations of connections, fast busy signals, and other strange results from dialing randomly generated numbers.

4. TELEPHONE SURVEY NONRESPONSE RATES

Given these qualifications on the computations of response rates, it is useful to discuss what levels of response rates are typical in telephone surveys. Levels of response rate on the telephone are affected by many of the same factors as rates in face to face surveys: the length of the questionnaire, the topic of the survey, the survey organization, the length of the survey period, the administrative rules on callbacks and refusal conversion, as well as other design features. It is important therefore in reviewing telephone survey response rates to keep these effects distinct from the effects of mode.

Cold telephone surveys in the United States tend to obtain response rates that are lower than comparable face to face surveys (Groves and Kahn, 1979; Mulry-Liggan, 1983; Fitti, 1979; Steeh, 1981). This result was found early in methodological research on telephone surveys and appears to be still applicable at this writing, nearly 10 years later. Despite this, there are some important exceptions. The works of Drew et al. (this volume, Chapter 14) and Cannell et al. (1987) show that high response rates are possible when the combination of topic and respondent rule is appropriate. Health and unemployment surveys have easily communicated importance. The use of a family or household informant increases the likelihood of participation.

There are two differences between telephone survey nonresponse and face to face survey response. First, a larger share of the nonresponse is associated with refusal cases on telephone surveys than on face to face surveys.[6] This arises because of the relative ease with which noncontacts in telephone surveys can be eliminated with repeated dialings. Second, relatively more of the refusals take place after only a few moments of interaction in telephone surveys. Although there have been few controlled studies of this, there is evidence that the majority of refusals may take place in the first minute of the conversation, after the interviewers have introduced themselves but before any substantial explanation can be given of the purpose of the call (see this volume, Chapter 16). This fact poses a large challenge for efforts at decreasing the refusal rate.

5. CORRELATES OF TELEPHONE SURVEY NONRESPONSE

Since nonresponse error is a function of both nonresponse rates and the characteristics of the nonrespondents, it is important to review current knowledge of the attributes of telephone survey nonrespondents. As with the field of nonresponse in general, our understanding of the correlates of nonresponse is limited, because of the need for external data sources to measure characteristics of the nonrespondents. There are, however, some studies which offer insight.

Age. There is repeated evidence that elderly persons disproportionately refuse to be interviewed in telephone surveys (Cannell et al., 1987; Weaver et al., 1975; Brown and Bishop, 1982). Cannell et al. show response rates of 57 percent among those 65 or older in a health survey, with overall response rates of 75 percent, using a random respondent rule. A similar phenomenon is evident in face to face surveys, but the telephone results reveal a stronger tendency toward elderly nonresponse. The possible causes of higher nonresponse among the elderly include (a) hearing loss leading to reluctance to use the telephone,

---

[6] Sebold's work (this volume, Chapter 15) shows that this may not be true with very short survey periods.

(b) social disengagement leading to hesitancy to interact with strangers, (c) fear of crime leading to fear of strangers, and (d) distinctive socialization regarding telephone use. The hearing loss argument has little appeal because the onset of hearing loss appears in early age groups. The social disengagement hypothesis has been criticized in the gerontological literature [see for example, Glenn (1969)]. The fear of crime argument does have support, but there have been no specific links to survey nonresponse. The socialization argument states that the current cohort of elderly began to use the telephone in a time when long distance telephone calls were relatively expensive and limited to very short interactions. The use of the telephone for unsolicited, long conversations with strangers may be deemed inappropriate. Again, there is no evidence to support this hypothesis, but there may be variation over countries in this tendency.

This volume presents the first documented failure to replicate the findings of higher nonresponse tendencies among the elderly in the telephone mode. Collins et al. (this volume, Chapter 13), in an experimental design in Great Britain, show higher nonresponse among the elderly, but no difference between the two modes of data collection. Failures at replication are generally an opportunity at theoretical advance, and such may be the case here. Speculation among survey researchers familiar with both cultures suggests that the relatively late onset of widespread telephone subscription may be related to lower response rates among the nonelderly in Great Britain (nullifying the age effect present in the United States). Some assert that telephone conversations with strangers are a rarer phenomenon in Great Britain than in the United States. Others note that the higher relative cost of long distance calls in Great Britain leads to a tendency to use the telephone only for pressing business and to rely on face to face conversations for a larger portion of communication needs. Whatever combination of factors produces the difference, new research into the causes of nonresponse in telephone surveys in the two societies is clearly warranted.

Education. The second finding with some replication is the higher nonresponse among lower education groups (Cannell et al., 1987). This finding is not independent of higher nonresponse among the elderly because of the lower educational attainment of the elderly. However, there are no multivariate studies to determine whether this is a spurious result.

Urban-Rural Differences. Finally, a common finding in face to face surveys is higher nonresponse in large urban areas than in others (see Steeh, 1981). There is some evidence that the urban-rural difference in response rates is diminished in telephone surveys. Groves and Kahn (1979) obtained that result in a U.S. national survey and showed that the relative proportion of urban apartment dwellers and occupants of other multiunit structures was higher on the telephone than in a companion face to face survey. Locked apartment buildings and multiunit structures with security guards often present insurmountable problems for face to face survey interviewers. In addition to those problems most U.S. survey organizations tend to encounter higher turnover among face to face interviewers in large urban areas. This phenomenon, representing another source of lower response rates, can be avoided in centralized telephone surveys.

Other correlates of nonresponse that have been reviewed in the literature refer to attributes of the interviewers, not of the sample persons. The centralization of telephone surveys has permitted researchers to focus on attributes of interviewers as influences on nonresponse independent of the effects of sample person attributes. This has occurred because of the randomized assignment of sample cases to interviewers. There are two findings of interest from these studies.

Interviewer Voice Quality. First, various attributes of the interviewer's voice may affect levels of cooperation obtained. Oksenberg and Cannell (this volume, Chapter 16) review in detail these results.

Interviewer Gender and Experience. Second, Groves and Fultz (1985) investigated the effects of interviewer gender on response rates to a consumer attitude survey. They found higher individual response rates among female interviewers than among male interviewers. There was some tendency for males to have lower response rates when the selected respondent was a female. When, however, analytic control of differences in levels of experience was introduced, the gender differences disappeared. That is, it was found that male interviewers tended to have less experience in interviewing than did female interviewers. More experience did lead to higher response rates, for both male and female interviewers. Controlling for experience, males had similar response rates as did females.

Length of the Survey Period. A common finding is that the number of calls and the number of days of calling required to achieve a specified response rate are higher for cold telephone surveys than for cold face to face surveys (Groves and Kahn, 1979; Mulry-Liggan, 1983). It is likely that this difference is an effect of the relative paucity of information that is obtained on unsuccessful dialings. In contrast, after an unanswered visit at a sample household, the interviewer often observes characteristics of the household or its neighborhood or seeks information from neighbors to guide the time of the next visit. This makes return visits more effective than repeated dialings.

The result of this difference is disproportionate nonresponse for quick telephone surveys, if similar interviewer resources are used. Figure 12.2 gives an idealized version of this result. Very short survey periods are more easily handled with face to face contacts. The empirical evidence comes from several organizations but most dramatically from RDD tests at the U.S. Census Bureau.

Interview Length. Research on "respondent burden" in the 1970s demonstrated that the length of the interview was not always related to the perceived burden of the interview by respondents (Frankel and Sharp, 1981). However, there is a strong belief among survey researchers that a telephone interview should be shorter than a comparable face to face interview, in order to achieve similar "quality." Unfortunately, there are very little data to support that belief. By and large, survey organizations in the United States have tended to use shorter telephone interviews than face to face surveys. However, Swedish researchers encounter less resistance to long telephone interviews. This volume contains a formal experiment in Great Britain (see this volume, Chapter 13), which found a 5 percentage point higher refusal rate on a 40-minute telephone interview as compared to a 20-minute telephone interview. More research of this type would be useful, but the issue is complicated by the fact that burden seems to be a function of topic, population, format of questions, interviewer behavior, as well as a variety of other design attributes. Hence, the research must vary those features as well in order to provide answers to a wide variety of surveys.

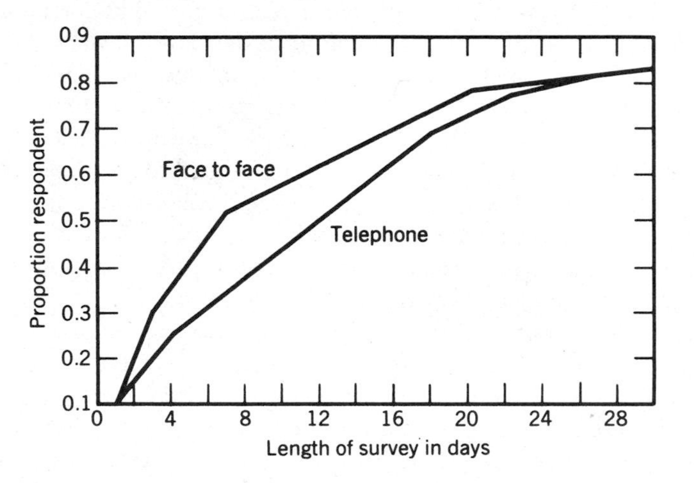

Figure 12.2. Response rates by mode.

6. EFFORTS TO INCREASE PARTICIPATION IN TELEPHONE SURVEYS

Efforts at increasing the response rates of telephone surveys have utilized two tactics. First, there have been a variety of efforts to alter the introductory comments of the interviewer in an attempt to reduce the immediate refusals often obtained. Second, there have been efforts to use advance letters to alert the sample persons to the upcoming survey request.

Introductions. The experiences in altering the initial interaction of the interviewer are mixed. O'Neil et al. (1979) report a series of experiments using survey like questions immediately after mentioning that the call concerned a research project, providing full description of the survey organization and the project, and providing verbal feedback to the respondent. No important differences in response rates were obtained with the experimental changes. Dillman et al. (1976) ran two experiments imbedded in telephone surveys. The manipulations of the introduction in the first survey included (a) the use of the residents' name (obtained from the telephone directory) and (b) a description of the study, sampling technique, and length of interview. Another treatment group in the experiment added to those features an offer of the study results and a description of the social utility of the project. No significant differences in response rates were obtained.

Position of Household Roster. Monsees and Massey (1979) changed the position of the household roster in a study using a family informant respondent rule. In one half of the sample, the roster was first obtained and then the respondent was asked to answer questions. In the other half, data on the telephone answerer was obtained first, then the roster was obtained for information on the other persons in the household. The authors found higher response rates in the condition of the later placement of the household roster. This result fits the beliefs of many interviewers that the household roster is one of the more difficult sets of information to obtain from sample persons early in the interaction.

Streamlined Respondent Selection Procedures. A similar desire to begin the interview as quickly as possible motivates a variety of selection procedures when a single respondent is sought from a sample household. They all are contrasted with a full household listing followed by a random selection of one eligible person. Instead, a two question procedure has been used, for example, first asking how many adults were resident, then how many female adults were resident. Based on the answers, the chosen respondent is identified by gender and relative age (e.g., "the youngest female," "the second oldest male"). This procedure can be developed to be equivalent to that of Kish (1949) or in a nonprobability version, like that of Troldahl and Carter (1964). Groves and Kahn (1979) studied this method and found a within-household coverage problem in about 9 percent of the sample cases, relative to a full household listing taken after the interview was complete. Another streamlined procedure is a nonprobability method — the last birthday or next birthday method. This asks one question, for example, which person 18 or older last celebrated a birthday, and chooses that person as a respondent. If within households, birthdays were randomly assigned to persons or the date of the survey contact were randomly determined *and* the household informant provided the information correctly, a probability sample would result. Experiments with this method, however, show a disproportionate number of women being selected compared to probability methods. There is, however, some evidence of higher response rates with the method. Thus the streamlined methods force a decision on the relative importance of coverage errors versus nonresponse errors.

Advance Letters. Another common tool to increase survey participation is the advance notification of the sample household of the survey request. This has been found effective in both face to face and telephone surveys (Dillman et al., 1976). Experimental results show that the contents of the letter seem to have smaller effects than its aspects of personalization. The effects of a letter are also diminished by the failure to recall its receipt (or to have received it) among persons contacted [a typical result being that about 75 percent acknowledge receipt and of those about 75 percent recall reading it (see Cannell et al., 1968)]. There are two hypotheses about the source of effects of advance letters: (a) the letter, if sent by a credible, authoritative source, tends to legitimize the later survey request, and (b) the knowledge that an advance letter was sent reduces anxiety felt by interviewers about the cold contact and improves their performance at the first contact. Groves and Snowden (1987), in a telephone study reinterviewing persons initially interviewed face to face, attempted to test these two hypotheses. In an experiment in which interviewers were told that some cases received letters, when they had not, the response rates for the group resembled those that were not sent letters.

The use of advance letters in telephone surveys is infrequent in the United States because of RDD sample designs, which do not supply names and addresses for sample numbers. The use of list assisted sample designs (this volume, Chapter 5) permits letters in a portion of the sample. The results of experimental uses of advance letters in cold telephone surveys suggest that response rates might be increased by 5 to 13 percentage points (see Traugott et al., 1987).

Refusal Conversions. Almost every telephone survey organization has procedures for persuading reluctant sample persons to provide the interview. Sometimes sample cases initially refusing are assigned to more experienced interviewers or to an interviewer of a different sex. Many survey organizations report that from 25 to 40 percent of the sample cases initially refusing later provide an interview. There is evidence that the success at conversion is a function of the "softness" of the refusals. The success at conversion is also a function of whether another household member is contacted, how well developed the rationale is for the initial refusal, and how well the second interviewer addresses the concerns of the reluctant respondent.

In some organizations, many refusal conversion attempts are used, in an effort to obtain high response rates. Unfortunately, there has been no open discussion among researchers about the ethical implications of repeated attempts to persuade a person who initially refused. In the extreme it is clear that such behavior tests the limits of "informed consent" (i.e., any failure to consent merely leads to another call).

7. NEEDED RESEARCH ON TELEPHONE SURVEY NONRESPONSE

Efforts to increase response rates place great emphasis on only one term in the nonresponse bias expressions above. In addition to the proportion of sample persons not measured, nonresponse error is a function of the characteristics of the nonrespondents relative to those of the respondents. For a simple linear statistic, if the efforts to reduce nonresponse affect different sample persons differently, it is possible that overall nonresponse error might increase with higher response rates. For that reason, some understanding of the characteristics of nonrespondents and reluctant respondents is needed to choose correctly survey design features affecting nonresponse.

The same comment can be made about the various nonresponse adjustment procedures used in telephone surveys (see this volume, Chapter 8), including imputation and weighting schemes. All of these procedures require the analyst to make assumptions about the equivalence of a set of nonrespondents to some subset of respondents. The respondent cases are then used to supply information about the nonrespondents.

There is a common weakness to current efforts at increasing response rates and at postsurvey adjustment for nonresponse — the absence of a theory of survey participation guiding the procedures. Dillman (1978) does offer concepts of social exchange to motivate various personalization efforts at advance letters and alternative introductions to the survey. However, additional concepts seem necessary.

There are three literatures in psychology that have been ignored by most survey research practice related to nonresponse. They are the literatures on compliance, altruism, and persuasion. From mostly experimental data, they have identified a large set of factors that affect decisions of persons to behave in a manner consistent with a formal request to do so made by another. These include notions of "reciprocation," the tendency to comply with requests from those who have provided some service in the past (Cialdini et al., 1975). Advance letter, incentives, even solicitous behavior of interviewers may fit this concept. Another factor is the perceived scarcity of opportunities to perform the behavior requested. Interviewers instinctively use this factor when they say, "This survey only has two more days before it's over, so you won't have an opportunity to respond if you wait." Conformity to behavioral norms also influence compliance. Some respondents clearly consent because of belief that they have a duty as citizens to participate in efforts seeking information about the society. This no doubt partly explains the high response rates for health surveys also. These literatures have also found that compliance is reduced when the subject has no unique abilities to provide the service, when many others could also help the requester. Here, telephone surveys face a unique challenge. Interviewers tell respondents that they represent thousands of others (the power of random sampling) but they often do not communicate why one of the others in the group of thousands could not respond.

Theoretical principles are useful in guiding research not because they have immediate implications for operations, but because they may suggest the underlying causes of the behaviors of interest. Most of the past research in telephone survey nonresponse has lacked a set of guiding theoretical principles that seek to explain why people participate in surveys. Without that, the field is likely to remain a set of scattered findings, with repeated failure at replications of results.

8. THE FUTURE

We find the temptation to speculate on the future irresistible in this chapter. Basically, we are pessimistic unless changes in research directions are made. The reasons for pessimism are clear. First, technological changes in telephone systems offer more barriers to respondent contact. Telephone answering machines are becoming more prevalent in the United States and other countries, mobile phones present problems of simple linking rules of persons to sample numbers, the capability of displays of the caller's number on the recipient telephone permits filtering of calls, and other changes are making survey use of telephone systems more complex. Second, there are long term societal trends in the decline in survey participation in the United States. Telephone surveys are not immune to these trends. Third, although the United States has not yet experienced such problems, other Western societies have experienced real shocks to participation, from governmental sources (e.g., the Swedish Data Act) or from political debate about the use of survey and census data (e.g., the West German experience in the middle 1980s).

In the face of these observations, new thinking about response rates in telephone surveys is required. As the last section notes, this thinking needs to be theoretically motivated. The construction of theories of survey participation is necessary. Efforts at increasing response rates face increasing opposition. Adjustment for the effects of nonresponse therefore needs greater attention. All adjustment requires some assumptions about the characteristics of nonrespondents. Those assumptions should be theoretically based.

There are clear targets for this work. The first is the correlation between age and response propensity (and its variation across countries). The second is the promise of centralization of the telephone interviewing to use new administrative and research designs. These might include the selection of refusal conversion efforts in a way to increase response rates *and* decrease the difference between respondents and nonrespondents, by attempting conversions of those cases not well represented already among the respondent group. The third is the investigation of links between nonresponse error and response error. There is little information about whether reluctant respondents tend to provide data of similar quality as those respondents easily interviewed. The concern of the investigator should be not only the reduction of nonresponse error but the reduction of the total error in the survey statistics, which means that putting too much effort and costs into refusal conversion might lead to larger total errors. This is a highly controversial issue that needs to be addressed. But without data we are all at a loss. It would be a relief for survey conducters to know when to terminate data collection. The fourth nonresponse research issue is the use of multiple modes of data collection (see this volume, Chapters 11 and 32), that might be designed to tailor the mode of data collection to the sample person's needs, and thereby reduce nonresponse error.

Business as usual is not likely to bring success to telephone survey researchers of the future. Ideas already exist to improve the chances of success but deliberate efforts to test them must proceed quickly.

CHAPTER 13

NONRESPONSE: THE UK EXPERIENCE

Martin Collins and Wendy Sykes
Social and Community Planning Research

Paul Wilson and Norah Blackshaw
Office of Population Censuses and Surveys

This chapter describes the problem of unit nonresponse to cold contact telephone interview surveys of the United Kingdom (UK) general population. We first examine the incidence of nonresponse in such surveys. The available information is limited but shows that response rates to telephone surveys are lower than to equivalent face to face interview surveys: at best at the level suggested by US experience — five percentage points lower. We go on to show that, where there are large differences between telephone and face to face response rates, they arise from a greater incidence of refusals to participate over the telephone. But when differences between these two modes are small, refusal rates are also similar and the difference arises surprisingly from a higher noncontact rate and from miscellaneous sources other than refusal to participate.

The crucial category of refusals to participate is examined in more detail. In particular, we explore the timing of refusals and the reasons for refusing. Finally, we look at the results of some strategies which have been adopted in attempts to reduce the problem: timing of calls; the deferment of resistance through making appointments; advance letters explaining the survey; attempts by a second interviewer to convert some of those who initially refuse to take part; the selection and training of interviewers; and shorter interviews.

We draw on results from four studies comparing telephone and face to face survey methods. All four studies involved attempts to interview probability samples of the UK adult population. The first three were designed explicitly as comparative investigations, two conducted by Social and Community Planning Research (SCPR) as experiments alongside the annual British Social Attitudes Survey (referred to here as BSA 1 and BSA 2) and one conducted by Marplan on behalf of the Market Research Development Fund ("Lifestyle in the 80's"). In each case we are able to compare telephone and face to face approaches for samples drawn from the same population: those whose addresses appear in the Electoral Registers and for whom a telephone number can be traced in a directory.

In none of these studies was nonresponse the initial focus. There were more basic doubts about the feasibility and usefulness of telephone surveys. Telephone coverage in the UK is still low by international standards — only around 80 percent — and as in other countries it is unevenly distributed. Researchers have been slow to adopt telephone interviewing, especially in social surveys which demand coverage of, or even focus on the less privileged groups where telephone coverage is low. In such surveys, if telephone interviewing is to be used, it must be in a dual mode setting, with nontelephone households interviewed face to face. Then a crucial consideration is whether or not the two interview modes will yield comparable and compatible data. This aspect of the comparative experiments is reported by Sykes and Collins (this volume, Chapter 19) and elsewhere (Sykes and Collins, 1987; Collins and Sykes, 1987).

One reason for expecting lack of comparability is the perception of the limited role of the telephone in the culture of the UK. Even the current coverage has been attained only quite recently: as recently as 1969, only one third of the UK households had telephones. Many people will therefore have come to telephone ownership only at an age when their communication habits were well established. The perception is thus that the telephone is reserved for short "business like" conversations and will be too unfamiliar to be used as an interviewing medium. It is at this point that concern about nonresponse arises. Perhaps, at the extreme, people will see the survey interview as an inappropriate use of the telephone and will refuse to take part.

The three experiments did much to reassure researchers on the basic issue of comparability (to the extent, for instance, that the UK Labour Force Survey now makes use of the dual mode approach). But they raised some more detailed issues: asking complex questions and — our interest here — the issue of nonresponse. The fourth study concentrated on these issues. Carried out by SCPR in collaboration with the Office of Population Censuses and Surveys (OPCS) Social Survey Division, it was a survey of "Attitudes towards Alcohol." Again it used a sample selected from Electoral Registers followed by a directory search for telephone numbers. In this instance, however, there was no directly parallel face to face survey. Comparisons are drawn instead with a face to face interview survey carried out by SCPR about a year earlier.

Further technical details of all four studies are given in the Appendix to this chapter.

1. THE INCIDENCE OF NONRESPONSE

1.1 Response Rates

Information on response rates in cold contact surveys of the UK general population is sparse and confusing. At one extreme, some market researchers quote rates as high as 90 percent. But they refer to quota sample surveys, where substitution is allowed for unanswered numbers and where no attempt is made to select a respondent at random when a number is answered. They also tend to refer to surveys involving only very short interviews. These very high response rates thus say little more than that only about 10 percent of those who answered the telephone refused to go on to answer a short questionnaire. At the other extreme, when dealing with telephone numbers in a more rigorous way, market researchers have referred to response rates of only around 50 percent.

The four experimental studies examined here involved probability samples of addresses and numbers, and random selection of a respondent for interview, with no substitution allowed at either stage; and the interviews were, by market research standards, comparatively long (in one case as long as 40-45 minutes). The response rates reported are conservatively calculated as the ratio of completed interviews to the total eligible sample. As Table 1 shows, these rates are low: in the range of 45 to 65 percent.

Response rates by telephone in the first three studies were unacceptably low, at only around 50 percent. The first two of these studies were carried out by SCPR, with no previous experience of such telephone interviewing, but the third was carried out by an experienced market research agency. For these studies, especially the first, completion rates in face to face interviewing were also low by usual standards. These were all very general studies of attitude and opinion, covering a wide range of topics and lacking a central theme. Such studies do seem to be difficult to 'sell' to respondents — what is their purpose; by whom and how will the answers be used? Perhaps this problem is accentuated in telephone contact, resulting in the much lower response rates for telephone interviewing (averaging about 50 percent) than for face to face interviewing (averaging about 65 percent).

The fourth study yielded a rather more encouraging response rate of 65 percent, still eight points lower than that achieved in the face to face survey. This study had a more precise topic (although still one that

Table 1. Response Rates in Four Comparative Studies for Completed Interviews as Percent of Total Sample Attempted

	BSA 1	BSA 2	Lifestyle	Alcohol
Face to face	60%	68%	67%	73%[a]
Telephone	53	46	47	65

[a]Face to face result for both telephone owners and nonowners.

yielded a relatively low response rate face to face). It used interviewers who were experienced with telephone surveys (although not necessarily with cold contact surveys or interviews as long as were involved here), and interviewing was carried out from a government office (known to be an advantage in gaining cooperation). The response rate could have been increased still further by possibly justifiable exclusion of a few unanswered numbers from the calculation and by minor changes in procedure, bringing it closer to that achieved face to face. But the latter rate may itself be an underestimate. Since we have no information about telephone ownership for nonrespondents to the face to face survey, we can quote only the response rate for the total sample, including nontelephone households. General experience shows that response rates are marginally higher in the population subgroups where telephone coverage is also above average. Thus, the rate of 73 percent may slightly understate the response among telephone households. On balance, we have to conclude that UK response rates in telephone surveys could be at much the same level, relative to face to face surveys, as that suggested by Groves and Kahn (1979) — about five percentage points lower — but that the difference may well be closer to ten points.

1.2 Categories of Nonresponse

Table 2 provides more detail of response rates to our four comparative studies, breaking the nonresponse down into three categories: refusals, by or on behalf of the respondent; noncontacts, including both cases where no contact at all was achieved and cases where contact was made at the number but not with the required respondent; and a miscellaneous "other" category.

The first result to emerge is that where there were major discrepancies between telephone and face to face interviewing in terms of overall response rates (in the second and third studies), the difference arose mostly from a much higher refusal rate over the telephone — about 40 percent, compared with 20 percent face to face. This fits in with

Table 2. Analysis of Response and Nonresponse

		Completed	Refusal	Noncontact	Other
BSA 1:	Telephone	53%	25%	6%	16%
	Face to face	60	25	4	11
BSA 2:	Telephone	46	38	11	5
	Face to face	68	21	6	5
Lifestyle:	Telephone	47	43	9	1
	Face to face	67	18	8	7
Alcohol:	Telephone	65	19	7	9
	Face to face	73	19	4	4

popular expectations and with some evidence from the US (e.g., Hochstim, 1967; Henson et al., 1974; Groves and Kahn, 1979) that refusals are more likely in telephone than in face to face contact.

The other two studies, however, where response rates over the telephone were closer to those achieved face to face, contradict these expectations. In these cases, refusal rates over the telephone were the same as those met face to face. Here, the higher nonresponse rate in telephone interviewing arose from rather more noncontacts (two or three points) and from the "other" category (five points).

We do not expect higher noncontact rates in telephone surveys. In face to face interviewing, the constraints of interviewer availability and cost tend to limit us to three or four attempts to make contact. In telephone interviewing, additional calls can be made without incurring travel time and expense and at a near zero cost. (In the studies reported here, as many as 20 calls were made to some numbers in an attempt to make contact.) The expectation, then, is that telephone surveys will show lower noncontact rates than similar face to face interview surveys.

Our four comparative studies showed – if anything – fractionally higher noncontact rates over the telephone. The explanation may well lie in the degree of inflexibility in the timing of calls introduced by a centrally located telephone interviewing facility. Differentials in connect charges (and interviewer productivity) greatly encourage the use only of evening shifts – from say 4 pm to 9:30 pm. (And it has to be remembered that the existence of only a single timeband in the UK prevents some of the flexibility that can be introduced in the US, with its time differences.) Some reduction in the noncontact rate might be achieved by supplementary daytime interviewing, but this could be at considerable extra cost.

Other possibly contributory factors include the presence of some ineligible addresses (e.g., empty and second homes) not eliminated from the telephone sample because of the greater problem of identifying them; and failures to make enough calls to certain addresses. The latter shortcoming can arise from another inflexibility in central location interviewing: the need to close one survey and free the facility for the next. There may also have been some limitations in the administration of our studies. Face to face interviewers take individual responsibility for a particular batch of sample addresses, keeping a close check on the outcome at each and usually making considerable effort to at least make contact. In a central location telephone survey the responsibility is shared with some or all other interviewers, and records are centrally kept. Administrative procedures have to replace the more focussed efforts of the face to face interviewer. In this respect, computerized sample management offers a major advantage, but close supervision and emphasis in interviewer briefing on sample management may also provide solutions.

The miscellaneous "other" category of nonresponse includes a wide variety of causes, none of them individually of any great significance: long-term absences, sickness, senility, language problems, administrative failures, rejected questionnaires, and so on. Most such problems will not have differential effects on telephone and face to face surveys.

The greater incidence of nonresponse in this category for two of our telephone surveys may reflect a problem which has been discussed elsewhere (e.g., Groves and Kahn, 1979): that it may be more difficult to establish over the telephone precisely what has happened to prevent an interview. Some doubtful or even unclassifiable nonresponse will then arise. At the extreme, it will be difficult to decide in some instances whether a case should be treated as one of nonresponse or one of ineligibility. This is not as great a problem with our studies as it would be with studies using random digit dialing, but some instances do arise. For example, in the fourth study about 1 percent of numbers were assigned to the nonresponse category when no connection was achieved or when the number was always engaged. These could, in a less conservative measure of nonresponse, be treated as ineligible. In another 2 percent of cases, contact was made at a number but it was found to be not at the address for which it had originally been traced. In these cases no interview was attempted and the number was assigned to the residual category of "other" nonresponse. Alternative treatments would have been to attempt an interview anyway or to regard the number as ineligible for the survey. Although only tiny numbers of cases are involved here, less conservative approaches treating them as ineligible could lift the reported completed rate from 65 percent to 67 percent.

1.3 Respondent Characteristics

Our concern is not, of course, simply with the incidence of nonresponse but with the associated risk of bias. Are nonrespondents different from respondents in terms which matter to the survey objectives?

In any survey the possibilities for approaching this question are limited since, by definition, direct information about nonrespondents will not be available. The most common approach is to compare the achieved sample of respondents with the survey population in terms of characteristics available from a reliable external source such as census data. With telephone surveys — at least in the UK —there is the added problem that the external comparative data for the appropriate population, the 80-85 percent with telephones, will themselves be survey based and hence already subject to nonresponse and sampling error.

An indirect approach adopted for the studies reported here is to compare the profiles of respondents yielded by the two interviewing modes. Thus, respondents to telephone interviews can be compared with telephone owning respondents to face to face interviews. This tells us only about nonresponse differentials between the two modes (ignoring possible response error differences in the two modes). But if we find no differences, we can with some confidence accept that the findings of comparisons between all respondents to face to face surveys and population data (e.g., Lievesley, 1988) apply also to telephone surveys.

This comparison has been made for the three studies carried out by SCPR, with the overall finding of little difference between respondents to the two modes. In the first study (BSA 1) no statistically significant differences were found between the two achieved samples for a range of demographic and socioeconomic variables. In the second study (BSA 2) some statistically significant but still small differences were found. The sample achieved by telephone included slightly higher proportions of childless couples below retirement age and of people in some nonmanual occupation groups. It included slightly lower proportions of couples with children and of "homemakers" not in paid employment.

The latest study, that of attitudes to alcohol, provides the soundest basis for this comparison, having the largest sample size and a more acceptable response rate in the telephone survey. Here the samples (limited to telephone households in the face to face interview sample) were compared in terms of eight variables: employment status, partner's employment status, housing tenure, number of adults, number of persons under l8, age of respondent, marital status, and social class. For only the first two of these variables were significant differences found: small but apparently in line with those found earlier. The sample interviewed by telephone included more people in paid work (62 percent compared with 56 percent face to face) and fewer homemakers of working age (14 percent compared with 18 percent). The same pattern occurred in terms of the status of the respondent's partner (if any). The telephone sample included more people with working partners (63 percent against 58 percent) and fewer with partners involved solely in homemaking (16 percent against 22 percent).

These small differences cannot be expected to have much impact on survey data but, taken with the earlier results, they suggest that telephone interview samples include relatively more working couples. Given that this is a group which tends to be somewhat underrepresented in face to face interview surveys telephone interviewing seems to offer a slight advantage in this respect, presumably because of the greater number and more appropriate timing of contact attempts. (Certainly these are the respondents whose households prove more difficult to contact: 85 percent of our nonworking respondents and 87 percent of those with a nonworking partner lived in households which were contacted at the first attempt, compared with 81 percent of working respondents and of those with either a working partner or no partner at all.)

Despite this clear and readily understandable pattern, we should reiterate that the main finding is a lack of difference between the samples achieved by the two modes and, hence, our belief that there is nothing unusual about nonrespondents to telephone surveys compared with those to face to face interview surveys.

None of these comparisons, of course, addresses the central question as to whether nonrespondents would have differed in terms of their answers to the survey questions themselves. This has been a matter of crucial concern in all four comparative studies, since they sought to investigate the effects of mode of interview on responses to survey questions. In each case, researchers had to be satisfied that they were not looking at effects confounding mode of interview with differential nonresponse: hence the comparisons described here. Confidence that such confounding did not occur is supported by the general finding of lack of response differences between the two modes. If there were a differential nonresponse effect, it must have been confounded with a mode response effect which balanced it in order to yield a zero effect overall. The more plausible explanation is that neither effect occurred.

Again the lack of differential nonresponse effects does not deny the possibility of such effects common to both modes of interviewing. Indeed, the very fact that response rates are found here and elsewhere to vary so much with the survey topic leads one to believe that widespread nonresponse effects must exist. But this is not a new conclusion nor one peculiar to telephone surveys.

2. REFUSALS TO PARTICIPATE

In the remainder of this chapter we concentrate on one source of nonresponse: refusals to participate. This is not to say that the problem of noncontact can be ignored: indeed the results in the previous section indicate that it is more serious than might be expected. But, as already mentioned, all four comparative studies were conducted in a limiting framework of interviewing largely confined to weekday evenings, with only a small amount of weekend work. Our findings on contact rates are thus not of great interest. They confirm earlier results for face to face surveys: that contact rates are quite high on weekday evenings from about 6 pm onwards (when about 10 percent of calls are unanswered) and rather higher earlier in the week. Calls at other times, while valuable in contacting those not available at the peak time, will show substantially lower productivity (e.g., nearly 40 percent of calls on Saturday afternoon are unanswered) as well as being more expensive in terms of call charges. Clearly more work is needed, especially in the area of conditional probabilities of making contact with those not initially contacted in the evening.

2.1 Reasons for Refusing

Effective solutions to the problem of refusals to participate will depend on knowing how and why they occur. Only then can we judge whether and how they might be overcome and the costs and benefits of specific techniques for doing so.

Only two of our surveys confirmed the findings of others that higher refusal rates contribute substantially to the lower telephone response rates compared with those obtained face to face. But there remains concern that the telephone approach is less acceptable to the general population than is face to face contact, as discussed in our introduction. The reasons given by those who refused to take part in the Alcohol Study, however, do not support such fears. Only 8 percent gave reasons that referred specifically to the method of interview: that the respondent was tired of unsolicited calls (mostly direct selling), objected to telephone surveys, or did not like talking on the telephone. Another 5 percent said that a 25 minute interview was too long. Certainly, interview length is rarely given as the main reason for refusal in face to face surveys (Lievesley, 1988). Apart from these very few references to the telephone, the reasons given for refusing to take part were much the same as those encountered in face to face surveys.

2.2 Immediate Refusals

In all our studies, including those showing little difference between the interviewing modes, refusals over the telephone tended to occur in the initial moments of contact, as shown in Table 3. Across the four telephone surveys, one-third of the refusals occurred even before the interviewer could establish which individual should be interviewed. A further one-fifth were proxy refusals given on behalf of the selected respondent by the person who had answered the telephone. Only the remaining 44 percent of refusals were received directly from the person chosen for interview. This is in marked contrast with the pattern of refusals in face to face contact, where relatively few refusals occurred prior to selection of a respondent or were delivered by a proxy. About 70 percent of refusals were received directly from the selected respondents.

This difference between the two modes confirms reports elsewhere that a large proportion of refusals to telephone surveys occurs early in the interaction between interviewer and respondent (e.g., Oksenberg et al., 1986). The contrast with a face to face survey suggests that direct contact allows the interviewer more opportunity to avoid outright refusal to take part, even if this may serve only to postpone the event.

We should not be too hasty in judging the telephone approach to be inferior because of this tendency to obtain early refusals. It may simply be that hardcore refusers or refusers who would be very difficult to convert

Table 3. Timing of Refusals

	Stage		
	Before Selection	Proxy	Selected Respondent
Telephone	34%	22%	44%
Face to face	12	19	69

give a firm negative answer earlier rather than later. Some support for this hypothesis is found in the analysis of reasons for refusal. Over half of refusals before selection were clear refusals (not interested, don't want involvement, don't like surveys, hung up with no reason given, invasion of privacy, object to telephone surveys). Refusals of this sort were somewhat less common at a later stage (40 percent). The problem may even have been exacerbated by the advance letter to telephone sample members. This gave forewarning of the survey and enabled people to make their decision and prepare their response before they were contacted.

In spite of this comforting reasoning, it is of concern that such a high proportion of refusals preclude the opportunity to carry out the selection procedure, to speak to the sample member in person, and to apply any persuasive techniques. In order to explore how those refusals are made, the reasons given, the opportunities available for interviewers to deal with them, and their success in doing so, we tape recorded a small sample of survey introductions. Of 80 recordings, 67 were audible and 50 of these complete (i.e., they recorded every stage of the introduction, including callbacks, from first contact to final outcome). Only 11 of the recordings were of early refusals — a very small number, allowing only tentative assessments.

Having established that the call had reached the correct address, the interviewer's standard introduction was: "My name is _____ , phoning from the Government's telephone interviewing unit in Hampshire about a survey of attitudes to alcohol. A letter was sent to your address explaining a bit about the survey. Did you see it?"

In a few cases the refusal occurred as soon as the interviewer introduced herself, but mostly it came when the interviewer asked if the letter had been received. The refusals were firm and certain, leaving little opportunity for interviewers to negotiate:

> Yes. Forget it. We're not interested and we don't want to do it.
>
> Yes we did. We don't want to get involved with it, thank you very much.
>
> Yes, we don't want to take part, thank you.
>
> Yes. I haven't the time. I'm looking after someone who is ill. (HUNG UP)

These reactions seem to support the view that immediate refusals come from hard core refusers, who would be very difficult to convert under any circumstance.

3. METHODS FOR REDUCING REFUSALS

The theory of social exchange, adapted by Dillman as a theoretical framework for tackling nonresponse to surveys, asserts that "the actions of individuals are motivated by the return these actions are expected to bring and, in fact, usually do bring from others... It is assumed that people engage in any activity because of the rewards they expect to receive" (Dillman, 1978; p.12). The implications for survey research follow directly: "minimize the costs of responding, maximize the rewards for doing so, and establish trust that those rewards will be delivered." Putting this strategy into effect may not be easy: in what areas and how can we minimize costs and maximize rewards; what might we have to give up in the process?

In this section we consider a range of strategies for reducing refusals to telephone surveys. We are not concerned here with an exhaustive exploration of the variables which might affect refusal rates (many of which are in any case fixed for a given survey, such as the topic or sponsor) but with a review of potential fieldwork tactics.

In approaching the problem it is useful to distinguish between two main kinds of reasons for refusal: those which can be explained in situational terms (arising from circumstances at the time of the request); and those which can be explained in terms of more principled attitudes towards surveys or some aspect of the survey in question — the sponsor, topic, or methodology (see Goyder, 1986). In practice, people often have mixed reasons for refusing, and sometimes a given reason is not the true one. For example, "too busy" is often a polite concealment of a more fundamental objection to participating or, simply, a sign of resistance to taking part which is unformulated in the respondent's mind. Nevertheless, the situational versus principled distinction provides a useful framework for considering strategies for dealing with the problem of refusal if only because it indicates the need for a range of solutions.

3.1 Timing of Calls

Goyder (1986) discusses two studies in which respondents were asked about the importance of a variety of factors in deciding whether or not to grant an interview: interest in the topic of the survey; whether the topic of the survey is socially important; initial impressions of the interviewer; and so on. The highest scoring factor was "What you are doing when the interviewer calls": 74 percent of respondents rated this as either extremely important or very important.

It might be expected that timing of calls would have an impact on such situational refusals: that broad patterns of time use would lead to differences between days of the week or times of day in the incidence of refusals. Information on this is scarce and often conflicting (this volume, Chapter 12). Our own survey of attitudes towards alcohol showed little difference between days of the week or times of day in the incidence of refusals. Of calls where contact was made at a number, calls on Monday evenings met with a slightly higher incidence of refusals (18 percent of calls) than calls on other days (14 percent), but this does not fit in with any obvious pattern of domestic behavior. There was no systematic variation in terms of time of day, but this proves little since our main interviewing shifts did not begin until 4 pm.

3.2 Appointments

Appointment systems represent an approach to situational refusals. When faced with a potential situational refusal (genuine or apparent), continued efforts to persuade the sample members may only increase resistance. Even if such efforts are successful, the chances of obtaining only a partial interview may be increased. To avoid this interviewers can be instructed to "back off" and try to arrange a more convenient time to call back. Such a strategy was used on the Survey of Attitudes towards Alcohol, and although no experimentation was built in to test the efficacy of the system, we can look at the outcomes at addresses where appointments were made and speculate upon its success.

Of the 526 addresses at which appointments were made in the call sequence, 81 percent resulted in productive interviews. This compares with 58 percent of addresses where appointments were not made. Although a difference of this sort is to be expected since appointments will mostly be made only when any initial resistance has been broken down, that is, not with hardcore refusers, it is substantial. We are optimistic that the strategy functioned as expected.

Appointments can be used in a more systematic way with all sample members. The appointments are arranged at the first contact with the individual — who may, of course, agree to an interview there and then. Such a course of action may be desirable when the interviews are long or if the target population is deemed to be particularly busy. For general population surveys the additional cost may outweigh the benefits but to our knowledge this has not been systematically evaluated.

3.3 Advance Letters

It has been argued that the "shock" element of a cold call and and a request for an interview can contribute to refusal rates (Dillman, 1978). An advance letter removes this element of surprise in a telephone survey and also, according to Dillman, "enhances the importance of exchange considerations." It provides an opportunity to stress the rewards of participation (boosting both intrinsic and extrinsic motivation to take part); to make reassurances about confidentiality, and to establish the respondent's trust in the survey organizers. There is evidence that advance letters can improve response rates — particularly on general population surveys (Dillman et al., 1976; Clarke et al., 1987). The latter experimented with advance letters on the British General Household Survey. The response rate from those who were sent such letters was significantly higher (84 percent) than from those who were not (79 percent). The difference between the two groups was largely due to the difference in the proportion of households who refused to cooperate. The advance letter was popular with interviewers, who felt that it helped allay respondents' initial suspicions and was a more professional way of seeking an interview. The main drawback was that some sample members were already armed with a refusal when the interviewer called, creating a difficult situation for interviewers to retrieve.

Experiments were carried out on both our British Social Attitudes telephone experiments, advance letters being sent to a random half of each sample. The letters notified a household of the impending call and when it was likely to come. They gave notice of the need to select one person at random from the household, and hence asked that the letter be shown to every household member. A concise description of the survey was given, avoiding too detailed information which might encourage some people to feel that the survey did not apply to them. The research director's name and telephone number were provided in case the recipient had questions. Finally, respondents were encouraged to set a later appointment with the interviewer if the call occurred at an inconvenient time.

In both experiments the subsample which received a letter yielded a marginally higher response rate than the subsample which did not (56 versus 50 percent and 48 versus 43 percent). Differences between the subsamples in terms of their refusal rates were less marked although, once again, the anticipated pattern emerged: 23 percent of those sent letters on BSA 1 refused compared with 27 percent of those not sent a letter. The equivalent statistics for BSA 2 were 37 and 38 percent respectively.

3.4 Refusal Conversion Attempts

Most survey organizations recognize the potential value of reissuing refusals for another attempt, although identifying which refusals should be tackled can present difficulties. For face to face surveys a main determinant is the interviewer cost. By telephone, this presents less of a problem since travelling costs are negligible, but this still leaves the difficulty of deciding which refusals can be attempted without alienating those involved or causing undue anxiety.

In our alcohol study, interviewers for reissues were selected for their high success rates on the survey and their willingness to take on the task. About half of the initial refusers were rejected on the basis of the comments written down by the interviewers at the time. Examples of refusals eliminated are: addresses from which a written or telephoned refusal was obtained in response to the advance letter; addresses where the selected person or contact was elderly or clearly worried about the call; addresses where strong objections were voiced, accompanied by threat to complain to a Member of Parliament, the Press, or OPCS; addresses where the person reached was clearly a hard core refuser (e.g., he expressed dislike of surveys). Despite this selectivity, only 13 percent of reissued addresses resulted in interviews, boosting the total final response rate by 2 points. This is well below the levels encountered by other researchers, who have reported 30 to 40 percent conversion rates (e.g., Alexander et al., 1986).

Interviewers involved in this refusal conversion effort were questioned about the experience. According to them, the easiest nonresponses to convert were those where the selected person had not been reached before; elderly people who needed a slower, simpler explanation; and those who had said originally that they were too busy or tired.

3.5 The Effect of the Interviewer

Differences exist between interviewers in terms of the refusal rates they obtain. Table 4 shows refusal rates for the 16 interviewers who worked on the Survey of Attitudes Towards Alcohol ranging from 28 percent down to 10 percent. While this range is "stretched" by the impact of sampling variability, the pattern is not consistent with a hypothesis of no differences between interviewers.

Such variation suggests that a reduction in overall refusal rates might be achieved in three ways: through selection; training and experience; and "weeding." "Weeding" refers to the dismissal of interviewers with consistently high refusal rates and depends on routine monitoring and

Table 4. Refusal Rates By Interviewer

Interviewer	Percent Refusals
A	28%
B	27
C	27
D	27
E	25
F	24
G	24
H	22
I	22
J	21
K	19
L	18
M	16
N	15
0	14
P	10

evaluation of performance over several surveys. Eastlack and Assael (1966) reduced refusal rates on their market survey of 25 to 35 percent in the first three months to 12 percent over the remaining seven months using this strategy. It is, however, extravagant compared with better initial selection and training.

In terms of selection, the main concern has been to identify interviewer characteristics which explain variations in refusal rates. For example, it has been suggested that men are likely to obtain higher refusal rates than women because calls from strange men are more threatening and likely to arouse anxiety. There is, however, no generalized evidence to support this suggestion (e.g., Groves and Fultz, 1985). We were unable to test it in our own survey since all the interviewers were female. In addition no correlation was found between refusal rates and either the interviewer's age or her level of education.

The importance of auditory stimuli in forming initial impressions of the interviewer in a telephone survey has led to research into the impact on refusal rates of interviewers' voice characteristics (see this volume, Chapter 16). Their results may have implications not only for the selection of interviewers (most survey organizations assess potential interviewers' telephone voices but there are few systematic guidelines to help in this), but also their training. Other aspects of interviewer training have received more attention. For example, efforts have been made to identify successful introductions — testing the length, structure, and content of this early part of the interaction (Dillman et al., 1976; O'Neil et al., 1979). Such studies have failed to come up with any generally applicable findings. More promising, perhaps, is the work of Morton-Williams and Young (1986), concentrating on the style of interaction. They find, for example, that successful face to face interviewers tend to be flexible and responsive to the reluctance expressed by respondents, addressing the problem directly. Less successful interviewers tend to be less adaptive.

Our recordings allow us to add our own impressionistic findings. Interviewers with lower refusal rates seemed confident and professional; they show an easy command of responses to queries and reluctance and, especially, an ability to avoid dead end situations. The less successful interviewers performed as well as the more successful when things went as expected. They did not appear to create resistance. But when it arose, they showed a lack of confidence and a tendency to panic; they seemed unprepared for problems, gave in too easily, and failed to avoid "dead-ends."

It seems that interviewers can develop persuasive skills through practice and experience. In spite of the difficulties involved in separating cause and effect (the worse interviewers may be dropped or leave), there is evidence that more experienced interviewers obtain lower refusal rates (e.g., Groves and Fultz, 1985). None of our interviewers had been employed as OPCS telephone interviewers for longer than 27 months at the time of the survey. But the eight who had been employed for a year or more included five of the six with the lowest refusal rates. Similarly, most of the interviewers had little experience of cold contact surveys. But of the four who had worked on more than one such survey, three were among the six interviewers with the lowest refusal rates.

3.6 Interview Length

Finally, we can look at the impact of interview length. It is a commonly held view in the UK that long interviews are not viable over the telephone: long telephone calls are thought to be unconventional and liable to arouse resistance, and it is thought to be uncomfortable to communicate by telephone for any length of time.

An experiment with interview length on the first of our studies resulted in refusal rates of 14 percent for a 40 minute questionnaire and 9 percent for a 20 minute version. (Prospective respondents were told of the interview length before being asked to continue.) Although this result supports the view that long interviews are less likely to be accepted, the difference is less than some might expect. And it is not clear that the problem arises entirely from the respondent: interviewers may well share the expectations of reluctance and be tentative in their approach.

APPENDIX. TECHNICAL DETAILS OF THE EXPERIMENTAL STUDIES

1. The First British Social Attitudes Study (BSA 1)

The annual British Social Attitudes Survey is carried out face to face with a sample of the population of Great Britain aged 18 and over, living in private households. The survey uses a multistage design in which (i) areas are selected with probability proportional to electorate; (ii) addresses are selected with probability proportional to the number of electors listed; and (iii) one eligible individual is selected at random.

For the telephone interview experiment, 798 addresses were drawn in the same way, covering the same 114 areas. These addresses (and the names of all residents) were submitted to British Telecom for a directory search, yielding listed numbers for 515 addresses (65 percent). Of these numbers, 429 were issued to interviewers during the survey period, the remaining 86 (a random subset) being withheld. At 53 percent of the numbers attempted, an interview was completed successfully with an adult, randomly selected as in the main survey. The interview was scheduled to last either 20 or 40 minutes (in a systematic split half experiment) with questions drawn in sequence from the hour-long face to face interview.

To provide a comparison, a random subset of 570 addresses from the main survey sample was submitted to the same directory search procedure. The results used here are those achieved by a face to face approach to the 313 (55 percent) addresses for which listed numbers were found. A face to face interview had been completed successfully at 60 percent of these addresses.

2. The Second British Social Attitudes Study (BSA 2)

For this study a sample of 3,600 addresses was selected from the Electoral Register and submitted to the British Telecom search, yielding listed numbers for 2,207 addresses (61 percent). These were systematically assigned to three subsamples: 631 to be approached face to face; 787 by telephone using "paper and pencil" interviewing; and 789 by CATI. (The face to face sample was smaller in anticipation of a higher response rate.)

The CATI experiment was a total failure, telling us more about the problems of inadequately planned adoption than about the method itself. The results discussed here are restricted to the subsamples approached face to face or by "pencil and paper" telephone interviewing. In the face to face subsample all 631 addresses were attempted, yielding 68 percent completed interviews. In the telephone subsample 730 numbers were attempted, with a 46 percent response rate. In each case one individual was randomly selected for interview using a standard 25 minute questionnaire.

3. Lifestyle in the 1980s

This study, carried out by Marplan on behalf of the Market Research Development Fund, used a similar methodology. An initial sample of 2,500 addresses was systematically assigned to either face to face or telephone interviewing subsamples. For the latter, numbers were sought in published directories, 764 being traced (61 percent). Following early indications of a low response rate by telephone a supplementary sample of 1,250 addresses was drawn; again numbers were sought in published directories, 721 being traced (58 percent). Addresses for which numbers were not traced were issued to face to face interviewers in order to establish whether a telephone existed and, if so, to obtain the number. This yielded a further 322 numbers.

In comparing response rates we are limited to the published comparison between a telephone approach to all 1817 telephone households (including both listed and unlisted numbers) and a face to face approach to the original 1250 addresses (including nontelephone households). In assessing interview mode effects (this volume, Chapter 19) we are more fortunate, being able to compare telephone and face to face approaches to subsamples of addresses with listed telephone numbers. All interviews used a standard questionnaire (apart from some split ballot experiments).

4. Attitudes towards Alcohol

For the telephone survey a sample of 2700 addresses was selected from the Electoral Register and submitted to the British Telecom search, yielding numbers for 1645 addresses (61 percent). All numbers were attempted, interviews being completed with a randomly selected adult at 65 percent of them.

Two questionnaire versions were used, the sample being systematically assigned between them. In Version 1 questions were identical to those asked in a prior face to face survey; in Version 2 a number of question form variations were introduced. In examining nonresponse we do not distinguish between the two versions.

The results obtained in this telephone survey are compared with those obtained in a face to face survey carried out about a year earlier. This indirect comparison brings some limitations, which have to be accepted on budgetary grounds, as noted in the text. In looking at nonresponse they involve also the problem that information on telephone ownership is not available for nonrespondents to the face to face survey, so that we are not comparing approaches to the same target population. In both areas there are some added concerns about possible variations between survey organizations: the telephone survey was carried out by the Government's Social Survey Division, the face to face survey by an independent research institute.

This research was carried out within the SCPR Survey Methods Center, funded as a Designated Research Center by the Economic and Social Research Council.

CHAPTER 14

NONRESPONSE ISSUES IN GOVERNMENT TELEPHONE SURVEYS

J. Douglas Drew, G. Hussain Choudhry,
and Lecily A. Hunter
Statistics Canada

Major demographic surveys conducted by government statistical agencies are typically large scale recurring surveys designed to yield precise estimates for national and subnational levels of aggregation. They often follow panel designs in which respondents are interviewed on several occasions, and in many cases the surveys were established before the advent of telephone survey methods. For these long established surveys, historical comparability of the data and preservation of the time series are very important.

The combined effect of these constraints is that the adoption of telephone survey methods has proceeded in a slower, more evolutionary fashion among government statistical agencies than has been the case in other survey taking sectors.

As Trewin and Lee (this volume, Chapter 2) point out, among government statistical agencies telephone methods are currently pretty much restricted to warm telephone interviewing, where initial interviews are face to face but reinterviews (that is, interviews on subsequent occasions) are by telephone.

Exceptions are Sweden and Denmark where current methods are based on cold telephone interviewing (that is, the conduct of initial interviews by telephone) for samples selected from other frames. Similar approaches are currently being investigated in Great Britain (Collins and Sykes, 1987) and in Canada as described later in this chapter. In these instances, the telephone interview is generally preceded by an introductory letter.

The use of the telephone additionally for sampling purposes is not as common among statistical agencies, although it has been the subject of research for some of the major demographic surveys in Canada and the

United States. In Canada, some of the smaller demographic surveys conducted by Statistics Canada have been based on telephone samples.

In Section 1 we highlight the research and developments at statistical agencies in the United States and Canada in the use of telephone survey methods for demographic surveys, and we examine how considerations relating to nonresponse have had an important bearing on the directions of research currently being pursued.

Sections 2 and 3 focus on nonresponse and response issues for two studies of telephone survey methods conducted by Statistics Canada. The first study involves local interviewers doing a mixture of warm and cold telephone interviewing for a sample drawn from a nontelephone frame, while the second study involves cold telephone interviewing from regional office sites for samples drawn from telephone frames.

1. DEVELOPMENTS AT AGENCIES IN CANADA AND THE UNITED STATES

The Current Population Survey conducted by the United States Bureau of the Census and the Canadian Labour Force Survey conducted by Statistics Canada were both among the first large scale recurring probability sample surveys in the world. Initially both surveys were based on area frames and followed rotating panel designs with face to face interviews of households on repeated occasions.

The Current Population Survey follows a rotation pattern in which households remain in the sample for two spells of four consecutive months, with eight months between the conclusion of the first spell and the start of the second. A warm telephone interviewing procedure was introduced in 1954 in which face to face interviews were carried out in the first, second, and fifth months that households were in the sample, and interviewing was by telephone where possible in other months. Studies prior to the introduction of warm telephone interviewing indicated there did not appear to be material differences in the labour force estimates, as compared with face to face interviewing. Since then, the use of telephoning has been extended to second-month cases (U.S. Bureau of the Census, 1978).

A similar warm telephone interviewing procedure was introduced into the Canadian Labour Force Survey (LFS), in the early 1970s in urban areas, in the early 1980s for the rural areas. Under the LFS design, households remain in the sample for six consecutive months, and the procedure consisted of telephone interviewing for households in their second through sixth months in the sample. Testing and analysis indicated that the telephoning had a favourable impact on cost and nonresponse rates, without any detectable impact on the estimates of labour force characteristics. (Muirhead et al., 1975, Choudhry, 1984)

Testing of warm telephone interviewing procedures was also carried out in the United States for the National Crime Survey (Bushery et al., 1978), with similar findings.

During the early 1980s, the U.S. Bureau of the Census carried out research into dual frame methods combining samples from telephone and area frames. This research included testing of random digit dialing (RDD) sampling methods and cold telephone interviewing. Two surveys examined were the Current Population Survey and the National Health Interview Survey, the latter in joint research with the National Center for Health Statistics.

Findings from both studies were that nonresponse rates were significantly higher for cold telephone interviewing than under either the warm telephoning used for the Current Population Survey or the face to face interviewing used in the National Health Interview Survey. For the Current Population Survey, the overall nonresponse rate for the telephone sample was 16 percent compared with nonresponse rates in the 5-6 percent range for the ongoing survey (Mulry-Liggan, 1983).

These levels of nonresponse were viewed as problematic. As Groves and Lyberg (this volume, Chapter 12) show, the magnitude of nonresponse bias is a function both of the difference in characteristics of respondents and nonrespondents, and of the nonresponse rate itself. Hence higher nonresponse rates carry with them the risk of higher nonresponse bias. Biemer (1983), for the case of national estimates of unemployment from the Current Population Survey, showed that for a dual frame mixed mode design, under the assumption of no bias in the area sample, the optimal proportion of sample to allocate to the telephone frame would drop from 23 to 3 percent as the bias of the telephone sample (that is, the nonresponse bias plus response bias) increased from 0 to 5 percent.

Taking the National Crime Survey, Lepkowski and Groves (1986) showed that under the same bias assumptions as considered by Biemer, the optimal allocation of sample to the telephone frame would drop from 85 to 14 percent for estimation of total crimes. However, again for the same bias assumptions, they also showed that the optimal allocation of sample to the telephone frame would remain virtually constant at 82 to 83 percent for estimation of the total number of robberies, demonstrating how dramatically the situation can differ not only across surveys, but from statistic to statistic within the same survey.

As a consequence of the increased risk of nonresponse bias under cold telephone methods, and the implications such bias would have on mean squared error of survey estimates, the Bureau of the Census is channeling its telephone survey research efforts into investigation of warm telephone procedures. The approaches being investigated involve initial face to face interviewing by local interviewers and reinterviews in later months conducted from central sites using CATI (Marquis and Blass, 1985).

In Canada, although nonresponse rates for cold telephone interviewing have been found to be higher than under warm telephone interviewing, the gap is much smaller than that experienced in the United States, and research into cold telephone methods is continuing, as described in the next two sections.

2. COLD TELEPHONE INTERVIEWING FOR AN AREA FRAME SAMPLE

2.1 Description of the Test

The "telephone first interview" test has been embedded into the ongoing Labour Force Survey since October 1985 in urban areas of Ontario and Quebec.

Under the area frame design followed by the LFS, the primary sampling unit (cluster) in urban areas is a city block. Within selected clusters, up-to-date lists of all dwellings are maintained. Every six months, a new sample of dwellings is selected within each block, staggered across blocks so that one-sixth of the clusters have newly selected dwellings each month.

A billing list of published telephone numbers is purchased from the telephone company. The civic addresses of newly selected dwellings are matched with the telephone billing addresses to obtain telephone numbers. From among a sample of clusters with two or more successfully matched addresses, one dwelling per cluster is randomly selected for the test sample, and another is selected for the control sample.

Newly selected dwellings in the test group have their telephone number preprinted on their questionnaire to permit a first month interview by telephone. Those in the control group do not have their telephone numbers preprinted, and so are treated in the usual manner with face to face first month interviews. In conformance with standard practices for the LFS, both test and control dwellings receive an introductory letter one week prior to interview. Most of the LFS interviewers in urban areas of the two provinces are employed in the test, with an average of two out of the 10–12 dwellings in a sample cluster for the first time receiving the test treatment. For analysis, comparisons have been made between the paired test and control samples.

For the test sample, face to face first month interviews are permitted where operationally expedient. For example, in cases where no contact has been made in telephone attempts during the first two days of the week in which interviewing takes place, a face to face interview is attempted on the third day, when the interviewer is normally in the neighbourhood doing other face to face interviews. This approach has lead to telephone first interviews for 75 percent of the test sample.

The mixed methodology permitting cold or warm telephone interviewing for the test sample was adopted since it was felt to correspond most closely to what could eventually be implemented under the area frame. A second test found a higher instance of telephoning where entire assignments were converted to the experimental procedure, with telephoning in 85-90 percent of households for which telephone numbers could be obtained. Hence, given the 70 percent success rate in matching to obtain telephone numbers, such a mixed methodology could result in approximately 60 percent of first month interviews being done by telephone in urban areas, with the remainder being face to face. The proportion of telephoning under such an approach could be increased, for example, if the area frame were to be replaced by a list frame either containing published telephone numbers, or at least constructed so as to facilitate matching to telephone company lists. The feasibility of such a list frame has been studied at Statistics Canada by Drew et al. (1988).

It is worth noting that the test design did not permit a comparison of cold versus warm telephone interviewing since, for a portion of the test sample, particularly the no contact cases, both telephone and face to face interviews were attempted.

2.2 Results on Nonresponse and Response

The nonresponse rates for the test and control samples, by month in the sample, appear in Table 1 along with t-statistics for the differences. No significant differences were detected in nonresponse rates. This is not surprising since the test procedure defaulted to a mixture of face to face interviewing and further telephoning when initial telephone attempts were unanswered.

Table 2 presents the estimate for the test sample as a percent of that for the control sample, for several characteristics of interest. Also, t-statistics for the difference between the test and control samples are given, with flagging of significant differences at the 5 percent level.

Table 2 reveals only two significant differences between the test and control treatments for Ontario: fewer unemployed males 25 years of age or

Table 1. Telephone First Interviews Test: Nonresponse Rates by Month in Sample (October 1985 – February 1987)

Month in Sample	Quebec			Ontario		
	Test	Control	t-Statistic	Test	Control	t-Statistic
1	4.6	5.2	−0.87	4.7	4.6	0.23
2	3.4	3.2	0.35	3.3	3.2	0.14
3	2.9	3.2	−0.64	3.4	2.8	1.43
4	2.8	3.5	−1.33	2.8	2.8	0.07
5	3.7	3.1	1.05	3.2	3.2	−0.07
6	3.2	3.2	0.00	2.9	2.3	1.48
All months	3.4	3.6	−0.40	3.4	3.2	0.79

Table 2. Telephone First Interview Test: Estimate for Test Treatment as a Percent of Estimate for Control Treatment

Characteristic	Ontario		Quebec	
	Percent	t-Statistic	Percent	t-Statistic
Employed	98.9	−0.62	95.0	−2.42*
Unemployed	93.5	−1.23	117.8	2.57*
Persons 15+	99.5	−1.03	97.5	−2.00*
No. of 1 person hhlds	105.2	1.51	100.7	0.03
No. of 2 person hhlds	97.3	−1.07	106.7	2.27*
No. of 3+ person hhlds	99.6	−0.18	88.2	−2.57*
Employed in 1 pers. hhlds	118.1	2.83*	111.6	1.27
Employed in 2 pers. hhlds	97.2	−1.00	101.6	0.33
Employed in 3 pers. hhlds	97.5	−0.59	85.0	−2.80*
Unemployed in 1 pers. hhlds	82.3	−1.06	119.2	0.99
Unemployed in 2 pers. hhlds	91.0	−0.92	135.1	3.03*
Unemployed in 3+ pers. hhlds	97.8	−0.41	102.6	0.28
Employed male 15–24	98.9	−0.02	75.6	−3.60*
Employed male 25+	97.6	−1.28	99.8	−0.02
Employed female 15–24	99.7	−0.07	103.8	0.37
Employed female 25+	100.5	0.01	92.6	−2.12*
Unemployed male 15–24	104.7	0.04	113.3	0.86
Unemployed male 25+	80.8	−1.99*	103.3	0.57
Unemployed female 15–24	110.0	0.60	133.0	1.64
Unemployed female 25+	92.1	−0.81	132.7	2.47*
Population male 15–24	100.5	0.16	85.8	−2.31*
Population male 25+	97.8	−1.41	99.2	−0.34
Population female 15–24	102.9	0.66	104.7	0.58
Population female 25+	98.3	−1.41	97.4	−1.83

*$p < .05$

older for the test treatment, and more persons employed in one person households for the test treatment.

For Quebec, however, several significant differences between the test and control samples were detected. Estimates that were significantly lower for the test treatment included: persons 15 years of age or older, males 15–24 years of age, total persons employed, employed males 15–24 years of age, employed persons in three or more person households, employed females 25 years of age or older, and the number of three or more person households. Estimates that were higher for the test sample included: total persons unemployed, unemployed females 25 years of age or older, the number of persons unemployed in two person households, and the number of two person households.

These results for Quebec seem to point to the underenumeration of persons in the telephone treatment since the alternative of overreporting of persons with face to face interviewing seems unlikely. In particular the underreporting seems to be concentrated among employed males 15–24 years of age in larger households. Some of these households are then turning up as 2 persons households with higher instances of unemployment among the reported persons than is the case for actual 2 person households. At the same time, unemployment is being overreported relative to the control treatment, particularly for females.

We are unsure what is leading to the underreporting of employed young males in larger households in Quebec. A relatively benign possibility is that such persons have looser ties to the household and are simply more likely to be forgotten when rostering is done over the telephone. If this were the case, then the problem may be resolvable by placing more emphasis during the rostering on ensuring that all household members are included. Another possibility is that of intentional underreporting. If this were the case, then better assurances of the confidentiality and purely statistical uses of the data may help.

3. TELEPHONE SAMPLING TEST

3.1 Details of Test Design

The telephone sampling test was also conducted for the provinces of Ontario and Quebec, but unlike the telephone first interview test, the telephone sampling test used a separate sample from that for the regular Labour Force Survey. Two sampling approaches were investigated: (i) a list sampling approach, and (ii) an RDD approach. A list containing the addresses and telephone numbers of the published residential telephone subscribers was purchased from the telephone company on a quarterly basis. Counts of nonpublished telephone numbers by hundreds banks were also obtained. (A hundreds bank is constituted by the area code and the first 5 digits of the telephone number within which there are potentially one hundred numbers represented by the last 2 digits.)

In order to stratify the telephone sample, maps of telephone exchanges were obtained and their boundaries were geocoded in order to link them to subprovincial areas of interest to the Labour Force Survey. The list of published residential numbers was then sorted by subprovincial area and telephone exchange, and a systematic sample of telephone numbers was drawn. Sample cases were sent a prior introductory letter about the LFS. Since the nonpublished telephone numbers and the business telephones reaching households were not included in the list sampling, these were covered by an RDD sampling in which the hundreds banks were selected by probability proportional to size sampling using counts of nonpublished telephone numbers as size measures, and in which screening for households was restricted to numbers not appearing on the residential published list. An offer to mail the introductory letter about the Labour Force Survey was made at the time of screening, which was done during the week prior to interviewing.

In the RDD sampling approach, the banks were stratified into two strata: sparse and nonsparse, based on the counts of residential numbers (both published and nonpublished) in the bank. The cutoff for defining the nonsparse banks was 30 working residential numbers in the bank. The allocation of the sample between the two strata was proportional to the counts of residential numbers. The sample within the strata was selected in two stages: (i) a probability proportional to size selection of banks where the size measure was the count of residential numbers in the bank, and (ii) screening to select one residential telephone number within each sparse bank and two per nonsparse bank. No introductory letter was sent. Under the RDD sampling approach, hit rates (i.e., the percentage of numbers called reaching households) of 71.3 percent in Quebec and 64.0 percent in Ontario were obtained. These hit rates should be analogs of secondary hit rates under a two stage Waksberg (1978) design. This compared with secondary hit rates of 68 percent in Quebec and 54 percent in Ontario for an earlier survey — the Health Promotion Survey — conducted by Statistics Canada using a Waksberg design, and overall hit rates under the Waksberg design of 54 percent for Quebec and 47 percent for Ontario (Health and Welfare Canada, 1985).

3.2 Results on Nonresponse and Response

The nonresponse rates and refusal rates for the RDD and list sampling treatments and for the telephone households in Ontario and Quebec for the Labour Force Survey are presented in Tables 3 and 4, along with t-statistics for the differences between these three sets of rates. The tables demonstrate clearly that both nonresponse and refusal rates are significantly higher for the two test treatments than those for the ongoing Labour Force Survey. While for most months there were no significant differences at the 5 percent level in nonresponse rates between the list and RDD treatments, for the combined sample over all months the nonresponse rate of 6.9 percent for the list treatment was significantly different from the 8.5 percent for the RDD treatment. For the refusal component of nonresponse, with the exception of the first two test months where the RDD refusal rates were higher, differences between the test treatments were not significant at the 5 percent level. The introductory letter for the list sample cases would appear to offer at least a partial explanation for the observed differences in nonresponse rates.

The nonresponse rates were also monitored for the RDD procedure during screening and were found to be higher than those during interviewing (12.1 versus 8.5 percent). The nonresponse rates were also obtained for the systematic and the RDD components of the list sampling method (Table 5). We observe that the nonresponse rates are higher for the RDD component. The major portion of the RDD component of the list treatment is comprised of nonpublished telephone numbers. Therefore, we can conclude that the nonresponse rates are higher for the households with nonpublished telephone numbers. In fact the higher nonresponse rates are directly attributable to the higher refusal rate for the RDD component. For example, in Ontario the nonresponse rate was 11.1 percent for the RDD, versus 2.3 percent for the systematic component. This would seem to support the case for approaches involving warm telephone interviewing of households with nonpublished numbers, which, for example, in Ontario and Quebec account for 9 percent of households.

The labour force characteristics of responding households with nonpublished telephone numbers were also examined. Due to small sample size for nonpublished numbers in the telephone test, the sample of those with nonpublished numbers from the regular Labour Force Survey was used. The Labour Force Survey sample records with telephone numbers from the November 1986 survey were matched with the October 1986 and January 1987 telephone company lists of published telephone numbers. If the telephone number could be matched to either of the telephone company lists, the telephone was classified as published,

Table 3. Telephone Sampling Test: Nonresponse Rates and *t*-Statistics by Survey Month and Sample Design

	Nonresponse Rates			*t*-Statistics for Difference		
Survey Month	RDD	List	LFS	RDD-List	RDD-LFS	List-LFS
Dec 1985	10.0	6.4	4.1	2.67	4.98	3.21
Jan 1986	9.4	5.8	4.0	2.93	5.25	2.49
Feb 1986	10.2	8.6	4.5	1.13	5.28	4.81
Mar 1986	10.3	8.5	4.4	1.34	5.46	4.75
Apr 1986	9.0	6.9	4.2	1.71	5.10	3.15
May 1986	6.4	5.2	4.2	1.13	2.81	1.39
Jun 1986	7.2	6.6	3.9	0.53	3.95	3.28
Jul 1986	10.0	8.5	5.5	1.12	4.54	3.17
Aug 1986	7.9	6.5	3.4	1.17	5.04	3.73
Sep 1986	5.5	5.6	3.2	−0.10	2.82	2.90
Overall	8.5	6.9	4.1	2.49	8.54	6.53

Table 4. Telephone Sampling Test: Refusal Rates and *t*-Statistics by Survey Month and Sample Design

	Refusal Rates			*t*-Statistics for Difference		
Survey Month	RDD	List	LFS	RDD-List	RDD-LFS	List-LFS
Dec 1985	4.5	2.1	1.2	2.61	3.95	2.05
Jan 1986	4.8	2.6	1.3	2.54	4.69	2.66
Feb 1986	4.7	3.3	1.2	1.57	4.66	3.79
Mar 1986	4.6	3.1	1.4	1.61	4.31	3.25
Apr 1986	4.9	3.6	1.5	1.48	4.79	3.28
May 1986	2.4	2.6	1.4	−0.31	1.97	2.27
Jun 1986	2.8	2.4	1.1	0.55	3.16	2.53
Jul 1986	2.1	1.9	0.8	0.35	2.72	2.31
Aug 1986	1.9	2.0	0.8	−0.08	2.53	2.58
Sep 1986	1.7	2.0	0.8	−0.44	1.95	2.37
Overall	3.4	2.6	1.2	1.76	6.13	4.71

otherwise it was classified as nonpublished. The average household sizes (in terms of persons 15 years of age or over) and the unemployment rates for the households with published versus nonpublished telephones are given in Table 6, and as can be seen they are very similar.

Table 5. Telephone Sampling Test: Overall Nonresponse Rates and Refusal Rates for RDD and Systematic Components of List Sampling Approach

	Nonresponse Rates		Refusal Rates	
Province	RDD Component	Systematic Component	RDD Component	Systematic Component
Quebec	12.8	7.5	7.0	2.4
Ontario	14.2	5.9	11.1	2.3

Table 6. Telephone Sampling Test: Average Household Size and Unemployment Rates for the Households with Published and Nonpublished Telephone Numbers by Province (1986 LFS Sample)

	Published		Nonpublished	
Province	Average Size	Unemployment Rate	Average Size	Unemployment Rate
Quebec	2.13	9.5	2.10	9.9
Ontario	2.21	5.6	2.22	6.0

Table 7 gives the average number of calls over all months in the sample by final response status, and it shows little difference between the list and RDD treatments. For respondents the average number of calls required was just over two in both cases. Refusals were only marginally more difficult to contact than respondents since the average of just over three calls included a followup attempt by a senior interviewer in the roughly 50 percent of cases where the initial refusal was judged not to be firm. For other nonresponse, an average of 12 calls were made. Figure 14.1, which shows the response rates for the list and RDD treatments as a function of the number of calls, further illustrates the importance of multiple attempts in attaining high response rates. Response rates of over 50 percent were obtained with just one call, and these increased to 86.9 percent for the list treatment after up to 5 calls. However, it took up to 9 calls to achieve a list response rate of 91 percent, and the final list sampling response rate of 93 percent was attained only after up to 17 calls. It is of interest to observe how steady the gap between RDD and list treatment response rates was.

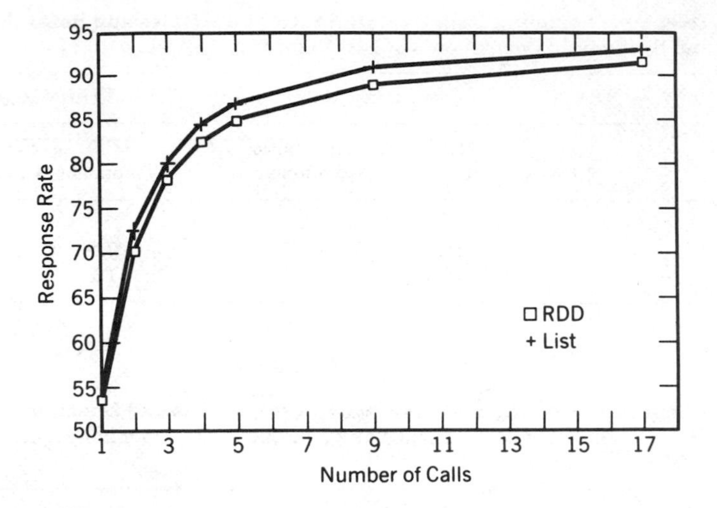

Figure 14.1 Telephone sampling test. Effect of number of calls on response rate.

Table 7. Telephone Sampling Test: Average Number of Calls per Household by Final Status

	Average	
Final Status	RDD Sampling	List Sampling
All Households	2.7	2.6
Respondents	2.2	2.1
Nonrespondents	8.7	8.6
Refusal	3.0	3.3
Other Nonresponse	12.3	11.8

3.3 Frame Changes and Implications for Panel Surveys

Our test design in which interviews were attempted for sampled numbers over a six month period permitted us to observe the rate at which the telephone frame became out of date. As part of the survey operations, unanswered numbers were checked with the telephone company to determine their status as working or nonworking. For the list sample, at the time of the first month's interview, when the list was on average three months old, 5.2 percent of numbers were no longer working, while the

RDD sample, 2.1 percent of numbers identified as working residential numbers at the time of screening were similarly found to be nonworking two weeks later at the time of the first interview, with an average over all panels of 3.9 percent.

If it is assumed that the size of the telephone universe is relatively stable over short durations of time, then the proportion of no longer working numbers in a sample of working numbers selected at an earlier point in time should be the same as the proportion of new working numbers not represented in the sample. In order to study the characteristics of those with new telephone numbers, respondents to the January 1987 Labour Force Survey were classified as having new telephone numbers if their number appeared on the January 1987 telephone company list, but not the October 1986 list. Sizable differences in labour force characteristics were found; for example, for Quebec unemployment rates of 20.2 percent for new numbers versus 10.9 percent for old numbers, and for Ontario corresponding rates of 9.6 and 6.6 percent. These findings illustrate the potential importance of frame changes — potential in the sense that it may depend on the subject matter — as an issue in the application of telephone methods to panel surveys.

4. Discussion

Earlier we reviewed how current research into telephone survey methods by government statistical agencies in the United States is being directed towards use of warm telephone methods, due in large part to higher nonresponse for cold telephone methods (16 versus 6 percent).

In Canada, nonresponse rates for methods involving a mixture of cold and warm telephone interviewing are comparable to those for warm telephone methods, and research is now focusing on response comparisons of the mixed versus warm methods. In one of two provinces studied there were no differences in survey estimates. However, in the second province differences were found, and further work is required to see if the cause of the difference can be identified and eliminated. Also in Canada, while nonresponse rates to cold telephone methods are higher than those for warm telephone interviewing (7–8 versus 4 percent), the gap is smaller than was the case in the United States. While nonresponse rates are not viewed as problematic for the cold telephone methods, more research is needed on methods for dealing with rapid deterioration of telephone sampling frames, and on response comparisons with warm telephone methods.

It is unclear why the American and Canadian experiences on nonresponse rates to cold telephone methods differ, though one can speculate that the differences may not be as great as they initially seem. While Mulry-Liggan (1983) reported an overall nonresponse rate of 16 percent for cold telephone methods over seven months of testing, the rate for the last three months was only 11.7 percent, which is strongly suggestive of a learning effect. If one accepts a learning effect hypothesis, then the remaining gap is not that great and might be attributable to procedural differences. For example, the Canadian methodology made use of introductory letters for telephone list sample cases, and there is evidence in the literature that introductory letters lead to improved response rates (Dillman et al., 1976). Also in the Canadian methodology, RDD sample cases were screened in advance of the survey at which time an introductory letter was offered, which, according to the "foot-in-the-door" theory advanced by Reingen and Kerman (1977), could have a positive impact on nonresponse rates although evidence on this is mixed (Groves and Magilavy, 1981). The U.S. methodology differed in that it was based on RDD sampling, with the first interview proceeding as soon as households were screened in, along with an offer to mail out material describing the survey. Also as we have seen, nonresponse rates for telephone surveys are quite sensitive to operational procedures, so that even minor variations in the number or scheduling of calls or in other factors could give rise to remaining differences. The differences in nonresponse could also be due to different attitudes of the two populations in general or certain subgroups of the populations towards telephone interviewing.

CHAPTER 15

SURVEY PERIOD LENGTH, UNANSWERED NUMBERS, AND NONRESPONSE IN TELEPHONE SURVEYS

Janice Sebold
U. S. Bureau of the Census

Telephone surveys exhibit two distinctive administrative problems, each arising from the absence of visual and physical contact with sample units. Both of these are generally absent in face to face surveys and hence have caused difficulties for researchers changing modes of data collection in an ongoing survey.

The first problem, addressed both by Weeks (this volume, Chapter 25) and Groves and Lyberg (this volume, Chapter 12), concerns the relative inefficiency of repeated calls on sample telephone numbers. In face to face surveys, interviewers at the time of the first call on a sample address often obtain valuable information which can guide the timing of future calls. In telephone surveys, in contrast, although the cost of repeatedly dialing a number is much lower, the amount of information obtained about unanswered numbers is minimal. The relative paucity of information leads to the need to make repeated callbacks on units. This problem is exacerbated in two stage random digit dialing (RDD) sample designs (like that of Mitofsky-Waksberg). In these designs, when a sample number is determined to be nonresidential (perhaps after several days of calling) another sample number is selected to replace it in the sample. The replacement numbers thus receive their first call in the middle of the survey period. Hence, other things being equal, more calling on more days may be required to complete a typical sample. Short survey periods may risk lower contact rates.

The second problem, most common in RDD sample designs in the United States, is that arising from unanswered numbers. Unanswered numbers are not completely analogous to unanswered rings of a door bell or unanswered knocks on a door. As Groves and Lyberg (this volume,

Chapter 12) note, some nonworking telephone numbers in the United States, will ring when dialed, forever if the caller has the patience to hear it. During the telephone survey administration, therefore, some unanswered numbers should be discarded (because they are nonworking) and others should be pursued (because they lie in units with no one answering at that moment). Although some organizations attempt to gain information on the status of such numbers from local telephone business offices, in general, this is not possible. Hence, there is some indeterminacy in the eligibility of sample cases. This produces difficulty in setting callback rules, estimating response rates, and designing postsurvey adjustment procedures for nonresponse.

This chapter reports the results of an investigation into the effect of lengthening a survey period on noncontact rates in an RDD survey and into the characteristics of sample numbers that repeatedly rang without answer during that survey. The studies were part of methodological research on telephone interviewing in the U.S. National Crime Survey, conducted by the U.S. Census Bureau.

1. DESIGN OF THE NATIONAL CRIME SURVEY EXPERIMENT

At present the U.S. National Crime Survey (NCS) uses a two-stage stratified cluster sample of addresses (from address lists, maps, and building permits) to select households for interview. Rotating panels of approximately 60,000 households (10,000 per month), are interviewed seven times at six month intervals. All persons age 12 years and over are interviewed individually and asked about crimes that may have occurred to them in the previous six months. Only the first interviewed respondent is asked questions about "household crimes," such as burglary. All respondents are asked about "personal crimes," such as assault. Under the current design the first and fifth interviews are conducted in person. Second, third, fourth, sixth, and seventh interviews are conducted by telephone. Most calls are made from the interviewers' homes.

RDD has been proposed as a potentially efficient method for obtaining part of the NCS sample. A dual frame design, with an address sample component, would be maintained to account for households which could not be reached by telephone. Between April 1985 and June 1986 the Census Bureau conducted an RDD experiment using the NCS. Sample telephone numbers were selected using the Mitofsky-Waksberg two stage method (Waksberg, 1978). A cluster of four units was used at the secondary stage of selection. Primary screening for each month's sample was done during the last two weeks of the previous month. To correspond to the face to face interview survey period, a two week interviewing period was specified for the RDD test, with interviewing done in a centralized telephone facility with CATI.

For the RDD experiment, 4100 cases were included in the sample. Two versions of the questionnaire were tested. One version had the interviewers collect demographic information on the entire household at the beginning of the interview. The second version had the interviewers collect the information at the end of the interview with the first person in the household.

2. LENGTH OF SURVEY PERIOD

In an RDD two stage sample survey on cigarette smoking Massey et al. (1981) found that by lengthening the survey period by 30 days, the response rate increased from 74 percent to 81 percent. This increase in the survey period involved efforts to persuade initial refusals to respond and included repeated callbacks on numbers unanswered on initial dialings. Brehm (1985) describes the probability of obtaining an interview with each passing day that a sample number receives calls in an 18 day survey period. He notes that after the first seven days, each additional day yields less than five percent increase in response rates. He estimates that increasing the survey period from 11 to 18 days might result in a five percentage point increase in response rates.

Using a two week survey period for the NCS experiment, the average response rate for both versions of the questionnaire for the April through December, 1985, period was 76 percent. The base for this rate includes units known to be eligible occupied plus all those which might have been eligible. The average refusal rate for the same period was 11 percent and the "other noninterview" rate was 13 percent.

In an effort to decrease the nonresponse rate, in further testing conducted in 1986, the Census Bureau lengthened the survey period and conducted a followup of all the unanswered telephone numbers. The usual two week survey period for the NCS was extended to four weeks. Only the second version of the questionnaire was used for the extended interview period since this version had a slightly higher response rate (74 percent versus 77 percent) during the earlier phases of the NCS/RDD experiment.

One way to measure the effects of the extended interviewing period is to compare results from the two week period in 1985 to those of the four week period in 1986. As shown in Table 1, the average response rate increased 3 percentage points, from 77 percent for the 1985 two week period to 80 percent for the 1986 four week period. Notice that the refusal rate went up 3 percentage points, and the other noninterview rate was cut from 13 percent to 7 percent. Some of the nonrespondents from the two week interview period became refusals when recontacted. It appears also that the interviewers were not as productive during the first two weeks of the extended interview period as in the two week period as a whole (a 74.1 percent response rate versus a 77.2 percent rate). Interviewers did know that they had four weeks instead of two to complete the survey and may have "stretched out" the work.

The largest single change evident in Table 1 is the reduction of the "other noninterview" rate by moving from two weeks to four weeks (from 14.5 percent to 6.6 percent). This demonstrates that the lack of information about unanswered numbers is not a fatal property of RDD sample surveys and can be somewhat overcome by continued dialing on sample numbers. Sample numbers assigned to residences tend to be answered after repeated calls on different days of the week, times of the day, and weeks of a month.

Obviously, however, a decision to extend the survey period in order to increase the response rate also raises the cost of the survey. Unlike personal visits where the interviewer can ask a neighbor for the best time to contact the respondents, RDD gains little information from each dialing and a great deal of time can be wasted on a household away for extended periods. Thus, many calls are made on a number with no chance of being answered even with an extended interviewing period.

Table 1. NCS/RDD Response and Refusal Rates by Length of Interview Period

	Interviewing Period	
Category[a]	April-December 1985	January-March 1986
After 2-week period:		
Average response rate	77.2%	74.1%
Average refusal rate	10.0	11.4
Average other-NI rate	12.8	14.5
After 4-week period:		
Average response rate		80.1
Average refusal rate		13.3
Average other-NI rate		6.6

[a]NI=noninterviewers

3. UNANSWERED TELEPHONE NUMBERS

After the interviewing on an RDD survey has ended, there always remains a set of sample numbers that were never answered when dialed. Without knowing which numbers correspond to households that were not reached and which are nonresidential, it is unclear what the denominator of the survey response rate should be. Further, in the Mitofsky-Waksberg design, differential probabilities of selection may be introduced into the design from unanswered numbers, depending on how they are classified. If an unanswered number is indeed nonresidential, then a replacement number should have been generated and called, in order to maintain equal probabilities of selection across clusters. If an unanswered number is residential, no replacement should have been generated, in order to maintain the equal probability nature of the sample.

Groves and Kahn (1979) found that in a single stage sample, only 22 percent of the sample numbers randomly dialed were working household numbers. They also found that the proportion of unassigned numbers in the national telephone system was between 69.5 percent and 72.4 percent depending on whether or not all unanswered numbers are counted as working or nonworking telephone numbers. They were able to identify 93 percent of working household numbers with 10 calls or less in a single stage sample. About 2 percent of the working residential numbers required 17 or more calls to complete the case. Hence, the response rate ranged from 59 percent to 70 percent, depending on whether or not the unanswered numbers were included in the base.

As shown in Table 2 about a fifth (21.4 percent) of the NCS/RDD interviews were taken in one or two calls. However, about 9 percent of the cases required more than 25 dialings to complete. Each household member 12 years of age and older was interviewed individually and these results include all calls made to the sample numbers. The large number of cases that remain incomplete after repeated dialings suggests that substantial numbers of dialings are required in RDD surveys to contact the household numbers and complete an interview. Given the problem of nonworking numbers yielding ringing tones, even greater attention to the number of calls on sample numbers is required. Given the results in Table 2, it is clear that the proportions of unanswered numbers that are residential and nonresidential will vary depending on how many dialings have been made to the numbers.

Massey et al. (1981) conducted a followup of unanswered numbers and initial refusals from an RDD survey on cigarette smoking. The nonresponses had received a total of 11 dialings before receiving a final disposition, and the fifth call was made to a telephone company business office to determine if the number was a working residential number. Of

Table 2. Number of Calls to Complete a Interview

No. of Calls	Interviews	Percent	Cum. Percent
1	369	9.2	9.2
2	494	12.3	21.4
3	406	10.1	31.5
4	393	9.8	41.3
5	261	6.5	47.8
6	228	5.7	53.4
7	216	5.4	58.8
8	201	5.0	63.8
9	161	4.0	67.8
10	134	3.3	71.1
11	119	3.0	74.1
12	114	2.8	76.9
13	80	2.0	78.9
14	66	1.6	80.5
15	53	1.3	81.8
16	65	1.6	83.4
17	54	1.3	84.8
18	40	1.0	85.8
19	48	1.2	87.0
20	25	0.6	87.6
21	31	0.8	88.4
22	29	0.7	89.1
23	33	0.8	89.9
24	18	0.4	90.3
25	27	0.7	91.0
25+	362	9.0	100.00

the unanswered numbers that were followed up, 38 percent were identified as nonhousehold numbers, 36 percent remained in the unanswered category, and 28 percent resulted in a household contact.

Groves (1978) followed up two small samples of RDD generated numbers. The followup involved both calls to local telephone business offices and repeated dialings of the numbers, in order to determine their residential status. For one group of 20 numbers dialed at least 12 times, 19 (95 percent) were nonworking numbers. For a group of 32, dialed over 17 times each, 15 were nonresidential numbers (47 percent).

As part of various RDD experiments, the Census Bureau conducted three followup studies of unanswered telephone numbers. In two 1984 studies, followup telephone calls were made to 266 unanswered numbers. About 80 percent of those cases were resolved during followup and approximately 40 percent of the previously unresolved unanswered numbers were nonresidential numbers. Most of these cases required less than 10 attempts to resolve their status. For a 1985 followup study of 239 cases, 83 percent of the unanswered cases were resolved. Unlike the earlier studies, only 38 percent of the resolved cases were residential and 44 percent were nonresidential or nonworking telephone numbers. These cases on average had received 20 calls prior to the followup efforts.

The differences across the studies in the percent of telephone numbers found to be working residential numbers may be due to differential lags between the original survey period and the followup of answered numbers. For the 1984 Census study, the followup was conducted a month after the original survey period. For the 1985 Census study, up to six months passed before the followup was conducted. Groves' study was conducted immediately after the survey period ended. The followup of Massey et al. occurred from one to four weeks after the period. In order to gain knowledge about the statuses of cases *at the time of sampling*, followup efforts close to the survey period are to be preferred.

In addition to dialing the numbers repeatedly to obtain their status, the Census effort involved attempts to learn why the number was not answered during the initial survey period. As shown in Table 3, among households reached in the followup the reason most frequently given for not answering the phone during the original survey period was that household members were "busy" or "never home." Approximately 44 percent responded that they were absent during the entire survey period or indicated that the unanswered number belonged to a seasonal residence. Additional effort to reach those people on vacation or with telephones at seasonal residences during the original survey period would not have been productive. The answer category, "in the process of moving," demonstrates a limitation of followup studies — the fact that nonworking numbers can change to working household numbers over time and be assigned to a household which recently moved into the housing unit.

Nonresponse bias arising from unanswered numbers that are residential is a function of both the rate of nonresponse and the characteristics of the persons in households never contacted. The discussion above gives some notion of the rate of household numbers among those unanswered after several dialings. Insight into the other component of the nonresponse bias can be obtained by examining the characteristics of persons reached during the followup efforts.

Massey et al. (1981), in their followup study, compared the initial respondents to the respondents contacted in the followup. They found the two groups to be similar except for a larger proportion of people 65 years old or older among the followup respondents. They also found that the household nonresponse rate for the recycled unanswered numbers was larger than the household nonresponse rate for households contacted on

Table 3. Reasons of Noncontact During Original Survey Period Reported in Followup Contact

Reason	Percent of Total
Busy households/work/never home	33
On vacation/absent entire survey period	32
Seasonal residence	12
In the process of moving	7
At school	3
Unplugs phone	2
Don't know/no reason	10
Total	100
n	86

the initial interview attempt (30 percent versus 22 percent). One explanation was that persons who are more difficult to contact by telephone appear to be slightly less likely to agree to participate in a survey interview.

The NCS/RDD experiment also attempted to learn the characteristics of persons contacted through the followup. Table 4 provides a demographic comparison of the households contacted during the followup with those contacted during the original survey period. By and large the two groups are similar. However, the unanswered numbers had proportionally more never married persons and fewer married persons. There is also evidence that the households which did not answer their telephones were smaller in size than those interviewed in the original survey period. In these households, fewer people would be at home at any given point, other things being equal.

Finally, the followup group contains relatively fewer persons over 65 years of age than the original survey. (These results were statistically significant at the .05 significance level.) Recall that Massey et al. found *more* people 65 years old or older in the followup group. The difference between the two studies may be due to several factors. First, Massey et al. include refusals and partial interviews in their followup study, whereas the NCS/RDD effort included only unanswered numbers. Other data show that elderly persons are disproportionately nonrespondents in telephone surveys (see this volume, Chapter 11). Also, the lag time between original interview and followup was different. The 1985 Census Bureau study allowed up to seven months between original and followup; Massey et al. allowed from one to four weeks.

The followup of unanswered numbers also had a larger proportion of missing demographic data than the original RDD survey. The main

Table 4. Demographic Characteristics of Persons Contacted During Initial and Followup Interviewing Period

	Time of Contact	
	Initial Period	Followup Period
Marital status of head of household		
Married	61%	51%
Widowed	13	9
Divorced	10	3
Separated	2	10
Never married	13	27
Sex of head of household		
Male	56	56
Female	44	44
Race of head of household		
White	87	91
Black	9	6
Other	4	3
Age of head of household		
Under 20	2	1
20–34	30	28
35–49	29	28
50–59	13	20
60–64	7	13
Over 65	20	11
Average household size	2.6	2.0
n	86	

objective of the followup was to obtain the residential status of a telephone number. After the status was obtained and the interviewer confirmed that the telephone number reached belonged to the respondent during the original survey period, the followup case was counted as complete even if the respondent refused to provide demographic information on the household. No attempt was made to recall those households in which no demographic information was obtained. In the original RDD survey, however, a case could not be completed until demographic information was obtained for the household and all refusals received at least one refusal conversion attempt.

4. CONCLUSIONS

Doubling the length of an RDD survey period from two to four weeks increased the response rate by 3 percentage points, but it is not clear that the slight rise in the response rate is offset by the increased costs associated with extending the survey period. The lengthening of the survey period effectively reduced the number of noncontacted cases, but also generated a higher refusal rate. The tendency for those households reached after repeated dialings to refuse an interview request is a finding worthy of further study by other organizations and for other topics.

Although for surveys conducted over many weeks only a small portion of the sample numbers remain unanswered, short surveys face a more severe problem of noncontact. What proportion of the unanswered numbers at the end of a survey are eligible residential cases depends both on the number and pattern of calls. Short survey periods, which generate larger proportions of unanswered numbers, also tend to have a larger portion of those numbers that are residential. In the Census experiments approximately 50 percent were working residential numbers. The demographic characteristics of the persons in these households were similar to those reached during the survey period, indicating that the bias due to nonresponse may not be severe for demographic characteristics. For statistics measuring social characteristics related to "at home" patterns (e.g., employment rates), greater bias due to noncontact may arise.

CHAPTER 16

EFFECTS OF INTERVIEWER VOCAL CHARACTERISTICS ON NONRESPONSE

Lois Oksenberg and Charles Cannell[1]
The University of Michigan

Almost all attention to sources of interviewer effects in survey research has focused on the techniques that interviewers use and the content of what is said. Psycholinguists and social psychologists, however, recognize that a large amount of communication is nonverbal. While survey researchers have studied some effects of nonverbal factors such as interviewer appearance or dress in face to face interviews, in general those factors have not received much attention. Nonverbal factors likely to play important roles in face to face interviews include those arising from visual and auditory stimuli. In telephone interviews, visual cues are eliminated, leaving auditory stimuli as the sole source of nonverbal communication. Psycholinguistic research suggests that these auditory stimuli are likely to be an important influence on respondents' attitudes and behaviors in telephone interviewing. This chapter reports our efforts to study nonverbal stimuli in telephone interviewing—in particular, their relationships to interviewer refusal rates.

The starting point was the discovery that refusal rates differed significantly among telephone interviewers. To give one example, SRC (Survey Research Center, The University of Michigan) interviewers working on a continuing monthly survey showed marked consistency in refusal rates over time, while individual interviewer rates ranged from 6 to 42 percent. Such findings stimulated us to attempt to locate the bases for

[1]The authors are indebted to Dr. Donald Sharf and Mark Lehman for providing the acoustical measurement, and to Dr. Lerita Coleman for her valuable contributions to the first stages of the research. This chapter is based on work supported by the National Science Foundation under Grant No. SES-8112801 and by the U.S. Bureau of the Census under Contract No. BC-SAC-66276.

the differences. It was found that most refusals to be interviewed occurred early in the interview introduction. For example, an analysis of refusals to an SRC telephone survey showed that about 40 percent occurred during the first few sentences, and an additional 50 percent followed these sentences but were prior to the formal beginning of the interview, with under 10 percent occurring after the interview had started.

The first idea was that refusal rates might vary with the content of the interviewers' introductions. To investigate this, we had all interviewers use a standard introduction based on introductions used by the most successful interviewers. However, refusal rate differences among the interviewers persisted.

It appeared that differential refusal rates were unrelated to the content of the introductory statements. This led to examining characteristics of interviewers' speech apart from content; that is, to examining paralinguistic characteristics such as pitch, speech rate, loudness, etc., as potential influences. It is evident from the history of research in paralinguistics and language attitudes that listeners rely heavily on vocal characteristics to form impressions and attitudes about speakers (Scherer, 1979). For example, people who speak rapidly, loudly, less haltingly, or with greater variability in pitch are perceived as more persuasive than those who speak more slowly, softly, more haltingly, or with less variability in pitch (Mehrabian and Williams, 1969; Packwood, 1974; Apple et al., 1979). Rapid and loud speakers also are perceived as more competent and credible (Pearce and Conklin, 1971; Brown et al., 1973). Increases in pitch produce judgments of less competence and less benevolence (Brown et al., 1973). Findings such as these are based largely on laboratory studies with male speakers.

We wondered whether survey respondents share a sensitivity to voice characteristics of interviewers, perceiving voice characteristics similarly and reacting similarly to them. Further, we wondered whether voice characteristics lead to the attribution of positive or negative personal characteristics to the interviewer that in turn affect the willingness of people to be interviewed.

1. RESEARCH ON THE RELATION OF INTERVIEWER VOICE CHARACTERISTICS TO REFUSAL RATES

Because refusals to be interviewed occur early in the interviewers' introductions, our research focused on interviewer voice characteristics during the first few introductory sentences. We have explored vocal characteristics and their relation to refusal rates for three groups of female telephone interviewers. One group included 34 U.S. Census

Bureau telephone interviewers working on the National Crime Survey (NCS) or on a special survey using NCS procedures and a short version of the NCS questionnaire.

The other two groups were SRC interviewers working on the Consumer Attitudes Surveys (CAS). The SRC groups included interviewers with especially high refusal rates or with especially low refusal rates and excluded those with moderate refusal rates. One SRC group (SRC 1) included 10 interviewers, while the other group (SRC 2) included 12 interviewers. Each of these groups included both high and low refusal rate interviewers.

Recordings were made of all the interviewers giving actual telephone interview introductions. The recordings were used to analyze the interviewers' vocal characteristics. SRC interviewer recordings included about the first one-half minute of the standard CAS introduction. Census interviewer recordings were from the standard NCS introduction. To comply with legal restrictions, the recordings included only the interviewer statements.

A tape was prepared for each of the three groups with the selected segments of each interviewer's introduction included. Care was taken to maintain similar vocal intensity levels among interviewers on a tape. Two types of measures were obtained for each voice. One used trained listeners to rate various voice characteristics. The other included machine measurements based on acoustical analytic procedures. The measures were based on similar ones developed in our earlier research with the SRC 1 interviewers (Oksenberg et al., 1986; Sharf and Lehman, 1984).

Most of the measures were designed to capture voice characteristics that the previous research found to be related to refusal rates. With regard to listener ratings, the measures concerned physical characteristics of the voice such as average pitch, variation in pitch, loudness, rate of speaking, and standardness of pronunciation. The latter was defined in terms of similarity to pronunciation of national radio announcers. Also included were ratings of how natural and conversational the interviewer sounded, which was called "conversational delivery," and ratings of how pleasant the voice was to listen to.

Several personal evaluations were included. These were how competent the interviewer sounded, how confident she sounded, how friendly she sounded, and how interested in the task she sounded.

Special equipment was used to obtain the acoustic measures. The equipment included a frequency analyzer, a computer, and special computer programs. With the aid of this equipment, measures of fundamental frequency were obtained for each interviewer. Fundamental frequency (a characteristic of speech sound related to listener perception of pitch) was measured repeatedly at extremely short intervals, throughout the interviewer's introduction. The measurements for each interviewer were used to calculate her mean fundamental frequency, which represents the frequency value that best characterized her voice. The standard deviation of the fundamental frequency measurements for an interviewer also was calculated. The standard deviation was used to indicate how variable the interviewer was in producing fundamental frequency. Mean fundamental frequency and fundamental frequency standard deviation capture characteristics of speech sound related to listener perceptions of average pitch and variation in pitch.

Several acoustic measures relating to specific aspects of the rate of speaking were also included. However, acoustic measures of intensity (a measure of physical energy of speech sound related to perceptions of loudness) were not included, because intensity levels of the recordings had been controlled. Finally, following indications in the earlier research of their importance, we included measures of intonation contours in saying certain key words in the introductions. Intonation contour measures, described below, were available only for a subset of Census interviewers.

1.1 Rating Procedures and Reliability

Four raters were trained to rate the interviewers' voices from the recorded introductions. The rating method was an adaptation of a method used widely in psychophysical research called the *Method of Magnitude Estimation*[2]. The procedure involved comparison of each recorded voice to that of a "sample speaker." The sample speaker was also recorded giving the same introduction. The comparisons to the sample speaker were made for each characteristic being rated. To illustrate the method, as used to rate loudness, a rater assigns a number to each of the voices, reflecting its loudness relative to that of the sample speaker. The sample speaker is arbitrarily assigned a loudness value of 10. The rater is told to assign a number from zero through 20 to the voice. Numbers greater than 10 indicate voices louder than the sample speaker, while numbers less than 10 are for softer voices.

The first concern was whether the raters would agree among themselves on how they rated the voices. Table 1 gives the mean inter-rater reliabilities of the voice characteristic scales for each of the

[2] Listener ratings of the first group of interviewers were made by another method in the earlier research (Oksenberg et al., 1986). For research reported in this chapter, these interviewers were rerated following the procedure described here.

Table 1. Rater Reliabilities of Interviewer Characteristic Measures by Interviewer Group

	Reliability		
Scale	SRC 1	SRC 2	Census
Average pitch	.91	.83	.77
Variation in pitch	.80	.67	.73
Rate of speaking	.51	.86	.83
Loudness	.81	.80	.85
Pronunciation	.66	—	.77
Conversational delivery	.73	.78	.50
Pleasant to listen to	.72	.35	.28
Competent	.68	.68	.57
Confident	.67	.80	.69
Interested in his/her task	.54	.71	.55
Friendly	.69	.75	.80

Note: For SRC 1, n=10; for SRC 2, n=12; for Census, n=34.

interviewer groups.[3] As the table shows, agreement among raters was quite high for some characteristics, such as average pitch and loudness. In general the reliability levels were adequate for the group comparisons made in this research. The reliability of ratings of how pleasant the voice was to listen to is a notable exception.

1.2 Relations among Ratings and Acoustic Measures for 34 Census Interviewers

As has been found in other paralinguistic studies, a number of acoustic measures were significantly correlated with the listener ratings. For all 34 Census interviewers, the acoustic measure of mean fundamental frequency correlated $r = .86$ ($p < .01$) with listener ratings of average pitch. Similarly, the standard deviation of fundamental frequency correlated $r = .65$ ($p < .0l$) with listener ratings of variation in pitch. Of the acoustic measures designed to capture various aspects of the rate of speaking, two were strongly correlated with comparable listener ratings. They were speech rate and articulation rate ($r = .74$ and $r = .64$, respectively; $p < .01$). Speech rate was defined as the ratio of the number

[3]The analysis of variance method with repeated measures giving mean inter-rater reliabilities as described by Winer (1971) was applied to obtain the rating scale reliabilities.

of syllables to the total duration of the introduction. Articulation rate was the ratio of the number of syllables to the total duration of speech. Number of pauses was moderately but significantly correlated with listener ratings of rate of speaking ($r = .39, p < .05$).

1.3 Interviewer Voice Characteristics and Refusal Rates

The major question in this research was whether the voice characteristics show any relationship to refusal rates. Three of the listener ratings and one acoustical measure were significantly related to the refusal rate among the Census interviewers (Table 2).[4] Table 2 is based on the 25 Census interviewers for whom refusal rates were available. These rates ranged from .05 to .37 with a mean of .17. Listener ratings of these interviewers that correlated significantly ($p < .05$ or better) with refusal rates were rate of speaking ($r = -.49$), loudness ($r = -.42$), and confidence ($r = -.37$). Of the acoustic measures only the mean fundamental frequency was significantly related to the refusal rate ($r = .38$). Interviewers with low refusal rates were rated as speaking relatively faster, louder, and with more confidence. They also had lower mean fundamental frequencies.

In addition to assessing separately the relation between each vocal characteristic and the refusal rate for the Census interviewers, several forward stepwise regression analyses were performed with refusal rate as the dependent variable. A liberal rule for inclusion of a characteristic in the regression model was used for these exploratory analyses: a maximum attained significance level of .15 was allowed for a measure to be included in the model.

The regression analyses (data not shown) added little information to that provided by the correlations in Table 2. In an analysis for which all listener ratings were available for inclusion in the regression model, rate of speaking ($p \leq .005$) and average pitch ($p \leq .08$) were identified as the best predictors of the refusal rate. Loudness had a correlation with the refusal rate nearly as high as that for rate of speaking, but was not selected because of the intercorrelation of loudness and rate of speaking ($r = .36$). Had the rate of speaking not been available for inclusion in the regression model, loudness would have been the first characteristic selected.

A similar regression analysis was undertaken using just the last six characteristics in Table 2 (thereby excluding physical characteristics of the voice and standardness of pronunciation). Confidence was the one

[4] Intonation contour measures were not analyzed here because they were available for only a subset of 12 Census interviewers with extreme refusal rates.

Table 2. Correlations between Rating and Acoustic Measures and Refusal Rate for Census Interviewers

Measure	Correlation
Rating Measures	
Average pitch	.24
Variation in pitch	.11
Rate of speaking	−.49[b]
Loudness	−.42[c]
Pronunciation	−.25
Conversational delivery	−.10
Pleasant to listen to	−.15
Competent	−.27
Confident	−.37[d]
Interested in the task	−.09
Friendly	.05
Acoustic Measures	
Mean fundamental frequency	.38[a]
Fundamental frequency standard deviation	.22
Mean pause duration	.06
Number of pauses	.14
Articulation rate	.11
Speech density	−.09
Speech rate	−.21

Note: The correlations are based on the 25 Census interviewers for whom refusal rates were available.
[a] $p < .06$
[b] $p < .02$
[c] $p < .05$
[d] $p < .07$

characteristic selected ($p \leq .07$) as a predictor of the refusal rate. Had confidence not already been included in the model, competence would have been selected. (The correlation between competence and confidence was .85.)

In a similar regression analysis using only the acoustic measures to predict the refusal rate, mean fundamental frequency was the only measure included in the regression model.

1.4 Comparisons among Census and SRC Interviewers on Listener Ratings

The next question addressed was whether the findings for Census interviewers held for the SRC interviewers. However, the findings for the Census group, which included the full range of refusal rates, could not be directly compared to findings for the SRC groups, which included only interviewers with especially high or especially low refusal rates. The six Census interviewers with the highest and the six with the lowest refusal rates were selected for additional analysis. This was comparable to the SRC groupings. In the Census group, the high refusal rates ranged from .22 to .37 and the low refusal rates ranged from .05 to .10. High refusal rates ranged from .24 to .46 among SRC 1 interviewers and from .40 to .55 among SRC 2 interviewers, while low refusal rates ranged from .03 to .14 among SRC 1 interviewers and from .05 to .17 among SRC 2 interviewers.

For each group (Census, SRC 1, SRC 2), relations between vocal characteristics and refusal rates were examined.[5] For each group of interviewers, the η^2 were calculated, giving the proportion of variation in ratings attributable to the difference between the interviewers with high refusal rates and those with low (Table 3).

As Table 3 shows, there are differences among the three groups of interviewers. The findings for the Census interviewers and SRC 2 interviewers are highly similar. There are major differences, however, between these interviewers and SRC 1 interviewers. For SRC 1 interviewers, all the rated voice characteristics except pronunciation differed between interviewers with high and low refusal rates.[6] In contrast to this, fewer of these characteristics were related to the refusal rate for SRC 2 interviewers and Census interviewers. For them, results for the rate of speaking, loudness, and confidence (as well as results for competence for Census interviewers) were significant; interviewers with low refusal rates spoke fast, loudly, and with apparent confidence and competence. A standard American pronunciation (as judged by the raters) was significantly associated with low refusal rates for Census interviewers. (Pronunciation was not rated for SRC 2 interviewers because informal assessment indicated little variation among them.)

The mean ratings on scales for interviewers with high and low refusal rates (data not shown) suggest some reasons why results for SRC 1 interviewers differed from those for the other interviewers. SRC 1

---

[5] Evaluations were by planned comparisons tested within the context of the analysis of variance used to determine the reliabilities of the listener ratings.

[6] Except for pronunciation ($p \leq .07$), $p \leq .05$ for all characteristics.

Table 3. η^2 for Comparisons of Ratings of Characteristics for Interviewers with High and Low Refusal Rates among SRC 1, SRC 2 and Census Interviewers

Characteristic	SRC 1	SRC 2	Census
Average pitch	.25***	.05**	.04*[a]
Variation in pitch	.26***	.03	.04
Rate of speaking	.04*	.12**	.26***
Loudness	.34***	.28***	.18***
Pronunciation	.07	—[b]	.07**
Conversational delivery	.17***	.000	.00
Pleasant to listen to[c]	.27***	.01	.00
Competent	.11***	.02	.08***
Confident	.11***	.07***	.18***
Interested in his/her task	.14***	.02	.00
Friendly	.09**	.00	.00

Note: Comparisons for SRC 1 interviewers were based on five with high refusal rates and five with low refusal rates. Comparisons for SRC 2 and for Census interviewers each were based on six with high refusal rates and six with low refusal rates. Except as noted, high scores on characteristics were associated with low refusal rates where η^2 was significant at $p \leq .05$.

[a]Low average pitch was associated with low refusal rates.

[b]*Pronunciation* was not rated for SRC 2 interviewers.

[c]*Pleasant to listen to* was very unreliably rated for SRC 2 and Census interviewers.

* $p < .05$.

** $p < .025$.

*** $p < .01$.

interviewers with high refusal rates had markedly low average pitch and variation in pitch levels compared to all other groups of interviewers. Perhaps there is a range of pitches and of variation in pitch that is equally acceptable. Levels beyond this range, as may be the case with the high refusal rate SRC 1 interviewers, may be less acceptable, and lead to higher refusal rates. In addition, SRC 1 interviewers spoke slowly in comparison with the other interviewers and showed less variation in speaking rate, making it statistically difficult to obtain a strong relationship between speaking rate and refusal rate for SRC 1 interviewers. Where there was more variation in speaking rate, as there was among SRC 2 and Census interviewers, stronger relationships with refusal rate occurred.

In general, compared to interviewers with high refusal rates, interviewers with low refusal rates were heard as speaking relatively rapidly, loudly, and with a standard American accent, and as sounding more confident and more competent. Average pitch level or amount of variation in pitch does not seem importantly related to refusal rates, although there is some evidence that very low pitch levels and very low levels of variation in pitch may increase the refusal rate.

1.5 Comparisons among Census and SRC Interviewers on Acoustic Measures

Analyses for the three groups of interviewers also were undertaken for all acoustic measures except the intonation contour measures. There were no significant differences on any of the other acoustic measures between interviewers with high refusal rates and those with low refusal rates among either Census or SRC 2 interviewers. Consistent with findings based on listener ratings, however, SRC 1 interviewers with low refusal rates had higher mean fundamental frequencies ($t = 2.02$, $p < .08$) and higher fundamental frequency standard deviations ($t = 4.13$, $p < .01$).[7] Note that for the SRC 1 group, low refusal rate interviewers had relatively high fundamental frequency scores whereas the effect reported earlier for the full group of Census interviewers went in the opposite direction. While the importance of fundamental frequency is unclear, in general the acoustic measures included in these analyses were not useful predictors of the refusal rate.

There is an intriguing finding, however, concerning intonation patterns. An analysis was made of the intonations of Census interviewers for several key words in their introductions. The analysis involved plotting the fundamental frequencies used by the speaker from the first sound to the last sound made in pronouncing the word. The plotting of the frequencies created an intonation contour. Various properties of this contour were studied. For example, the word "hello," a greeting used in nearly all introductions, was analyzed and several intonation contour patterns were observed. One measure of the patterns was to compare the fundamental frequency at the start (onset) of saying "hello" with the end frequency (offset). The significance of this comparison was that the ratio of onset to offset frequency correlated significantly with the refusal rate. For the full group of Census interviewers the correlation was $-.50$ ($p < .05$).[8] With regard to the high and low refusal rate Census interviewers, the following patterns were found. The pattern for low refusal rate interviewers was a high frequency onset and low frequency offset. For high refusal rate interviewers the pattern was reversed, or the two frequencies were essentially equal.

Similar intonation contours were plotted for four other key words in the introduction. The same pattern was found for two of them near the beginning of the introduction. They included the interviewer's name and

---

[7] All tests were two-tailed *t*-tests.

[8] Intonation contour measures for the full group of Census interviewers were available only for "hello."

"Bureau." The remaining two ("survey" and "older"), occurring later in the introduction, did not show pattern differences between high and low refusal rate interviewers.

2. DISCUSSION

Although there were some inconsistencies among the interviewers studied, a general pattern emerged. Overall, interviewers rated as speaking rapidly,[9] loudly, and with a standard American pronunciation, and perceived as sounding competent and confident, had lower refusal rates than those with opposite patterns. With regard to intonation patterns, interviewers using a falling tone on key words early in the introduction had lower refusal rates than those using a rising tone.

It seems likely that the speech characteristics of low refusal rate interviewers in introductions—speed, loudness, standard accent, falling tones—contribute to listener perceptions of them as competent and confident. These findings are consistent with psycholinguistic research mentioned earlier, showing that people who speak rapidly and loudly are perceived as more persuasive than those who speak slower and softer (Mehrabian and Williams, 1969; Packwood, 1974; Apple et al., 1979). The findings are also consistent with the research showing that rapid and loud speakers are perceived as more competent and credible (Pearce and Conklin, 1971; Brown et al., 1973). For the Census interviewers in this study — with the exception of one pair of variables — the rate of speaking, loudness, and pronunciation ratings all correlated highly and significantly with both competence and confidence ratings (data not shown).

As for intonation patterns, falling intonation generally is considered appropriate for ending declarative statements, whereas a rising intonation is considered appropriate for ending questions. It is reasonable to expect that the use of rising intonation for declarative statements (as happened in the introductions analyzed here) tends to produce perceptions of the speaker as diffident and as seeking reassurance from the listener. In contrast, use of the falling intonation appropriate to declarative statements appears likely to produce perceptions of the speaker as sure of himself/herself.

---

[9] While a fast rate of speaking for introduction statements may help persuade respondents to participate, asking questions at a slower pace is probably better because it facilitates comprehension and attention to reporting tasks.

This picture of the successful (low refusal rate) interviewer as appearing competent, confident, and, more generally, potent contrasts somewhat with the belief of some interviewers and others that the way to induce cooperation is to be very friendly, ingratiating, and nonthreatening. Our research suggests that the latter, somewhat submissive, stance is not effective. Some, however, may fear that the more forceful approach is unpleasant, unfriendly, or threatening for the respondent. This does not appear to be so. For the Census interviewers, ratings of friendliness were not significantly related to ratings for rate of speaking, loudness, pronunciation, competence, or confidence. This means that interviewers with a more forceful approach generally were perceived as just as friendly as those with a less forceful approach.

The view of the successful interviewer—one that influences the respondent to be interviewed—as having a relatively forceful approach during the introduction is consistent with large bodies of research and theory about how to affect other people's behavior. Within social psychology, research on leadership, compliance, and persuasion all address this issue. In broad outline, competence, trustworthiness, and dominance (the tendency to lead, direct, influence, and control others) are seen as important attributes of people able to affect the behavior of others. Nonverbal characteristics carried by the voice, however, have received scant attention in this research literature as factors in persuasiveness, gaining compliance, or asserting leadership.

Our findings with regard to the particular vocal characteristics related to the refusal rate are tentative. If the characteristics we have identified turn out to be important influences on the refusal rate in general, a possible explanation can be found within a framework that integrates paralinguistic research findings with social psychological research on leadership and persuasion. The framework centers on the concept of dominance. In outline, dominance is indicated in speech at least in part by speaking loudly, rapidly, and with certain intonation patterns. Dominant people are perceived as competent, and also are likely to succeed in attempts to direct and influence others. Thus, assured-sounding interviewers are more successful in gaining respondent cooperation and, therefore, have lower refusal rates.

The role of a standard American accent is unclear. Standard speakers may be perceived as more competent because they are judged in some sense to speak "correctly." However, the issue is confused by the existence of regional and other subgroup pronunciation patterns—rated as nonstandard for this analysis—that carry additional connotations, particularly with regard to social status. If such speakers are perceived to have low social status, they may be less successful in eliciting cooperation.

3. IMPLICATIONS OF THE RESEARCH FOR TELEPHONE INTERVIEWING PRACTICE

There appear to be two interviewer roles. At the beginning, the immediate goal is to persuade the respondent to cooperate. For this, a forceful, self-assured approach appears to be useful. In contrast, during the interview the interviewer assumes a second role. Then the issue is how best to communicate clearly with the respondent. A slow pace here facilitates understanding of the questions and attention to reporting tasks.

There is evidence that speech patterns and voice characteristics can be changed. If successful voice characteristics are identified, training in speaking can be introduced into interviewer training programs. In fact, interviewers already are selected to avoid divergent speech patterns, so interviewers entering training would be within the normal range. We have tried some voice training on a small scale. The training included acquainting trainees with speech characteristics that our research has shown to be related to refusal rates, attempts to sensitize the trainees as to how they themselves sounded, and practice in the desirable speech characteristics. This limited experience indicated that such training is feasible. The next step would be to develop a full-scale training program or to find voice training programs already in existence that could be adapted for telephone interviewing.

The information presented here is by no means conclusive. At times the results are conflicting or unclear. These studies should be viewed as preliminary explorations into the complex but crucially important issue of response rates in telephone surveys. The results do suggest, however, that further research in voice characteristics and speech patterns may lead to effective techniques for significant increases in response rates.

SECTION D

DATA QUALITY IN TELEPHONE SURVEYS

CHAPTER 17

MEASURING DATA QUALITY

Paul P. Biemer
New Mexico State University

How does the quality of data collected in centralized telephone interviewing facilities compare with that of data collected by face to face interviewing or by mail? Considerable attention has been focused on this question in the last decade, as the bibliography to this volume will confirm. This chapter discusses the main methods which have been used to assess the quality of telephone surveys relative to surveys conducted using other modes of interviewing. It is not intended to be an exhaustive description of the numerous techniques for evaluating the quality of survey data. Rather, emphasis is given to two methods — mode comparison studies and record check studies — since these techniques are used later in Section D.

1. MODE COMPARISON STUDIES

Ideally, the "mode comparison study" involves conducting two surveys of the same population, usually simultaneously, which differ only in the mode of interview. Estimates from the surveys are then compared and inferences regarding the relative quality of the two sets of data are made. In what follows, we describe the theoretical justification for inferences based on these studies as well as a number of cautions in interpreting the studies' results.

1.1 Description of the Method

A sample survey is an integrated system of activities and operations for data collection which are closely tied to a particular mode of interviewing.

Many features of the design, such as the sampling frame, sample design, questionnaire design, interviewer recruitment and training procedures, respondent rules, interviewing procedures, etc., which are ideal for face to face interviewing, may be inefficient or even infeasible for telephone interviewing. If the objective of a study is to compare the quality of data from an efficiently designed face to face (or mail) survey to an equally well designed telephone survey, the mode of interview will typically not be the only design factor that will differ between the two surveys in the study. Rather, the comparison is really between two systems of data collection, each with design parameters and procedures that may be broadly equivalent, yet particularly adapted for efficiency in the given mode of interviewing. This consequential confounding of other design factors with the interview mode means that, in practical situations, the estimation of a "pure mode effect"[1] is not possible. If estimation of the pure mode effect is the objective, a laboratory type experiment may be necessary so that factors such as the interviewer, type of respondent, and refusal and noncontact rates can be strictly controlled. However, this is rarely the objective of the survey practitioner, whose main interest lies not just with the pure mode effect, but also with the combined effects of the mode of interview with other design factors. For example, if the survey practitioner conducting an ongoing survey is contemplating a change from using face to face interviewing to using centralized telephone interviewing, he or she will be concerned with how changes to the data collection system will affect data quality. Similarly, if a new or one time survey is being planned, the survey practitioners' interest is in how the data collection *systems* based on alternative interviewing modes will affect the quality of the data for this survey. Thus, in this section, unless otherwise stated, a "mode comparison study" will refer to a study comparing the quality of two *systems* of data collection developed for different modes of interviewing.

This discussion of the objectives of mode comparison studies has important implications for their design. In most cases, the designs of both survey systems in the comparison should, wherever possible, mirror those that would be implemented in a production setting. This strategy argues for efficiency in the survey designs rather than striving for equivalence of design parameters for the two systems. An example of this strategy is the design of the survey questionnaires. In a face to face interview, show cards might be used for some questions; however, this interviewing aid is

---

[1] By "pure mode effect" is meant, as an objective of a theory of communication, the effect on the responses to a particular question, or set of questions, of changes in the mode of communication between interviewer and respondent, where the only change in the interview situation is the mode of communication.

impossible for the telephone interview. Nevertheless, to avoid the use of show cards in the face to face interview simply to maintain comparability with the telephone interview would be a poor design decision, since this feature is an important component of the face to face system and one that would be used in actual practice. Another example is the survey period length, the time period beginning when the first interview is conducted and ending when the last interview is accepted. A two week period may be quite efficient for a face to face survey but too short for a random digit dialing survey (see this volume, Chapter 15). If a longer survey period would be within the time constraints for reporting the results in actual production, then a longer survey period should be implemented for the RDD survey in the mode comparison study.

The proper design of mode comparison studies can be quite difficult, and this increases the risk that the study will yield data that may lead to false inferences. Without careful thought and attention to the underlying assumptions, the comparisons of estimates may reflect nothing more than designer incompatibility. Further, there are risks in extending the results of mode comparison studies to other survey systems, especially those whose designs differ markedly from the systems in the original study. These problems and others are discussed later in this chapter.

Every design decision in a data collection system can have an impact on the total error of the resulting estimates — affecting either the bias, the variance, or typically both the bias and the variance. Since variance (even nonsampling variance) can usually be manipulated and controlled (for example, by adjusting the sample size), biases are considered to be the most problematic components of total survey error and the focus of most mode comparison studies. The following simple model will suffice to illustrate the basic methodology employed for these studies.

Consider two surveys for measuring a common set of population characteristics — one survey to be conducted by telephone and the other to be conducted either by face to face interviewing or by mail. Denote by $\hat{y}$ the corresponding estimate for some characteristic from the telephone survey and by $\tilde{y}$ the estimate from the comparison (i.e., mail or face to face) survey. Assume that the expected values of $\hat{y}$ and $\tilde{y}$, in the absence of nonsampling error, are identical, i.e., both surveys are measuring the same characteristic of the same population. Further assume that, as a consequence of nonsampling errors, $\hat{y}$ has a bias $B(\hat{y})$ and $\tilde{y}$ has a bias $B(\tilde{y})$. Typically, interest is on the difference between these two estimates, i.e., $d(\hat{y}, \tilde{y}) = \hat{y} - \tilde{y}$, referred to as the estimated mode effect. Then, the difference $d(\hat{y}, \tilde{y})$ is an estimator of $B(\hat{y}) - B(\tilde{y})$, i.e., the "differential bias" between the two surveys. If it can be assumed that $B(\tilde{y})$ is very small relative to $B(\hat{y})$, then $d(\hat{y}, \tilde{y})$ may be interpreted as an estimator of $B(\hat{y})$, the bias of the telephone estimate. This situation is plausible, for example, when the mode of interview for the comparison survey is regarded as the "preferred survey method"[2] for estimating the characteristic of interest. Quite often, however, this assumption is not tenable and our interest lies in determining which of the following three situations exist:

A. There is no statistically significant difference between the estimates from the two surveys, which implies that $B(\hat{y}) = B(\tilde{y})$.

B. The telephone survey estimate is significantly larger than the comparison survey estimate, which implies that $B(\hat{y}) > B(\tilde{y})$.

C. The telephone survey estimate is significantly smaller than the comparison survey estimate, which implies that $B(\hat{y}) < B(\tilde{y})$.

For large scale experiments, these determinations can be made at both national and subnational levels (e.g., the age, race, and sex demographic domains, and the region, state, and urban-rural geographic domains).

Outcome A is an indication that both estimates have the same bias (which, of course, may be zero) and are therefore of equal quality; or at least, there is no difference in quality that is discernible with the given precision of the comparison. Outcomes B and C by themselves tell us nothing about the relative accuracy of the two estimates since we have no information regarding the relative magnitudes of the biases. For example, B indicates that the bias in $\hat{y}$ is larger than the bias in $\tilde{y}$, however, the bias in $\hat{y}$ may be small and positive while the bias in $\tilde{y}$ is large and negative. The next section describes some ways to deal with this problem.

1.2 Interpreting the Results of Mode Comparison Studies

Statistical significance of a comparison alone is not necessarily indicative of a data quality differential between two surveys, and three other factors must be considered:

F1. The effect size, i.e., the magnitude of the difference

---

[2] The "preferred survey method" is discussed further by Hansen, Hurwitz, and Bershad, (1961). It is the method of data collection for which conditions are ideal (for example, nonresponse rates and response errors are as small as possible) and for which the survey budget is essentially unrestricted (for example, the best interviewers, the best training, as many callbacks as are warranted are used). The method may involve interviewing, accessing administrative records, or other forms of data collection.

F2. The direction of the difference, i.e., B or C

F3. Violations of the underlying assumptions for the mode comparison study which could explain the magnitude of the difference.

Regarding F1, the important question is whether users of the telephone survey results would reach different conclusions from those reached using the comparison survey's results. If the answer is "no," then the study analyst would have justification in concluding that both modes produce data of similar quality, and for recommending whichever mode of interview is the most cost effective, timely, and operationally efficient. Indeed, in many investigations, the conclusion of "no mode effect" is desirable since it leads directly to the conclusion "use the cheaper alternative," and little further investigation, except perhaps replication of the study results, is indicated. However, if the significant difference is important, the study analyst faces the burden of determining which estimate (or set of estimates) is preferred. For some surveys, there is a history of research that helps to determine the more accurate data. Unfortunately, in many cases fairly accurate knowledge of the true population (or domain) target parameters is not available, and a record check study (Section 2) is required.

For some characteristics, there may be substantial evidence (for example, from the literature) of a tendency of respondents to consistently misreport in one direction, regardless of the mode of interview. An example of this phenomenon is the reporting of some health characteristics in the National Health Interview Survey. It is a widely held belief that respondents tend to underreport such events as "doctors visits" and "days lost as a result of illness" (see Cannell et al., 1987). Factors such as respondent burden and recall loss have been cited as potential causes of the phenomenon. In terms of the model, this implies that both $B(\tilde{y})$ and $B(\hat{y})$ are negative. Thus, the telephone estimate being greater than the comparison survey estimate leads to the conclusion that the telephone estimate is less biased than the comparison survey estimate (this is the "more is better" theory). In statistical terms, $d(\hat{y}, \tilde{y}) > 0$ implies that $B(\hat{y}) - B(\tilde{y}) > 0$ or $B(\hat{y}) > B(\tilde{y})$. Since $B(\hat{y}) < 0$ and $B(\tilde{y}) < 0$ we have $|B(\hat{y})| < |B(\tilde{y})|$, i.e., the magnitude of the bias for the telephone survey is less than the magnitude for the comparison survey. Of course, other factors may contribute to make $d(\hat{y}, \tilde{y})$ significantly larger than zero. For example, more frequent reporting of health events on the telephone may be the consequence of either item or unit nonresponse, undercoverage of some segment of the population in one or both surveys, or greater displacement of events into the reference period for the telephone mode. A careful analysis of the data could reveal peculiarities in demographic characteristics such as sex ratios, age, or household size distributions that might also explain the discrepancy between the estimates.

Another situation where $d(\hat{y}, \tilde{y})$ being significantly larger (or smaller) than zero may indicate superior data quality in one of the modes under investigation occurs with "sensitive topics." Hochstim (1967) and Colombotos (1969) have argued that respondents will report more honestly on intimate and personal subjects — such as alcohol consumption, income, and marital relations — in telephone surveys than in face to face surveys. Apparently this effect, referred to as the "social desirability effect," stems from a desire by the respondent to avoid embarrassment or to appear favorable in the eyes of the interviewer, and this tendency is greater when the interviewer is present, as in a face to face interview. Here, the sign of the bias is expected to be in the direction of the socially desirable response. For example, a greater frequency of reports of high alcohol consumption by one mode of interview compared to another may be evidence of a smaller social desirability bias in the mode having the greater reporting. However, as mentioned before, the researcher must still contend with the difficulties of numerous confounding factors, which may also contribute to the difference between the two surveys, and caution must be exercised in interpreting the result of such studies.

In some cases the more is better and/or the social desirability effects assumptions are not appropriate for the characteristics of interest. Further, it may not be feasible to obtain the true values of the characteristics for the sample respondents through administrative records or other sources in order to estimate $B(\hat{y})$ and $B(\tilde{y})$ directly. Here, there may be little the survey methodologist can say about the relative data quality associated with two modes of interview when important differences between the estimates from the two test surveys are observed. However, if estimates of other indicators of nonsampling error are available, these may be compared between the two surveys to substantiate or support the claim of accuracy of one modality over the other. Examples of these indicators are: estimates of interviewer effects and other indicators of response variance or reliability, response rates, edit failure rates, coding error rates, agreement of the estimates with other external estimates, and so on.

Finally, regarding F3, a major underlying assumption of mode comparison studies is that both modes are measuring the same characteristic of the same population, referred to as the "homogeneity assumption." However, as mentioned earlier, it may be necessary to alter the wording of questions used for the face to face mode in order to be suitable for the telephone, and the risk of violating the first part of the assumption (i.e., same characteristics are measured) is increased. Further, the sampling frame, the nonresponse patterns, or the time periods for conducting the surveys may differ enough to raise serious concerns about whether the same populations are being measured. When $d(\hat{y}, \tilde{y})$ does not differ significantly from zero, the conclusion of "no mode effect" may still be plausible since violations of the homogeneity assumption should tend to reject the hypothesis. The real problem arises in the interpretation of significant mode effect. Here, the possibility arises that the system-level effects associated with the mode of interview explain less of the difference than the cumulative effect of violations of the aforementioned underlying assumptions.

There are a number of variations of the basic mode comparison study technique. For example, one or both surveys in the comparison may be conducted by a combination of modes ("mixed mode" designs); thus, telephone/face to face, self-administered/face to face, and self-administered/telephone combinations are possible. Rotating panel studies that alternate the mode of interview each time a household is interviewed have been used. So have designs that use face to face interviews followed by telephone reinterviews of the same respondents (and vice versa). An excellent discussion of these and other methods is provided in Groves (in press).

Three chapters in this section of the book rely on the use of mode comparison studies to compare the quality of data collected by telephone, face to face, or self-administered survey systems.

The first, by de Leeuw and van der Zouwen, is a meta-analysis of 31 mode comparison studies of telephone and face to face interviewing which were published between 1952 and 1986. This chapter provides evidence, gleaned from a systematic review of studies worldwide, of a small but worrisome bias for the telephone mode evidenced by five quality indicators.

The next chapter, by Sykes and Collins, summarizes the results of four mode comparison studies conducted at the Centre for Social and Community Planning Research in London. Three of the studies provide estimates of $B(\hat{y}) - B(\tilde{y})$ for demographic characteristics and attitudinal questions. The social desirability assumption is used for a number of items where this assumption is reasonable. The fourth study compares the social desirability biases of the two modes associated with questions on alcohol consumption. Three years of data are accumulated in this chapter to provide one of the most thorough investigations of telephone mode effects available in the literature.

The third chapter, involving the mode comparison study technique is by Bishop et al., which compares telephone data collection with the self-administered mode. Two parallel mode comparison studies were conducted — one in the United States and one in Germany — to investigate question order effects, open and closed form effects, and tone of wording effects. The U.S.-Germany replications provide the basis for broader generalizations of the results of the two studies than would be possible had the studies been conducted in only one country.

2. RECORD CHECK STUDIES AND OTHER METHODS

Because of the unavailability or inaccessibility of relevant administrative data for the characteristics typically measured in surveys, and the expense of obtaining records from official sources, record check studies occur relatively infrequently in the survey methodology literature. Further, the use of this method is, of course, restricted to factual items, for example, age and income. An implicit assumption of the method is that the records contain accurate information (i.e., the true values) for the survey characteristics of interest for the individuals in the study. With this assumption, separate estimates of $B(\hat{y})$ and $B(\tilde{y})$ can be computed and compared. From the perspective of evaluating the relative accuracy of the data collected by two modes of interview, the record check study is the ideal method, since direct estimates of accuracy are possible. However, three problems typically plague the method and may cause the records data to be less valuable for evaluating quality:

(a) The time periods for the record data and the survey data may not coincide

(b) The characteristic(s) being reported in the record system may not exactly agree with the characteristic(s) being measured in the surveys

(c) The information available for some individuals in the sample may not be adequate to allow an exact match to the administrative record; for example, the characteristics to be used in matching may only be available for respondents of the survey and not nonrespondents.

The effect of these problems is to bias the estimates of $B(\hat{y})$ and $B(\tilde{y})$ to an unknown extent, and hence comparisons of the two bias estimates could be misleading.

In Körmendi's study (this volume, Chapter 21), the results of a record check study conducted in Denmark are reported. Nearly ideal conditions prevailed for this study, which compared the accuracy of income information collected by face to face interviewing with that of telephone interviewing.

Körmendi used information from government tax records to directly estimate $B(\hat{y})$ and $B(\tilde{y})$. Her sampling frame contained enough information on nonrespondents in both surveys to obtain tax records for nonrespondents as well as for respondents. Thus, both the response and the *nonresponse* bias components are estimated. The study is a rare and important contribution to the literature on mode effects for income data.

Other methods for evaluating the quality of survey data involve the use of *reinterview surveys*, recontacting the households in a survey in order to obtain a second response for the same characteristics obtained in the initial contact of the households (see for example, Hansen, Hurwitz, and Pritzker, 1964); or *interpenetration of interviewer assignments*, randomly assigning the households in a survey to the interviewers available to conduct the survey (see, for example, Hansen, Hurwitz, and Bershad, 1961). Reinterview surveys are used primarily to assess the response reliability of survey data. However, very few studies have been conducted that compare the response reliability of telephone data with that of other modes [see Biemer and Wolfgang (1985) for one study]. Much more work is needed in this area. Interpenetration of assignments is typically used to estimate the contribution of interview variance to the total variance of a survey estimate. Several studies have been conducted which compare the effect on total variance of telephone interviewers and face to face interviewers (see, for example, Groves and Kahn, 1979; Groves and Magilavy, 1986). The evidence in the literature indicates that interviewer variance may be much smaller for telephone surveys. However, Stokes (in press) points out that previous studies may not be valid as a result of using an inappropriate estimator for the interviewer variance for categorical data items.

In the last chapter in this section, Stokes and Yeh use the idea of interpenetrating interviewer assignments to develop a new methodology for identifying telephone interviewers who contribute maximally to the interviewer variance of survey estimates. Their procedure, which is based on earlier work by Anderson and Aitken (1985) and Morris (1983), may allow survey managers to improve the quality of survey results by eliminating the "poor" interviewers, who have a large impact on total survey error, and emulating the "good" interviewers, who reduce the total error.

The techniques described in this chapter are similar in that they all aim at providing a direct estimate of one of the components of nonsampling error. They also share a major disadvantage: they are costly and/or inconvenient to use. Consequently, too few studies are reported in the literature that use the techniques to estimate and compare components of nonsampling error between alternative modes of interview. Yet this type of information is essential in order to extend the knowledge of the quality of data produced by alternative modes of interview. Mode comparison studies, though powerful and informative, are often inconclusive without additional information regarding the relative sizes of the nonsampling error associated with each mode. Thus, future research on evaluating and comparing the quality of telephone interviewing with other modes of interview should focus on providing more direct information on the components of nonsampling error.

CHAPTER 18

DATA QUALITY IN TELEPHONE AND FACE TO FACE SURVEYS: A COMPARATIVE META-ANALYSIS

Edith D. de Leeuw and Johannes van der Zouwen[1]
University of Amsterdam and Free University, Amsterdam

In the last two decades telephone surveys have become an increasingly popular alternative to traditional face to face surveys. Major advantages of the telephone survey are the comparatively low costs and fast results, the greater potential of quality control through stricter supervision and monitoring of the interviewer staff, and the possibilities of reaching respondents in difficult to visit or dangerous neighborhoods and late at night (Groves and Kahn, 1979).

The first concern of researchers in the field of telephone surveys has been with improving sampling techniques and reducing nonresponse (Dillman, 1978; Frey, 1983). However, better sampling techniques and a high response rate, and the resulting reduction in coverage, sampling, and nonresponse error, do not necessarily imply a high validity of the data (Jones and Lang, 1982; de Leeuw and Hox, 1987). Other sources of error, for instance the interviewer, the questionnaire, and the respondent, can threaten the validity of survey research too (Rosenthal and Rosnow, 1969; Phillips, 1971; Dijkstra and van der Zouwen, 1982). In short, the utility of a data collection technique does not only depend on costs, speed, and response rate, but also on the quality of the data collected (Herman, 1977).

Recently, the interest of researchers has focused on the issue of data quality. This has resulted in an increasing amount of literature on mode

---

[1] The authors thank Prof. Dr. G.J. Mellenbergh, Prof. Dr. T.F. Pettigrew, and Dr. W Sykes for their comments on earlier versions of this chapter, and Dr. J.J. Hox for his practical help in writing a computer program. Special thanks are due to those friendly and helpful librarians, who sent us relevant articles from all over the world.

comparison, and the influence of the survey method on data quality, and thereby in a growing need for a comprehensive review on this subject.

To integrate research findings on mode differences and provide a review on this topic, we used principles of meta-analysis (Glass, McGaw, and Smith, 1981). This method makes it possible to present both an overview of mode differences found with respect to data quality, and an estimate of the (effect) size of those differences. In this chapter we will first of all present a systematic overview of empirical findings on differences between telephone and face to face surveys with respect to data quality. Furthermore, we will propose theoretically based explanations for the differences found empirically.

1. METHOD

1.1 On Meta-Analysis

Though the name "meta-analysis" deceptively suggests otherwise, meta-analysis is not one method or one type of analysis. Meta-analysis or integrative analysis, as it is often called, is a coherent set of quantitative methods for reviewing research literature (Hedges and Olkin, 1985; Glass, McGaw, and Smith, 1981; Light and Pillemer, 1984). The primary aim of meta-analysis is inferring (noncausal) generalizations about substantive issues from a set of studies directly bearing on those issues (Jackson, 1980). In meta-analysis, quantitative study outcomes from known research on a particular, well defined question are statistically combined.

The methods used in meta-analysis are not new. The principles of meta-analysis are the same as those governing "ordinary" survey research. Typical steps taken in meta-analysis are: a precise definition of the research problem, data collection (i.e., collection of relevant articles or papers), coding of the variables of interest, and statistical analysis (Wolf, 1986). In general, the dependent variable "study outcome" is operationalized in two ways: both the significance level (p-value) and an effect size measure are coded. Furthermore, background variables such as year of publication and source of publication are routinely coded, just as age and sex are routinely asked in a survey. Also, several research design characteristics of each study are coded (e.g., sampling method, type of subjects). This coding process results in a data matrix in which the cases (or rows) are the research studies of interest for the meta-analysis. Standard statistical procedures can then be used. In other words, the basic idea is to apply statistical methods, with the published statistics in previous studies of interest as the data (Waelberg and Haertel, 1980).

This distinguishes meta-analysis from the more traditional, narrative forms of literature review (Bangert-Drowns, 1986).

1.2 Retrieval and Selection of the Studies Reviewed

We started our meta-analysis with an on-line computer search of the relevant literature. The abstracting services used were: Psychological Abstracts (1967-1986), Sociological Abstracts (1963-1986), Dissertation Abstracts (1861-1986), and Dialog/SSCI (Social Sciences Citation Index, 1972-1986). The following keywords were used, both single and in combination: artifact, bias, comparison, data collection method, face to face, interview, personal, response, response bias, response effect, response style, social desirability, survey, and telephone. In addition the abstracts of SRM, a Dutch documentation center in the field of social research methodology, were searched for the period 1979 to 1986. The reference lists of the studies found in this way were then searched to uncover additional studies.

Most studies found (81 percent) were done in the United States. This could partly be a result of the data bases available for the computer search. In order to avoid "retrieval bias," we published an appeal for research articles in three European newsletters.

Studies were included in the meta-analysis when they empirically compared telephone and face to face surveys. Articles that only reviewed past literature, without presenting any new material, were not included. A second important criterion for inclusion was that the study reported at least one indicator for "data quality." Therefore, papers reporting response rates only were not included.

In total 31 articles and reports were found in which telephone and face to face surveys were compared on the quantity and the quality of the data. Sixteen different journals in the domains of psychology, sociology, marketing and opinion research, medicine, and criminology provided the relevant literature. The oldest reference was published in 1952, the most recent one in 1986. Some studies were (partly) reported in more than one article or paper. In order to avoid introducing dependence between the cases in the statistical analyses, the unit of analysis or case in this meta-analysis is a study, and not an article or paper (Bangert-Drowns, 1986; Rosenthal and Rubin, 1986). This explains why the number of cases reported in the analysis is fewer than 31.

1.3 The Coding of the Studies

Data quality is a complex and fuzzy concept (Bailar, 1984; O'Toole, Battistuta, Long, and Crouch, 1986). Especially in a study of subjective phenomena (i.e., attitudes, beliefs or other attributes, which cannot be observed directly), it is difficult, if not impossible, to assess the correctness of the answers (cf. Turner and Martin, 1984). In those cases various proxy variables, or indicators for the quality of the data, have been used (Groves, 1978). As a result, a large variety of different indicators of data quality can be found in empirical comparisons of face to face and telephone surveys. In order to make a useful selection of these indicators, a content analysis was conducted on a subsample of 20 articles. Only those indicators for data quality used in at least two studies were retained and coded for the meta-analysis. These indicators are:

1. Accuracy, or response validity; for this indicator the answer of the respondent is checked against the "true" value as found in official records (e.g., the possession of a driver's license). This indicator is only applicable when validating information is available (cf. Sudman and Bradburn, 1974).

2. Absence of S(ocial) D(esirability) Bias; inversely proportional to the number of socially desirable answers on a particular question. An answer is said to be socially desirable when that specific answer is more determined by what is acceptable in society than by the real situation (cf. DeMaio, 1984).

3. Item Response, inversely proportional to the number of "no answer" or "missing data" per question (excluding "do not know" responses).

4. Amount of Information, indicated by the number of responses given in response to an open question or a checklist.

5. Similarity of response distributions obtained by different modes. Indicated by no significant difference between the proportions obtained under the different modes. This indicator, though often used, is only a very rough indicator for data quality.

Besides these five indicators of data quality, a sixth dependent variable was used: Response Rate, defined by the number of completed interviews divided by the total number of eligible sample units (cf. Groves and Kahn, 1979; Kviz, 1977).

A coding schedule, partially based on Sudman and Bradburn (1974), was used. We included background variables relating to the research report (e.g., journal, year and country of publication), and the study itself (e.g., type and size of sample, subject of the research and its saliency for respondents, equivalence of samples and questionnaires used in the study).

For each indicator of data quality the size of the mode effect was computed, using several indices, depending on characteristics of the indicator involved. For the indicator Amount of Information, the standard effect size *d* was computed (Glass, McGaw, and Smith, 1981); *d* expresses the difference between two means in terms of the (within-group) standard deviation. A *d* value of 0.1 indicates that the means differ by a tenth of a standard deviation. When comparing modes of data collection with respect to the other four indicators of data quality, one actually makes a cross-classification of the mode of data collection used and the type of answers given (e.g., a 2 x 2 cross-classification of face to face versus telephone with socially desirable versus not socially desirable, or with accurate versus inaccurate). In these cases Cohen's (1969) effect size *e* for cross-classifications is used. In the case of a 2 x 2 cross-classification, *e* equals the proportion of variance accounted for by the mode effect (Cohen, 1969).

To make the different estimates of effect size comparable, a standard norm for effect size (SNES) was constructed. This norm is based on Cohen's (1969) definitions of what is to be considered a small, medium, or large effect size. For instance, Cohen defines a *d* of 0.2 as a small effect. This is equated with an *e* (2 x 2 table) of 0.01, which Cohen also defined as small. To determine intermediate values, linear interpolation was used.[2]

Furthermore, for each indicator the statistical significance level of the difference was coded. We also coded the direction of this difference (i.e., which mode offers data of better quality). For the indicator Similarity, coding of the direction of the difference was only done when the authors of the original study gave a convincing argument and a decision in terms of better quality could safely be made (e.g., more reporting of undesirable behavior such as taking oral contraceptives as a Roman Catholic).

Since the unit of analysis is a study (Bangert-Drowns, 1986; Wolf, 1986), when a particular study used more than one measure of the same construct (e.g., Item (non)Response for several questions), effect sizes

---

[2] This resulted in a nine point (0–8) SNES scale. Key points are: 0 = no effect (e, $d = 0$); 2 = small effect ($e = 0.01$, $d = 0.2$); 4 = medium effect ($e = 0.09$, $d = 0.5$); 6 = large effect ($e = 0.25$, $d = 0.8$); 8 = maximum effect ($e = 0.8$, $d = 1.4$). Other values were interpolated, e values were rounded to three and d-values to two decimals before translation to the SNES scale (de Leeuw and van der Zouwen, 1987).

were combined by taking the mean effect size prior to the coding (Rosenthal and Rubin, 1986). All the studies were coded by one of the authors; problem cases were discussed with the coauthor and with a statistician.

2. RESULTS

In total, 28 comparisons between face to face and telephone interviews were coded. In three cases relevant and reliable information was not available; those cases were excluded from further analysis. The remaining 25 cases have been published between 1952 and 1986. For a short description of these cases see Table 1. The most frequently used outlets for research reports were the *Public Opinion Quarterly* (5 cases) and the *Journal of the American Statistical Association* (3 cases).

A variety of subjects was covered in the questionnaires, with a predominant role for questions about health (10 cases). In 8 cases, the surveys studied dealt primarily with questions about behavior, 6 dealt with questions about (biographical) facts, 6 with questions about attitudes, 2 with questions about feelings, and 1 dealt with questions about cognitions of the respondents.

The samples used varied strongly with respect to the number of respondents (std. dev. = 1484), with an average sample size of 1532. In most cases (20 cases or 80 percent), a random sample was drawn; in the remaining cases a convenience sample (3 cases) or a panel (2 cases) was used.

The response rate of the face to face interview is higher than that of the telephone interview: for the face to face interview a mean response rate of 75 percent is reported (std. dev. = 18 percent) versus a mean response rate of 69 percent for the telephone interview (std. dev. = 19 percent); $p < 0.02$.

For each of the five indicators of data quality the p values were combined over the cases, using the z transformation (Cooper, 1979; Rosenthal, 1978). For each indicator the combined p value was less than 0.01. However, for the indicator Accuracy this result is rather spurious. In one case an extremely large and significant effect size for Accuracy was found. This case was found in an early publication (Larson, 1952) and it concerned a very special topic (i.e., knowledge about the content of leaflets dropped from a plane). It can be argued that this specific case is an

Table 1. Major Characteristics of the 28 Studies Reviewed. Author, Year of Publication, Subject, Indicators Coded for Meta-Analysis, and Summary Conclusion as Given in the Original Article.[a]

First Author	Year	Subject	Indicator	Conclusion
Aneshensel	1982	Health/depression	Similarity	No significant mode effects
Assael	1982	Consumer/business	Accuracy	Telephone less accurate
Cahalan	1960	Consumer/newspaper	Similarity	No differences
Colombotos (also Colombotos 1965)	1969[a]	Health/opinion	S.D. bias, Similarity	Essentially no differences
Colombotos	1969[a]	Health/attitudes	Similarity	Essentially no differences
Dillman	1984	Housing	Item response, Similarity	Some evidence telephone more extremeness
Groves (also Groves 1979)	1978	Several topics	Item response, Amount info, Similarity	Tel. tends to yield fewer and faster answers
Henson	1978	Health/moods	S.D. bias, Similarity	Tel. fewer symptoms and higher on S.D. scale
Herman	1977	Voting	Accuracy, S.D. bias, Amount info	In general, no mode effects, tel. less willing to reveal sensitive info

Table 1 (Continued)

First Author	Year	Subject	Indicator	Conclusion
Herzog	1983	Reanalysis of earlier surveys; older subjects as focus of interest		Elderly gen. underrepresented, little evidence for mode by age interaction
Hinkle	1978	Health/mental	Similarity	Both methods yield comparable data
Hochstim (also Hochstim 1962)	1967[b]	Health/general	Item response, Similarity	Data collection strategies used proved to be practically interchangeable
Hochstim	1967	Health/cervical cytology	Accuracy, Item response, Similarity	
Janofsky	1971	Health/self disclosure	Similarity	In both modes, resp. equally willing to express feelings
Jordan (also Jordan 1978)	1980	Health/attitude	Item response, Amount info, Similarity	Tel. more missing income data, more extremeness, acquiescence & evasiveness
Kersten	1985[a]	Travel	Similarity	Small differences
Klecka	1978	Crime/victimization	Item response, Similarity	RDD tel. can replicate face to face interview with complex sampling
Larson	1952	Leaflet messages	Accuracy, Item response, Similarity	Serious doubt on validity of tel. responses

Table 1 (Continued)

First Author	Year	Subject	Indicator	Conclusion
Locander	1976	Personal facts	Accuracy	None of the methods differed significantly
Mangione	1982	Drinking	Similarity	Results equivalent
Oakes	1954	Consumer/free resp	Amount info	Average of suggestions less in tel.
O'Toole	1986[a]	Health/veterans	Accuracy, Item response	No mode differences observed
Rogers	1976	Housing/ city services	Accuracy, Item response, Similarity	Quality of data collected is comparable
Schmiedeskamp	1962	Consumer/finances	Similarity	Tel. some avoiding of definite positions
Siemiatycki	1979[a]	Health/ community needs	Accuracy, Item response	No evidence quality is superior with home interviewing
Wiseman	1972	Several topics	Similarity	Responses not always independent of method
Woltman	1980	Crime/victimization	Similarity	Number of reported victimizations less in procedure with tel. as major mode
Yaffe	1978	Health/utilization	Accuracy	In person strategies higher accuracy

[a]Country of origin of the studies was the United States, with the exception of Kersten (Holland), O'Toole (Australia), and Siemiatycki (Canada).
[b]Two separate studies are reported in one article. Some studies are partly reported in more than one article, the first author and year of publication of the additional article is then given in parentheses.

Table 2. Comparison of Data Quality in Face to Face and Telephone Surveys; Mean Effect Size, Mean, Standard Deviation, Minimum and Maximum SNES Value, Number of Comparisons in Analysis

Indicator	Mean *e*	Mean *d*	SNES[a] Mean	Stnd. Dev.	Min.	Max.	Number of Cases
Accuracy	0.00	—	0.4	0.7	0	2	8
Absence of Social Desirability Bias	0.01	—	1.3	0.6	1	2	3
Item (non)Response	0.01	—	0.8	1.2	0	3	9
Amount of Information	—	0.12	2.5	0.6	2	3	4
Similarity	0.01	—	0.7	0.9	0	2	16

[a]SNES (Standard Norm Effect Size) is measured on a nine point scale: (0) no effect, (4) medium effect, and (8) maximum effect.

"outlier," and we reanalyzed the data without this case.[3] This resulted only in a change for the estimates of the indicator Accuracy: without the 1952 case the combined p value for this indicator is 0.18. For the indicators Absence of Social Desirability Bias, Amount of Information, Item (non)Response, and Similarity, the difference found between telephone and face to face interviews was statistically significant and in favor of the face to face interview.

Table 2 summarizes the results for the mode effects on data quality. In most comparisons only one or two indicators of data quality were used. As a consequence, the data points for each indicator are limited and differ in number. When we inspect Table 2, we must conclude that the effects, though significant, are small. The largest effect sizes are found for the indicators Amount of Information (SNES = 2.5) and Absence of Social Desirability Bias (SNES = 1.3); however, even these are, according to Cohen's (1969) definition of effect size, small. In this sense the results are promising (Jordan, Marcus, and Reeder, 1980).

Going back to the individual studies (as summarized in Table 1), we see that in the majority of the comparisons no statistically significant mode effects are found. When differences are found, they are in favor of the face to face interview. This is in agreement with the findings of Smith

[3]With the 1952 case the combined p value for Accuracy is 0.00, the mean e value is 0.05 (std. dev. = 0.07), and the mean SNES is 1.0 with a standard deviation of 2.

(1984), who concluded in a review of the literature on telephone and face to face interviews that 'the tilt of research findings is towards better quality data from personal interviews.'

In the articles coded, sometimes additional indicators for data quality were reported. For instance, Jordan et al. (1980) compared response styles in a telephone and face to face interview. They found more acquiescence (agreement), more evasiveness (do not know and no answer), and more extremeness (extreme answer category) response bias in the telephone interview. A tendency for the telephone respondent to choose the more extreme point on a scale was also noted by Groves (1979). This result is partly corroborated by Dillman and Mason (1984) who investigated extremeness bias and report that 'there is some evidence to support the telephone extremeness response..., but it is neither strong nor completely consistent.' Also, aspects of (psychometric) reliability were investigated by several authors. Aneshensel, Frerichs, Clark, and Yokopenic (1982) found no differences between modes in the internal consistency (Cronbach's alpha) of a depression scale. Concerning consistency over time (test-retest reliability), no differences were found between the telephone and face to face interview (Herman, 1977; O'Toole et al. 1986; Rogers, 1976). Furthermore, Herman (1977) reported that when the initial interview was in person and the second by telephone, or vice versa, attitudes tended to be less consistent than when both interviews were of the same type. This last finding, when replicated, is of importance for the users of mixed mode techniques (this volume, Chapter 32).

Several authors (e.g., Groves, 1979; Jordan et al., 1980) have worded the hypothesis that the relative inexperience with telephone surveys is the cause of the better performance of the face to face interview, and that the differences between the modes should become less over time. The data of the meta-analysis presented here provide the means to investigate this idea. We computed the correlation coefficient between the year of publication and the indicators for data quality used in this study, of course without the extreme 1952 case. The correlations found between year and effect size are all negative, and range from −0.21 for Item (non)Response and −0.47 for Accuracy. Differences between the face to face and telephone interviews are indeed becoming less over the years.

3. A THEORETICAL EXPLANATION OF THE RESULTS

3.1 The Explanatory Model

To explain the mode differences found, we used a theoretical model on data collection (de Leeuw and van der Zouwen, 1987; van der Zouwen and de Leeuw, 1987). In this model earlier models of the interview process (e.g., those of Nowakowska, 1970; Sudman and Bradburn, 1974; Cannell, Miller, and Oksenberg, 1981; Dijkstra and van der Zouwen, 1982; van der Zouwen, Dijkstra, and van de Bovenkamp, 1986) are integrated. The model is further elaborated, using earlier attempts to explain mode effects (e.g., Groves and Kahn, 1979; Singer, 1979; Sykes and Hoinville, 1985). Figure 18.1 gives a graphical summary of the model. Dependent variables of the model are the performance characteristics of the modes (Dillman, 1978): Response Rate, Amount of Information, Item Response, Accuracy, and Absence of Social Desirability Bias. The independent variables characterize the method of data collection used. For instance, interviewing by telephone is more associated with centrally located interviewers and direct supervision of interviewers. This enhances the control over interviewers' performance in the telephone interview. The traditional face to face interview has fewer limitations with regard to social customs (i.e., the acceptability of conversations with strangers and the acceptability of long speech acts), more possibilities to mail introductory letters (as compared to telephone surveys using random digit dialing), and a larger "channel capacity." These relationships are represented by arrow 1 in Figure 18.1.

The difference in channel capacity may affect the performance characteristics through various intervening variables. Without visual support, both the understanding of and the responding to questions are more difficult for the respondent, and the interviewer has less opportunity to notice inadequate response behavior. This will lead to a greater cognitive complexity of the task for the respondent, and to lesser control over respondents' behavior in the telephone interview. Also, without visual contact between interviewer and respondent, the affective distance may be larger and the rapport between them smaller. This may lead to less motivation of the respondent, but also to less need for approval. Since the incoming signal in telephone conversations is nearly inaudible for bystanders, responses are incomprehensible to other persons, especially when closed questions are used. This reduces the potential influence of bystanders (arrows 2 to 6).

The large amount of control over interviewers, which characterizes the modern telephone survey, will probably have a beneficial effect on the adequacy of interviewing: the supervisor can listen in on the interviews

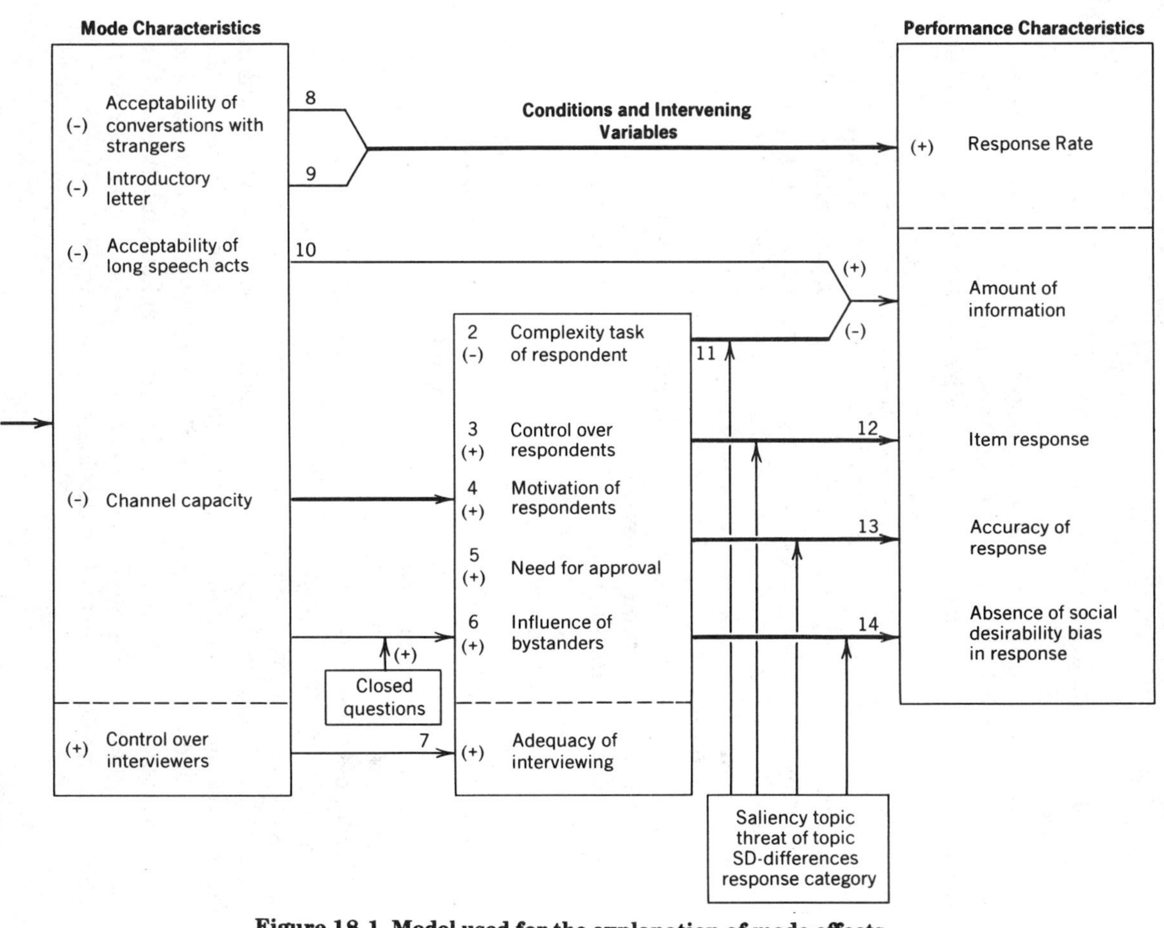

Figure 18.1 Model used for the explanation of mode effects.

and correct inadequate formulations of the questions or inadequate interviewer behavior (arrow 7).

3.2 An Explanation of the Effects Found

The amount of information given in response to open questions and checklists is higher in face to face interviews. Two factors are important for understanding this mode difference. First of all, there is a higher degree of acceptability of long speech acts in face to face interviews (see Figure 18.1, arrow 10). Secondly, face to face interviews are characterized by a larger channel capacity. This gives the interviewers more opportunities to motivate respondents when a speech act dries up, and to help them by making their tasks less complex (arrows 2, 3, 4, 11).

A small mode difference was found concerning the number of "no answers" and "missing data." The somewhat higher item response in face to face interviews can be explained by the larger channel capacity of this mode, and the resulting possibilities for a higher control over respondents, for a stronger motivation of respondents, and for less complexity of respondents' task (arrows 3, 4, 2). This effect is somewhat tempered by the lower potential for control over interviewers (arrow 7) in face to face interviews, resulting in a small difference.

A small and nonsignificant mode effect was found for the accuracy, or (checked) validity of the responses. Former research on response effects (Dijkstra and van der Zouwen, 1982) has shown that characteristics of the topic, such as saliency and threat, play an important role here. Depending on the topic of the question, the factors "need for approval" and "influence of bystanders" will have a negative influence on the accuracy in face to face interviews, and counteract the positive influence of more motivation and control over respondents.

Finally, a small difference in socially desirable answers was found in favor of the face to face interview. This effect cannot be explained by the theoretical model. In fact, considering the small need for approval and the small possible influence of bystanders in telephone interviews, less social desirability bias would be theoretically expected in telephone interviews. We will come back to this issue in the discussion.

4. DISCUSSION

Only small differences were found between telephone and face to face interviews. Furthermore it was found that these differences have become smaller over time. Since mode effects are small, other factors such as preferences and accessibility of respondents and costs (Groves, 1979; Woltman et al., 1980; Nicholls and Groves, 1986) may be more important for an individual researcher in deciding which method is best (cf. Dillman, 1978).

The finding that only small differences exist does not necessarily imply that our methods of data collection are good. For instance, although no significant difference in accuracy was found between telephone and face to face interviews, both methods show some bias (Assael and Keon, 1982; Locander et al., 1976). Especially with threatening questions, no data collection method is perfect (Bradburn, Sudman, et al., 1979). Consequently, it is important to keep improving data collection techniques, and to experiment with new methods.

Reviews are always one step behind the actual state of the art. In the articles reviewed here no explicit reference was made to the use of computer-assisted telephone interviewing (CATI). As the acceptance and the use of CATI are increasing rapidly (Spaeth, 1987; this volume, Chapter 29), it is of paramount importance that further research in this field be done, and that controlled experiments and comparisons with CATI be performed.

Considering the higher degree of anonymity in telephone interviews, more honesty and less social desirability may be expected. However, the results of this meta-analysis suggest a slight advantage for face to face interviews in this respect. Other authors (e.g., Smith, 1984; Bradburn, 1983) conclude that results are equivocal at best.

Why telephone interviews are not as effective as expected in asking sensitive questions is not known, but some possible reasons suggest themselves. Johnson, Hougland, and Clayton (1987) commented on a certain "unease" of respondents when discussing sensitive subjects by telephone. Groves (1979) found a marked respondent preference for the face to face interview. Major reasons were "know who you are talking to" and "liking to see who one is dealing with." Woltman et al., 1980) also report that being able to see the interviewer and check credentials are reasons cited for preferring face to face interviews. Dillman (1978) emphasizes the importance of trust in order to obtain valid answers. It is conceivable that the beneficial influence of the greater anonymity in telephone interviews only works when respondents are convinced that they are dealing with a worthy organization, which can be trusted with socially undesirable answers. Whether this posthoc explanation is correct should be investigated empirically.

Hopeful for researchers interested in sensitive topics are the results of two recent European studies. Körmendi (this volume, Chapter 21) had in Denmark the unique opportunity to check responses with income data registered by tax authorities. She did find a higher item response in the face to face interview and more "do not know" answers and refusals on income questions in the telephone mode. Her major conclusion is encouraging: the method of data collection has no influence on the checked accuracy of answers. There even is a slight tendency toward higher accuracy for the telephone in certain subgroups. Sykes and Collins (this volume, Chapter 19) describe three direct experimental comparisons between telephone and face to face interviews in Great Britain. Their main finding in all three studies was the similarity of the answers obtained by the different modes. Though not statistically significant, across questions they detect a tendency to give more socially desirable answers in face to face interviews. In a fourth nonexperimental comparison, they found more reported drinking in a telephone interview than in a face to face interview held a year earlier.

There is reason for optimism about the further improvement of telephone interviewing. Not only are the differences found between the modes small, the differences are becoming even smaller over the years. Let us concentrate on further increasing the possibilities and the quality and of telephone interviews!

APPENDIX

Articles coded for the meta-analysis. Some studies were partly reported in more than one article or paper. These are marked with an asterisk.

Aneshensel, Caro S., Frerichs, Ralph R. Clark, Virginia A. and Yokopenic, Patricia A. Measuring depression in the community. A comparison of telephone and personal interviews. *Public Opinion Quarterly*, 1982, 46, 110–121.

Assael, Henry and Keon, John. Non sampling versus sampling errors in survey research. *Journal of Marketing*, 1982, 46, 114–123.

Cahalan, Don. Measuring newspaper readership by telephone: two comparisons with face-to-face interviews. *Journal of Advertising Research*, 1960, 1, 1–6.

*Colombotos, John. The effects of personal versus telephone interviews on socially acceptable responses. *Public Opinion Quarterly*, 1965, 29, 457–458.

*Colombotos, John. Personal versus telephone interviews: effect on responses. *Public Health Report*, 1969, 84, 773-782.

Dillman, Don A. and Mason, Robert G. *The influence of survey method on question response*. Paper presented at the annual meeting of the American Association for Public Opinion Research, Delavan, Wisconsin, 1984.

*Groves, Robert M. On the mode of administering a questionnaire and responses to open-ended items. *Social Science Research*, 1978, 7, 257-271.

*Groves, Robert M. Actors and questions in telephone and personal interview surveys. *Public Opinion Quarterly*, 1979, 43, 190–205.

*Groves, Robert M. and Kahn, Robert L. *Surveys by telephone; a national comparison with personal interviews*. New York: Academic Press, 1979.

Henson, Ramon, Cannell, Charles F. and Roth, Aleda. Effects of interview mode on reporting of moods, symptoms, and need for social approval. *Journal of Social Psychology*, 1978, 105, 123–129.

Herman, Mary B. Mixed-mode data collection: telephone and personal interviewing. *Journal of Applied Psychology*, 1977, 62, 399-404.

Herzog, A. Regula, Rodgers, Willard L. and Kulka, Richard A. Interviewing older adults: a comparison of telephone and face-to-face modalities. *Public Opinion Quarterly*, 1983, 47, 405-418.

Hinkle, Andrew Lee and King, Glen D. A comparison of three survey methods to obtain data for community health program planning. *American Journal of Community Psychology*, 1978, 6, 389-397.

*Hochstim, Joseph R. Comparison of three information gathering strategies in a population study of sociometrical variables. Proceedings of the Social Statistics Section of the American Statistical Association, 1962, 154-159.

*Hochstim, Joseph R. A critical comparison of three strategies for collecting data from households. *Journal of the American Statistical Association*, 1967, 62, 976-989.

Janofsky, A. Irene. Affective self-disclosure in telephone versus face-to-face interviews. *Journal of Humanistic Psychology*, 1971, 11, 93-103.

*Jordan, Lawrence A., Marcus, Alfred C. and Reeder, Leo G. Response styles in telephone and household interviewing: a field experiment from the Los Angeles health survey. *Proceedings of the Section on Survey Research Methods of the American Statistical Association*, 1978, 362-366.

*Jordan, Lawrence A., Marcus, Alfred C. and Reeder, Leo G. Response styles in telephone and household interviewing: a field experiment. *Public Opinion Quarterly*, 1980, 44, 210-222.

Kersten, H.M.P. and Moning, H.J. Differences in estimates due to changes in methods of data collection. *Kwantitatieve Methoden*, 1985, 19, 31-47.

Klecka, William R. and Tuchfarber, Alfred J. Random digit dialing: a comparison to personal surveys. *Public Opinion Quarterly*, 1978, 42, 105-114.

Larson, Otto N. The comparative validity of telephone and face-to-face interviews in the measurement of message diffusion from leaflets. *American Sociological Review*, 1952, 17, 471-476.

Locander, William, Sudman, Seymour and Bradburn, Norman. An investigation of interview method, threat and response distortion. *Journal of the American Statistical Association*, 1976, 71, 269-275.

Mangione, Thomas W., Hingson, Ralph and Barrett, Jane. Collecting sensitive data. *Sociological Methods and Research*, 1982, 10, 337-346.

Oakes, Ralph H. Differences in responsiveness in telephone versus personal interviews. *The Forum*, 1954, 19, 169.

O'Toole, Brian I., Battistuta, Diana, Long, Ann and Crouch, Keitha. A comparison of costs and data quality of three health survey methods: mail, telephone and personal home interview. *American Journal of Epidemiology*, 1986, 124, 317-328.

Rogers, Teresa F. Interviews by telephone and in person: quality of response and field performance. *Public Opinion Quarterly*, 1976,40, 51-65.

Schmiedeskamp, Jay W. Reinterviews by telephone. *Journal of Marketing*, 1962, 26, 28-34.

Siemiatycki, Jack. A comparison of mail, telephone, and home interview strategies for household health surveys. *American Journal of Public Health*, 1979, 69, 238-245.

Wiseman, Frederich. Methodological bias in public opinion surveys. *Public Opinion Quarterly*, 1972, 36, 105-108.

Woltman, Henry F., Turner, Anthony G. and Bushery, John M. A comparison of three mixed-mode interviewing procedures in the national crime survey. *Journal of the American Statistical Association*, 1980, 75, 534-543.

Yaffe, Richard, Shapiro, Sam, Fuchsberg, Robert R., Rhode, Charles A. and Corpeno, Helen C. Medical economics survey methods study; cost-effectiveness of alternative survey strategies. *Medical Care*, 1978, 16, 641-659.

CHAPTER 19

EFFECTS OF MODE OF INTERVIEW: EXPERIMENTS IN THE UK

Wendy Sykes and Martin Collins
Social and Community Planning Research, England

As we have discussed elsewhere (this volume, Chapter 13), our research into the effects of interviewing by telephone has arisen in a hostile, or at best doubting, environment. In Britain, telephone surveys seem to break taboos about private telephone etiquette (for example, against unsolicited calls) and to challenge ideas about what the telephone can and should be used for: mainly short, purposeful conversations. Despite changes over the last decade, the British public is thought to be resistant to telephone interviewing, unsophisticated in the use of the telephone, and unpracticed in carrying out complex cognitive tasks through the medium. Our conclusions deny almost all of these beliefs except perhaps that of general reluctance to be interviewed over the telephone.

This chapter is based on results from four comparative studies carried out between 1983 and 1986. The first three — two studies of British social attitudes carried out by the Survey Methods Centre at Social and Community Planning Research (SCPR), and a study of lifestyle in the 1980s carried out by Marplan Ltd. for the Market Research Development Fund (MRDF)[1] — provided direct comparisons of telephone and face to face interviewing. In each case samples of people accessible by telephone were selected, using electoral registers to generate names and addresses and directories to generate telephone numbers where available. Addresses were then systematically allocated to either telephone or face to face interviewing. (For further details, see the Appendix to Collins et al., (this volume, Chapter 13.)

[1] These three studies are referred to hereafter as BSA 1, BSA 2, and Lifestyle, respectively.

Full results from these three studies have been reported elsewhere (Sykes and Collins, 1987; Market Research Development Fund, 1985). Here we summarize in Section 1 their findings in the context of possible psychological and sociological explanations.

The fourth study — of attitudes towards alcohol — was a collaborative venture between SCPR and the Social Survey Division of the Government Office of Population Censuses and Surveys (OPCS). This also used a sample selected from the electoral register followed by a directory search for telephone numbers. Here, however, comparisons are drawn not with a parallel face to face experimental sample but with an interview survey carried out by SCPR about a year earlier.

This study, discussed in Section 2, was designed to build on the generally encouraging results of the three earlier experiments. It sought to follow up some of the isolated instances of mode effects found earlier and to explore, through split ballot tests of question form and wording, approaches to obtaining information by telephone. (Again further details of the study design will be found in the Appendix to Collins et al., this volume, Chapter 13.)

1. THREE DIRECT COMPARISONS

The first three studies described above yielded direct comparisons between telephone and face to face interviewing over a wide range of descriptive behavioral and attitudinal variables and for a variety of question types. The results, described in detail by Sykes and Collins (1987), were generally encouraging. The main finding in all three studies was the similarity of the answers obtained by the different interviewing modes. This result is consistent with those of similar studies in other countries, notably the United States. (Colombotos, 1969; Wiseman, 1972; Lucas and Adams, 1977; Groves and Kahn, 1979; Jordan et al., 1980; Körmendi et al., 1986).

1.1 Nonresponse

In all three studies, efforts made during selection to ensure comparability between the samples interviewed by alternative modes were threatened by differential nonresponse. The telephone subsamples yielded consistently lower completion rates: 53 percent versus 60 percent on BSA 1; 46 percent versus 68 percent on BSA 2; and 45 percent versus 67 percent on the lifestyle survey. (A full analysis of response to the survey is given in Collins et al., this volume, Chapter 13.)

Fortunately there seems to have been relatively little impact on sample profiles. For both BSA 1 and BSA 2 the samples to be compared were tested for differences on a number of demographic and socioeconomic variables of relevance to the survey. These were: age within sex, marital status, household composition, economic status, socioeconomic group, and geographical location. No statistically significant evidence of differential nonresponse was found on BSA 1. Statistically significant but still small differences were found on BSA 2: the telephone sample had more childless couples under 60 and fewer couples with young children and teenagers. It also had a higher proportion of intermediate and junior nonmanual workers and fewer homemakers.

The possible effects of these small differences would be worrying if the studies *had* suggested differences on substantive items between interviewing modes. In their absence, it is of course possible that there were two offsetting effects, interviewing mode effects working to balance out the effects of differential nonresponse. A far less strained explanation is that neither effect occurred, that the sample differences arising from nonresponse were too small and/or weakly correlated with the survey variables to affect the results.

1.2 Mode Effects

The marginal distributions of response yielded by the different modes of interview were compared and differences tested for statistical significance using chi squared tests, even though the data arose from a complex (not a simple random) sampling procedure. Much has been written about the effect of complex designs on the sampling distribution of the chi square statistic (e.g., Holt et al., 1980) and about the adjustments that should be made in order to compensate for the clustering in the design. In essence, with designs that are clustered, the *true* estimates of variability are usually higher than those based on the assumption of a simple random sampling design. This will lead, ultimately, to test statistics that are too *large*, and hence to the false rejection of null hypotheses (i.e., to anticonservative tests). However, with respect to the social attitudes survey, much research has focused on the calculation of *true, complex* standard errors. This has revealed that, for attitudinal variables, the *design factor* (defined as the ratio of the complex standard error to the simple random sampling standard error) with the design used in these surveys rarely exceeds 1.2. It was felt that this provided sufficient justification for using standard rather than adjusted chi squared tests, particularly since the literature argues that in a 2-way test of independence the consequences of clustering are unlikely to be severe (Holt et al., 1980). It was not felt that the large amount of computation necessary to calculate corrected chi squared statistics was justified.

The overall pattern of results for the two social attitudes studies is shown in Table 1. The table shows the incidence of statistically significant mode effects across 95 questions (or parts of questions) in BSA 1 and 69 questions in BSA 2.

Given a degree of lack of independence between the questions in a single survey — even one which covers a variety of topics —our *overall* conclusion is that these results are approximately consistent with a hypothesis of no mode effects. That is, the percentage of questions showing mode effects is approximately that which would be expected to occur by chance (i.e., if the items in the survey were considered as independent).

There was no obvious explanation for those differences which (did) occur. We found no consistent patterns and no apparent single or set of characteristics linking the questions. For example, questions asked with a show card face to face were no more likely to show mode effects than those asked without. Similarly, items identified in advance as likely to pose particular response problems for respondents interviewed by telephone did not, in the event, appear to do so. These included questions incorporating one or more potentially difficult concepts, long questions and questions with large numbers of response options.

The overall conclusion from the lifestyle comparative study was very similar: "We can be reassured that for most straightforward market research questions the telephone will give the same answers as a face to face interview" (McDonald, 1985).

This conclusion is probably readily acceptable in the United States, with its background of methodological research and familiarity with

Table 1. BSA 1 and BSA 2 Percentage of Questions Showing Mode Effects

Number of Questions	BSA 1 (95)%	BSA 2 (69)%
No significant difference	91	87
Significant at 5%	7	9
Significant at 1%	2	4

BSA 1 sample sizes: telephone 310, face to face 229.
BSA 2 sample sizes: telephone 354, face to face 360.

telephone interviewing. It is becoming accepted in the United Kingdom but still faces the kinds of prejudice against telephone interviewing mentioned in our introduction. It should, perhaps, come as less of a surprise given the results of earlier psychological investigations of the effects on communication of physical separation and the absence of visual cues.

Argyle (1969) identified six functions for nonverbal cues during face to face interaction:

1. Mutual attention and responsiveness (to indicate that attention is being paid)

2. Channel control (to determine turn taking in the interaction)

3. Feedback (to convey reactions to other participants' utterances, e.g., pleasure, surprise, comprehension)

4. Illustrations (accompanying gestures that illustrate what is being said and provide emphasis)

5. Emblems (gestures used instead of words)

6. Interpersonal attitudes (to convey the attitudes and intentions of the protagonists toward one another).

The effects of removing the visual channel of communication depend to some extent on the type of interaction taking place and the relative importance of these different functions, and partly on the substitutability of verbal for nonverbal signals. In a review of small group laboratory experiments comparing telephone and face to face conversations, Reid (1977) concluded that the effects of the medium are, for the most part, "subtle, small, and elusive." In particular, the withdrawal of vision had no measurable effect of any kind on the outcome of communication involving information transmission and problem solving tasks.

1.3 Sensitive Questions

Within the overall pattern of lack of mode effects, we can look at some specific types of question where there are occasional suggestions of effects or where earlier comparative research has suggested such effects might arise. One example is the "sensitive" or "threatening" question.

We adopt a very broad definition of such questions. These are questions dealing with issues of a personal or private nature, and/or to which certain responses are clearly more socially desirable. Included are questions about the respondent's circumstances and behaviour (income, dealings with the police) as well as attitudes and beliefs (e.g., sexual mores, religious conviction).

Few of the items identified on any of the studies as potentially sensitive or threatening showed evidence of mode effects. For example, there was only one such item on BSA 1, out of a possible 15 (over the telephone 42 percent of respondents admitted having no religion, compared with 28 percent of those interviewed in person). There was, however, a consistent *pattern* across the questions: a tendency to give more socially desirable answers face to face. Thus, a higher percentage of respondents to BSA 1 interviewed by telephone admitted to having been questioned by the police over the past two years in connection with a crime. Similarly, a variety of questions covering crime and sexuality asked on the survey of lifestyle yielded apparently more honest answers over the telephone.

These findings — of weak support for the telephone approach —were ones which evidence from psychological research had led us to expect. They are also supported elsewhere in the literature (Schmiedeskamp, 1962; Colombotos, 1969; Hochstim, 1967; Rogers, 1976; Cannell et al., 1987) although not in all cases (Wiseman, 1972; Henson et al., 1974; Groves et al., 1979).

Of particular interest here is the work of Short and his colleagues (Short et al. 1976). Their work in small group experiments examines the effects of *social presence* on the content, characteristics and outcome of an encounter. *Social presence*, which combines the effects of physical presence and visual communication, is conveyed least strongly in audio-only communication and most strongly when people meet face to face. Some intermediate level is achieved with audiovisual media such as videos.

A weak social presence confers a sense of anonymity, which, according to Short et al. (1976), decreases the magnitude of conforming response in a wide range of situations. Since social presence is felt less strongly over the telephone, it is arguable that in certain situations this mode of communication is less likely to arouse anxiety or tension in an encounter — ingredients in the tendency to conform. [This may account for the fact that some forms of stuttering disappear over the telephone (Boettinger, 1977).] Telephone selling to industry and retail trade has capitalized on the apparent weak position of the telephone salespeople whose questions "sound harmless" and who "elicit ready answers from their prospects" (Wage, 1973). Anxiety such as that frequently expressed by face to face respondents (particularly the elderly) that they might not be giving the right answers, or in other respects performing well (Young, 1987), may be less prevalent in telephone interviews.

Finally, where the participants in an encounter are able to see one another as well as hear one another, many more clues about the social characteristics of the other are available to evoke the cultural norms and stereotypes that can influence the opinions expressed. For survey interviews "social expectation" effects are a well documented phenomenon. For example, both black and white respondents tend to give more racially liberal responses to black than to white interviewers (Hatchett and Schuman, 1975–76; Schaeffer, 1980; Campbell, 1981), especially on questions pertaining to race (Sudman and Bradburn, 1974). Thus, for items or topics particularly susceptible to such effects, the telephone may hold certain advantages over face to face methods. [Although in at least one study, race of interviewer effects were found to persist even over the telephone (Cotter, Cohen, and Coulter, 1982).]

1.4 Open Questions

The results of the experimental surveys also supported a handful of fairly stable findings on the interaction of mode and question form reported in the literature on telephone interviewing. The most significant of these is a tendency for respondents to give truncated answers to open ended items over the telephone (e.g., Groves and Kahn, 1979). Although neither of the social attitudes studies carried any open ended items, the lifestyle study included a number of spontaneous awareness measures. A higher percentage of telephone respondents gave no answers to these questions, and of those that did answer, the average number of responses given over the telephone was significantly lower than face to face.

The tendency to obtain truncated response has been associated with the faster pace of telephone interviewing (Williams, 1977). Certainly this characteristic was noticeable on BSA 2 where the same questionnaire was used for both modes of interviewing. Ten percent of the telephone interviews designed to take 25 minutes were conducted in under 20 minutes, compared with 5 percent of face to face interviews. And less than a third, compared with 41 percent, took more than half an hour.

Analysis of tape recorded telephone conversations and face to face conversations carried out by Wilson and Williams (1977) demonstrated not only that telephone conversations tended to be shorter, but also that the length of each utterance (i.e., the number of words between speaker switches) and the mean number of speech acts tended to be shorter. This was an uncontrolled experiment: the content or purpose of the conversations was not randomly assigned to a mode of communication, and the authors concur that the characteristics identified could be highly correlated with the nature of the encounters. However, Davies (1971) also found that the participants in telephone conversations concerning problem solving tasks took significantly less time than in face to face contact and exchanged fewer suggestions. Chapanis et al. (1972) likewise found that participants exchanged more messages in the face to face than in the telephone condition. Wilson and Williams (1977) attribute such findings to the lack of certainty of speakers regarding the reception of their utterances, resulting from the absence of continuing feedback.

Another reason why telephone conversations may be shorter than those carried out in person is that there tend to be significantly more pauses in face to face encounters (Williams, 1978). Again this may be related to the need for feedback; those interacting over the telephone may rely on a verbal flow to provide continuity in the absence of other cues about what is taking place. The telephone conversationalist feels compelled "to participate by taking an active part in maintaining the flow of talk" (Ball, 1968). In telephone interviews this might take the form not only of respondents providing answers more quickly but also interviewers allowing less time between items. Because of this, answers requiring more processing by respondents (i.e., those involving more difficult cognitive tasks such as recalling, giving multiple responses, or formulating full answers in their own words) might be expected to be vulnerable to media effects.

1.5 Questions Involving Response Scales

Scale items have also received some attention in the methodological literature on telephone interviewing. In most cases, the interest stems not from a concern with scale items *per se*, but from debate about how to handle questions which face to face would probably be administered with a show card. Rating scales fall into this category — visual aids being routinely employed in the face to face interview both to help respondents remember the distinctions between scale points and to alleviate interviewer and respondent fatigue on scale batteries.

Alternative ways of presenting scale questions to telephone respondents include reading out each of the scale points and asking the question in two stages as "unfolding." For example a five point "satisfaction-dissatisfaction" scale would first be presented with three response options (*Satisfied, Dissatisfied* and *Neither Satisfied nor Dissatisfied*). "Satisfied" would then be asked: "Is that 'Very Satisfied' or just 'Satisfied'?" Those endorsing "Dissatisfied" would be asked: "Is that 'Very Dissatisfied' or just 'Dissatisfied'?"

Findings from research on the topic are difficult to evaluate. For example, Groves and Kahn (1979) recommended the two step (or "unfolding") alternative on the basis of results which demonstrated a tendency toward extreme positive responses and lower inter-item correlation with the one step version. Miller (1984), controlling for some of the potentially confounding elements in the former study, found the opposite: the one step scale was recommended because it produced *fewer* extremely positive endorsements and showed *higher* inter-item correlations. It also produced less missing data and was preferred by interviewers who found the two step version too long and repetitious. In both these studies the comparisons were between numerical scales with labelled end and middle categories (asked in one step) and two step scales with all seven categories labelled. Though perhaps justifiable on pragmatic grounds, (these may be the practical alternatives for long scales) such experiments make it difficult to evaluate and understand the processes which are actually at work.

The lifestyle study tested a number of 5 point agree-disagree scales using an unfolding method over the telephone and both one step (with a show card) and two step versions face to face. The two step versions in either mode were identical, allowing the researchers to control for either mode or form effects in assessing any differences in the answers obtained under the experimental conditions. In the event, no significant differences were found between any of the mean scores. Nor was there any consistent pattern across the test items of higher or lower mean scores from any of the approaches.

These issues were also pursued further in the survey of attitudes toward alcohol discussed in the next section.

2. THE SURVEY OF ATTITUDES TOWARDS ALCOHOL

2.1 Objectives

The fourth study which we will discuss has to be seen against the background of the earlier comparative experiments. As described in the introduction, it did not involve a directly comparative subsample interviewed face to face but relied on a more indirect comparison with a previously conducted face to face survey. The aim in this case was not to show for a fourth time that telephone interviewing was generally acceptable but to build on the results of the earlier studies. Thus, as we have mentioned, the issue of "sensitivity" was tested more thoroughly through a choice of survey *topic* involving large numbers of items that might be regarded as sensitive or threatening.

On the subject of question form, the objective was to consider not only mode effects but also the effects of *different* questioning approaches over the telephone, using a series of split ballot experiments. These were in the two areas identified above — open questions and questions involving response scales. In all, about half the questions in the survey involved split ballots. Question forms considered to be more similar — if not identical — to those asked face to face were combined in one questionnaire version, with variants combined in a second version.

While the main criterion by which these question variants are to be judged is their effect on response distributions, the study incorporated facilities for other judgments. These included a survey of the interviewers employed.

2.2 Overall Mode Effects

From the above comments it will not be surprising that this study in fact yielded many more examples of differences between telephone and face to face interview data. In fact, statistically significant differences were found for as many as 68 out of 97 questions (70 percent). This should not be seen as contradicting the earlier results. Quite apart from the deliberate choice of topic and of question forms expected to be liable to mode effects, a number of complicating factors should be taken into account.

First, as in the previous studies, response rates differed somewhat between the two interviewing modes: 65 percent by telephone compared with 73 percent. This *could* have led to differences between the sample profiles, although given our previous findings it seemed unlikely. A further potential source of such differences was a difference between the two modes in terms of the sample base or survey population. The telephone survey was conducted only with people whose numbers were traced in directory searches. The comparative face to face data are drawn from a national sample. People who said they had no telephone were excluded, but the telephone owning subsample will include people with unlisted numbers (around 12 percent of U.K. telephones) and possibly people with access to a telephone outside their living accommodation.

The combined effects of differential nonresponse and differences of definition were not large, however. The samples were compared in terms of eight variables: employment status, partners' employment status, housing tenure, number of adults, number of under 18s, age of respondent, marital status, and social class. For only two of these variables were significant differences found: small but apparently in line with those found earlier. The sample interviewed by telephone included rather more people in paid work (62 percent compared with 56 percent face to face) and rather fewer homemakers (14 percent compared with 18 percent). The same pattern occurred in terms of the status of the respondent's partner (if any). The telephone sample included more people with working partners (63 percent against 58 percent) and fewer with partners involved solely in home making (16 against 22 percent).

A second set of limitations on the mode comparisons arose from the decision to compare our telephone results with the earlier face to face survey rather than with a coincidental experimental subgroup. There was an interval of over a year between the two studies. More importantly, perhaps, the face to face survey was carried out in the Spring (of 1985); the telephone survey was carried out in the two months before Christmas 1986 —a time of a year at which alcohol becomes a more salient topic (more people are drinking, and there is an annual campaign against drinking and driving). In addition, the telephone data were collected by interviewers from social survey division of OPCS, while the face to face data were collected by SCPR interviewers. Differences could arise in terms of house style, training or past experience.

An illustration of *possible* interference by these factors is an explicable but not predicted difference between the two modes in the incidence of "don't know" responses. These were more frequent in face to face interviewing than they were in telephone interviewing (of 58 questions yielding "don't know" answers from either interviewing mode, 42 gave a higher percentage face to face), especially at questions concerning knowledge and beliefs about alcohol and its effects. It could be that the weaker sense of social presence over the telephone makes respondents less bound by the answers they give and more willing to guess or to express weak beliefs. This may be worthy of further investigation, but the lower incidence of "don't know" answers over the telephone is just the kind of difference we might expect to arise from differences in interviewer background or from the increased salience of the topic.

Perhaps *most* important, however, in considering the incidence of statistically significant differences we should have in mind two aspects of the testing process itself. First, in contrast with the earlier multitopic surveys, all the questions here relate to the single topic of alcohol. This must reduce the independence between different questions: one significant result will be associated with others. (This may have been aggravated by the presence in the questionnaire of several sequences of very similar scale questions.) Second, the sample sizes involved here are quite large — around 1500 (telephone owners) interviewed face to face and around 1000 interviewed by telephone (500 in each version of a split ballot). As a result, statistical significance will often be attached to differences that are of little or no substantive importance. Two examples of such significant but unimportant differences are shown in Table 2.

2.3 Sensitive Questions

Turning to the first of the specific areas chosen for further investigation in this study — sensitivity and threat — we have, in alcohol, a sensitive topic overall. Within the questionnaire, however, we identified 24 questions as being "sensitive." These covered drinking patterns and practices, amounts of alcohol consumed, the likely effects of alcohol on respondents and questions on drinking and driving. Examples are given below:

- When *you* have been drinking alcohol, how likely is it that you will start to feel sexually excited? Is it *very* likely, *fairly* likely, fairly *unlikely*, or *very* unlikely?

- I'm going to read out some descriptions of the amount people drink and I would like you to tell me which one fits you best. Would you say you. . . . hardly drink at all, drink a little, drink a moderate amount, drink quite a lot, or drink heavily?

Table 2. Statistically but Not Substantially Significant Mode Effects

	Telephone	Face to Face
Drinking as a cause of work accidents	(1002) %	(1307) %
Very common cause	7	5
Fairly common cause	29	26
Not very common cause	41	46
Unusual cause	24	22
Likelihood of respondent becoming upset and depressed after drinking	(478)[a] %	(1404) %
Very likely	1	—
Fairly likely	2	2
Fairly unlikely	24	27
Very unlikely	73	70

[a]A split ballot question, so only version one is included.

- Would you say in the past 12 months, you have ever been a little bit drunk? IF YES: About how often? Please stop me when I get to the category which applies to you. READ OUT: Every day, most days, 3–4 times a week, 1–2 times a week, 1–2 times a month, 3–4 times in the year, 1–2 times in the year?

Statistically significant differences between the telephone and face to face methods were identified on 11 of the 24 questions. Generally speaking, where such differences did occur they lay in the expected direction. Although not dramatically so, the telephone method elicited marginally "better" data — data that were presumed to be less affected by social desirability considerations. For example, Table 3 shows the frequency with which respondents claimed to have drunk various beverages over the past year. Generally, a smaller percentage of telephone than face to face respondents reported very infrequent drinking (three or four times a year or less) or total abstention, and a higher percentage endorsed "1–2 days per month." No consistent differences were found for the higher frequency categories.

Reports of the *amount* of alcohol drunk were also higher over the telephone. Average amounts reported for each type of alcoholic drink were around 10 percent higher over the telephone. A measure combining frequency of consumption and usual amount of alcohol drunk was used to classify respondents as heavy, moderate, frequent light, infrequent light, or occasional drinkers: averaging across the categories of alcohol, 57 percent of telephone respondents were classified in the lower category of occasional drinkers, compared with 63 percent of face to face respondents.

These differences are statistically significant, fit with prior expectations, and confirm the rather tentative findings of the earlier experiments. But we have to say again that they are small and would probably have little impact on most substantive interpretations and uses of the data.

2.4 Open Questions

The well documented tendency for open questions to obtain less detailed answers over the telephone has been explained in terms of conventional *styles* of telephone communication (e.g., few pauses, shorter utterances). One strategy for dealing with the problem is to devise ways of counteracting convention, encouraging both interviewers and respondents to take more time in dealing with open questions. Perhaps the most obvious single method is to actually *tell* respondents (and incidentally interviewers) what is required of them. Previous work in this area has

Table 3. Frequency of Consumption of Various Alcoholic Drinks in the Past Year

Frequency		Shandy	Beer	Spirits	Port	Wine
Infrequent or	T	75%	43%	53%	61%	46%
abstention	F	80	45	58	73	49
1-2 days per month	T	15	17	25	16	31
	F	10	15	20	13	25
1-2 days per week	T	10	38	20	12	23
to 'most days'	F	9	37	20	13	26
Every day	T	—	3	2	1	1
	F	—	3	2	1	—

been carried out with some measure of success (Miller and Cannell, 1982), and we decided to explore the possibilities in this survey.

Split ballot question wording experiments were conducted on four open questions. The experimental version of each carried instructions to respondents to take their time in answering. For example:

Version 1.

You sometimes hear talk about people who have a "drinking problem." What do you understand by a "drinking problem"?
PROBE FULLY. RECORD VERBATIM

Version 2.

You sometimes hear talk about people who have a "drinking problem." Bearing in mind that you can take as much time as you like to answer. . .What do *you* understand by a "drinking problem"?
PROBE FULLY. RECORD VERBATIM

Before looking at the effects of these variations in question *form* it has to be said that the overall comparisons between *telephone* and *face to face* interviewing were generally not dramatic. Substantive interpretations of the data would not be affected.

In this survey there was no evidence that the telephone responses were truncated. In fact, telephone interviewing obtained, on average, *more* answers than the face to face approach. A likely explanation of this result is differences in the probing skills of interviewers. Evidence that the OPCS telephone interviewers were more persistent in probing for details arises at a number of questions. For example, mode effects on an item dealing with the diseases caused by alcohol related mainly to the balance of responses between two categories: one general ("liver disease") and one specific ("liver cirrhosis"). Over the telephone, 26 percent of responses were in the specific category, 14 percent in the general one. Face to face, fewer (22 percent) were in the specific category and more (21 percent) in the general.

This suggestion of especially careful handling of open questions by our telephone interviewers may explain the failure of the experimental approach to obtain fuller answers. The question versions with instructions to respondents to take their time yielded almost exactly the same number of answers per respondent as did the standard versions. (Averaged over the four questions, each version yielded 1.7 answers per respondent.)

One other possibility is that the experiment worked but not in quite the way anticipated. All the telephone interviewers handled *both* question versions, and it is possible that the message apparent in the *experimental* version affected their handling of *both* types of open questions asked over the telephone. So, we had no differences *between* the telephone variants and, marginally, more data over the telephone compared with the face to face approach.

Interviewers' opinions on the alternative question forms were divided: six preferred the standard form, nine the version with instructions, and one interviewer had no preference. The most common reason for preferring the experimental version was it persuaded respondents to talk more slowly by telling them that their answers were being written down in full. So, a marginal advantage of the experimental approach may be that it makes the interviewer's recording task rather easier. On the other hand, interviewers who preferred the standard approach felt that it sounded less patronizing and was shorter and more simple.

2.5 Scale Questions

Scale questions were heavily used in the original (face to face) survey of attitudes towards alcohol. They took a variety of forms: Likert type agree/disagree items (for example, concerning the effects of alcohol); items asking whether drinking was more or less prevalent than in the past among particular subgroups of the population or under particular circumstances; and several scales of frequency, of events, or of forms of behaviour. In a number of cases, items formed repetitive batteries.

In these circumstances substantial use was made of show cards in the face to face interview: to aid differentiation between response categories that might *sound* similar and/or to encourage a more considered reaction to repetitive questions. Of the 48 scale questions also used in the telephone interview questionnaire, 37 had made use of showcards in the original interview.

The concerns that lead to the use of showcards in the face to face interview are just those that researchers expect to cause problems in a telephone interview. Hence, we might expect to find more mode effects on those questions adjudged to need showcards in the face to face interview. Our results support this expectation, but not conclusively: 30 of the 37 showcard questions (81 percent) yielded significant mode effects, compared with 5 of the 11 questions asked without showcards (45 percent).

In a telephone interview survey one can consider mailing out showcards or other stimulus material prior to the interview. But this can be a costly and unreliable process. Under many circumstances — where addresses and/or respondent identity are not known — it is simply unfeasible. One alternative that we examined here was the use of two stage unfolding questions, intended to reduce confusion between categories, to counter the memory problems involved in reading out long lists of categories, and to slow down the response process. This approach was tested against the alternative of simply reading out the response categories.

While the unfolding method aims to tackle the same problems as are addressed by the use of showcards, we cannot necessarily expect it to yield the same results. (There is limited evidence that unfolding questions in face to face interviews yield different answers from those obtained face to face in single stage questions, by reading out or by using showcards.) In this experiment, we found no clear-cut pattern. Indeed the results follow almost every possible pattern! In some instances we found both mode effects and question form effects, in some mode effects without question form effects and, in others neither type of effect.

A case where we found both effects is a battery of ten Likert type agree/disagree scales. In the face to face interview respondents were asked to choose an answer from a card (*Agree Strongly*, *Agree*, *Disagree*, *Disagree Strongly*). In the telephone interview these response categories were either read out as printed on the card or were approached in two stages: "Agree" or "Disagree," followed by either "Strongly agree" or just "Agree" or "Strongly disagree" or just "Disagree" depending on the answer in Stage 1.

The results are summarized in Table 4, showing the distribution of responses averaged over the ten items in the battery. The comparison between the face to face questions and the read out telephone questions shows the extremeness bias mentioned earlier.

Table 4. Average Responses to Ten Agree/Disagree Items

	Agree Strongly	Agree	Disagree	Disagree Strongly
Face to face	9%	34%	48%	9%
Telephone:				
Read out	14	26	41	19
Unfolded	18	22	35	25

Table 5. Average Responses to Six Likelihood Items — Other People

	Very Likely	Fairly Likely	Fairly Unlikely	Very Unlikely
Face to face	14%	33%	28%	24%
Telephone:				
Read out	19	35	24	22
Unfolded	19	33	25	22

Face to face, 18 percent of unions were said to be strongly held, compared with 33 percent over the telephone. The division between agreement and disagreement was not, however, affected. The unfolding approach, rather than reducing the effect, actually increases it: with this approach, 43 percent of opinions were said to be strongly held. (Again the division between agreement and disagreement was unaffected.)

An example of mode effects but *no* question form effects is found in another battery of items dealing with the likely effects of drinking on other people of the respondent's sex and age. In the face to face interview, respondents were asked to choose an answer from a card (*Very Likely, Fairly Likely, Not Very Likely, Not at All Likely*) to say how likely drinking was to make *other people* relaxed, aggressive, and so on. As with the agree/disagree items, one version of the telephone interview used the simple approach of reading out the response categories while the alternative version used an unfolding technique: "Likely" or "Unlikely," followed by "Very" or "Fairly."

The results in Table 5 are again a summary — the average of seven such items. Here we see a mode effect, with telephone respondents being more likely to believe that alcohol will affect people, but no difference between the two approaches to the telephone interview.

In an immediately following question, respondents were asked to apply the same likelihood battery to *themselves*. In this case, as Table 6 shows, we find *neither* mode effects nor marked question form effects.

We might suggest that mode effects will be greater for less salient topics, e.g., when talking about other people rather than oneself. But this would actually be an extension of suggestions (e.g., Miller, 1984) that *question form* differences will be greater under such circumstances. It does not seem to explain why the question about other people should show mode effects but no question form effect.

One topic that has particularly exercised researchers in the area of telephone questionnaires is that of income. In this study, an income question was asked in two alternative forms. In one version, the interviewer read out income bands starting from the lowest, until the respondent indicated his or her income had been reached. The alternative used a form of unfolding, first asking whether the respondent's income was in, above, or below the band in the middle of the scale, then going on to read out the relevant categories, again starting from the lowest. This variation in question form — expected to reduce underestimation —in fact had no effect on the distribution of answers.

These very mixed findings emerging from comparisons of response distributions are echoed in interviewers' reactions to the different forms. Where the unfolding version was simple, as for example in the agree/disagree items, it was preferred — by 11 of our 16 interviewers. Those who preferred it found it less confusing than the read out version (especially when interviewing the elderly). It involved less repetition of the question; in particular, it avoided the need for (or formalized) the probing for strength of opinion so often needed in the read out version when responses were unclear.

Table 6. Average Responses to Six Likelihood Items — Self

	Very Likely	Fairly Likely	Fairly Unlikely	Very Unlikely
Face to face	8%	20%	22%	50%
Telephone:				
Read out	10	20	20	51
Unfolded	9	19	19	54

More complex unfolding questions — like the income question variant mentioned above — were not popular with interviewers. The straightforward read out version was felt to flow better, to be more concise and, hence, to be less confusing to respondents. In the specific case of the income question, the unfolding form was felt to increase the sensitivity of the question.

2.6 Mode Effects on Older Respondents

One main area in which reservations about telephone interviewing remain strong in the United Kingdom is that of interviews with older respondents. Hoinville (1983) summarized some commonly recognized features of such interviews. Older respondents may find the interview more threatening than younger people, may find it more difficult to concentrate, and may have problems with hearing, comprehension, or recall. Over the telephone, some of these difficulties are likely to be exacerbated, and we might expect to find larger mode effects among older respondents.

To explore this, a number of questions were identified that we felt might cause problems for older respondents. These were mainly sensitive items dealing with drinking behaviour or questions that might be confusing, or pose difficult response tasks. Response distributions were tested for significant interactions between respondents' age (under 65/65 and over) and mode of interview.

Of 40 questions examined only 11 demonstrated interaction effects between age and mode of interview. And some of these were not clear-cut examples of the effect we were looking for: larger mode effects among older respondents. One example of where this pattern *did* emerge is the question battery asking about the effects of alcohol on other people. Table 7 shows the pattern for the average item.

Here we see mode effects among older respondents which are in the same direction as, but larger than, those among the under-65s. It seems that the factors contributing to mode differences on these items were of exaggerated importance for the older age group.

In some cases, however, no such clear conclusions can be drawn. Then, the significant interaction tended to arise from variations between the two age groups in the *direction* of the mode effect. Further research is needed to explain the extent of the problem. Is it, for example, confined to certain kinds of question topics or to certain question forms? Explanations will also be needed if the problem is to be tackled effectively.

Table 7. Age Mode Effects on a Question Battery

	Very Likely	Fairly Likely	Fairly Unlikely	Very Unlikely
Age 65+				
Face to face	11%	32%	27%	29%
Telephone, read out	22	30	17	30
Absolute difference	11	2	10	1
Aged under 65				
Face to face	14	35	29	22
Telephone, read out	18	36	25	21
Absolute difference	4	1	4	1

CHAPTER 20

A COMPARISON OF RESPONSE EFFECTS IN SELF-ADMINISTERED AND TELEPHONE SURVEYS

George F. Bishop
Institute for Policy Research, University of Cincinnati

Hans-Juergen Hippler and Norbert Schwarz
Center for Surveys, Methods, and Analysis

Fritz Strack
University of Mannheim[1]

Many experiments have shown that the results of social surveys can be significantly affected by the way in which the questions are worded, the form in which they are presented, and the order or context in which they are asked. Nearly all of this evidence, however, has come from survey interviews conducted either face to face or by telephone (see Bishop 1982, 1983, 1984, 1985, 1986, 1987; Kalton et al., 1978, 1980; Krosnick and Alwin, 1987; Schuman et al., 1983, 1986; Schuman and Ludwig, 1983; Schuman and Presser, 1981; Smith, 1987; see also Dijkstra and van der Zouwen, 1982). With one exception (Hippler and Schwarz, 1986), none of the well known response effects in this literature have, to our knowledge, been replicated in a self-administered or mail questionnaire, in which respondents typically have more time to think about each question and the implications of their answer to one question for their answer to

[1] The authors would like to thank Paul Biemer, Edith Desiree de Leeuw, and Johannes van der Zouwen for their comments and suggestions on revising the previous draft of this chapter.

another.[2] It would thus be useful to know which response effects generalize to self-administered surveys, which do not, and why.[3]

Theoretically, we should expect some response effects to occur in both self-administered surveys and telephone or face to face interviews, but others should either disappear or become much less pronounced in magnitude in a self-administered situation. Question order effects, for example, should either vanish or become negligible in a self-administered survey because, unlike in a telephone or face to face interview (without a show card), respondents can consider all the questions and response alternatives before answering. So, for the same reason, should most response order effects be reduced in magnitude, if not eliminated, in a self-administered instrument, with the exception of long lists or item scales that respondents may inspect too hastily (see Schuman and Presser, 1981, pp. 72–74). In contrast, question form and wording effects should be just as likely to occur in a self-administered survey as in a telephone or face to face interview because the information presented to respondents in all of these modes of data collection (e.g., the presence or absence of a middle response alternative; the word "forbid" or "allow") is essentially equivalent. Because respondents in a self-administered survey have more time to think about the meaning of the questions, however, subtle variations in how they interpret them may well occur, resulting in significant differences between this and other modes of data collection.

To test these hypotheses, we designed a cross cultural experiment and replication that compared the response effects of variations in question form, wording, and context in a telephone survey with those in a self-administered survey, the two modes of data collection that we thought

---

[2] Since drafting this chapter, we have learned of a recent study by Ayidiya (1987), suggesting that one well known question order effect involving the Communist/American newspaper reporter items used by Schuman and Presser (1981) does replicate in a mail survey, but that another involving the same abortion item used in our experiment does not, confirming the findings shown in Table 3 below. The data from Ayidiya's study also show that a variety of recency response order effects reported in the literature are either eliminated or significantly reduced in magnitude in a mail survey, confirming too the results presented in Table 1 below. Furthermore, Ayidiya has found that acquiescence response effects due to question format are just about as likely to occur in a mail survey as in a telephone or face to face interview, confirming our findings on question form effects as well. So there is now convergent evidence for our hypotheses from an independent investigation.

[3] There are, of course, other inherent differences between self-administered and telephone or face to face surveys (e.g., interviewer effects, task related effects, etc.) which might account for some of the variation in response effects by mode of data collection, but we suspect that they are relatively minor. Though we have no conclusive evidence that mail survey respondents tend to look ahead at the questions and answers, this assumption would seem quite plausible.

were most different from one another.[4] The experiment and the replication were done at about the same time, with the same questions, with similar populations: college students at the University of Cincinnati in the United States and students at the University of Mannheim in West Germany. The principal reason for using student subjects, other than limited resources, was that we did not think a sufficiently high response rate could have been achieved with a self-administered survey of the general public, whether done by mail or face to face delivery. Any differences we might find between the two modes of data collection with the general public might then be due to a difference in response rates. The use of West Germany and the United States as cultural settings for the experiment was purely a matter of convenience, an outgrowth of a visit by the first author to the survey research center, ZUMA, in Mannheim. Though the authors assumed cultural differences between the two societies would affect the marginal distribution of responses to various questions (e.g., a more "conservative" pattern in the United States), they expected that response and mode of data collection effects, if they were truly universal, should replicate from one cultural setting to another.

1. RESEARCH DESIGN

1.1 The Experiment in the United States

The data for this experiment were collected in February, March, and April of 1986 from a systematic random sample of 724 graduate and undergraduate students selected from a current telephone directory for the University of Cincinnati.[5] Half of these students were randomly assigned to be interviewed by telephone; the other half received the questionnaire in a self-administered form that was personally delivered to their residence and returned either by mail or by having it picked up by

---

[4] Our assumption is that respondents have the least amount of time to think about the question and their answer in a telephone interview, somewhat more time to stop and think in a face to face interview, and the most time in a self-administered survey where they can look over all the questions and responses before answering. Thus we would expect to find the largest difference in response effects between telephone and self-administered surveys.

[5] We are unable to evaluate any bias due to nonresponse because the directory service on campus does not provide any population statistics or documentation on the accuracy of the directory. The response rate for the two surveys as reported below, however, was relatively high, indicating that nonresponse bias was probably not a significant source of error.

the person who delivered it, typically the latter.[6] The response rate for the telephone survey was 83.9 percent; for the self-administered survey, with an intensive followup, it was 76.8 percent.[7] In both the telephone and the self-administered surveys, respondents were randomly assigned to receive either form A or form B of the questionnaire (see Appendix).

Table 1 outlines the various experiments reported in this chapter.[8] A majority of the items were exact replications of questions from previous split-ballot experiments conducted by Schuman and his associates or by Bishop and his coworkers. The authors constructed the question on left/right political identification as a substitute for an item on liberal/conservative identification used by Schuman and Presser (1981) because the latter terms did not have much currency in the political culture of West Germany, whereas the concepts of "the left" and "the right" seemed more comparable in meaning across the two cultures. Similarly, we designed the question about forbidding or allowing smoking in public places such as restaurants because we thought this issue would be somewhat more salient and comparable in meaning across the two societies than those used in previous investigations of the forbid-allow effect (Hippler and Schwarz, 1986; Schuman and Presser, 1981, pp. 276–283). Finally, we created the "International Trade Act of 1986" as a comparable, cross cultural surrogate for the fictitious "1975 Public Affairs Act" invented by Bishop and his colleagues (1980, 1986) and the real, but obscure, legislative acts used by Schuman and Presser (1981). So, as with the questions about left/right political identity and forbidding or allowing smoking in public places, the experiment with the International Trade Act

---

[6] The first author would like to thank Susan Ackerman, Shirley Frayer, and Andy Smith for their conscientious efforts in helping him to collect these data. He owes them all a big debt of gratitude.

[7] Most of the nonrespondents in both the telephone and the self-administered surveys were students who could not be located because the directory numbers listed for them had been disconnected. Only 10 potential respondents (3 men, 7 women) refused to be interviewed, all of them in the self-administered condition in which the questionnaire was to be personally delivered. Fourteen potential respondents in the self-administered condition did not return their questionnaires even after repeated reminders to do so by telephone. Otherwise, cooperation was unusually high compared to surveys of the general public.

[8] In addition, the questionnaire in both the German and the U.S. studies included two experiments on the effects of question length on responses to questions about respondents' memories of the wars in Afghanistan and the Falkland Islands. These data are part of a separate study on memory effects and so they are not included here. The questionnaires in both the self-administered and the telephone surveys in Cincinnati also included a question order experiment with items from the NORC General Social Survey concerning the death penalty and the harshness with which criminals are treated by the courts, as well as various measures of involvement with the death penalty issue. These items were part of an independent study by Susan Ackerman and are therefore not analyzed here.

item represents a conceptual, rather than an exact, replication of previous split-ballot studies.

1.2 The Replication in West Germany

The conditions under which the data were collected in the German setting were somewhat different, though the wording and sequence of the questions in this study were, with one minor exception, identical to those in the U.S. experiment.[9] First of all, the data for the German study were collected during the first two weeks of April 1986, whereas the U.S. data were gathered in February and March as well as early April of 1986. To our knowledge, however, there were no international or domestic events in February and March of that year, nor for that matter in April, that might have differentially affected responses to any of the questions in either of the two experiments.

While the U.S. study included graduate and undergraduate students from a variety of disciplines, the subjects for the German investigation were all undergraduates at the University of Mannheim, majoring in law and business administration. Also, unlike the U.S. study, these students were initially contacted in the classroom and then asked to participate in a survey. After they agreed to participate, they were randomly assigned to one of the four conditions: the telephone or the self-administered survey and either Form A or Form B of the questionnaire. Subjects assigned to the self-administered condition received one of the two versions of the questionnaire and were asked to fill it out immediately, whereas subjects assigned to the telephone condition completed a one page questionnaire for a separate study in which they were asked on the second page for their telephone number, the best time for contacting them, and their first name. Most of the questionnaires were completed within 12 minutes. All of the questionnaires were then collected and each participant got a mechanical pencil as a reward.

During the following three days the subjects in the telephone condition were interviewed by five professional interviewers. Because of a concern over the relatively small sample sizes in the German experiment, the interviewers were instructed and trained to go through the interview

---

[9] On the question about nuclear power plants (see Appendix), the U.S. version reads, "Some people say that the United States needs to develop *new* power sources from nuclear energy...[emphasis added]," whereas the English translation of the German version reads, "...*alternative* power sources [emphasis added]." We doubt that this minor change in wording made an important difference in the meaning of the item, though we really cannot be sure without performing still another experiment.

Table 1. List of Experiments in Question Form, Wording, and Context

	Form A	Form B
Response Order		
Divorce	Middle alternative presented in *last* position	Middle alternative presented in *middle* position
Nuclear power plants	Middle alternative presented in *last* position	Middle alternative presented in *middle* position
Question Order		
Abortion	Question on abortion if birth defect, asked *before* question about abortion if woman wants no more children	Question on abortion if birth defect, asked *after* question about abortion if woman wants no more children
Trade with Japan	Question about trade restrictions by Japan asked *before* question about trade restrictions by U.S./Germany	Question about trade restrictions by Japan asked *after* question about trade restrictions by U.S./Germany
Middle Response Alternative		
Marijuana penalties	Middle alternative omitted	Middle alternative offered
Left/right pol. ID	Middle alternative offered	Middle alternative omitted
Defense spending	*Middle* alternative offered	*No opinion alternative* offered instead of middle alternative
No Opinion Alternative		
Arab/Israel relations	No opinion alternative *offered*	No opinion alternative *omitted*
International Trade Act of 1986	No opinion alternative *omitted*	No opinion alternative *offered*
Tone of Wording		
Smoking in public	Allow	Forbid
Open vs. Closed Form on Work Values	Closed	Open

as quickly as possible, reinforcing the tendency of respondents to give the first answer that comes to mind, whereas in the self-administered survey respondents had been encouraged to take their time in answering.[10]

Each of these procedures was designed to strengthen the response effects of the manipulation of the question forms, making the differences between the two modes of data collection large enough to detect statistically with small samples and categorical variables.

Only eight of 163 subjects assigned to the telephone condition could not be contacted, resulting in a response rate of 95.1 percent. A total of 194 self-administered questionnaires and 155 telephone interviews were completed. From the self-administered group 11 people who reported not having a telephone were eliminated from the analysis to make the two samples more comparable. This resulted in a total of 183 subjects in the self-administered condition, a relatively small number as compared to that in the U.S. experiment.

The replication in West Germany, then, was more like a typical laboratory experiment, conducted partly under field conditions, whereas the U.S. experiment was more of a field study similar to the usual survey. These variations in implementing the experiment, however, make it all the more valuable if the results should replicate from one setting to the other.

2. FINDINGS

2.1 Response Order Effects

Divorce Issue. Schuman and Presser (1981, Chap. 2) have discovered that when respondents are asked whether a divorce should "be *easier to obtain, more difficult to obtain*, or *stay as it is now*," they are significantly more likely to select the middle response alternative, *stay as it is now*, when it is offered in the last, rather than the second (or middle) position in the response sequence. We had hypothesized that response order would make a significant difference in the results for the divorce item in our telephone surveys, as it did in Schuman and Presser's studies, but that it would have no significant effect on the results of the self-administered surveys. Surprisingly, however, we found that the order in which the middle alternative was presented in the divorce item made no significant

---

[10] In the self-administered survey the manipulation was strengthened, in both experiments, by the instructions to "...take your time and read each question carefully before answering the questionnaire."

difference in the results, in either the telephone survey or the self-administered survey, in either the U.S. or the German experiment (data not shown here)[11]. This response order effect may not replicate with college students because the issue of divorce does not have the same psychological significance for them as it does for older adults, many of whom have had to make a decision about divorce in their lives, or who are contemplating such a decision. Probing respondents who select the answer, "stay as it is now," with a followup question about why they chose that alternative, might reveal differences in the meaning of divorce for these two populations.

Nuclear Power Issue. When respondents are asked, "Are you in favor of *building more nuclear power plants*, would you prefer to see *all nuclear power plants closed down*, or would you favor *operating only those that are already built?*," they are significantly more likely to choose the middle alternative, *operating only those that are already built*, when it is presented in the last rather than the second (or middle) position in the response sequence (Bishop, 1987). As with the divorce item, we had hypothesized that response order would make a significant difference in the results for the nuclear power item in our telephone surveys, and little or no difference in the findings from the self-administered surveys. The figures in Table 2 tend to confirm this hypothesis. When interviewed by telephone, respondents in the U.S. experiment were noticeably more likely to select the middle alternative if it was presented in the last, rather than the middle, position. The German data showed the same pattern: the difference between the two question forms in the telephone survey was sizable (15 percent) and close to being statistically significant, despite the small subsample sizes. When respondents were given the self-administered form, however, the order in which the middle alternative was presented in the nuclear power question made no significant difference in the percentage of respondents who chose it in either of the two experiments. Though there was a tendency for German respondents to select the middle alternative more often when it was presented in the middle rather than the last position in the self-administered survey, this pattern is probably due to chance since it did not replicate in the U.S. experiment.

---

[11] To conserve space, this and several other tables of data have been omitted from this chapter, but are available from the authors on request.

Table 2. Response to Nuclear Power Plants Item by Question Form and Mode of Data Collection

	Telephone		Self-Administered	
	Middle Alt. Offered in *Last* Position	Middle Alt. Offered in *Middle* Position	Middle Alt. Offered in *Last* Position	Middle Alt. Offered in *Middle* Position
Nuclear plants (U.S.)				
Build more	31.9%	32.6%	32.5%	38.7%
Close all	17.8	26.7	21.7	18.8
Operate only those built	50.3	40.6	45.8	42.5
	100.0	99.9	100.0	100.0
	(185)	(187)	(166)	(181)
	X^2 (middle vs. other responses) = 3.10, df=1, p=.078		X^2 (middle vs. other responses) = 0.25, df=1, p>.25	
	Three-way interaction (response by form by mode): X^2=0.74, df=1, n.s.			
Nuclear plants (Germany)				
Build more	14.5%	18.7%	25.3%	20.9%
Close all	15.8	26.7	23.1	15.4
Operate only those built	69.7	54.7	51.6	63.7
	100.0	100.1	100.0	100.0
	(76)	(75)	(91)	(91)
	X^2 (middle vs. other responses) = 3.67, df=1, p=.055		X^2 (middle vs. other responses) = 2.73, df=1, p<.098	
	Three-way interaction (response by form by mode): X^2=6.40, df=1, p<.05			

2.2 Question Order

Japanese Trade Issue. As expected, when respondents in the U.S. experiment were interviewed by telephone, the order in which the questions about Japanese-American trade relations were asked made a sizable and significant difference in the results (see Table 3). Indeed, the results were remarkably similar to those reported by Schuman and Ludwig (1983): respondents were significantly more likely to favor limiting Japanese imports to the United States (69.4 percent) than they were to favor limiting U.S. exports to Japan (53.8 percent) when each question was asked in the first position ($\chi^2 = 9.6$, df = 1, $p < .01$). Similarly, we found that support for limiting U.S. exports to Japan (67.9 percent) increased significantly when respondents were asked about it immediately *after* the question about trade restrictions on Japanese imports by the United States, presumably because a norm of even handedness had been evoked by the sequence of the questions. Furthermore, we found, as did Schuman and Ludwig, that the norm of even handedness does not necessarily operate with equal force in both directions: support for limiting Japanese imports (67.2 percent) did not decline significantly, if at all, when respondents were asked about it immediately *after* the question about limiting U.S. exports to Japan. This asymmetry, as Schuman and Ludwig suggest, may be the result of American perceptions that the "unfair" Japanese competition for the U.S. market is what needs to be righted by restrictions on imports.

But when respondents were asked these same questions in the self-administered form, the order in which they were presented had, as predicted, no significant effect on the results. This does not mean, however, that the norm of even handedness had no influence on responses to the questions about Japanese-American trade relations in the self-administered form. To the contrary, because respondents were able to look at both of the questions about trade restrictions simultaneously, they could not help but realize that a norm of even handedness was called for in answering the questions. And that is why we find, unlike the results of the telephone survey, that respondents were *not* significantly more likely to favor trade restrictions by the United States (68.1 percent) than they were to favor restrictions by Japan (64.5 percent) when each question was asked in the first position ($\chi^2 = 0.53$, df = 1, n.s.). Indeed, the absence of an order effect on the responses to these questions in the self-administered form is precisely what Schuman and Presser's hypothesis would predict.

The same question order effect occurred in the German experiment, even though the marginal distributions of responses to the trade items were quite different from those in the United States (i.e., more favorable

Table 3. Response to Japanese Trade Items by Question Form and Mode of Data Collection

	Telephone		Self-Administered	
	Limit U.S. Item Asked *Before* Limit Japan Item	Limit U.S. Item Asked *After* Limit Japan Item	Limit U.S. Item Asked *Before* Limit Japan Item	Limit U.S. Item Asked *After* Limit Japan Item
Should Japan limit U.S. imports				
Yes	53.8%	67.9%	64.5%	67.6%
No	46.2	32.1	35.5	32.4
	100.0	100.0	100.0	100.0
	(186)	(187)	(166)	(182)
	χ^2=7.26, df=1, p<.01		χ^2=0.25, df=1, p>.25	
	Three-way interaction (response by form by mode): χ^2=2.17, df=1, .10<p<.25			
Should U.S. limit Japanese imports				
Yes	67.2%	69.4%	68.1%	68.1%
No	32.8	30.6	31.9	31.9
	100.0	100.0	100.0	100.0
	(186)	(186)	(166)	(182)
	χ^2=0.11, df=1, p>.25		χ^2=0.00, df=1, p>.25	
	Three-way interaction (response by form by mode): χ^2=0.09, df=1, p>.25			

	Limit Germany Item Asked *Before* Limit Japan Item	Limit Germany Item Asked *After* Limit Japan Item	Limit Germany Item Asked *Before* Limit Japan Item	Limit Germany Item Asked *After* Limit Japan Item
Should Japan limit German imports				
Yes	12.8%	30.7%	30.0%	25.0%
No	87.2	69.3	70.0	75.0
	100.0	100.0	100.0	100.0
	(78)	(75)	(90)	(92)
	χ^2=7.34, df=1, p<.01		χ^2=0.57, df=1, p>.25	
	Three-way interaction (response by form by mode): χ^2=6.58, df=1, p<.02			
Should Germany limit Japanese imports				
Yes	24.4%	36.5%	41.1%	33.7%
No	75.6	63.5	58.9	66.3
	100.0	100.0	100.0	100.0
	(78)	(74)	(90)	(92)
	χ^2=2.65, df=1, p=.103		χ^2=1.07, df=1, p>.25	
	Three-way interaction (response by form by mode): χ^2=3.64, df=1, p=.056			

toward "free trade"). When respondents were interviewed by telephone, the sequence of the questions made a significant difference in the results, but it made little or no difference when the respondents were given a self-administered questionnaire. Unlike the results of the U.S. experiment, however, a norm of even handedness appears to have influenced responses to both questions more equally. Support for limiting Japanese imports to Germany declined (24.4 percent) when respondents were asked about it immediately *after* the question about trade restrictions by Japan, though the difference was not statistically significant. The effect of the norm may be more symmetrical in the German case because trade relations with Japan are probably not viewed as unbalanced as they are in the United States. In other words, a necessary condition for the norm to operate with equal force in both directions may be a perception that both parties (e.g., nations) are presently engaged in fair and equal competition. Otherwise, the effect of the norm will be asymmetrical, acting to equalize the "unfair" competition, as in the U.S. case.

Abortion Issue. Here too, as hypothesized, the order of the questions made a difference in the results, but only when respondents were interviewed by telephone (cf. Schuman and Presser, 1981; Bishop et al., 1985). In both experiments, respondents were more likely to approve of an abortion for a woman who is married and does not want any more children when they were asked about it on the telephone, and *before* the question about abortion in the case of a possible birth defect, than when they were asked about it *after* the latter question (see Table 4). In the self-administered questionnaire, however, the sequence of the questions made little or no difference in the results. Though the evidence for the hypothesis was statistically significant only in the German experiment, the pattern in the two studies was sufficiently similar that it is highly unlikely to be the result of chance.

2.3 Middle Response Alternatives

As in previous studies (Bishop, 1987; Schuman and Presser, 1981, Chap. 6) we found that respondents were much more likely to select a middle response alternative if it was explicitly offered to them than if it was not (data not shown here). This pattern occurred on both the question about marijuana penalties and the item on left/right political identification, and in both experiments. And, as predicted, this question form effect was just as likely to occur in a self-administered survey as in a telephone survey, and to about the same degree in each experiment.

We also discovered that respondents were significantly more likely to select the middle response alternative on Form A of the question about

Table 4. Response to Abortion Item by Question Form and Mode of Data Collection

	Telephone		Self-Administered	
	Women's Right Item Asked *After* Birth Defect Item	Women's Right Item Asked *Before* Birth Defect Item	Women's Right Item Asked *After* Birth Defect Item	Women's Right Item Asked *Before* Birth Defect Item
Abortion if woman does not want any more children (U.S.)				
Yes (allow)	51.9%	59.3%	51.2%	47.5%
No	48.1	40.7	48.8	52.5
	100.0	100.0	100.0	100.0
	(185)	(189)	(166)	(181)
	χ^2=1.77, df=1, .10<p<.25		χ^2 = .34, df=1, p>.25	
	Three-way interaction (response by form by mode): χ^2=2.22, df=1, .10<p<.25			
Abortion if woman does not want any more children (Germany)				
Yes (allow)	42.1%	69.3%	49.5%	58.2%
No	57.9	30.7	50.5	41.8
	100.0	100.0	100.0	100.0
	(76)	(75)	(91)	(91)
	χ^2=11.50, df=1, p<.01		χ^2=1.42, df=1, .10<p<.25	
	Three-way interaction (response by form by mode): χ^2=2.98, df=1, p=.084			

defense spending than the no opinion alternative on Form B of the question (data not shown here), clearly indicating that these two response alternatives are not psychologically equivalent. As would be expected, however, this question form effect was observed in both the telephone survey and the self-administered survey, and in both experiments. In other words, this form effect, just like those in the previous experiments with the marijuana and left/right items, did not interact significantly with the mode of data collection in either of the two experiments.

2.4 The No Opinion Alternative

Not surprisingly, as with middle responses, both experiments showed that respondents were significantly more likely to choose a no opinion alternative if it was explicitly offered to them than if it was not, and not only on the real issue of Arab-Israeli relations, but also on the fictitious "International Trade Act of 1986" (data not shown). Again, as expected, this form effect was just as evident in the self-administered survey as it was in the telephone survey, and in both investigations. Question form effects, then, whether the result of the presence or absence of a no opinion alternative or a middle response alternative, do not appear to depend upon the mode of data collection, though we have obviously not tested this hypothesis fully through comparisons with data from face to face interviews (cf. this volume, Chapters 18, 19, and 21).

2.5 Tone of Wording

The figures in Table 5 on the forbid/allow effect are difficult to explain. As Schuman and Presser (1981) would predict, the effect does not generalize very well, if at all, to a concrete subject such as regulating smoking in public places. But their prediction seems to apply only when the data are collected by telephone. In both experiments, we found no significant differences by question form in the telephone survey. The data for the self-administered surveys, however, are much harder to interpret. In the U.S. experiment we discovered, contrary to the results of all previous research on the forbid/allow effect, that respondents were significantly more likely to say that something, such as smoking in public places, should be *forbidden* than they were to say that it should *not be allowed*, whereas in the German experiment the results were exactly the opposite: respondents were much more likely to say that smoking in public places should *not be allowed* than they were to say it should be *forbidden*, the pattern we would expect to have found, if any. Suffice it to say that a further replication is in order.

2.6 Open vs. Closed Question Form

A comparison of the responses to the open and closed form of the work values question in Table 6 shows the following:

1. As in previous experiments by Schuman and Presser (1981, Chap. 3), most responses to the closed form of the question fell within the first five precoded categories, whereas responses to the open form spread much more widely beyond these five categories. This pattern occurred in both the self-administered and the telephone surveys, and in both experiments, though it was somewhat sharper in the U.S. data than in the German data. For whatever reason, the German students were significantly more likely than U.S. students to volunteer more than one response to the question on both the open and closed forms, and in both

Table 5. Response to Forbid-Allow Items by Question Form and Mode of Data Collection

	Telephone		Self-Administered	
	Allow Form	Forbid Form	Allow Form	Forbid Form
Smoking in public places (U.S.)				
Yes (allowed, not forbidden)	51.1%	47.6%	53.0%	38.7%
No (not allowed, forbidden)	48.9	52.4	47.0	61.3
	100.0	100.0	100.0	100.0
	(186)	(189)	(166)	(181)
	χ^2=0.32, df=1, p>.25		χ^2=6.61, df=1, p<.02	
	Three-way interaction (response by form by mode): χ^2=2.18, df=1, .10<p<.25			
Smoking in public places (Germany)				
Yes (allowed, not forbidden)	66.7%	72.4%	47.1%	67.4%
No (not allowed, forbidden)	33.3	27.6	52.9	32.6
	100.0	100.0	100.0	100.0
	(78)	(76)	(85)	(92)
	χ^2=0.59, df=1, p>.25		χ^2=7.53, df=1, p<.01	
	Three-way interaction (response by form by mode): χ^2=1.50, df=1, .10<p<.25			

the self-administered and the telephone surveys. Notice too that respondents in both studies who received the open form of the question in the self-administered survey were more likely to have given more than one response to the question (by writing it in) than those who received the open form of the question in the telephone survey, most likely because of the inability to probe and clarify such responses in the self-administered condition.

2. If we examine the data for just the first five categories common to both forms, it appears that there was a substantial difference between the open and closed forms in the percentage choosing a *feeling of accomplishment* as the most important work value. As Schuman and Presser have discovered, respondents were much more likely to select the *feeling of accomplishment* category when it was explicitly offered to them on the closed form than to volunteer it on the open form. This response pattern was evident in both the telephone and self-administered surveys in the U.S. sample, but only in the telephone survey in the German sample. Some of the difference between the open and closed form on the Accomplishment category, however, may be due, as Schuman and Presser

have suggested, to the fact that many respondents who are coded into the Satisfaction category on the open form (see Table 6) would, if properly probed, end up in the Accomplishment category. Many of the respondents who gave more than one codable response, particularly on the open form of the self-administered survey, might also have selected Accomplishment as the most important work value if forced to choose. So the apparent difference between the two forms may represent primarily variations in coding and probing procedures on the open form of the question. Other differences between the two forms on the common categories seem to be relatively minor, especially given the small subsample sizes on which they are based, especially in the German study.

3. On the *closed* form of the question respondents in the U.S. sample were significantly more likely to select one of the last three of the five common response alternatives (control of work, pleasant work, security) when they were interviewed on the telephone than when they were given the self-administered form ($\chi^2 = 4.27$, df $= 1$, $p < .04$). But this response order effect did not replicate in the German sample. So the results are somewhat ambiguous, pending a further replication, preferably with a much larger sample.

3. CONCLUSION

Though the results of this crosscultural experiment and replication were not as unequivocal as we might have liked, they clearly suggest that question order and response order effects are significantly less likely to occur in a self-administered survey than in a telephone survey, whereas question form and wording effects are probably just as likely to occur with one mode of data collection as another. To the extent that such response effects are regarded as unwanted systematic sources of error in survey measurement, our findings on question order and response order effects would indicate that the quality of data gathered through self-administered surveys may, other things being equal (e.g., response rate, respondent literacy), be better than that obtained by telephone surveys [see this volume, Chapters 18, 19, and 21 for comparisons of data quality in telephone and face to face interviews]. Further replications of these findings with similar, as well as different, populations would certainly be useful, as would extensions to other topics, response effects, and variations in modes of data collection (face to face interviews and standard mail surveys). For we now know that generalizations about response effects in surveys are even more conditional than we thought they were.

Table 6. Response to Work Values Item by Question Form and Mode of Data Collection

	Telephone		Self-Administered	
	Open	Closed	Open	Closed
Most prefer in a job (U.S)				
Pays well	8.4%	8.6%	6.7%	14.5%
Feeling of accomplishment	16.3	49.5	22.2	53.6
Control of work	8.4	16.1	3.3	12.0
Pleasant work	11.1	21.5	8.9	14.5
Security	8.4	4.3	1.1	3.6
Liking/satisfaction	30.5	0.0	8.3	0.0
Promotion opportunity	3.2	0.0	5.6	0.0
More than one response	3.7	0.0	41.1	1.2
Other	10.0	0.0	2.8	0.6
	100.0	100.0	100.0	100.0
	(190)	(186)	(180)	(166)
Most Prefer in a Job (Germany)				
Pays well	0.0%	1.3%	1.1%	2.2%
Feeling of accomplishment	15.5	42.1	15.6	11.1
Control of work	2.8	21.1	3.3	22.2
Pleasant work	11.3	14.5	2.2	23.3
Security	1.4	7.9	0.0	3.3
Liking/satisfaction	19.7	0.0	11.1	0.0
Promotion opportunity	1.4	0.0	1.1	0.0
More than one response	42.3	6.6	56.7	36.7
Other	5.6	6.6	8.8	1.1
	100.0	100.0	100.0	100.0
	(71)	(76)	(90)	(90)

APPENDIX. WORDING OF THE QUESTIONS IN THE SELF-ADMINISTERED SURVEY

Form A	Form B
1. In your opinion, should divorce in this country be... 1. easier to obtain 2. more difficult to obtain 3. stay as it is now	1. In your opinion, should divorce in this country be... 1. easier to obtain 2. stay as it is now 3. more difficult to obtain
2. Do you think that the Japanese government should be allowed to set limits on how much American industry can sell in Japan?	2. Do you think that the American government should be allowed to set limits on how much Japanese industry can sell in the United States?
3. Do you think that the American government should be allowed to set limits on how much Japanese industry can sell in the United States? 1. Yes 2. No	3. Do you think that the Japanese government should be allowed to set limits on how much American industry can sell in Japan? 1. Yes 2. No
4. Do you think it should be possible for a pregnant woman to obtain a *legal* abortion if there is a strong chance of serious defect in the baby? 1. Yes 2. No	4. Do you think it should be possible for a pregnant woman to obtain a *legal* abortion if she is married and does not want any more children? 1. Yes 2. No
5. Do you think it should be possible for a pregnant woman to obtain a *legal* abortion if she is married and does not want any more children? 1. Yes 2. No	5. Do you think it should be possible for a pregnant woman to obtain a *legal* abortion if there is a strong chance of serious defect in the baby? 1. Yes 2. No
6. Do you think that smoking in public places, such as restaurants, should be allowed? 1. Yes 2. No	6. Do you think that smoking in public places, such as restaurants, should be forbidden? 1. Yes 2. No

Form A	Form B
7. Some people say that the United States needs to develop new (alternative) power sources from nuclear energy in order to meet our needs for the future. Other people say that the danger to the environment and the possibility of accidents are too great. What do you think—do you... 1. favor building more nuclear power plants 2. prefer to see all nuclear power plants closed down 3. favor operating only those that are already built	7. Some people say that the United States needs to develop new (alternative) power sources from nuclear energy in order to meet our needs for the future. Other people say that the danger to the environment and the possibility of accidents are too great. What do you think—do you... 1. favor building more nuclear power plants 2. favor operating only those that are already built 3. prefer to see all nuclear power plants closed down
8. In your opinion, should penalties for using marijuana be... 1. more strict 2. less strict	8. In your opinion, should penalties for using marijuana be... 1. more strict 2. less strict 3. about the same as they are now
9. Some people believe we should spend less money for defense. Others feel that defense spending should be increased. How about you—do you think defense spending should be.. 1. increased 2. decreased 3. continued at the present level	9. Some people believe we should spend less money for defense. Others feel that defense spending should be increased. How about you—do you think defense spending should be... 1. increased 2. decreased 3. no opinion
10. The *United Nations* has been considering the *International Trade Act of 1986.* Do you... 1. favor the passage of this act 2. oppose the passage of this act	10. The *United Nations* has been considering the *International Trade Act of 1986.* Do you... 1. favor the passage of this act 2. oppose the passage of this act 3. no opinion
11. On most political issues, would you say you are on the left, on the right, or in the middle? 1. Left 2. Right 3. Middle	11. On most political issues, would you say you are on the left or on the right? 1. Left 2. Right

Form A	Form B
12. Do you think the Arab Nations are trying to defeat Israel, trying to work for a real peace with Israel, or do you not have an opinion on that? 1. Trying to defeat Israel 2. Trying to work for a real peace with Israel 3. No opinion	12. Do you think the Arab Nations are trying to defeat Israel, or are they trying to work for a real peace with Israel? 1. Trying to defeat Israel 2. Trying to work for a real peace with Israel
13. [Question about the Falkland Islands]	13. [Question about the Falkland Islands]
14. [Question about Afghanistan]	14. [Question about Afghanistan]
15. The next question is on the subject of work. People look for different things in a job. Which *one* of the following five things would you most prefer in a job... 1. work that pays well 2. work that gives a feeling of accomplishment 3. work where there is not too much supervision and you make most decisions yourself 4. work that is pleasant and where the other people are nice to work with 5. work that is steady with little chance of being laid off	15. The next question is on the subject of work. People look for different things in a job. What would you *most prefer* in a job?

CHAPTER 21

THE QUALITY OF INCOME INFORMATION IN TELEPHONE AND FACE TO FACE SURVEYS

Eszter Körmendi[1]
The Danish National Institute of Social Research

1. PROBLEMS OF ASKING INCOME QUESTIONS IN SURVEYS

There appears to be a great deal of agreement in Western industrialized countries on the type of questions respondents perceive as being threatening or embarrassing.

"Threatening questions" encompass questions where the answers may cause the respondents to fear a lowering of their esteem in the eyes of others, i.e., questions relating to social desirability.

Such questions can either involve activities which are considered embarrassing, strictly private, illegal, etc., or conversely, activities regarded as desirable, leading to higher social status, being "with it," etc. A classic example, and the type of question most often treated in textbooks, is income questions (Sudman and Bradburn, 1982). Difficulties can arise from both cognitional and emotional circumstances.

First, a memory factor is involved, especially when the question concerns income earned the previous year. This difficulty is somewhat lesser in face to face interviews by means of a show card with relatively large income brackets which help the interviewee to find the correct bracket. Furthermore, the large intervals on the card may reassure the interviewee that only an approximate rather than an exact knowledge is required.

[1]The author thanks soc.drs. J. Noordhoek, Danmarks Statistik, for his comments on an earlier version of this paper and for his practical help that made the comparison between the self-reported and register data possible.

Second, income questions contain a knowledge factor especially in relation to personal net income or family incomes. Specifying their income can be a complicated matter for certain groups of persons such as the self employed as compared with wage earners. Likewise, nonresponse on the part of certain subgroups is probably more a result of lack of knowledge than of unwillingness to answer. Several studies show item nonresponse to be much higher, for example, among married women than among men (Andersen and Christoffersen, 1982; Shih, 1983).

Third, the response to income questions is influenced by social acceptability. This factor can also result in a high proportion of "don't know" responses as well as in a high refusal rate, especially in groups with very high or very low incomes (Locander and Burton, 1976; Andersen and Christoffersen, 1982; Shih, 1983; Hippler and Hippler, 1986).

Another difficulty derives from the fact that personal income, along with other subjects such as party affiliation, religion, and sexual habits, belongs in the private sphere.

The show card, which enables the respondent just to pick the letter representing the proper category and thus avoid naming the actual figure, is a particularly advantageous method in situations where interviewer and respondent are not alone. This kind of assistance, however, is not available with regard to telephone interviews. The questions may therefore place greater demands on memory and be felt to be a greater intrusion on privacy.

Problems of information quality connected with sensitive questions are usually centered on two phenomena — namely, a higher rate of nonresponse and a reduced accuracy or validity of the answers.

2. PREVIOUS RESEARCH

A long series of studies (Schmiedeskamp, 1962; Rogers, 1976; Groves, 1979; Siemiatycki, 1979; Jordan, 1980; Aneshensel et al., 1982) have shown that it is especially difficult to obtain answers to income questions in telephone interviews.

The significance of question wording for response willingness and, to a degree, response accuracy, has been intensively treated in the international literature on methodology. Some researchers consider it absolutely necessary that only simple questions be used in telephone interviews. Dillman (1978) is of the opinion that respondents cannot grasp more than 4 to 5 response categories in situations where both questions and answer categories must be read aloud to them. Horton (1978) suggests that telephone questions should be simple. Complex questions can be broken up into several lesser and simpler ones.

Especially in income questions, various other question forms have been used in an attempt to adapt to the telephone medium.

For example, Locander and Burton (1976), who (experimentally) tested the effect of the question form by gathering income data via telephone, reported that respondents are in fact sensitive to the form used when answering income questions. Their findings suggest that nuances in question form may either reduce or increase the threatening impact of the question.

Locander and Burton noted that question wording had an influence not only on response willingness but also on response accuracy. They have found that with a general population or an upper income group starting at the top income bracket and working down produced the highest response rate and the most accurate estimate of income. For low income groups it was better to start with the lowest income levels and work upwards. Starting in the middle (the zeroing technique) requires on the average the fewest questions. By the zeroing technique, it was assumed that the direction of the expectation would not be established in the mind of the respondent because the categories are not read in one direction, as in the other forms.

In 1983, Shih attempted to verify Locander and Burton's findings and discovered that the proportion of "don't know" responses and refusals was higher when the zeroing type of questions were used.

Other studies also indicate that the question technique seems to have an influence on the distribution of responses to income questions. For example, Sykes and Hoinville (1985) find significant differences in household income assessments resulting from face to face and telephone interviews. They note a significant underreporting of household income over the telephone but believe that these results can just as well be ascribed to differences in question techniques as to the interview method itself. However, in later research Sykes and Collins did not find similar differences between the income information obtained in telephone and face to face interviews (this volume, Chapter 19).

According to Hippler and Hippler (1986), recent literature on the design of questionnaires recommends that income questions be given an open-ended form, i.e., without any answer categories, partly in order to achieve more precise results and partly in order to avoid those problems usually associated with the analysis of categorized data. The tendency to answer prematurely, thereby underreporting income, can also be avoided in this way. However, a systematic examination of several nationwide surveys in West Germany shows (Hippler and Hippler, 1986) that open-ended questions tend to have higher levels of refusal than closed versions. It is assumed that one of the reasons for this is that income questions in open-ended form are experienced as more threatening and offensive.

3. THE PURPOSE OF THIS STUDY

The purpose of this study is to compare the quality of income information obtained by telephone and in face to face interviews. The concept of quality is described by the size of item nonresponse, by the structure of item nonresponse, and by the accuracy or validity of the obtained income information. In other words the aim of the project was to study the nature and extent of effects of interviewing mode on data quality.

4. RESEARCH DESIGN

4.1 The Sampling Frame

As a public research institute, the Danish National Institute of Social Research has access to a large number of central administrative registers for research purposes. The Central Population Register (a computerized register of the total population with permanent residence in Denmark) is used as a sampling frame in connection with the majority of representative surveys conducted by the institute.

Consequently, in addition to being able to draw samples from the register, the institute has access to information on several demographic characteristics, such as gender, age, marital status, citizenship, and community of residence for both respondents and nonrespondents.

The sample for this study was drawn from the Central Population Register in 1984 among 16- to 99-year olds as two simple random selections. It consisted of 1,000 individuals selected for telephone interviewing and 2,000 persons selected for face to face interviews. The difference in the size of the two samples was due to technical reasons.

4.2 Field Procedures

The data presented here were collected during the omnibus survey of August of 1984 by the interviewing staff of the institute. Telephone numbers were found for almost 90 percent of those drawn for the telephone interviewing. An attempt was made to interview the remaining approximately 100 persons face to face. In a few cases, the method originally decided upon was replaced by the other interview form at the request of the interviewee. A few days before the start of the field work both samples received a short letter of introduction.

4.3 Testing Validity

Information on gross and net income for all respondents was obtained from the tax authorities. The data were compared with the information obtained from respondents in order to examine the extent to which the two interviewing methods resulted in systematically different answers. Also access to financial information on the nonrespondents' income permitted an analysis of the effect of nonresponse on the answers provided. Finally, it was possible to compare the incomes of respondents with those of nonrespondents.

4.4 Income Questions

Four income questions were included in the survey: personal gross and net income, and family gross and net income. A show card with 14 income categories was used for personal interviews. An attempt was made to compensate for the absence of a show card on the telephone by specially coaching the interviewers in question technique. For example, it was stressed that in cases where answers were not immediately forthcoming, the words "approximately" or "not to the penny" could be added. In addition, the interviewers were instructed that income brackets appropriate to the respondent's position could be suggested if the interviewee was reluctant to answer the open-ended income question. Furthermore, the importance of speaking clearly and slowly was accentuated. Answer categories were only used in cases where the respondent was extremely reluctant. In other words the question structure was virtually open-ended. This question technique was chosen in order to use the best approach for each mode. This means that the study is testing a question effect and a mode effect at the same time and a possible confounding cannot be excluded. Table 1 shows the wording of the four income questions.

5. FINDINGS

5.1 Response Rates

As mentioned in Section 4.2, telephone numbers were found for almost 90 percent of those selected for telephone interviewing. The remaining were attempted as face to face interviews. This procedure resulted in total response rates of 77.9 for telephone interviews supplemented with

Table 1. Missing Information for Four Income Questions by Interview Methods

Question Text	Face to Face			Telephone		
	Don't Know	Refusal	Total Nonresponse	Don't Know	Refusal	Total Nonresponse
How large was your personal gross income in 1983, i.e., before taxes and deductions?	5.7%	1.5%	7.2%	10.3%	2.5%	12.8%[a]
How large was your personal net income, i.e., income after taxes?	9.7	1.6	11.3	13.1	3.1	16.2[b]
How large was your family's gross income in 1983, i.e., before taxes and deductions?	16.3	1.8	18.1	23.2	2.6	25.8[c]
How large was your family's net income in 1983, i.e., income after taxes?	23.2	1.9	25.1	30.1	3.1	33.2[d]
Total persons		1,421			651	

[a] $\chi^2 = 17.23$, df = 1, $p < 0.05$.
[b] $\chi^2 = 8.81$, df = 1, $p < 0.05$.
[c] $\chi^2 = 15.86$, df = 1, $p < 0.05$.
[d] $\chi^2 = 14.51$, df = 1, $p < 0.05$.

personal interviews and 76.7 for the personal interview sample, i.e., the rates did not differ significantly.

In order to compare telephone and personal interviewing methods, one must compare those individuals interviewed personally with the group interviewed by telephone.

Only those individuals who completed the interview in the predetermined manner have been included in the study. The two groups were tested for a number of characteristics which may be presumed to influence the distribution of responses. The tests showed that there were no significant differences between groups as regards gender, age, and vocation. The study material consists of responses from 1,512 persons selected for and interviewed face to face and 651 persons selected for and interviewed by telephone. Ninety-two persons without telephone among the face to face respondents were furthermore excluded from the analysis.

5.2 Item Nonresponse

As can be seen in Table 1, the tendency to avoid answering income questions, through either "don't know" responses or refusals, was systematically greater among those interviewed by telephone than among those interviewed face to face. The differences are statistically significant at the 5 percent level for all four questions.

Willingness and ability to answer were in the analysis defined as the simple presence or absence of income information, regardless of whether the missing information was the result of a "don't know" response or a refusal.

The analysis showed that occupational status is the background factor correlating most strongly with capability and willingness to answer these questions. In all four questions there are significantly higher incidences of gainfully employed persons who either will not or cannot answer these questions on the telephone. A multidimensional analysis revealed that the tendency to avoid answering the question about gross personal income on the telephone was significantly higher among 25- to 59-year olds of both genders with jobs.

The results seem to indicate that the same groups of persons who are most negatively disposed to answering income questions in face to face interviews also refuse to furnish this information on the telephone.

The fact that this information is obtained from a significantly larger number of individuals in face to face interview situations should probably be seen in the light of the opportunity for persuasion uniquely present in this situation. According to statements from interviewers themselves, occasions where a "don't know" or refusal can be converted to the desired information in a face to face situation are common. However, these procedures require time and personal interaction. The quicker tempo of the telephone interview, combined with the fact that contact between the two parties is maintained exclusively through conversation, makes it extremely difficult, if not impossible, to take advantage of the usual strategies.

5.3 Response Differences

The information obtained in the two different interview situations revealed no difference in personal income between the two groups. This was the case with both gross and net income. The answers showed identical income distribution despite the greater proportion and different pattern of nonresponse in the telephone interviews. This might be a consequence of the adaption of the question technique to the telephone medium and the spontaneous nature of most answers, which were elicited without the use of answer categories.

As opposed to personal income, significant differences were found at a 5 percent level between the family incomes of the telephone and face to face groups. This applies to both gross and net income in the income brackets 0 to 79,999 kroner and 80,000 to 199,999 kroner.

Table 2 shows that the proportion of incomes in the three lowest categories is 5 to 6 percentage points higher in the telephone group than among those interviewed face to face. The same applies to high incomes, with 53 percent of the telephone group reporting gross family income of 200,000 kr. or over compared to barely 50 percent of those interviewed face to face.

The calculation of median scores did not, however, reveal any significant difference between the two distributions.

The concentration of both gross and net family incomes in the lowest income brackets, and a somewhat greater proportion in the highest bracket, might be the consequence of the fact that it is mainly individuals with moderate incomes who avoid answering income questions on the telephone. The differing rates of nonresponse across income groups could support this hypothesis.

The more correct size of incomes reported over the telephone presents another possible explanation. One analysis of face to face interview data from 1976 (Andersen and Christoffersen, 1982), involving a comparison between personally reported information and the tax authorities' final tax assessments, revealed that respondents had overreported their incomes by 2.5 percent. However, this seemingly moderate deviation concealed large differences between various subgroups. The analysis showed a significant

Table 2. Family Income by Interview Method

	Face to Face		Telephone		
Amount (kr.)	Individual Categories	Grouped	Individual Categories	Grouped	*n*
Gross income					
0 or negative	0.26%		0.00%		3
1 – 39,999	3.52		6.21		71
40,000 – 79,999	9.36		12.84		171
		13.1%		19.1%[a]	
80,000 – 119,999	12.29		10.56		194
120,000 – 149,999	9.45		8.28		150
150,000 – 199,999	15.29		9.32		223
		37.0		28.2[b]	
200,000 – 249,999	18.38		20.50		313
250,000 – 299,999	13.23		16.56		234
300,000 –	18.21		15.73		288
		49.8		52.8[c]	
Total	100.00	100.0	100.00	100.0	
Total persons	1,164		483		1,647
Net income					
0 or negative	0.38		0.00		4
1 – 39,999	6.86		10.11		117
40,000 – 59,999	9.21		11.72		149
		16.5		21.8[d]	
60,000 – 79,999	11.47		10.80		169
80,000 – 99,999	12.31		11.49		181
100,000 – 119,999	13.35		7.59		175
		37.1		29.9[e]	
120,000 – 149,999	16.26		16.09		243
150,000 – 199,999	19.55		22.53		306
200,000 –	10.62		9.66		155
		46.4		48.3[e]	
Total	100.00	100.0	100.00	100.0	
Total persons	1,064		435		1,499

[a] $\chi^2 = 9.24$, df $= 1$, $p < 0.05$.
[b] $\chi^2 = 11.70$, df $= 1$, $p < 0.05$.
[c] $\chi^2 = 1.17$, df $= 1$, n. s.
[d] $\chi^2 = 5.58$, df $= 1$, $p < 0.05$.
[e] $\chi^2 = 6.89$, df $= 1$, $p < 0.05$.
[f] $\chi^2 = 0.47$, df $= 1$, n. s.

tendency among the lowest paid towards overreporting income and, among the highest paid, a tendency to underreport. According to Koolwijk (1969) a linear relationship often exists between discomfort and the frequency with which a respondent engages in the relevant activity. The function for income, on the other hand, is curvilinear.

The fact that the telephone medium is more impersonal may reduce the need for social acceptability. When interviewer and interviewee are not in visual contact, the inclination to under- or overreport income is perhaps somewhat reduced.

Finally, the differences in the distribution of personal and family incomes could also be the result of the use of different questioning techniques. As shown in Table 1, substantially more individuals in both interview situations are either unwilling or unable to answer questions on family income than questions on personal income. These four questions are usually put right after each other in the order shown in Table 1. The increasing difficulty of these questions results in many respondents' initially attempting to avoid answering. In such cases the interviewer will most likely begin to read out the response categories, starting with the lowest intervals, in order to encourage willingness to answer. The differences could derive from changes in the question technique rather than from the effect of the interview method.

5.4 Response Validity

Table 3 shows the correlation between self-reported gross and net incomes and the information from tax authorities. For the purpose of the analysis, the income information from the tax authorities has been divided into the same categories as in the questionnaire.

A relatively large measure of agreement can be observed between reported and registered data for the respondents considered as a whole. However, this similarity conceals large differences, especially for employment categories. Regardless of whether interviewing was carried out face to face or over the telephone, a comparison reveals extremely large discrepancies for the self-employed.

This is not surprising since both gross and net incomes are especially difficult to estimate for this group. The larger measure of agreement for personal interviews could be a result of the availability of tax returns and other documents in the face to face situation.

For the other subgroups Table 3 does not reveal any remarkable differences in accuracy between the answers obtained by telephone and face to face. However, income correlations are somewhat higher for both wage earners and the unemployed who were interviewed by telephone.

Table 3. Correlation Between Self-Reported Income and Income According to Tax Authorities

	Gross Income			Net Income		
	Spearman Corr.[a]	ASE[b]	*n*	Spearman Corr.[a]	ASE[b]	*n*
Sex						
Men						
Telephone	0.85	0.03	294	0.78	0.04	281
Personal	0.87	0.02	672	0.81	0.02	643
Women						
Telephone	0.89	0.02	271	0.84	0.03	264
Personal	0.91	0.01	640	0.83	0.02	617
Employment						
Wage earners						
Telephone	0.92	0.01	317	0.85	0.02	302
Personal	0.90	0.01	752	0.80	0.02	716
Self-employed						
Telephone	0.57	0.14	39	0.46	0.15	37
Personal	0.69	0.08	97	0.61	0.10	84
Not employed[c]						
Telephone	0.86	0.03	209	0.80	0.03	206
Personal	0.82	0.03	461	0.78	0.03	458
Age						
16–24 years						
Telephone	0.94	0.01	111	0.92	0.01	111
Personal	0.90	0.02	199	0.87	0.03	196
25–39 years						
Telephone	0.86	0.04	176	0.84	0.03	163
Personal	0.86	0.02	438	0.79	0.03	412
40–59 years						
Telephone	0.87	0.04	156	0.76	0.05	151
Personal	0.86	0.03	392	0.79	0.03	371
60 and over						
Telephone	0.82	0.04	122	0.73	0.06	120
Personal	0.82	0.03	282	0.74	0.04	280
Total						
Telephone	0.90	0.02	565	0.85	0.02	545
Personal	0.90	0.01	1,312	0.85	0.01	1,260

[a]The Spearman correlation coefficient is a measure of rank correlation. Its interpretation is the same as the customary Pearson's product moment correlation.
[b]ASE is the asymptotic standard error.
[c]Includes students, schoolchildren, old age pensioners, housewives, unemployed, draftees, and persons ill for extended period of time.

The inclusion of gender and age in the analysis does not produce systematic differences in the validity of income data from the two methods.

Examining which subgroups exhibit a tendency to under- or overreport their incomes is another way of testing the respondents' own figures against data provided by the authorities. The relationship between the sets of information is summarized in Table 4.

Considered as a whole the respondents' data on gross income exhibit a surprisingly high degree of accuracy with a slight tendency to underreport in telephone interviews and an analogous slight tendency to overreport in a face to face situation.

Most noteworthy in Table 4 is the tendency among groups with an income above 100,000 kr. in 1983 to overreport gross income. With only a few exceptions this applies to all subgroups. Aside from the self-employed, whose considerable overreporting almost certainly arises from the difficulty of defining the concept, data exhibit modest but nevertheless interesting differences for those interviewed face to face and by telephone. The tendency to overreport income in higher income brackets seems to be somewhat less marked among respondents interviewed over the telephone. Table 4 reveals that statistically significant overreporting occurs only in face to face interview situations. This could be the result of respondents experiencing a social pressure to accentuate or overestimate the success represented by a relatively large income in a situation involving personal interaction. These differences may be the result of a tendency to overreport socially desirable traits and behavior in the face to face situation. This phenomenon has been observed in connection with other subjects linked to social desirability (Colombotos, 1969; Locander et al., 1976).

Another interesting characteristic of the data in Table 4 is the underreporting of gross income in both face to face and telephone interviews by almost all groups with yearly incomes under 100,000 kr. Part of the explanation for this may be the method of calculation, which involved a comparison of interval midpoints with precise figures from tax authorities, thus introducing a considerable source of error for incomes close to the upper limit of an interval. Another explanation might be that old age pensioners forget to report small additional sources of income and confine answers to the amount of their monthly pension payments. The same could be true for young students, who also often have several sources of income. However, the tendency to underreport the gross income is only significant at the 1 percent level for two subgroups (women, unemployed) and in both cases only in telephone interviews.

In spite of the modest differences in the tendency to over- and underreport in the two interview situations, the material could not document any significant differences in the accuracy of answers between the two modes of interviewing.

Table 4. Self Reported Gross and Net Incomes in Telephone and Personal Interviews Compared with Income Information from Tax Authorities

	Gross Income			Net Income		
Category	Self-Reported Average Income (1,000 kr)	Official Average Income (1,000 kr)	S.E.[a] (1,000 kr)	Self-Reported Average Income (1,000 kr)	Official Average Income (1,000 kr)	S.E.[a] (1,000 kr)
Sex						
Men						
Telephone	139	137	3.6	84***	93	2.5
Personal	138*	131	2.1	83***	89	1.6
Women						
Telephone	70**	74	1.4	47**	50	1.1
Personal	78	81	2.2	50**	54	1.3
Employment						
Wage earners						
Telephone	132	133	1.4	84**	88	1.3
Personal	135*	133	1.1	84***	89	1.0
Self-employed						
Telephone	187	155	22.9	90*	124	15.3
Personal	175*	149	11.8	92	106	9.6
Not employed)[b]						
Telephone	51**	57	2.1	36***	40	1.3
Personal	51	56	3.0	35**	39	1.5
Age						
16–24 years						
Telephone	56	58	1.5	37	39	1.1
Personal	57	57	1.3	36*	38	1.7
25–39 years						
Telephone	135	130	3.8	84**	90	2.0
Personal	128	124	2.0	81**	86	1.8
40–59 years						
Telephone	128	129	4.7	79*	89	4.0
Personal	139	135	10.1	81**	89	2.4
60 and over						
Telephone	81	88	3.6	52**	59	2.4
Personal	74	78	4.4	49*	52	1.4
Total						
Telephone	106	107	2.0	66***	72	1.4
Personal	108	107	1.5	67***	72	1.0

[a] Standard error of the difference between reported and official.
[b] Includes student, schoolchildren, old age pensioners, housewives, unemployed draftees, and persons ill for extended periods of time.
* Significant at 5 percent level.
** Significant at 1 percent level.
*** Significant at 0.1 percent level.

As regards net income, all subgroups without exception overestimate the amount of taxes paid, thereby underestimating the size of net income. Systematic qualitative differences between the two methods of data collection could not be uncovered in this case either.

5.5 Income for Nonresponse Groups

A frequent assertion in methodology literature is that many individuals do not wish to answer income questions because their answers would deviate from the usual norms and expectations, i.e., from the socially desirable. In fact several empirical studies show that the groups refusing to answer this type of question have average incomes far exceeding those who do answer them (Andersen et al., 1982; Hippler and Hippler, 1986). One problem with these studies is that missing responses can take the form of their direct refusal or "don't know" answers. As seen in Table 1, the first type of reply is relatively rare in Danish surveys. Since a "don't know" answer may in fact constitute a polite refusal, these two categories have been combined for purposes of this analysis. Table 5 shows gross income registered by the tax authorities and by interview result.

With the exception of one subgroup (60 years or more selected for face to face interviews) the nonrespondents are characterized by systematically lower incomes than the respondents. Nonresponse is known to be much greater in urban areas and is even concentrated in certain parts of cities. Marital status, age, and gender play a role as well. These demographic characteristics are usually related to income levels. Thus it is not surprising that average income is lower for nonrespondents. What is more surprising is that among those interviewed by telephone the interval between incomes of respondents and nonrespondents is much larger than among those selected for face to face interviews.

A comparison of those who answered the question on personal gross income with those not volunteering this information is ambiguous. Table 5 does show a tendency towards lower average incomes among the latter. The few cases of higher average incomes among nonrespondents occur in the group selected for face to face interviews. This could be interpreted as a sign of greater significance being attached to social desirability, i.e., exceptionally high income, in face to face situations. The size of the standard error of the mean, which is remarkably greater for personal interviews, supports this assumption.

All in all, the material suggests that, while nonresponse on income questions is significantly greater for telephone interviews and the composition of nonresponse is different, these factors have no significant effect on distributions. In the case of both interview methods

Table 5. Average Gross Income in 1,000's of Kr. According to Tax Authorities, by Interview Result

	Have Answered		No Answer to Income Question		Not Interviewed	
	Average	Standard Error	Average	Standard Error	Average	Standard Error
Sex						
Men						
Telephone	137	5.056	123	13.887	105	8.676
Personal	131	3.256	122	24.982	117	7.258
Women						
Telephone	74	3.063	72	6.728	65	5.450
Personal	81	2.932	80	13.789	75	4.932
Employment						
Wage earners						
Telephone	133	3.614	110	12.867		
Personal	133	2.237	116	11.290		
Self employed						
Telephone	155	22.584	117	16.018	84	5.149[a]
Personal	149	12.795	119	38.992	94	4.359
Not employed[b]						
Telephone	57	3.557	56	7.038		
Personal	56	3.673	71	16.506	56	
Age						
16–24 years)						
Telephone	58	4.663	54	14.688	44	7.480
Personal	57	3.632	59	15.000	53	5.387
25–39 years						
Telephone	130	4.716	110	13.234	107	10.204
Personal	124	3.063	143	51.440	110	6.657
40–59 years						
Telephone	129	7.337	130	16.043	112	11.718
Personal	135	4.406	130	29.353	110	6.948
60 years and over						
Telephone	88	7.009	61	7.593	63	7.582
Personal	78	5.779	59	7.430	88	10.777
Total						
Telephone	107	3.292	94	7.600	84	5.149
Personal	107	2.301	92	12.279	94	4.359

[a]Employment status is not available from the Central Population Register.
[b]Includes students, schoolchildren, old age pensioners, housewives, unemployed, draftees, and persons ill for extended periods of time.

nonrespondents have lower average incomes than respondents. Similarly, the group not reporting income, either in the form of refusal or "don't know" answers, is characterized by having lower average incomes than those supplying information. This naturally does not preclude the possibility of those who refuse being predominantly recruited among persons with much higher incomes than the respondents. Irrespective of the data collection method, the combined effect of missing responses on income questions is an overestimation of income for the sample as a whole and for most subgroups.

CHAPTER 22

SEARCHING FOR CAUSES OF INTERVIEWER EFFECTS IN TELEPHONE SURVEYS

Lynne Stokes and Ming-Yih Yeh
The University of Texas at Austin

1. REASONS FOR INTERVIEWER DIFFERENCES

It has long been believed that interviewers contribute a personal bias to the data they collect. The evidence for this comes from interpenetration experiments, whereby each interviewer receives as his/her assignment a random sample of the population (or some subset of it). If interviewer assignment averages differ by more than chance would predict, then that difference is attributed to an interviewer effect. Interviewer effects have been observed in both telephone and face to face surveys and for a variety of question types.

Among the reasons which have been suggested for these effects are the following:

1. Interviewers may not follow directions exactly, either purposefully or because those directions have not been made explicit enough. They may not ask questions exactly as worded or follow skip patterns correctly. They may make mistakes in coding or recording responses, or they may err in the way they record responses. However, research designed to test whether individual interviewer biases are correlated with frequency of incorrect interviewer behavior have generally not been able to detect relationships (Groves and Magilavy, 1986; National Center for Health Statistics, 1987).

2. Interviewers may vary in their inflection, tone of voice, or other personal mannerisms that are not controlled or discussed in training. It is difficult to measure these characteristics, though, so it hasn't been clearly established empirically whether or not they are related to interviewer

biases. Oksenberg and Cannell (this volume, Chapter 16) have begun to develop methods for measuring interviewer vocal characteristics and have found that they appear to be related to interviewer response rates, but have not investigated the relationship to interviewer biases. It has also been found that interviewer attitudes about a question (Bailar, Bailey, and Stevens, 1977) and interviewer expectations about the difficulty of the task for the respondent (Singer, Frankel, and Glassman, 1983), which may subtly influence interviewer behavior, are associated with response rates. The effect of response rate on interviewer bias will be discussed later.

3. Another potential source of interviewer effects is respondent reaction to characteristics of the interviewer that he or she cannot change, such as his or her race, age, or sex. For example, evidence of deference behavior by respondents, which is an attempt to give a socially acceptable answer to interviewers according to their race, have been found in both face to face (Campbell, 1981; Schuman and Converse, 1971) and telephone interviews (Cotter, Cohen, and Coulter, 1982; Reese et al., 1986) on items that are racially sensitive.

4. An observed interviewer effect may actually result from nonresponse, since the characteristics of an interviewer's nonrespondents may vary. This source of interviewer effects is fundamentally different from the other three. In this case, the cause of the observed differences is not that a single respondent would respond differently to different interviewers, but rather that interviewers have access to different subsets of the population. Since it is known that interviewer response rates often differ and that nonresponse is generally not at random in the population, this would seem to be a likely source for apparent differences among interviewers.

Interviewer differences are important because they have an impact on the mean square error of estimates of means and totals. If interviewers are seen as fixed features of the survey design, as they might be in an ongoing survey with little turnover, then a useful measure of total error for an estimate might be obtained by treating the interviewer effect as a bias. After all, if interviewers differ from each other, some must differ from any standard of truth as well. If, on the other hand, interviewers are seen as a random feature of the design, as they would be if interviewers for each survey realization were chosen randomly from a population of interviewers, then their impact on the estimate could also be described as an increase in its variance. This way, the mean square error would reflect the expected variability from one sample realization, with a particular set of interviewers, to the next realization, which used a different set of interviewers. The latter of these two approaches is the way interviewer effects are usually viewed, perhaps because bias is difficult to measure, except in the most unusual of circumstances, such as that enjoyed by Körmendi (this volume, Chapter 21) in her Danish study of income.

In this chapter, our goal will not be to assess the impact of interviewer effects on the estimators. Instead we develop a method designed to estimate each interviewer's relative bias, defined as the difference between an individual's and the group's average bias. This sometimes provides clues to the causes of interviewer effects, in addition to simply their presence. The purpose of having such a method would be so that the causes, once identified, could be reduced or eliminated.

In Section 2, we discuss differences in interviewer effects in face to face and telephone surveys, as well as special problems that arise when estimating the effects in telephone surveys. In Section 3, we describe a common model used to assess the impact of interviewer effects on the variance of the sample mean to demonstrate why interviewer effects should be of concern to the researcher. We also comment on why this model may be inappropriate for modeling interviewer effects in categorical data. In Section 4, we discuss a method for estimating interviewer relative biases based on an alternative to the model introduced in Section 3. Finally, in Section 5, we illustrate our method with an application to data collected in a random digit dialing (RDD) survey, and in Section 6, we discuss potential uses for the methodology.

2. COMPARISON OF EFFECTS IN TELEPHONE AND FACE TO FACE INTERVIEWS

Studies designed to detect mode effects, such as those reviewed by Sykes and Collins (this volume, Chapter 19), concentrate on examining differences in marginal distributions of responses to similar questions on similar populations for telephone and face to face surveys. Detection of mode differences in interviewer effects must be approached in a different way, since neither marginal differences nor lack of them are proof of the existence or nonexistence of differences in interviewer effects. To detect mode differences in interviewer effects, one must determine if individual interviewer biases are more or less variable in one mode than the other. Few attempts to measure the differences in this variability have been published. None have compared it for identical items from similar populations using the same survey conditions (such as interviewer pool and interviewer training). This would be a very difficult and expensive task to coordinate on a large scale.

Despite this, many researchers have speculated that the variability among interviewers will be smaller in telephone survey operations, especially those with centralized facilities, than for face to face surveys. If we examine the potential causes of interviewer effects discussed in Section 1, we can see where this belief comes from. First, if variability in interviewer biases is caused by the first mentioned source, which might be described as trainable interviewer behavior, then the better interviewer control available in a centralized telephone facility should reduce the effect. Interviewers can be more easily monitored and training can be more uniform and frequent there. Interviewers also have the opportunity to observe and learn from one another, potentially making their behavior more similar. The use of computer-assisted telephone interviewing (CATI) could make controlling differences in interviewer behavior even more complete as, for example, through more uniform probing and editing of responses. Furthermore, if interviewer differences could be identified, then measures to reduce them, such as changes in training or instructions to interviewers, could be more quickly implemented.

If the cause of the interviewer differences is the second source discussed in Section 1, a difference in delivery or tone, then there would seem to be less chance of benefitting from the better interviewer control of telephone facilities. Intuition would suggest that since these more subtle behaviors are related to personality, they could not be easily influenced. However, preliminary results reported by Oksenberg and Cannell (this volume, Chapter 16) suggest that training may be able to induce uniformity among interviewers on at least some of these behaviors.

If differences in interviewer biases are attributable to respondent reaction to the interviewer, they too are likely to be lessened when interviews are conducted by telephone. The respondent's ability to differentiate characteristics of interviewers, like race, age, or simply appearance, will be reduced or eliminated.

On the other hand, there are also some reasons to believe that interviewer effects may be more of a problem in telephone surveys than in face to face ones. Since response rates are typically lower by telephone, the interviewer differences that result solely from nonresponse, the fourth source discussed in Section 1, are potentially more pronounced. Secondly, Miller and Cannell (1982) and Sykes and Collins (this volume, Chapter 19) have suggested that the faster pace of telephone interviews leads to less thoughtful responses, which in turn may lead to greater susceptibility to interviewer effects. Finally, we will see in Section 3 that the increase in variance of the sample mean as a result of interviewer differences is proportional to interviewer workload size. Since interviewer workloads are typically larger in telephone surveys, there is a potential for a larger impact due to the interviewer effects.

A clear advantage of telephone over face to face surveys is the economy with which estimation of interviewer effects by interpenetration experiments can be accomplished there. In face to face surveys, interpenetration requires that the enumeration areas of two or more interviewers be combined and the sample units from the combined areas be randomly distributed to the interviewers (Mahalanobis, 1946). This ensures that the units in those interviewers' assignments would have the same expected values if there were no interviewer effects. However it also requires each one to travel over a larger area, which increases the cost and time to complete the survey. Since this cost may be substantial and since measurement of nonsampling errors has a low priority in most survey organizations, estimation of interviewer effects is rarely undertaken in face to face surveys.

By contrast, little additional cost is incurred by interpenetration in telephone surveys. The theoretically simplest way to design the procedure would be to randomly assign all sample units to the interviewers at the beginning of the survey and insist that they pursue those cases until it ends. Such a procedure would not be a good idea from a practical point of view, though, since interviewers would be forced to either work all shifts or accept high nonresponse rates, since some respondents in their assignments would not be reachable during their regular shifts. A better, and more common, method is to randomly assign all units available to be called on a given shift to the interviewers working on that shift. Since the expected value for units may vary by shift, only those interviewers working in the same shift may be compared directly in order that any difference in their assignment averages beyond what should occur by chance can be attributed to interviewer effects.

Other comparability adjustments may need to be made to meet practical requirements of the field operation as well. For example, if the caseload is large and the schedule is tight, it may be more efficient to exercise some control over the order in which telephone numbers are to be called. Even when these ordered cases are distributed randomly to interviewers, some inequities in their assignments may be induced. For example, fast interviewers will receive more cases further down the priority ranking than slow ones, again making direct comparisons of interviewer assignments reflect priority differences confounded with interviewer differences. A solution is to group cases according to priority. If only a few priority levels are used, priority can be thought of as another subgrouping of the population, like shift.

Frequently, the interviewers' work over several survey replicates are pooled for the purpose of measuring interviewer effects. If characteristics of the population are changing over time, then replicate should be considered as yet another subgrouping of the population, within which interviewer assignments are interpenetrated.

So although telephone surveys can make interpenetration of interviewer assignments relatively cheap, practical considerations of the fieldwork can make implementation of the designs difficult. More complex models are needed than those used for analyzing data from the simpler interpenetration designs used in face to face surveys.

A second difference in implementation of an interpenetration design in the two modes is the balance of interviewer assignments. In face to face surveys, the assignments are generally controlled to be nearly equal in size. In telephone surveys, the designer has less control over the assignment sizes, since only the number of hours an interviewer works is controlled. Estimators of parameters which measure the size of the interviewer effect therefore have different properties, due to these unequal assignment sizes.

3. MODELING INTERVIEWER EFFECTS

In this section we describe a simple model for interviewer effects, and use it to show that the impact of such effects on the variance of the sample mean can be overwhelming. We then discuss why the simple model may be inappropriate, especially for categorical items with complex interpenetration designs.

Let d_{tij}, $t = 1, ..., r$, $i = 1, ..., k$, and $j = 1, ..., n_{ti}$ be a random variable that denotes the response of the jth respondent in interviewer i's assignment in subgroup t of the population. The subgroups may be shift, priority, or shift $\times$ priority levels, or any other subgroup of the population within which interviewer assignments are interpenetrated. One can show that the variance of the sum of any random variables can be written as

$$\begin{aligned}
\mathrm{Var}\left(\sum_t \sum_i \sum_j d_{tij}\right)\\
&= \sum_t \sum_i \sum_j \mathrm{Var}(d_{tij})\\
&\quad + 2 \sum_t \sum_i \sum_{j<j'} \mathrm{Cov}(d_{tij}, d_{tij'})\\
&\quad + 2 \sum_{t<t'} \sum_i \sum_j \sum_{j'} \mathrm{Cov}\,(d_{tij}, d_{t'ij'})\\
&\quad + 2 \sum_t \sum_{t'} \sum_{i<i'} \sum_j \sum_{j'} \mathrm{Cov}(d_{tij}, d_{t'i'j'})\,.
\end{aligned} \tag{22.1}$$

Components of the second term of Eq. (22.1) denote the covariance between two responses in the same subgroup of the population and which are recorded by the same interviewer. Those in the third term of Eq. (22.1) denote the covariance between pairs of responses, one in one subgroup and one in another, but which are also recorded by the same interviewer. Finally, those of the fourth term denote the covariance between two responses recorded by different interviewers.

A common model for interviewer effects assumes that

$$\mathrm{Cov}\,(d_{tij}, d_{t'i'j'}) = 0 \text{ for } i \neq i' \text{ and any } t, t', j, j'; \tag{22.2}$$

that is, there is no correlation among units which are not recorded by the same interviewer. It also assumes

$$\mathrm{Var}(d_{tij}) = \sigma^2 \text{ for all } t, i, \text{ and } j \tag{22.3}$$

and

$$\mathrm{Cov}(d_{tij}, d_{t'i'j'}) = \rho\sigma^2 \text{ for } i = i' \text{ and any } t, t', j, j'\,. \tag{22.4}$$

Under these assumptions, one can show that

$$\mathrm{Var}(\bar{d}) = (\sigma^2/rkn)\,[1 + (n-1)\rho] \tag{22.5}$$

in the simple case of a completely balanced design, which is one in which $n_{ti} = n$ for all t and i, that is, all interviewer assignments are the same size n.

The parameter ρ in Eqs. (22.4) and (22.5) is called the *intra-interviewer correlation* since it is the correlation between any two units in the same interviewer's assignment under this model. An estimate of ρ is often reported as a measure of the severity of interviewer effects for an item because of its direct effect on the variance of $\bar{d}$. Although typical values of ρ are small, usually ranging between 0 and about 0.05, the impact on $\mathrm{Var}(\bar{d})$ can be seen from Eq. (22.5) to be great if the interviewer assignment sizes are large. For example, if $\rho = 0.01$ and $n = 101$, $\mathrm{Var}(\bar{d})$ would be doubled over its size for an equal sample size design in which each interviewer conducts a single interview.

Assumptions (22.2), (22.3), and (22.4) are strong ones, however, and untrue in certain circumstances. For example, Eq. (22.2) implies that there is no correlation between two units not in the same interviewer's assignment, even if the responses have been supervised by the same crew leader or coded by the same coder. Equation (22.3) implies that the variability of the responses must be equal over all r subgroups of the population. For some continuous responses, this assumption would be questionable. The situation is more serious if the item is a categorical one and d_{tij} records whether ($d_{tij} = 1$) or not ($d_{tij} = 0$) the respondent belongs to a particular category. Then Eq. (22.3) is impossible if the proportion of the responses in a given category, $E(d_{tij})$, differs from one subgroup to another. This is because the variance of such a variable, called a Bernoulli variable, depends upon its mean; specifically $\mathrm{Var}(d_{tij}) = E(d_{tij})[1 - E(d_{tij})]$.

Assumption (22.4) has two questionable implications. The first is that the covariance between two of an interviewer's responses in the same subgroup of the population must equal the covariance between two responses where one is in one subgroup and one in another. The second is that the covariance between two of an interviewer's responses in the same subgroup of the population is constant over all subgroups. Even if d_{tij} is continuous, these may not be true. For example, suppose d_{tij} denotes household income and the subgroupings of the population are replicates. If interviewer differences diminish with increasing interviewer experience, we would expect $\mathrm{Cov}(d_{tij}, d_{tij'}) > \mathrm{Cov}(d_{t'ij}, d_{t'ij'})$ if t' denotes a later replicate than t. For categorical items, the situation is again worse, since in that case, $\mathrm{Cov}(d_{tij}, d_{t'ij'})$ and $\mathrm{Cov}(d_{tij}, d_{t'ij'})$ again depend on $E(d_{tij})$ and/or $E(d_{t'ij'})$.

The problems with the assumptions leading to Eq. (22.5) have been recognized and discussed by Anderson and Aitkin (1985) and Pannekoek (1988). Each proposed a different model for the interviewer effect in a categorical item, the former using logistic regression and the latter a beta-binomial model. We choose to follow the development of Anderson and Aitkin because their model makes it easier to take into account interpenetration within subgroups of the population.

4. IDENTIFYING DISCREPANT INTERVIEWERS

In this section we propose a logistic regression model for describing the effects of the interviewers on a categorical response in a complex interpenetration design. Then the interviewer relative biases can be estimated from the model, with the intent of identifying those interviewers who obtain responses that are extreme or discrepant from the others. For this reason, we treat the interviewer as a fixed rather than a random effect, which is a departure from the usual point of view in interviewer variance studies (Kish, 1962; Anderson and Aitkin, 1985).

It should be noted that we cannot know if these discrepant interviewers are further or closer to the truth than the others, since we have no measure of such a standard. Nevertheless, our goal is to identify these interviewers so that they may be investigated.

Since we are restricting attention to categorical responses, we again let $d_{tij} = 1$ if the jth response in interviewer i's assignment in subgroup t of the population is in a specific category and let $d_{tij} = 0$ otherwise. We propose the logistic regression model

$$\begin{aligned} p_{ti} &= P(d_{tij} = 1) \\ &= \exp(\alpha_t + \beta_i)/[1 + \exp(\alpha_t + \beta_i)] \end{aligned} \tag{22.6}$$

for $t = 1, ..., r$; $i = 1, ..., k$. The logistic model is one of several commonly used for describing effects on categorical data; it seems to fit the data examined in Section 5 adequately. Now α_t and β_i are parameters which measure the effect of the tth subpopulation and the ith interviewer, respectively, on the category proportion. Having the α_t's in the model allows us to remove the effect of subpopulation on the responses in order that we can make fair comparisons among interviewers. Under the assumptions of this model, the ranking of the β_i's, $i = 1, ..., k$, is identical to the ranking of the p_{ti}'s for each subpopulation t. So any interviewer with an unusually high or low β_i is one with large relative bias.

Any of a number of software packages perform logistic regression and thus could be used for fitting Eq. (22.6). Several of them provide a choice of estimators for parameters of the model. Since maximum likelihood estimators are needed for a subsequent step in our analysis, we used SAS PROC CATMOD, whose output contains these estimates as well as estimates of their variances.

Our experience was, however, that when Eq. (22.6) was fit to data from our RDD survey, the interviewers whose estimated β_i values for an item were most extreme were often those who had recorded the fewest responses. It is possible, of course, that those interviewers actually had the most extreme relative biases. However, we also know that the variance of an estimator of β_i will be large when the amount of information available on that interviewer's performance is small. Extreme values of $\hat{\beta}_i$, the estimator of β_i, would then be more likely than extreme values of $\hat{\beta}_{i'}$, say, if interviewer i' had a larger assignment, even if $\beta_i = \beta_{i'}$.

A methodology which was developed for the problem of estimating many parameters of the same type when there is not a lot of information available on any one of them is the empirical Bayes method (Morris, 1983). Its advantage is that under certain conditions, it provides estimates which are better (in a sense to be discussed later) than those produced by classical competitors.

Its advantage comes from an assumption of its model that all parameters to be estimated (all β_i's in our case) are similar. This allows information about one interviewer's β_i to be gained from information about another's, an idea which has been referred to as "borrowing strength."

The assumption of similarity among the β_i's is stated mathematically in the empirical Bayes model by saying that the β_i's are random variables which are independent and identically distributed. For the analysis undertaken here, we further assume that the β_i's are from a normal population having mean β and variance A, which are both unknown; i.e.,

$$\beta_i \sim N(\beta, A). \tag{22.7}$$

The magnitude of A is then a measure of the similarity among the β_i's.

If β and A had been assumed known, this would be the type of assumption made in a Bayes analysis, where exact priors are specified for unknown parameters of the model. By contrast, the empirical Bayes model assumes only some characteristics of the prior (in our case, normality), but not a unique distribution.

A second assumption needed for the empirical Bayes analysis we perform is that

$$\hat{\beta}_i \mid \beta_i \sim N(\beta_i, V_i), \text{ for } i = 1, ..., k \tag{22.8}$$

and are independent, with the V_i's known. Then from Eqs. (22.7) and (22.8) the marginal distribution for each $\hat{\beta}_i$ can be shown to be

$$\hat{\beta}_i \sim N(\beta, A + V_i). \tag{22.9}$$

The empirical Bayes estimate of β_i is then

$$\beta_{EBi} = (1 - \hat{W}_i)\,\hat{\beta}_i + \hat{W}_i\,\hat{\beta}\,, \tag{22.10}$$

where $\hat{W}_i = V_i/(V_i + \hat{A})$, and $\hat{A}$ and $\hat{\beta}$ are estimates of A and β. Note from Eq. (22.9) that information about A and β is available from the $\hat{\beta}_i$'s. β_{EBi} is a weighted average of β_i, which describes behavior of the ith interviewer alone, and $\hat{\beta}$, which contains information about all interviewers. The weight $\hat{W}_i$ is determined by the relative magnitudes of V_i and $\hat{A}$; if V_i is small relative to $\hat{A}$, most of the weight is on the individual estimate β_i rather than $\hat{\beta}$ and vice versa. Empirical Bayes estimators are sometimes referred to as shrinkage estimators since they pull the $\hat{\beta}_i$'s toward the estimated mean of the prior.

Efron and Morris (1973) proved that if Eqs. (22.7) and (22.8) hold, then the β_{EBi}'s are on the average better than the $\hat{\beta}_i$'s as estimators of the β_i's in the sense that $E[\sum_i (\beta_i - \beta_{EBi})^2] < E[\sum_i (\beta_i - \hat{\beta}_i)^2]$ if $k \geq 3$, where the expectation is over the distribution given in Eq. (22.7). A practical beneficial outcome for our application is that the interviewers who are identified by the estimation process as being most discrepant are less likely to have been identified just because their estimate was based on a small amount of information.

In our application, there are reasons supporting the choice of assumptions (22.7) and (22.8), but their validity is not assured. For example, a normal prior for the β_i's, as required by Eq. (22.7), was chosen for our analysis because its theory has been most fully developed. It is believed, however, to enjoy a certain robustness property, which can be roughly described by saying that a conjugate prior (which the normal distribution is in this case) has the largest risk of any prior with the same mean and variance (see Morris, 1983, Theorem 1). The assumption of normality in Eq. (22.8) is justified in our application, at least for large samples, since the $\hat{\beta}_i$'s are maximum likelihood estimators. Although the V_i's are not known as required by Eq. (22.8), we can estimate them (estimates are provided by SAS PROC CATMOD) and replace V_i by $\hat{V}_i$ in β_{EBi}. The assumption of independence of the $\hat{\beta}_i \mid \beta_i$'s is clearly violated in our application. However, we do have estimates of their correlations (also provided by SAS PROC CATMOD) for our data, and they are fairly small. Therefore we ignore rather than try to adjust for this complication.

To implement the empirical Bayes procedure we have described, we first need estimates of β and A in order to compute β_{EBi}. We also need a method for obtaining confidence intervals based on the estimates. Several procedures have been proposed for these problems (Morris, 1983; Laird and Lewis, 1987).

We used Morris's method for the analysis presented in Section 5. It was implemented using software written in SAS PROC IML. (A listing of that program is available from the authors.) Its outputs include the estimates β_{EBi}, $i = 1, ..., k$, along with estimates of the standard error s_i for each. Morris (1983) proposes an approximate empirical Bayes $(1-\alpha)100$ percent confidence interval of the form $\beta_{EBi} \pm zs_i$, where z is the $1 - \alpha/2$ quantile of the standard normal distribution.

5. DATA ANALYSIS

In this section, we illustrate the use of the methodology developed in Section 4 for some data from a random digit dial (RDD) telephone survey. The data were collected by the U.S. Bureau of the Census in 1982. The questionnaire used was nearly identical to that of the Current Population Survey (CPS) and included the usual demographic items as well as those concerned with employment status. There was special interest in the measurement of response rate for certain sensitive questions, such as household income.

The data analyzed here were collected by 16 interviewers and contain results of interviews with 1414 households. The number of completed interviews per interviewer ranged from a high of 249 to a low of 14. Interviewers generally worked two concurrent shifts of 2-1/2 hours each, but were not consistent in the shifts they worked from one day or one week to the next. These shifts were collapsed for the purpose of analysis into four categories: morning (9 A.M. − 2 P.M. weekdays), afternoon (2 P.M. − 7 P.M.), evening (7 P.M. − midnight), and weekend (all day Saturday). Shift defined the only subpopulations which were found necessary to account for differences in proportions for the variables whose analyses are presented here. Although the cases were distributed according to five priority groups and contained data from three replicates, no significant differences among these groups were confirmed. The model, then, is that shown by Eq. (22.6), where $r = 4$ and $k = 16$.

The items considered in this analysis are household income and employment status. The item household income was asked as an unfolding question (Groves and Kahn, 1979), so its responses were collected in categories. Those categories, which were fairly fine, were pooled to form the four larger categories: low (<\$10,000); medium (\$10,000−\$25,000); high (>\$25,000) income; and a nonresponse (NR) category. The response to the employment status item was recorded in one of the categories: working, with a job but not at work, looking for work, in school, retired, keeping house, unable to work, or other nonlabor force. Several of these categories were rarely used, which made estimates

Table 1. Comparisons of Empirical Bayes and Maximum Likelihood Estimates of Interviewer Effects for the 1982 RDD Study: Income Item

	High		Medium		Low		Nonresponse	
Interviewer	$\hat{\beta}_i$ ($\sqrt{\hat{V}_i}$)	β_{EBi} (s_i)	$\hat{\beta}_i$ ($\sqrt{\hat{V}_i}$)	β_{EBi} (s_i)	$\hat{\beta}_i$ ($\sqrt{\hat{V}_i}$)	β_{EBi} (s_i)	$\hat{\beta}_i$ ($\sqrt{\hat{V}_i}$)	β_{EBi} (s_i)
1	−0.024	−0.268	−1.573	−1.393	−1.416	−1.487	−1.852	−1.692
	(.34)	(.21)	(.41)	(.26)	(.40)	(.19)	(.47)	(.31)
2	−0.735	−0.478	−0.886	−1.132	−1.875	−1.575	−1.110	−1.216
	(.34)	(.21)	(.36)	(.25)	(.45)	(.23)	(.37)	(.31)
3	−0.537	−0.412	−1.848	−1.444	−1.469	−1.498	−0.905	−1.105
	(.41)	(.22)	(.55)	(.31)	(.47)	(.22)	(.42)	(.35)
4	−0.721	−0.439	−1.060	−1.235	−0.709	−1.371	−2.652	−1.811
	(.60)	(.28)	(.60)	(.32)	(.57)	(.29)	(1.04)	(.60)
5	−0.464	−0.429	−0.911	−1.014	−1.210	−1.382	−2.177	−2.079
	(.14)	(.12)	(.16)	(.15)	(.17)	(.16)	(.22)	(.21)
6	0.351	−0.154	−2.288	−1.567	−1.384	−1.480	−1.969	−1.753
	(.33)	(.25)	(.54)	(.35)	(.39)	(.19)	(.49)	(.39)
7	−0.306	−0.356	−2.296	−1.534	−1.433	−1.492	−0.858	−1.098
	(.44)	(.22)	(.64)	(.31)	(.50)	(.22)	(.46)	(.31)
8	−0.664	−0.499	−1.165	−1.225	−1.929	−1.607	−0.917	−1.001
	(.22)	(.17)	(.25)	(.19)	(.32)	(.21)	(.23)	(.22)
9	−0.240	−0.308	−0.876	−1.041	−1.391	−1.470	−2.611	−2.270
	(.20)	(.15)	(.22)	(.19)	(.24)	(.15)	(.36)	(.33)
10	−0.335	−0.357	−1.205	−1.252	−1.104	−1.413	−2.038	−1.855
	(.26)	(.17)	(.30)	(.22)	(.34)	(.21)	(.38)	(.33)
11	−0.681	−0.499	−1.346	−1.320	−2.296	−1.663	−0.667	−0.826
	(.24)	(.18)	(.28)	(.21)	(.41)	(.28)	(.27)	(.26)
12	−0.705	−0.494	−1.678	−1.464	−1.198	−1.425	−1.061	−1.144
	(.27)	(.19)	(.32)	(.24)	(.28)	(.19)	(.28)	(.26)
13	−0.084	−0.232	−1.814	−1.562	−1.547	−1.518	−1.396	−1.406
	(.20)	(.16)	(.27)	(.23)	(.25)	(.15)	(.24)	(.22)
14	−0.027	−0.201	−1.224	−1.254	−1.640	−1.543	−2.023	−1.905
	(.19)	(.16)	(.23)	(.18)	(.26)	(.16)	(.28)	(.26)
15	−0.320	−0.344	−1.436	−1.380	−1.922	−1.628	−1.078	−1.127
	(.18)	(.14)	(.22)	(.18)	(.25)	(.20)	(.21)	(.20)
16	−0.853	−0.524	−0.947	−1.134	−1.713	−1.550	−1.140	−1.221
	(.31)	(.22)	(.31)	(.23)	(.36)	(.19)	(.32)	(.28)

of α_t and β_i for them unstable. Consequently, we had to pool responses into four categories for purposes of analysis: employed (including "working" and "with a job but not at work"), keeping house, other nonlabor force, and else (including "looking for work," "in school," "retired," and "unable to work").

Table 1 displays for each interviewer the maximum likelihood $\hat{\beta}_i$ and empirical Bayes estimates β_{EBi} of β_i, along with estimates of their standard deviations $\sqrt{\hat{V}_i}$ and s_i, respectively, for each category of the item household income. Table 2 displays the same information for the item employment

Table 2. Comparison of Empirical Bayes and Maximum Likelihood Estimates of Interviewer Effects for the 1982 RDD Study: Labor Force Item

	Working		Keeping House		Other Nonlabor Force		Else	
Interviewer	$\hat{\beta}_i$ ($\sqrt{\hat{V}_i}$)	β_{EBi} (s_i)	$\hat{\beta}_i$ ($\sqrt{\hat{V}_i}$)	β_{EBi} (s_i)	$\hat{\beta}_i$ ($\sqrt{\hat{V}_i}$)	$\beta_{EB(i)}$ (s_i)	$\hat{\beta}_i$ ($\sqrt{\hat{V}_i}$)	β_{EBi} (s_i)
1	0.703	0.586	−1.288	−1.704	−2.560	−2.019	−3.085	−2.373
	(.34)	(.24)	(.39)	(.33)	(.51)	(.28)	(.63)	(.46)
2	0.178	0.338	−2.588	−2.283	−1.485	−1.794	−1.518	−1.691
	(.33)	(.24)	(.61)	(.38)	(.39)	(.23)	(.43)	(.33)
3	−0.435	0.141	−1.918	−2.045	−1.354	−1.773	−1.401	−1.640
	(.43)	(.32)	(.52).	(.34)	(.44)	(.26)	(.46)	(.35)
4	0.523	0.496	−1.724	−2.003	−2.961	−2.048	−1.685	−1.822
	(.56)	(.31)	(.68)	(.40)	(1.05)	(.43)	(.67)	(.42)
5	0.247	0.294	−2.195	−2.176	−1.988	−1.928	−1.391	−1.465
	(.15)	(.13)	(.24)	(.21)	(.21)	(.15)	(.19)	(.18)
6	1.190	0.795	−2.296	−2.195	−2.115	−1.935	−3.224	−2.354
	(.36)	(.28)	(.51)	(.34)	(.46)	(.23)	(.74)	(.50)
7	0.173	0.370	−2.037	−2.094	−2.595	−2.011	−1.166	−1.516
	(.44)	(.28)	(.55)	(.35)	(.64)	(.31)	(.46)	(.36)
8	0.169	0.278	−1.927	−2.010	−1.427	−1.741	−2.019	−1.980
	(.22)	(.19)	(.33)	(.26)	(.27)	(.23)	(.33)	(.28)
9	0.502	0.496	−2.106	−2.114	−2.347	−2.020	−1.574	−1.654
	(.20)	(.17)	(.30)	(.24)	(.31)	(.23)	(.26)	(.23)
10	0.917	0.710	−2.030	−2.083	−1.739	−1.856	−3.465	−2.528
	(.30)	(.23)	(.43)	(.31)	(.44)	(.22)	(.63)	(.49)
11	0.262	0.350	−2.622	−2.345	−1.306	−1.740	−1.711	−1.774
	(.24)	(.20)	(.46)	(.33)	(.35)	(.26)	(.32)	(.27)
12	0.214	0.325	−1.764	−1.916	−1.764	−1.853	−2.024	−1.982
	(.25)	(.20)	(.33)	(.27)	(.31)	(.18)	(.34)	(.28)
13	0.420	0.440	−2.281	−2.222	−1.968	−1.914	−1.636	−1.700
	(.20)	(.17)	(.31)	(.25)	(.26)	(.17)	(.26)	(.23)
14	0.784	0.684	−2.248	−2.198	−2.071	−1.941	−2.273	−2.153
	(.21)	(.18)	(.33)	(.26)	(.30)	(.19)	(.32)	(.28)
15	1.116	0.927	−4.017	−2.856	−1.802	−1.857	−2.264	−2.166
	(.19)	(.19)	(.53)	(.51)	(.24)	(.16)	(.28)	(.25)
16	0.482	0.483	−1.707	−1.898	−2.389	−2.001	−1.933	−1.922
	(.28)	(.21)	(.36)	(.28)	(.41)	(.25)	(.37)	(.30)

status. Several cases of changes in ranking from the $\hat{\beta}_i$'s to the β_{EBi}'s are apparent from Tables 1 and 2. For example, interviewer 4 is ranked among the most extreme on the high income category on the basis of his maximum likelihood estimate, while his empirical Bayes estimate is only the sixth smallest. Note also that $\hat{\beta}_4$ is the least precise of all the estimates, having an estimated standard error of .60.

A more enlightening way to display the empirical Bayes estimates shown in Tables 1 and 2 is graphically, which is done in Figures 22.1 and 22.2. Before examining these graphs, first note from Eq. (22.6) that the proportion of afternoon shift respondents who would be recorded by interviewer i as being

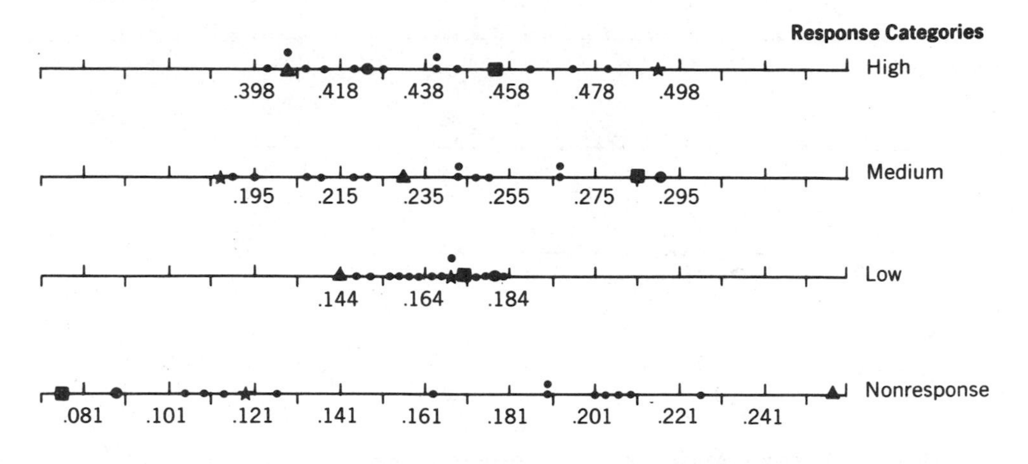

Figure 22.1 Model based estimates $\hat{p}_{Ai} = \exp(\hat{\alpha} + \beta_{EBi})/[1 + \exp(\hat{\alpha}_A + \beta_{EBi})]$ for each of the 16 interviewers and for each category of the item household income. ▲ = Interviewer 11; • = Interviewer 5; ■ = Interviewer 9; ★ = Interviewer 6.

in a particular category, according to our model, would be $p_{Ai} = \exp(\alpha_A + \beta_i)/[1 + \exp(\alpha_A + \beta_i)]$, where α_A is the effect of the afternoon shift and β_i is the effect of the ith interviewer for that category. Figure 22.1 plots estimates of these parameters, denoted by $\hat{p}_{Ai}$, for each category of the household income item, while Figure 22.2 shows the same for the employment status item. The scale for each category is the same, but each is centered at the estimate of its prior mean, $\hat{\beta}$.

In effect, we are comparing our model-based estimates of the proportion of respondents each interviewer would have recorded in a category if he or she had done all interviewing in the afternoon shift. The ranking of the estimates of p_{Ai} , $i = 1, ..., k$, will be the same as that among the β_{EBi}'s, or the same as if we had adjusted to any other shift. We do the adjustment just to aid interpretation; comparison of absolute magnitudes and differences between them for the parameters p_{Ai} are more meaningful than absolute magnitudes and differences among the β_i's.

Figure 22.1 clearly shows that the largest discrepancy among interviewers occurs in the nonresponse category. This is not unexpected, as the highest ρ values in interviewer variance studies are frequently reported for nonresponse categories. A more interesting observation from Figure 22.1 can be made by retaining the identity of some of the extreme interviewers. Then we can see that the interviewers whose relative biases are most extreme in the nonresponse category frequently are also extreme on one or more substantive categories as well. Interviewers 5 and 9, who have an extremely small proportion of nonrespondents for the income

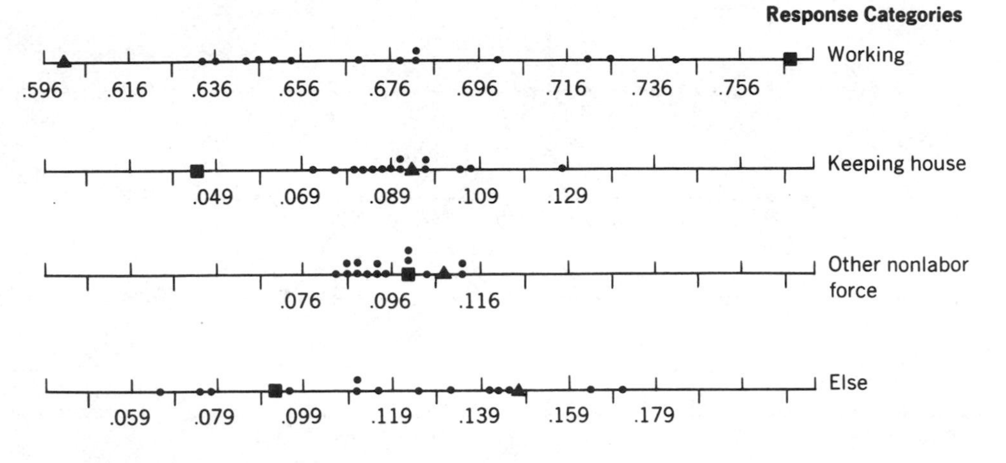

Figure 22.2 Model based estimates $\hat{p}_{Ai} = \exp(\hat{\alpha} + \beta_{EBi})/[1 + \exp(\hat{\alpha}_A + \beta_{EBi})]$ for each of the 16 interviewers and for each category of the item employment status. ▲ = Interviewer 3; ■ = Interviewer 15.

item, report an extremely large proportion of medium income respondents. On the other hand, interviewer 11, who has a large proportion of nonrespondents, has low proportions in the high and low income categories. This may suggest that interviewers 5 and 9 are successful in encouraging medium income respondents to answer the income question, while other interviewers are not. It also suggests that interviewer 11 seems to lose respondents from the high and low income categories, perhaps because of too slow or otherwise annoying delivery of the unfolding question. (The question starts with middle income categories and progresses toward either end, depending on the answers obtained from the respondent.) It may be, then, that a substantial proportion of the apparent variability among the responses in some categories of this item is due to nonresponse. The only obvious exception to this idea is interviewer 6, who seems to be classifying too many respondents in the high income category and too few in the medium, but has a typical proportion of nonresponses.

Figure 22.2 also suggests that some interviewers may be classifying respondents into certain categories according to different rules than other interviewers. For example, interviewer 15 seems to be mixing the employed and keeping house categories, perhaps suggesting a misunderstanding of the rules for classifying the employment status of a housewife who works irregularly. Interviewer 3 also seems to classify too few respondents as employed and too few in some of the nonlabor force categories. Figure 22.2 also suggests, for the keeping house category, at least, that most of the variability among interviewers can be attributed to only a couple of them. If the difficulties of those interviewers could be eliminated, then interviewer effects might cease to be a problem for that category.

6. CONCLUSIONS

We have made several suggestions concerning procedures for investigating interviewer effects in survey data which are not now standard. The first is that an alternative to the usual model for interviewer effects be used when the responses are categorical and the interpenetration design is complex. The second is that more information about the potential source of differences among interviewers can be obtained if estimates of individual interviewer effects are made and if all categories of an item are examined together rather than one at a time. And finally, when interviewer sample sizes vary greatly, as they may do in telephone surveys, an empirical Bayes estimation procedure may help in identifying interviewers who are discrepant.

SECTION E

COMPUTER-ASSISTED TELEPHONE INTERVIEWING

CHAPTER 23

COMPUTER-ASSISTED TELEPHONE INTERVIEWING: A GENERAL INTRODUCTION

William L. Nicholls II
U. S. Bureau of the Census

Computer-assisted telephone interviewing (CATI) employs interactive computing systems to assist interviewers and their supervisors in performing the basic data-collection tasks of telephone interview surveys.[1] In typical applications the interviewer is seated at a computer terminal or microcomputer, wearing a telephone headset. As survey questions are displayed on the computer screen, the interviewer reads them to the respondent and enters responses on the computer keyboard. In most systems, question wording and branching between items is computer controlled, based on prior entries or case input, and answers as entered can be checked to prompt correction of edit failures and reconciliation of apparent inconsistencies.

Current CATI systems range in size from a single personal computer serving as a standalone interviewer workstation to configurations of 100 or more interviewing and supervisory stations supported by a large mainframe. More commonly, CATI systems use a minicomputer and dumb terminals or networked microcomputers for 5 to 60 workstations. Current systems also vary widely in the amount and kind of computing assistance they provide to telephone survey operations. Some merely supply a blank online questionnaire to display questions and accept entries. The interviewer must enter case numbers and input data, while

[1]This definition of CATI and the listing of CATI capabilities in Table 1 are taken from Nicholls and Groves (1986) with permission of the *Journal of Official Statistics*. The history and uses of CATI summarized in Section 1 are based on more extensive coverage of these topics in the same source.

sample management and call scheduling are handled by offline procedures. Other systems automate virtually all survey interviewing, supervisory, and management functions, providing the range of computer assistance outlined in Table 1. The chapters in Section E focus on CATI systems with broad capabilities.

Table 1. Capabilities of Current CATI Systems

1. *Sample management*. The sample and associated case input data are stored on computer files. The system (rather than the interviewing staff) maintains the sample status of each case and links its input data to the interview and output record.

2. *Online call scheduling and case management*. The system selects the cases to be called as interviewers request another case. The priority, sequence, and timing of most calls is set by the system at least until the respondent is first reached.

3. *Online interview*. The online interview has the following features.
   a. The system displays interviewer instructions, survey questions, and response categories on the interviewer's computer screen.
   b. Screens may contain "fills" or alterations of the display text based on prior answers or batch input from case records.
   c. Answers to closed questions may be entered by numeric or alphanumeric codes; and those codes and other numeric entries may be edited by sets of permissible values, by ranges, or by logical or arithmetic operations.
   d. Edit failures may result either in an unaccepted entry (requiring another attempt) or in display of additional probes or questions to be asked.
   e. Extended text answers may be entered for open questions.
   f. Branching or skipping to the next item is automatic and may be based on logical or arithmetic tests of any prior entries or input data.
   g. Interviewers may interrupt and resume interviews in midcourse; review, back up to, and (if permitted) change prior entries; and enter interviewing notes at appropriate points.

4. *Online monitoring*. The system is able to reproduce any interviewer's screen at a supervisor's terminal where audio monitoring may also occur.

5. *Automatic record keeping*. The system maintains records (or keeps summaries) of online calls, their outcomes, response rates, and interviewer productivity and makes this information accessible to survey supervisors and managers in online or printed reports.

6. *Preparation of data sets*. The CATI output files are produced in a form ready for the next stage of processing. This may be post interview editing and coding, batch editing or imputation, weighting, and analysis.

1. HISTORY AND USES OF CATI

The first CATI systems were developed by U.S. market research organizations in the early 1970s (Fink, 1983). They found ready acceptance by meeting important needs in the market research field. By entering interview responses directly to the computer, tabulations could be produced much more quickly after the close of interviewing. By accumulating counts of key respondent characteristics, quota targets could be precisely and efficiently met. By adding visual to audio monitoring from supervisory (or client representative) terminals, sufficient quality and production control were provided to make separate verification surveys unnecessary. And by utilizing computer capabilities for randomization, rapid calculations, and table lookup, new types of market research studies were made possible. Through the 1970s and early 1980s, the use and sophistication of CATI in commercial market research grew rapidly (Dutka and Frankel, 1980; Fink, 1983). Today CATI is commonplace in market research agencies in the United States and other western nations where residential telephones are ubiquitous.

University survey research centers began their largely independent development of CATI in the mid 1970s (Shure and Meeker, 1978; Palit and Sharp, 1983; Shanks, Lavender, and Nicholls, 1980). They stressed computing support for probability sampling, such as clustered random digit dialing, and for call and callback scheduling to attain the high response rates probability sampling requires. They also extended the range of interviewer actions, such as backing to prior items, resuming interrupted interviews, and entering interviewer item notes, to more closely approximate those of personal interviewers in academic surveys (Shanks, 1983). University CATI developers also introduced CATI to the broader statistical community, emphasizing its value for survey documentation, the standardization of survey practice, and its potential contributions to survey methodology (Freeman, 1983; Groves, 1983; Nicholls, 1978). Today most university and private sector survey research organizations in the United States have CATI facilities or are planning to acquire them in the near future (Spaeth, 1987).

U. S. governmental agencies demonstrated an early interest in CATI but did not begin acquiring their own CATI capabilities until the early 1980s (Tortora, 1985; Nicholls, 1983b). In government applications, CATI frequently is employed in multimode data collection, such as telephone followup to mail questionnaire nonresponse or second and later visit interviews after an initial face to face interview. These uses have prompted the closer integration of CATI with computer support for personal interviewing and mail questionnaires. Introduction of CATI for production data collection is currently in progress at the U.S. Census

Bureau, the National Agricultural Statistics Service, the Bureau of Labor Statistics, the Centers for Disease Control, the Netherlands Central Bureau of Statistics, Statistics Canada, and Statistics Sweden, and is being considered by additional governmental agencies in the United States and abroad.

2. VARYING CONCEPTIONS AND TYPES OF CATI

The use of CATI for broadly differing survey applications has produced varying conceptions of CATI, some complementary and some competing, which differ in their objectives and requirements.

2.1 Functions of CATI

Some survey organizations view CATI primarily as a means *to facilitate or expedite telephone surveys*, making them easier and faster to complete. By building common survey procedures directly into CATI systems, or into prepackaged setup modules, surveys with similar designs can be conducted more efficiently, even by field staffs with limited survey experience. From this perspective, CATI's major benefits derive from: (a) fast and simple methods of questionnaire setup; (b) direct support of interviewers in selecting respondents, phrasing questions, and choosing the next question; (c) entry of responses in machine readable form to speed processing; and (d) simple methods of generating output files and administrative reports. These gains are most easily realized in relatively simple surveys with limited filling, branching, and edit checking.

Other survey organizations stress CATI's ability *to enhance and control survey data quality*. They view CATI's primary function as substitution of computer processes for clerical procedures believed to pose most difficulties for paper and pencil interviewers. From this perspective, CATI's major benefits derive from: (a) systematic control over the scheduling of calls and callbacks; (b) tailored wording of complex questions based on prior responses; (c) computer-controlled branching between questionnaire items and sections; (d) automatic range and consistency edit checking during the interview; and (e) careful monitoring of interviewer performance to ensure that intended procedures are followed.

A third function of CATI, which may be viewed as an extension of the second, is *to permit new types of surveys* not possible with paper-and-pencil methods. Computer-assisted interviews may: (a) randomize question sequences or question wording in complex factorial designs; (b) incorporate arithmetic calculations or logical checks not readily performed with paper methods; (c) utilize table lookup routines to match responses with lists of possible alternatives, such as makes and models of automobiles; and (d) use data from prior interviews (or from records) in dependent interviewing without disclosing that data to the interviewer in advance (Fink, 1983; Groves and Nicholls, 1986).

The first of these three functions of CATI is not necessarily consistent with the second two. Surveys that use many of CATI's quality enhancement features or collect data difficult to obtain with paper and pencil methods are not necessarily simple to set up and conduct. They may require extensive preparation, careful pretesting, and a thoroughly trained staff for proper execution.

2.2 Schools of CATI Questionnaire Design and Interviewing

Schools of CATI questionnaire design have emerged which reflect differences both in CATI systems and in the types of surveys for which they are used. CATI questionnaires may be item, screen, or form based; may vary in their use of scripting; and may be designed primarily for forward movement or for both forward and backward movement.

Item based CATI systems display one survey question and answer space at a time, perform edits after each entry, and typically erase the screen before the next item appears. Branching is computer controlled. When an entry is made to one item, the next item is displayed and the cursor moves to its answer space. Items cannot be answered out of sequence since each item requires an entry before the next is displayed.

Screen based CATI uses the screen rather than the item as the basic unit of questionnaire design. The screen may have one or several items; and in some systems the answer spaces are grouped in a separate answer section. As in item based CATI, branching is computer controlled, and the items must be answered in sequence. Screen based CATI is especially convenient for multiple answer questions and for sets of short, noncontingent questions on the same topic. Some item based systems may operate similarly by displaying multiple items concurrently or consecutively on the same screen.

Form based CATI takes a different approach to answer entry, movement between items, and editing. A screen typically contains answer spaces for many items, often in a table format, but the interviewer may use cursor control keys to complete them in any order. Editing is postponed until all the screen's items are answered. This design emulates one page paper forms, especially those with rows and columns of figures to be checked against marginal totals or displayed data from prior reports.

This format is especially useful in surveys of establishments, such as businesses, where the organization's record keeping practices may govern the order in which items are answered.

Some CATI systems permit item based, screen based, and form based displays in different parts of the same questionnaire. Systems also have been developed with split screens to accommodate both item and form based CATI concurrently. The item based section leads the interviewer through the questions in a standard order and handles contingent questions. But the interviewer can shift to the form based section when necessary to record data in a different order or to correct prior entries summarized there.

An item is called *scripted* when it provides the interviewer with the exact wording of the question and *unscripted* when it merely labels the needed entry, such as "Sex" or "Marital Status." CATI's filling capabilities permit unusually thorough scripting. A household CATI survey may ask: "IS JOHN now married, separated, widowed, divorced, or HAS HE never been married?" Or the same question may be partially scripted as in a paper questionnaire, reading: "[Are you/Is PERSON] now married, widowed, separated, divorced, or [have you/has he/has she] never been married?" The interviewer inserts the correct name, pronouns, and verb forms. Filled scripting requires more setup time, and organizations differ in their estimates of its cost effectiveness. Scripting also takes up more display space. Form based CATI frequently uses unscripted items to permit many variables per screen. Form based CATI screens, therefore, often resemble data entry screens. Some form based systems have adopted compromises by displaying the full text of each form based item at the bottom of the screen as it is reached, or by allowing the interviewer to toggle between scripted and unscripted versions as needed.

The first CATI systems were designed primarily for forward movement. Interviews proceeded from screen to screen without the option of backing up more than an item or two. Even current systems may erase the entries of backed to items or prohibit backing across roster boundaries or after randomized items. Systems limited to forward movement have some advantages for opinion surveys since they ensure that initial responses are not revised after later topics are introduced. Inhibited backing presents more problems for factual surveys or those with many contingent items. If entry errors at key branch points cannot be corrected, the interviewer may have no option but to abort the interview and start over. To circumvent these problems and to give interviewers the same backing options they have in paper questionnaires, systems have been developed which permit interviewers to back to, review, and change the answers of prior items without erasing intervening entries. They usually provide a quick return to the next needed item, even when an answer change revises the branching path.

Forward only systems require different approaches to interviewer training and questionnaire design than forward and backward systems. In forward only systems, special care may be required to avoid interviewer errors, to warn interviewers when the last chance to revise a key entry has been reached, or to add special screens where changes to key prior entries can be made. In forward and backward systems, interviewers usually require additional training in the use of interviewer movement commands and may need on-screen help choosing proper targets for jump back commands. The programming of forward and backward questionnaires also requires closer attention so they will function correctly both in forward movement and after backing and changing answers (Nicholls and House, 1987).

2.3 CATI as Facility or Function

The most common use of CATI at present is in *standalone CATI facilities* conducting telephone interviews for a number of surveys. Berry and O'Rourke (this volume, Chapter 29) present an organizational analysis of such facilities. The facility receives a questionnaire, sampling specifications, and survey specific instructions, and returns a data file of completed interviews and various field performance and administrative reports. Such facilities may incorporate integrated system modules for survey sampling, call and callback scheduling, coding of verbatim answers, and tabulation of results; or the CATI software may articulate with other programs which perform these tasks.

The development of CATI has also become *part of a broader movement toward computer-assisted data collection.* The same or closely related software and hardware may be employed for computer-assisted data entry (CADE), computer assisted self-administered questionnaires (CSAQ), and computer-assisted personal interviewing (CAPI). Further extensions may link CATI with modules for label printing and check in of mail questionnaires, computer-assisted coding of verbatim replies, analyst review of completed forms, failed edit followup, maintenance of sampling frames, and longitudinal data bases, and perhaps even desktop publishing of final tabulations. Current redesigns of CATI systems seem prompted more by their integration into broader systems than by needed enhancement in CATI *per se.* CATI may be seen in the future less as a standalone system than as one of many functions performed by larger computer based survey systems.

3. CHAPTERS ON CATI IN THIS VOLUME

The four remaining chapters in Section E, and two in Section F on the administration of telephone surveys, provide more detailed information about CATI and its consequences for telephone surveys.

In Chapter 24 Baker and Lefes discuss the design of current and future CATI systems. While this chapter should be of most interest to CATI systems designers, it also provides general readers with an overview of the architecture and functioning of CATI systems. Baker and Lefes find broad agreement across organizations in user requirements for CATI; and they develop a logical system design covering most known CATI systems. System differences, they argue, arise largely from variations in physical design.

In Chapter 25, Weeks presents a summary of CATI call scheduling methods based on a review of practices at several major survey organizations. Since CATI call scheduling procedures usually have clerical analogues and address such common survey goals as minimizing nonresponse and costs, this chapter should prove of almost equal interest to those who use clerical rather than computer-assisted telephone methods.

Chapter 26 provides a general introduction to CATI questionnaire design. The authors, House and Nicholls, propose six objectives for the design of CATI questionnaires, which recognize their dual nature as survey instruments and computer programs. They then provide general design strategies, checklists of procedures, and alternative techniques contributing to those objectives. The authors focus on item and screen based CATI systems which permit both forward and backward movement, the types of systems most commonly used in governmental and university household surveys.

In Chapter 27, Catlin and Ingram examine the effects of CATI on survey data quality and survey costs, based on a comparison of CATI and paper and pencil methods in the same survey. The large sample, extended duration, and controlled conditions of this study make it an excellent vehicle to test previous conclusions about CATI's effects. While a growing literature suggests that CATI generally can match or exceed paper telephone methods in data quality and cost effectiveness, a sufficient number of exceptions to these encouraging conclusions have been documented to recommend careful advance testing when a large survey or one providing an important data series is converted from paper to CATI methods.

Relatively little has been published on CATI field work procedures. Morton and House (1983) reported early experiences in CATI training; and Nicholls and Groves (1986) have summarized field impressions on

CATI interviewer training, supervision, and turnover. This volume's Section F on the administration of telephone surveys includes two chapters contributing importantly to this area. In Chapter 30, Cannell and Oksenberg review interviewer monitoring methods which generally apply to both CATI and paper and pencil telephone surveys. In Chapter 29, Berry and O'Rourke summarize changes in recruiting and training practices that survey organizations have experienced as they move from paper to CATI telephone methods. The latter chapter also analyzes the impact of CATI on the internal organization of survey centers.

CHAPTER 24

THE DESIGN OF CATI SYSTEMS: A REVIEW OF CURRENT PRACTICE

Reginald P. Baker and William L. Lefes[1]
National Opinion Research Center

This chapter describes in relatively nontechnical terms the general design of computer-assisted telephone interviewing (CATI) systems. It focuses on the design of the computer programs that drive the CATI interview. Other system functions such as scheduling of calls, quality control monitoring, and case management are not considered. CATI systems in use by university and governmental survey organizations form the basis for the discussion, but individual systems are neither described nor evaluated. Rather, the discussion is aimed at a general review of current CATI systems and emphasizes the interaction between design options and system capabilities. The first section provides a brief review of the history of CATI. The second looks at system requirements for CATI. The third presents a generic CATI system design. The chapter concludes with some brief speculation about future directions of CATI system development.

1. THE EMERGENCE OF CATI TECHNOLOGY

Four objectives have guided most of the development of CATI systems over the last 20 years: greater efficiency in interviewing, improved quality in survey data, fast access to survey results, and reduced cost (e.g., Groves, 1983; Fink, 1983; Shanks, 1983). As computers came into common use for

[1] The authors wish to thank the following individuals for their comments on an earlier draft of this chapter: Michael F. Weeks, Research Triangle Institute; Matthew Futterman, University of California, Los Angeles; William E. Connett, Institute for Social Research, The University of Michigan; and Arnold E. Levin, U. S. Bureau of the Census.

the tabulation of survey results and large scale, centralized telephone interviewing emerged as an important survey methodology, CATI appeared as a logical next step. The computer's ability to make decisions about which instructions to execute under variable circumstances, and to enforce a set of rules about acceptable entries, led to the concentration of several discrete activities — interviewing, data entry, and data editing — into the single activity of CATI. This concentration of activities, CATI supporters argued, would make telephone interviewing more efficient and less costly (Nelson, Peyton, and Bortner, 1972). It also would produce higher quality data, since errors could be identified and resolved at the time of the interview. With the proper system design, moreover, data could be ready for analysis almost as soon as the interview was completed.

The history of CATI system development can be divided into three broad phases. Systems developed in each phase differ both in the hardware they run on and in their software designs.

Phase I began in about 1972 and ended around 1977. The prototypical system of this era was the Survey Response Processor (SRP), a CATI system developed at Chilton Research Services in the middle 1970s (Fink, 1983; Shanks, 1983). The SRP was a COBOL based, parameter driven, transaction processor running on an IBM 370 mainframe computer. The system was designed so that the amount of programming required to bring up an instrument was kept to a minimum. Specially trained nonprogramming staff called "authors" broke questionnaires down into a set of transactions or data records describing question text, precoded answers, valid answers, logic skips, and even consistency checks between questions. These transactions were read by the SRP and used by the system to drive the interview.

Phase II, running from about 1978 to 1983, was characterized by greater use of minicomputers and so-called generative system designs. The system developed by Shure and Meeker (1978) at UCLA was typical. Their system abandoned the transaction based design of the SRP in favor of a design which used a specially formatted version of the questionnaire. CATI questionnaire designers edited a machine readable version of the questionnaire, adding special characters to indicate question text, responses, skips, and so forth. This edited questionnaire was read by the system and translated into FORTRAN code, which was then compiled and executed to drive the interview. Compared to the SRP, the UCLA system's setup procedure was more intuitive in that the input to the system looked more like a traditional questionnaire. It also ran on less expensive and more reliable hardware, a PDP-11 minicomputer.

Phase III began around 1983 and coincides with the emergence of microcomputing on a broad scale. Under the most common hardware design, the components of CATI — questionnaire modules and output data — are kept on a central file server to which interviewer stations in the form of standard IBM compatible microcomputers are connected. When an interview begins, the CATI application is loaded into the CPU of the microcomputer station and run. Software designs for such systems rely less on compilers and more on specially designed interpretive languages. CASS (Palit and Sharp, 1983; Palit, 1980) is one of the first and most widely used systems of this general type.

2. SYSTEM REQUIREMENTS

The basic requirements for a CATI system have been stated and discussed with remarkable unanimity by a number of authors including Rustemeyer et al. (1978), Freeman and Shanks (1983), Nicholls and Groves (1986), and again by Nicholls in the previous chapter of this volume. Figure 24.1 organizes these requirements into a system requirements model.

A system requirements model is the basic road map that a system designer uses to develop a design. The model classifies requirements into five groups:

1. The outputs the system must produce
2. The inputs needed to produce the outputs
3. The operations that must be performed to produce the outputs
4. The operational controls needed to assure that the outputs are correct
5. The resources the system will need.

Figure 24.1 shows three types of inputs to the CATI system: the survey questionnaire, sample or panel data that might be referred to in the questionnaire, and the actual responses keyed to the system by the interviewer. The basic operations of the system include the ability to display and/or modify question text, accept and store responses, permit nonstandard movement, repeat groups of questions to construct rosters, perform logical and arithmetic tests, replay all or part of an interview, and manage dialogue between interviewer and respondent so that errors or inconsistencies in the interview can be corrected or resolved.

The model also defines controls on the operations of the system to ensure that the data captured and stored by the system are correct in the context of a specific questionnaire. These controls include performing

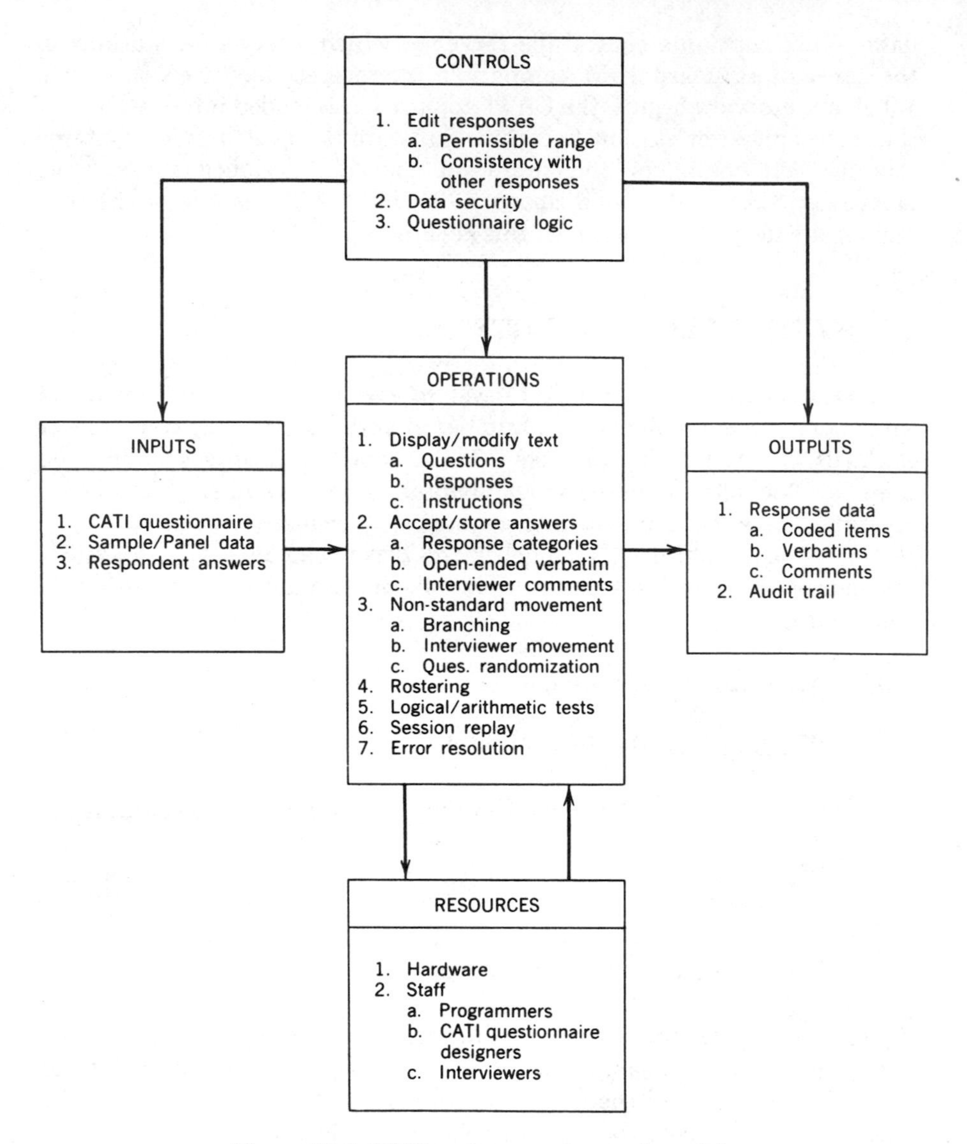

Figure 24.1 CATI system requirements model.

range checks on items, comparing responses from different questions for consistency, and enforcing the skip pattern of the questionnaire. The basic output of the system is the response data along with interviewer comments and an audit trail. Resources include the hardware on which the system will run and a staff of programmers, CATI questionnaire designers, and interviewers.

3. A GENERALIZED CATI SYSTEM DESIGN

Despite widespread agreement on the capabilities of a viable CATI system, current system designs differ widely. To use the language of the system designer, we might say that the general outlines of a *logical design* are clear. By this we mean that we can specify in broad conceptual terms what a system must do, what its components must be, and how those components must be related to one another for the system to operate properly. Despite this, the variety of *physical designs* is so great that generalization is difficult. In this context "physical design" means the details of the system's architecture such as the programming language it is written in, the function and structure of the individual modules that comprise it, the input/output methods used, its file and record structures, and the computer hardware on which it runs.

The following discussion emphasizes the logical design of CATI systems. Physical design is discussed anecdotally to help illustrate how different systems implement specific components or features.

3.1 Software Architecture

Software architecture refers to the basic components that constitute a software system and the relationships that exist among them. Figure 24.2 represents the software architecture of a generic CATI system. The logical design described there is sufficiently generalized to allow for a variety of physical designs. It also satisfies all of the requirements identified in Figure 24.1. While no particular system is depicted, most of the system designs we know of can be fit to the general model presented there.

The design in Figure 24.2 features two basic system components: an application generator and a runtime module.

The application generator is a questionnaire compiler. It reads a file or set of files describing the survey questionnaire. Inputs are examined for errors and a set of outputs produced. These outputs are then input to the runtime module, where the computer code required to conduct an interview is generated. The runtime module is executed to conduct the interview and write out the resulting data.

The design in Figure 24.2 shows only those components needed for conducting an interview. The figure does not include other system service

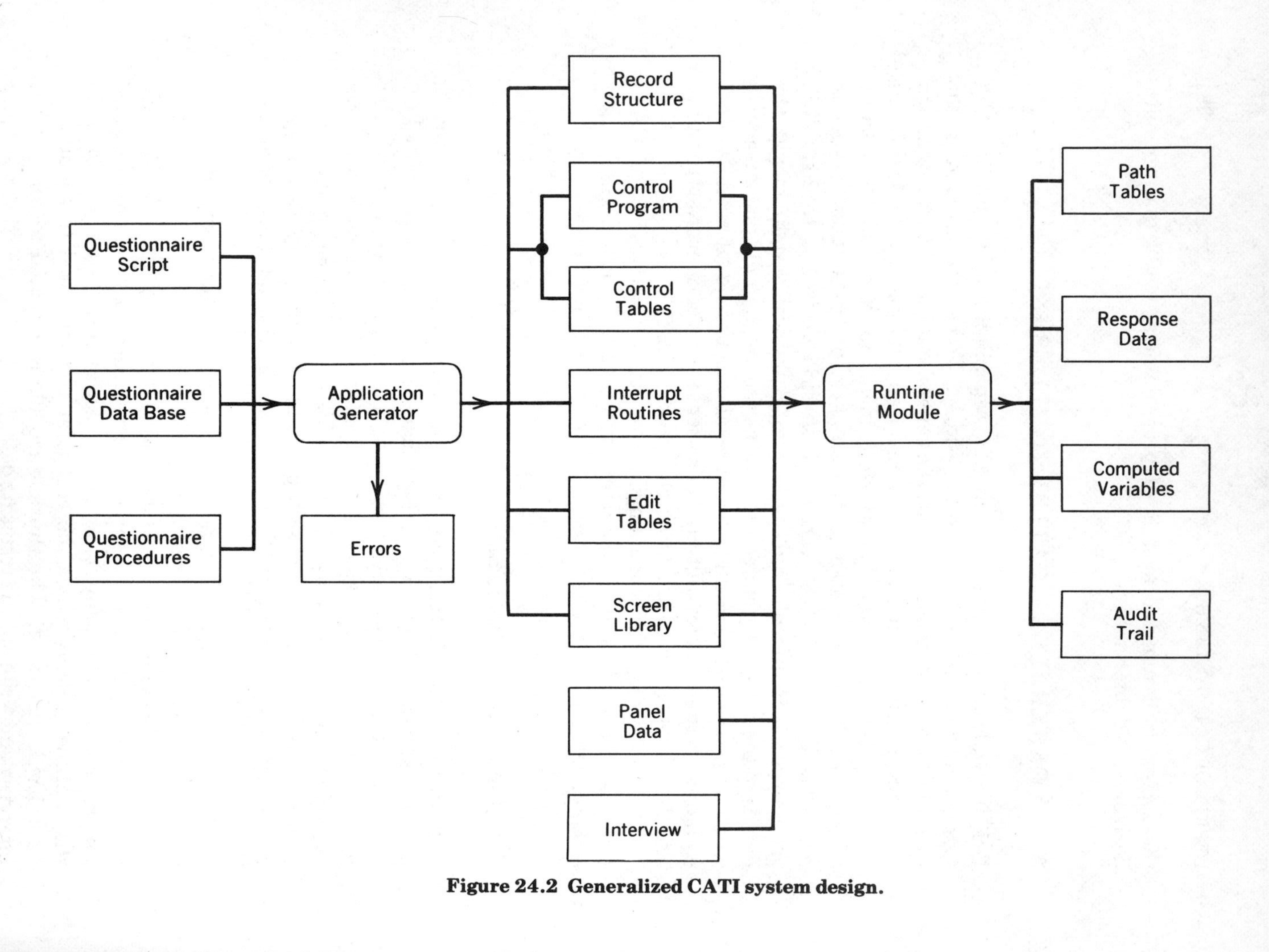

Figure 24.2 Generalized CATI system design.

modules or utilities, such as text editors to prepare the questionnaire, post processors to reformat data, or the case management components that are part of most CATI systems.

3.2 Questionnaire Input

One fundamental difference among CATI systems is the method used to enter the questionnaire into the system. The most common approaches include the following:

1. A questionnaire script in which a machine readable version of the questionnaire is annotated with a variety of symbols to identify question text, response values, length of the output data item, branching to be performed, and so forth. Scripts often lack the more powerful tools of a true programming language such as DO loops, complex statements to control branching, and the ability to create subroutines or program modules that can be called from different locations in the questionnaire.

2. A questionnaire data base in which question and answer text, valid values, branching instructions, and the like are stored.

3. A CATI procedural language in which the questionnaire is represented as a series of program loops and subroutines. We have seen systems that used one of the standard programming languages such as BASIC, PL/1, or FORTRAN, although they are less popular today than in the past. Systems more often use a CATI specific language with the ability to include subroutines written in another language.

4. Some combination of script, data base, and a procedural language.

Each of these approaches has advantages and disadvantages. On the one hand, questionnaire scripts and questionnaire data bases tend to be easy to set up and require little training or skill on the part of the CATI questionnaire designer. On the other hand, both have relatively weak programmability and lack the flexibility and power of a procedural language. Data bases have the added advantage of transportability — that is, they simplify the development of question banks from which questionnaire designers can build CATI applications using previously set up items.

In the end, the best approach probably will depend on the size and complexity of the surveys to be conducted. Large, complex questionnaires with intricate skip patterns are difficult to manage without the power and flexibility of a true CATI programming language. Either the script or the data base approach is probably better suited to relatively straightforward questionnaires on quick turnaround surveys. Our preference is for a hybrid approach that uses a questionnaire data base in concert with a powerful procedural language.

3.3 The Application Generator

Two broad types of application generators are commonly used: compiler based and interpreter based.

A compiler is a high level program that reads another program and translates it into machine language so that it can run on a specific computer. Compiler based CATI systems generally function in one of two ways. Some systems have application generators capable of reading annotated questionnaire script and rewriting it in a third generation language such as FORTRAN. In other words, the application generator translates the questionnaire into a FORTRAN program. This FORTRAN program is compiled using a FORTRAN compiler, which in turn creates a machine language program that can be run under the standard runtime system. A second, more direct approach is sometimes used for processing questionnaires written in a CATI procedural language. Such systems have a CATI language compiler that reads the CATI procedural language and directly produces a machine language program.

Interpreters, by contrast, translate their input into instructions that are executed immediately. In interpreter based CATI systems, the application generator reads script, procedural code, or even data base entries. This input is prepared in special formats so that it can be read by an interpreter at the time of the interview. Under this design, the application generator may write out a new program in the CATI interpretive language or it may simply assemble all of the code and files that are needed when the interview is run. The runtime module is an interpreter which reads the uncompiled CATI code that drives the interview.

Of course, not all systems have a single application generator into which all of the questionnaire files are input on a single pass. Some break the questionnaire into several different components, each of which is processed individually by the program or programs constituting the application generator. Since all parts of the questionnaire are logically related to one another, the system must also be able to reassemble them into a coherent whole.

While there are strong arguments in favor of both compilers and interpreters, the current trend clearly favors interpreters. Improvements in interpreter design and the rush toward microcomputers are the principal reasons. Compiler based systems with their relatively large memory requirements are mostly confined to mainframe and minicomputer environments.

Regardless of their physical design, CATI application generators generally perform two key functions: (1) syntax checking with error reporting and (2) production of outputs needed by CATI to conduct an interview. The level of error checking that most application generators perform is relatively low, being confined to the validity of commands and correct usage. Structural or logical errors in the questionnaire are usually not discovered by the application generator; these are more often searched for manually in a separate, time consuming debugging step.

One of the more fruitful areas for additional work by CATI designers is in the development of software to assess the correctness of CATI questionnaires automatically (Nicholls and House, 1987). Correctness in routing instructions and consistency across items should be enforceable by current programming techniques. Programs that read a CATI program and generate a flowchart of the CATI questionnaire should also be possible for comparison with a questionnaire analysis prepared by the questionnaire author. Optimization rules designed to streamline the runtime code and make it perform more efficiently comprise still another potential enhancement.

3.4 Components of the CATI Application

The outputs from the application generator together make up the CATI application. While virtually all systems have some technique for performing the functions of each of these outputs, not all systems have discrete components that do so. The trend in earlier systems was to consolidate all functions — branching, editing, nonstandard movement, display of text and answers — in a single CATI control program. The more recent trend has been to decompose the various functions into separate pieces to give the system designer greater flexibility, particularly in memory management.

Figure 24.2 divides the CATI application into six components: a control program, a set of control tables, a set of edit tables, a library of interrupt routines, a screen library, and a record structure.

Many, but not all, systems generate a main *control program*. Systems that produce third generation language code (e.g., FORTRAN or PL/1) or feature a CATI-specific compiler, often represent the entire questionnaire in the control program. This program performs all the necessary interview operations and controls (sometimes with the aid of external files) especially proper movement through the interview. In systems that use an interpreter as a runtime component, the control program is usually a set of survey-specific instructions that can be interpreted. In systems that use a compiler to produce object code (a machine language program), the control program is the object code.

In addition to a control program, or in place of it, some systems produce a set of *control tables* that govern movement through the interview. These tables specify the order of the questions as well as the branches to be taken based on respondent answers or other data loaded externally to the system. These tables may control simple branching based on a single response or complex branching requiring the evaluation of complex instructions that are separately stored but referenced in the control table. The tables may be left on disk or loaded to memory during the interview. The use of control tables is particularly suited to systems that employ a data base approach to questionnaire setup.

A third set of outputs from the application generator is some form of *edit tables*. The purpose of these tables is to provide the system with parameters that allow it to validate responses as they are keyed. In some systems, these edit specifications are embedded in the CATI control program. In others, they are stored in files where they can be accessed as needed during the interview. Again, depending on the system design and the hardware, these tables may be stored on disk where they are accessed as needed during the interview or they may be kept in memory.

Many systems also create a library of *interrupt routines*. These routines are used to execute complex skip instructions, resolve inconsistencies, and manage nonstandard interviewer movement. We call these interrupt routines because when they are called, standard movement through the questionnaire is suspended until the routine is executed and control is passed back to the control program.

The incorporation of relatively self-contained interrupt routines into CATI applications makes for a very flexible and powerful system. It allows the CATI questionnaire designer to guide the interviewer through elaborate probes, reviews of prior answers, reviews of definitions, entry of comments, and other digressions with the respondent outside of the main body of the questionnaire. Once a CATI application is up and running, the system's response to special circumstances can be changed or refined without affecting the main CATI program.

Another potential output from the application generator is a *screen library*. Some systems, especially older ones, do not use a screen library but display questions, instructions, and responses by writing lines of text directly to the screen. Over the last several years the trend has been toward storing complete and formatted items — question and response categories — in text files that are moved directly into screen memory rather than written to the screen as a series of individual lines.

Each approach has its advantages. In systems that write to the screen, it is often easier to design the display so that answers are keyed next to or directly after each question, much as one fills out a hard copy questionnaire. This technique also simplifies the substitution of nonstandard text, such as previous responses or panel data, into question text. With screen libraries, answers are often entered in an answer or dialogue box at the bottom of the screen, which can be troubling for interviewers. On the positive side, screen libraries tend to improve system performance and simplify questionnaire setup.

We are just beginning to see the emergence of CATI systems that use a more advanced screen design technique called *forms*. This technique is most often used by data base management systems. It allows for great flexibility in the location of items on the screen, including the definition of items as display only, updatable, or text substitution. With such an approach, several questions can be placed on the same screen and the interviewer can be permitted to move freely among them in any order. Because forms are "painted" and stored like screens in a library, they are most easily implemented in systems with screen libraries. Although the effect can be duplicated with systems that do not use screen libraries, some very tedious programming is usually required.

Finally, all CATI systems have some device by which the *output data record* is defined and constructed. The application generator creates an output format to tell the system where individual data items are to be stored. This definition may be developed by the CATI questionnaire designer and included as part of the CATI program or it may be created automatically by the application generator.

There is no recognized standard for the order in which items are stored. Some systems arrange them alphabetically by item name, others use the order of the interview. For those who prefer to minimize the amount of post-CATI processing, the latter seems preferable. If, however, one is concerned about the ability to add items to the output record after interviewing has begun, another approach may provide greater flexibility. The greatest flexibility is in systems that do not assemble the output record at the end of the interview, but use a post processor. These systems write each response out to a transaction file at the time of entry. This file can be reformatted into the desired format or formats at any time.

3.5 The Runtime Module

When an interviewer begins an interview, he or she invokes the runtime module for the questionnaire. This module may be a compiled object code that is loaded and run, or it may be an interpreter that reads the CATI application and executes without compilation.

The conventional view that a compiler makes for a better CATI system than does an interpreter is changing. While compiled code will often run faster than an interpreter, a well designed interpreter can be written to run at an acceptable speed, even on microcomputers. Interpreters are easier to write than compilers, and they can be designed to meet the precise needs of CATI with some efficiency. Finally, interpreters usually will run in less memory than compilers, another major advantage for microcomputer based CATI systems.

3.6 Runtime Outputs

Virtually everything discussed thus far has been general in that it applies to the way in which a questionnaire is set up and managed internally by the computer. The focus now shifts to the particular, that is, the outputs that are created as part of the interview with an individual respondent. These are also represented in Figure 24.2.

One key piece of information maintained by the system describes the *path of the current interview* through the questionnaire. This is important because it both documents which questions were answered and controls system movement when certain kinds of interrupts occur. These interrupts are usually of two types: (1) nonstandard movement under interviewer, rather than system control; and (2) breakoffs, that is, temporary or final terminations of the interview before its completion.

Nonstandard movement occurs when an interviewer goes back to verify or change a previously answered item. When this happens, the system must know which questions are on the current path through the questionnaire and the current valid responses. It also must use information about the current path to position itself at the correct point in the interview when the interrupt is concluded. In those systems that retain all responses even if the respondent has changed them in a later part of the interview, path information also is needed to discriminate between applicable (active) and inapplicable (inactive) responses before the final data record can be assembled.

Information about the current path is generally stored in one of two ways. In most systems an array or set of lists is maintained with information about each question including the current answer and its status — on or off the current path, applicable or inapplicable. When the interviewer backs up, the system references this array to determine whether a question has been answered and is on the current path. If both conditions are true, the system jumps back to the indicated question, finds the current answer, and displays it to the interviewer. If a response is changed, the path array is used to replay the interview so that any inconsistencies or changes in routing brought on by the new response are identified and resolved.

When an interview is broken off (e.g., the respondent cannot continue or a technical problem appears) the total system environment must be saved, typically on disk, for future use. The total system environment includes information about where the breakoff occurred, all valid responses, the active skip pattern, and the values of any computed variables. When the interview is resumed, this environment is recreated so that the interview can continue where it left off.

The ability of a CATI system to handle breakoffs and nonstandard interviewer movement is extremely important. Both are relatively frequent events in CATI surveys. Systems that have difficulty restarting interviews after breakoffs, erase responses when interviewers go back to previously answered questions, or allow interviewers access to questions off the current interview path pose numerous problems for all types of surveys. In our view, they should be avoided.

A second output from the runtime module is the *response data*. There is considerable variety in how this is managed, but all seem to be variants of three general approaches:

1. Responses are written to disk at the time of entry as transactions. Sometime after the conclusion of the interview, a post processor reads the transactions and constructs the data record.

2. Responses are kept in memory until the interview is complete. Either during the interview or at its conclusion, responses are collected into the correct record format. At the conclusion of the interview, the data record is written to disk.

3. Responses are maintained in a data base environment and updated directly as they are entered from the keyboard.

Writing to disk as responses are entered provides a good measure of protection against system failure, yet it may impact system response time. Unless the data are also retained in memory, there is the added disadvantage of limiting the ability of the interviewer to back up and review previously entered responses. Applications that require quick turnaround of the collected data may also be hindered by the need to run a post processor before tabulations can be run.

Future CATI systems will probably write responses directly to data bases. These systems will support more complex questionnaires and manage the resulting data more efficiently. Their most important advantage derives from the ability of data bases to maintain complex relationships among groups of entities. Rosters of family members and event histories are difficult for many systems. Data bases will help.

For now, the hardware on which the system will run probably should determine the output technique. Continually writing responses to disk ought to be the rule for mainframe or minicomputer systems. CPU failures can be relatively frequent and they impact all interviewers when they occur. Mainframes and minicomputers are also fast enough, both in CPU processing and in disk I/O, to avoid response time degradation while writing responses to disk. Microcomputer based systems, with their greater reliability and slower disk drives, may not require the extra protection, especially given the added cost of performance degradation. However, these systems have memory constraints that limit the amount of information that can be retained throughout the interview.

Most systems also include *verbatim responses* and *interviewer comments* as part of the basic output data. Virtually all systems of which we are aware write each of these to separate files where they can be coded and/or reviewed at a later date.

Another common output of the runtime module is the value of *computed variables* defined by the CATI questionnaire designer. Many of these variables are intermediate calculations used, for example, to control a particular respondent's path through the questionnaire or to check consistency among responses. All systems use them, but they have little value after the interview is completed. Only computed variables useful in analysis of the data are retained in the final output record.

Finally, most systems write some kind of *audit trail.* This may be a separate file or it may be incorporated in whatever path management technique is used. Audit trails are used for debugging, and therefore should be written to disk on a flow basis rather than assembled in memory and written out only when the interview is concluded.

Some systems that write every response transaction to disk throughout the interview use this transaction as an audit trail as well. A more effective approach is to write every character entered on the keyboard including responses, backup commands, and invocations of the help facility to an audit trail. Such files are extremely useful for debugging. They also provide a means by which interviews can be "replayed" and questionnaire design deficiencies identified through an analysis of interviewer movement. Since this constant writing of characters can slow response time, this feature is often treated as an option. It may be turned on during the early weeks of a new survey and discontinued once the application has stabilized.

4. FUTURE TRENDS IN CATI SYSTEM DESIGN

Choosing a good CATI system design requires both analytic skills and an ability to see the future. Matching system features to one's specific requirements is the easy part. The more difficult task is determining the direction of future CATI developments so that an upgrade path is available as both the technology and its use in surveys changes.

Survey organizations are now at a watershed in CATI system design. Those that have been using productive minicomputer based systems for many years are redesigning or replacing those systems to take advantage of major improvements in computing technology. The generally lower cost, greater reliability, and increased power of today's computers are luring many away from stable, established systems. Some organizations are turning away from hardware such as the PDP-11 and toward newer machines such as the Vax. Others are abandoning minicomputers completely to bring up microcomputer based CATI systems.

As in the past, today's minicomputers have much to offer CATI. In system performance and capacity, many are on a par with all but the largest mainframes. New operating systems with multitasking capabilities, high density, high speed disk drives, and the ability to serve a large number of terminals simultaneously make minicomputers even more attractive. Software tools to write and support CATI on minicomputers, such as compiled programming languages and data base management systems, are numerous and powerful. This increased computing power may not be necessary to accommodate the relatively modest demands of the interview software; but the performance of a minicomputer may be required for case management systems handling tens of thousands of cases and for tabulations used in quality control or quick turn around analysis.

Despite these advantages, many if not most survey organizations are moving away from minicomputer hardware toward microcomputers. One major reason is cost. Almost anyone can mount small scale CATI surveys on microcomputers. PCs are inexpensive and getting more so. Their reliability, compared to that of previous generations of computers, is a technological marvel, and there are many commercially available CATI systems that can be set up and run by almost any experienced PC user in a matter of days. By connecting microcomputers together to form a local area network (LAN), many of the advantages of a minicomputer can also be realized.

Of course, microcomputers have a great deal more to offer than their low cost and high reliability. Their portability combined with continued improvements in microcomputer based CATI systems make it possible to conduct computer controlled interviews virtually anywhere, including respondents' homes. The emergence of new data collection approaches such as CAPI and CSAQ is also a factor. A single system capable of running in a number of modes —CATI, CAPI, CADE, and CSAQ — is enormously attractive to organizations that use these different modes, especially within the same survey. Such a system need only be set up once on each survey. The advantages for economy and uniformity across methods of administration are considerable.

In our view, the future of CATI is bright. With continued improvements in both hardware and software there will be fewer and fewer tradeoffs in system capabilities. Systems will keep getting easier to use, have greater capacity, and provide still greater freedom and flexibility to the questionnaire designer and the interviewer. CATI already is a mature technology, but we have only begun to feel its impact in survey research.

CHAPTER 25

CALL SCHEDULING WITH CATI: CURRENT CAPABILITIES AND METHODS

Michael F. Weeks
Research Triangle Institute

A necessary component of every telephone survey is the call making effort to contact and interview eligible respondents and to identify and close out ineligible and nonresponding cases. The call making effort is a labor intensive process that typically consumes a significant portion of a survey's available resources. To structure, monitor, and optimize this process, survey researchers utilize some type of call scheduling system. Traditionally, call scheduling systems have been implemented manually by telephone survey personnel. With the advent of computer-assisted telephone interviewing (CATI), however, great strides have been made in automating call scheduling systems and improving their effectiveness and efficiency. This chapter describes current capabilities and methods for automating call scheduling in a CATI environment.

Although CATI has received considerable attention in the survey literature, the application of computer support to call scheduling has been largely overlooked. The literature in this area has focused on two topics: (1) general descriptions of the mechanics of a call scheduling system in a traditional paper and pencil environment (e.g., Dillman, 1978; Groves and Kahn, 1979); and (2) the timing of calls to respondents (e.g., Falthzik, 1972; Rogers, 1976; Fitti, 1979; Wiseman and McDonald, 1979; Vigderhous, 1981; Kerin and Peterson, 1983; Warde, 1986; Weeks et al., 1987). Some research has been done on the efficiency of automated call scheduling (this volume, Chapter 27; Nicholls and Groves, 1986), but with one exception dating from the early CATI era (Nicholls, 1978), substantive descriptions of CATI call scheduling have not previously appeared.

This inattention in the survey literature belies the fact that a CATI call scheduling system offers significant advantages over a traditional manual system and may well be a practical necessity for large telephone surveys involving thousands of cases and complex calling protocols.

This chapter attempts to update the literature by providing a description of current CATI call scheduling capabilities and methods at leading U.S. survey organizations. Seven such organizations were judgmentally selected and in-depth descriptions of their CATI call scheduling systems were obtained from contributors at each organization.[1]

We may broadly define CATI call scheduling as the overall system used to implement and control the call making effort in a CATI survey. The most obvious feature of a CATI call scheduling system is the use of a computer to support the call making effort, in contrast to the traditional non-CATI system that depends solely on the telephone survey staff to implement the calling rules and procedures manually. Upon closer inspection, however, there are essentially eight generic components that all CATI call scheduling systems share. These are defined briefly in Table 1. They are discussed individually in the next section. A concluding section summarizes the advantages of a CATI call scheduling system and suggests areas for future research.

1. CURRENT PRACTICE

1.1 Call Outcome Coding System

All call scheduling systems use a call outcome coding system to facilitate maintenance of a historical record (*call record*) of calls made to a sample case and the outcome of each effort. At the conclusion of each call, a code is entered into the system to indicate the appropriate outcome. The system then updates the online call record maintained for the case. The call record serves as input to the scheduler, which uses the recorded

---

[1] The participating organizations and contributors were: the Rand Corporation (Sandra Berry), Research Triangle Institute (James Knight), UCLA (Eve Fielder and Matt Futterman), University of Illinois (Johnny Blair), University of Michigan (William Connett and Sue Ellen Hansen), U.S. Bureau of the Census (William Nicholls II), and the U.S. Department of Agriculture (Gene Danekas). These descriptions formed the basis for this chapter, which attempts to paint a composite picture of current practice at these organizations. While the number of contributing organizations was necessarily limited for practical reasons, it is assumed that their shared features reflect state of the art methods, while procedures in selective use may represent alternatives for the future or for specialized applications.

Table 1. Eight Key Components of a CATI Call Scheduling System

1. *Call outcome coding system.* The system used to code the result of each call made on a sample case. Each call and the coded result are added to an online call record, which in turn is read by the scheduler.

2. *Scheduler.* A generic term used to describe the software system that reads the call record for sample cases, determines their current status, sorts them into appropriate queues, schedules calls on active cases, and delivers cases to interviewers for calling in accordance with a predetermined priority system.

3. *Calling protocols.* The rules and algorithms that specify the sequential actions to be taken on each type of case. The protocols are implemented by the scheduler.

4. *Priority system.* The rules used by the scheduler to determine the order in which cases are assigned to interviewers for calling.

5. *Assignment system.* The method of assigning cases to appropriate interviewers for calling.

6. *Case closeout system.* The procedures used to close out completed cases and assign an appropriate final code.

7. *Supervisory intervention.* Refers to the role of the telephone supervisor in the operation and control of the call scheduling system.

8. *Monitoring reports.* The system generated reports used to monitor operation of the call scheduling system and the progress of the call making effort.

information to determine the status of the case and to schedule the next appropriate action, based on the applicable calling protocol. The call record is also used by the interviewer to review the previous processing of the case before making the next scheduled call and by the supervisor in reviewing problem cases.

Most organizations maintain a record for each dialing. However, as a practical matter, some organizations distinguish between dialings and calls. For example, in the Census Bureau system the basic unit of the call scheduling system is the call — not individual dialings. Generally, a call requires only one dialing, and when this is so, the call outcome code represents both the dialing action and the result of the call. In certain circumstances, however, the applicable calling protocol may require more than one dialing per call, such as a second dialing to confirm a wrong number outcome. In such situations only the outcome of the last dialing is recorded in the call record, although counts of dialings are preserved elsewhere for evaluation purposes.

The codes used for a particular survey are determined in large measure by the characteristics of the survey design. The number and nature of the codes will vary for random digit dialing surveys, establishment surveys, and list sample surveys of households or individuals. Other design characteristics that affect the coding scheme include the use of primary and secondary respondents, followup interviews, a tracing component, and mixed mode strategies.

A call outcome coding system should satisfy at least three basic criteria. First, the codes should be discrete; if they overlap, the status of the case may be ambiguous. Second, they should provide for all outcome possibilities; otherwise, the interviewers will have to "force" a call outcome into an inappropriate category. Third, they should represent an appropriate level of specificity. This is important so that the information preserved on the processing of the case is sufficient for all uses. Generally, it is better to err on the side of too much specificity than too little, since codes can always be collapsed at a later stage of processing.

The University of Michigan uses a call outcome coding system that is fairly typical of the systems used by most organizations. Their call outcome classification system (omitting the actual numeric codes used) is shown in Table 2.

A CATI call scheduling system maintains a call record for each case in an online file that can be displayed by the interviewer or supervisor. A typical call record shows the date, time, interviewer, and outcome code for each call on the case. For appointment codes, the date and time of the appointment is also displayed. Some systems also provide a comment field for each call. In addition, a current status code and/or a code indicating the next action to be taken may be displayed. Numeric codes may be defined on the screen for the convenience of the interviewer.

1.2 Scheduler

The scheduler is a generic term that refers to the software that prioritizes, schedules, and delivers cases to the interviewing staff for calling. Although considerable variation exists in both operational details and terminology, the principal functions of the scheduler (or "case management system") are summarized below.

The *schedule construction module* (also known as the "batch scheduler" or "general scheduler") runs in batch mode at specified intervals, ranging from every few minutes to once a day. This module reads the relevant data in the call record for each case, determines the

Table 2. Typical Call Outcome Categories in Use by The University of Michigan

FINAL INTERVIEW
- Completed Interview
- Partial Interview

NO FINAL DISPOSITION – ANSWERED
- **No Contact**
  - First wrong connection (RDD)
  - First nonworking number, number not verified (RDD)
  - First wrong number for respondent, recontact only (List)
  - First respondent's number no longer in service, recontact only (List)
  - Answering service machine; directory assistance
- **Call Back**
  - Appointment with respondent
  - Call back (best time for respondent known)
  - Call back (best time for respondent not known)
- **Initial Refusal**
  - Initial refusal by respondent
  - Initial refusal, not by respondent
  - Initial refusal, respondent undetermined

NO FINAL DISPOSITION, NOT ANSWERED
- Ring, no answer
- Regular busy signal
- Complete silence
- Fast busy or strange noise

status of the case, assigns active cases to appropriate calling queues, and prioritizes the cases within selected queues (e.g., appointments by appointment time).

The *case assignment module* (or "dealer") delivers cases to appropriate interviewers. This module operates in real time rather than batch mode to provide immediate response to interviewer needs. When an interviewer indicates readiness for a case, the case assignment module selects the case with the highest current priority, based on a preprogrammed priority system, and transmits it to the interviewer's work station. In some systems, the program also determines the types of cases the requesting interviewer is eligible to receive and delivers the highest priority case appropriate for that interviewer. All systems use the case assignment module to select and deliver routine cases, such as those not previously attempted. However, many systems allow the interviewer to choose individual cases within the special queues employed for problem cases. All systems provide for manual override of the priority system when necessary.

Most systems include an *outcome evaluation module* (or "assessor") to perform certain evaluative functions in real time at the conclusion of a call that otherwise must be performed by the schedule construction module in batch mode. When the interviewer keys a call outcome code and exits the case, the outcome evaluation module determines its new status. If the case has been completed (e.g., an interview is obtained), the module will set the appropriate final outcome code and remove the case from the pool of active cases. If the case requires further action, the module sets a current status code and may also set a "next action to be taken" code. In some systems, nothing further happens until the next run of the schedule construction module. In other systems, the outcome evaluation module puts the case back into the appropriate queue if the controlling protocol specifies a relatively prompt callback (i.e., before the next run of the schedule construction module). Otherwise, the case may be relegated to a holding queue to await the next run of the schedule construction module. The outcome evaluation module offers the convenience of more fully automated call scheduling at the expense of operating in a real time mode.

Some schedulers include a *reserve mobilization module* that may be invoked when the interviewers begin to run out of work. This module liberalizes the standard calling protocols so that cases normally scheduled for a later time or shift can be scheduled earlier.

A key function of the scheduler is the maintenance of calling queues into which cases are sorted by current status to await assignment to the interviewing staff. The number and types of queues utilized vary widely by organization and sometimes even by survey within the same organization. Typical queues include those established for various types of appointment cases, appointment retries, busy retries, new cases, cold callbacks (cases for which earlier calls have not produced a contact), refusals, "supervisor hold" cases (those requiring supervisory review), and a variety of other nonroutine cases (e.g., language barrier, tracing required, etc.). Some organizations do not distinguish between new cases and cold callbacks but rather combine both types into a "search" queue.

1.3 Calling Protocols

Calling protocols are the rules and algorithms that specify the calling efforts to be made on each type of case. Most protocols are incorporated into the scheduler and implemented by it. This section describes current practice for selected protocols governing common types of cases.

Initial Calls. The general approach to scheduling initial calls on newly activated cases is to schedule the call in the earliest appropriate time slot. This places the case in the system promptly to weed out ineligible numbers and identify incorrect numbers at an early stage. The goal is to maximize the probability of a successful resolution of the case, including identification and deletion of ineligible cases, on the first call. The key issue is determining the most appropriate time slot for a particular type of survey.

For establishment surveys, initial calls are scheduled during normal business hours based on the respondent's local time. For panel surveys involving multiple interviews with study members, respondents are asked the best time to make the next call and recontacts are scheduled accordingly.

For first-time surveys of known households or household based populations, the literature suggests that the best time to place an initial call is on weekday evenings and weekends as opposed to weekday daytime slots (Wiseman and McDonald, 1979; Vigderhous, 1981; Weeks et al., 1987). Another motivation for focusing on weekday evenings and weekends is the potential for cost savings in telephone tolls and computer charges, since both typically are cheaper during evenings and weekends than during weekday daytime hours.

Current practice varies in scheduling initial calls for RDD surveys. Some organizations schedule the initial call for the first available time slot to get the case into the system promptly. Others schedule the initial call for the first available weekday daytime time slot to speed the identification and deletion of business and nonworking numbers and to save evening and weekend time slots for household calls. Still others schedule the first call for a weekday evening or weekend time slot. The rationale for this approach is that priority should be given to the household component of the sample since most nonhousehold numbers can be screened out during nonbusiness hours via operator intercepts or recorded messages. No published results have demonstrated which of these approaches is most effective.

Cold Callbacks. The standard approach to scheduling callbacks on cases for which earlier calls have produced no contact is to employ some sort of algorithm, or set of algorithms, designed to enhance the probability of an early contact. Once a call has been answered by the respondent, or someone acquainted with him or her, subsequent callbacks, if necessary, are usually based on information obtained during the contact.

In establishment surveys, cold callbacks are generally not necessary. Contact is usually established on the first call with someone who can either serve as the respondent or else suggest a callback time. If a cold callback is necessary, the same rules apply as for the initial call; that is, call during normal business hours based on the local time of the establishment.

Cold callbacks in list sample and RDD household surveys are more problematic. Organizations use a variety of algorithms to schedule such calls and frequently use the same algorithms for both types of surveys. If any distinction is made between the two, it usually involves only the timing of the first call.

The "every shift approach" simply schedules cold callbacks for each successive interviewer work shift. The case is rotated through each shift with few, if any, priority distinctions until someone answers or some maximum number of calls is reached. If still unanswered at the maximum, the case is either referred to supervisory review or automatically closed out as a final unanswered number. This approach ignores available research on optimal contact times in favor of an intensive effort designed to produce an early contact through sheer level of effort. It may be dictated by a short data collection period that precludes call scheduling optimization. Alternatively, it may be implemented near the end of the data collection period for all remaining noncontact cases; or it may be phased in for cases that have reached the end of a more optimal algorithm without contact. The advantage of the every shift (or even every hour) approach is that it usually produces good results within in a relatively short elapsed time. The main disadvantage is its expense, since more calls are required to complete the survey and a larger proportion of calls are made during weekday daytime time slots, when telephone toll charges and computer rates are at a premium.

The "scatter approach" is similar but somewhat more scientific. In this approach the interviewing workweek is divided into some number of day and time slots and algorithms are then defined that schedule calls in a systematic pattern throughout the set of time slots. If a case completes one algorithm without obtaining an answer, it reverts to the next algorithm, and so forth. Generally, this approach places heavy emphasis on scheduling calls at a variety of times but pays little attention to disproportionate probabilities among time slots.

The "contact probabilities approach" utilizes calling algorithms for cold callbacks based upon answered "hit" rates for various time slots. The hit rates used are either obtained from the literature or else developed internally from previous surveys of similar populations. Two variations of this general approach are widely used. The "static probabilities" approach evaluates time slots in terms of fixed probabilities, while the "conditional probabilities" approach attempts to adjust the probabilities of a hit based upon the timing of earlier no answer calls.

The most frequently used method of scheduling calls on noncontact cases is the "priority score approach." In this approach each noncontact case has a weight or priority score computed each time the scheduler is run. The priority score is based on a variety of factors considered relevant for the survey at hand. Some examples include: number of prior attempts made to date, number of attempts made in the current time period, number of attempts that day, number of elapsed days since the last attempt, distribution of prior attempts across a variety of time periods, time remaining to closeout, hit rate probability, and importance of the case for analysis purposes. Calls are made based on priority scores; the higher a case's priority score, the more likely it is that the case will be called during the current time slot.

The priority score approach is essentially a hybrid system that combines several features of other approaches with a variety of other factors that have intuitive or practical appeal. It is a flexible approach that can be easily adapted to the special requirements of a particular survey by simply changing the priority score formula. It should be noted that several organizations use this approach exclusively to schedule both initial calls to new cases and cold callbacks to cases in process. These organizations lump new and old cases together into a search queue, which contains all cases for which no previous contact has been attempted or made. New cases may receive priority points until attempted at least once, and perhaps additional points during optimal (weekday evening and weekend) time slots.

All call-scheduling algorithms for noncontact cases include some mechanism for controlling the level of effort expended on a case. For example, a maximum number of cold calls may be specified and the case closed out as a final noncontact when the ceiling is reached. Where the ceiling is set depends on a number of factors, including the time and resources available, desired response rate, and, in RDD surveys, degree of concern about noncontact cases of unresolved eligibility status. A fairly typical policy is at least 10 attempts on noncontact cases. Cases reaching the cold call limit are generally routed to the supervisor for review before the case is closed out.

Appointments. There are three categories of appointments in general use: (1) hard (firm, specific); (2) soft (general); and (3) estimated (guessed). Although definitions vary somewhat, hard appointments are generally defined as those that are suggested or agreed to by the respondent himself or herself. If the interview cannot be obtained on the first contact call, a hard appointment is the next best outcome. Soft appointments are those that are made with someone other than the respondent, such as a family member, and are less desirable because of the greater uncertainty about the respondent's willingness to keep it. Finally, an estimated appointment is one that is not agreed to by either the respondent or a secondary source. Rather, it is based on the interviewer's judgment as to the best time to call back.

Organizations differ in their use of estimated appointments. Some employ only the hard and soft categories and do not require the interviewer to schedule a callback if insufficient information was obtained to qualify as a hard or soft appointment. Under this approach, answered calls that do not result in an interview, a hard or soft appointment, a refusal, or other resolution, are returned to the cold callback queue and scheduled like a noncontact case. Other organizations require the interviewer to set an estimated appointment for such cases, even if no scheduling information is obtained. The rationale here is that the interviewer's best guess, based on a review of the call record and interchange with the informant, is probably better than simply returning control of the case to a noncontact algorithm.

When the interviewer sets an appointment, he or she keys the appropriate call outcome code and then enters the type of appointment and the date and time. As a general rule, the system also gives the interviewer the option of entering a comment about the appointment. The scheduler reads this information in the call outcome record, converts the appointment time to the time of the survey unit, and schedules the callback at the appointed time.

Interviewers are given allowable ranges of permissible appointment days and times. Appointments normally must be during the survey unit's operating hours and before survey closeout. Permissible days and times also may be programmed into the CATI system, so that the screen displays an error message if an unacceptable day or time is entered.

Protocols vary somewhat with respect to appointments that are not kept by the respondent. If the appointment callback is an unanswered number, the majority approach is to schedule the next callback one hour later. If this call also results in an unanswered number, the case is either scheduled for the original appointment time on the following day or else referred to the "supervisor hold" queue for review.

Refusals. Refusal cases typically receive some level of followup effort before they are finalized. The level of effort can range from a single conversion effort by any interviewer to an elaborate protocol providing for several conversion efforts by refusal conversion specialists. The emphasis placed on refusal conversion efforts depends on a number of factors, including the response rate goal for the survey, level of concern about potential nonresponse bias, proficiency of the regular interviewing staff in circumventing initial refusals, availability of skilled conversion specialists, and the survey unit's policy regarding the propriety of making repeated conversion efforts.

The most effective method of controlling refusal nonresponse is to minimize the incidence of initial refusals. Thus, in interviewer training programs, survey organizations emphasize methods of circumventing "gatekeepers" and motivating respondents to participate in the survey, including strategies for dealing effectively with questions, concerns, and objections likely to be encountered. One technique for assisting interviewers with this task is to post question/answer scripts prepared by survey management at each interviewer's work station for easy reference when challenged by reluctant sample members. Alternatively, such scripts are included on CATI screens and accessed by the interviewer through a "help" menu.

Once an initial refusal has been encountered, subsequent processing of the case is controlled by the refusal protocol, which specifies the number of conversion efforts to be made, who will make them, the timing of the conversion efforts, and how the conversion process will be monitored. The maximum number of conversion efforts permitted typically ranges from one to three, depending on the factors mentioned above. Most organizations utilize specially trained refusal converters. As a general rule, a different interviewer is used at each stage. Organizations typically require supervisory review after the initial refusal and after each conversion effort as a safeguard against harassment. Depending on the circumstances, the supervisor has the discretion of terminating the conversion process at any stage.

Other Protocols. The preceding paragraphs described current practice for several common types of cases. Depending on the survey, other types of cases may be defined with associated protocols. Additional types and their typical protocols are listed in Table 3.

1.4 Priority System

In a CATI call scheduling system, the majority of cases are assigned to interviewers by the scheduler's case assignment module in accordance with a preprogrammed priority system. The priority system helps ensure that interviewer labor and other resources are used efficiently. Efforts are directed to the most important or most promising cases first, with the less important or less promising cases receiving attention only if sufficient resources remain. A priority system may also be a factor in making interviewer staffing decisions. For example, the number of interviewers assigned to a shift may be determined by the labor required to call cases above some minimum priority level. Work on lower priority cases is deferred to later shifts when their priority increases. A priority system can be an important tool in managing project resources.

The priority system usually applies only to the "general" queues; that is, appointments, new cases, cold callbacks, and other queues that are eligible for assignment to the general interviewing staff. Special queues

Table 3. Typical Protocols for Various Other Types of Cases

1. *Regular busy signal.* Call back at specified interval (e.g., 15 or 30 minutes). After a prescribed number of busy outcomes, refer case to supervisor or close out automatically.

2. *Fast busy signal.* Schedule callbacks in different time slots. After a prescribed number of calls with same result, close out if RDD or refer to tracing mode if list.

3. *Answering machine.* Treat like an unanswered number and schedule the next call in accordance with the controlling protocol.

4. *Temporary nonworking.* Make a specified number of callbacks in different time slots. If the same result persists, close out if RDD or refer to tracing mode if list.

5. *Permanent nonworking.* If the prerecorded message or operator intercept confirms that the sample number was reached, close out if RDD or refer to tracing mode if list. Otherwise, redial to confirm.

6. *Wrong number.* Call back immediately to confirm. If same number is reached, close out as a "double wrong connection" if RDD or refer to tracing mode if list. [See Groves and Kahn (1979) for an explanation of double wrong connections.]

7. *Computer modem.* Call back immediately to confirm. If confirmed, close out if RDD or refer to tracing mode if list.

8. *No result from dial.* Same as type 7 above.

9. *Other type phone.* (Includes pay phones, mobile phones, prerecorded messages such as time, weather, etc.). If a person answers and confirms number, close out if RDD or refer to tracing mode if list. Otherwise, call back immediately to confirm.

10. *Friendly breakoff.* (Respondent terminates interview prematurely due to lack of time. Appointment is scheduled to complete interview.) Treat the case like a hard appointment.

11. *Hostile breakoff.* (Respondent terminates interview prematurely by refusing to continue. No callback is scheduled.) Treat the case like a refusal.

12. *Other result.* (No specific result code provided.) Key an explanation of the situation and refer the case to the supervisor.

established for refusals, supervisor hold, and other nonroutine cases are typically exempt from the priority system. They are normally assigned to specific interviewers or types of interviewers, such as bilingual interviewers and tracing specialists, who work independently of the routine caseload. The supervisor may monitor the various queue sizes, however, and impose a priority system among the special queues if staffing is inadequate to keep them all current.

The priority system prioritizes cases both across the general queues and also within selected queues. At the across queue level, the majority approach is to give top priority to the appointment queues — hard, soft, and estimated (if used), in that order. Appointment retries and busy retries also are usually given equal status with one of the appointment queues. New cases and cold callbacks follow the appointment queues either separately or as a combined search queue. At the bottom of the priority scale are other miscellaneous types of low priority cases, such as future appointments and noncontact cases scheduled for a future shift.

Within queues, the priority system arranges cases with specified callback times, such as appointments, appointment retries, and busy retries, by time of callback. Organizations that use a priority score approach for noncontact cases also prioritize the queues for new cases and cold callbacks (or the combined search queue) with case priority scores.

1.5 Assignment System

There are two basic methods of assigning cases to interviewers: by paper and online. In the paper approach, the scheduler (schedule construction module) produces an individual case assignment sheet for each interviewer, who summons a case by keying the case ID number. In the online approach, no paper is used. Instead, the interviewer simply indicates readiness to receive a case and the scheduler's case assignment module delivers the case with the highest current priority. Alternatively, the interviewer may choose a case from an online display of available cases, without the use of a hardcopy assignment sheet.

Some organizations utilize a fully online system. Typically, the general queues are under the control of the case assignment module, while the special queues established for nonroutine cases are under the control of the interviewer. Other organizations use a combination online/paper system. For example, noncontact cases may be assigned online, while appointments and the special queues are handled by paper assignment sheets. Advocates of the selective use of paper argue that it is less expensive, affords better control, facilitates review of case histories, and offers protection against system failure. None of the contributing organizations routinely uses a totally paper-based system, although this remains a viable option for small scale surveys or specialized applications.

There are three options in allocating individual cases to specific interviewers: (1) assign the case to any available interviewer; (2) assign it to a specific interviewer; or (3) assign it to a specific type of interviewer.

Option 1 is used by all organizations for noncontact cases since no information generally exists to suggest that one interviewer might be more appropriate than another. Option 2 may be used in circumstances where one specific interviewer may have a better chance of obtaining a successful outcome. For example, some organizations routinely assign appointment callbacks to the same interviewer who made the appointment, on the theory that this interviewer has already established rapport, and therefore should be more successful. Option 3 is utilized when the nature of the case requires a callback by a specific type of interviewer. For example, cases with language barriers will be assigned to the appropriate language queue and called by an appropriate bilingual interviewer. Other cases may be assigned to refusal conversion or tracing specialists.

Organizations that utilize interviewer specialists typically use the scheduler to help control the assignment of cases to appropriate interviewers. In these systems, the case assignment module maintains records of interviewers approved for each type of case and will only deliver cases to eligible interviewers.

1.6 Case Closeout System

This component performs the process by which cases are closed out and final codes assigned. There are four principal mechanisms by which this is accomplished.

The most common occurs when the interviewer obtains a final outcome, such as a completed interview, on a specific call. When the interviewer keys the call outcome code, the scheduler recognizes this as a final outcome, assigns a final outcome code, and removes the case from the pool available for assignment.

The second closeout mechanism is activated when the prescribed level of effort specified in a calling protocol is reached, such as the second conversion effort for a refusal. The scheduler recognizes that the calling protocol has been satisfied and closes out the case as described above.

The third way a case can be closed out is by supervisory action. The supervisor is generally responsible for reviewing various types of problem cases, such as perennial unanswered numbers, and for determining if additional actions should be taken. If the supervisor concludes that no further action is warranted, he or she will assign a final code.

Finally, a case can be closed out by special action of the CATI system or systems personnel. Examples include deleting a case as part of a sample reduction or when the respondent returns a mail questionnaire.

Like call outcome codes, final codes should be unique and exhaustive of all final outcome possibilities and should provide information at an

Table 4. Final Outcome Categories in Use by The University of Michigan

FINAL INTERVIEW
Completed Interview
Partial Interview

FINAL DISPOSITION – REFUSALS AND NONINTERVIEWS
Refusal, never reached for conversion
Initial refusal, breakoff, no conversion
Final refusal by respondent
Final refusal, not by respondent
Final refusal, respondent undetermined
Final refusal, breakoff on conversion
Final refusal, no conversion attempt
Noninterview, never answered
Noninterview, circumstantial
Noninterview, incomplete
Noninterview – nonsample, end of study replacements (RDD only)
Noninterview – grid filled

FINAL DISPOSITION – NONSAMPLE
RDD only
Final silence
Final strange noise or fast busy
Second wrong connection
Nonworking number
Nonresidential number
Out of primary area
Noninterview – grid filled, sample replaced
No eligible respondent
Nonsample, not regenerated
Recontact only
Second wrong number for respondent, no new number available
Second respondent's number no longer in service

appropriate level of detail. As with call outcome codes, it is better to err on the side of overly detailed codes since little used codes can always be collapsed later. The final codes used by The University of Michigan are shown in Table 4.

The Census Bureau also utilizes prefinal as well as final outcome codes. These prefinal codes are set by the outcome evaluation module and record the tentative final outcome of cases that are still in process. At any point in the survey, a tabulation of both final and prefinal codes will summarize what the distribution of final codes would be if data collection were terminated at that point.

1.7 Supervisory Intervention

The telephone supervisor plays a key role in the operation and control of a CATI call scheduling system. Although a full description of the supervisor's role is beyond the scope of this chapter, this individual's principal responsibilities typically include recruiting new interviewers, as necessary; assuming a lead role in interviewer training programs; scheduling interviewer labor; monitoring caseloads in the various queues, especially the appointment and special queues not controlled by the case assignment module of the scheduler; reviewing problem cases that have been routed to the supervisor hold queue; providing overall supervision of the interviewing staff; implementing quality control procedures such as audiovisual monitoring of interviewers' calls; taking incoming calls from respondents; and printing and analyzing a variety of system produced monitoring reports.

The CATI system typically provides the supervisory staff (and sometimes survey management) with a variety of computer utility programs that assist them in discharging their responsibilities. The Census Bureau divides these programs into four main sets: (1) those that review or control the general flow of work; (2) those that review the status or history of individual cases; (3) those that change the progress of the case or set a final code; and (4) other specialized programs.

1.8 Monitoring Reports

CATI organizations produce a wide variety of computer generated reports that are used by supervisory staff, systems personnel, and survey management to monitor the call scheduling system and the overall data collection effort. One of the greatest advantages of a CATI system over a paper and pencil approach is the wealth of readily available monitoring data. Although content, format, and terminology vary widely across organizations, there are several generic types of reports in common usage. These are briefly described in Table 5.

2. ADVANTAGES OF CATI CALL SCHEDULING AND SUGGESTIONS FOR FUTURE RESEARCH

A CATI call scheduling system of the kind described above offers numerous advantages over a traditional manual system. Automation reduces the potential for human error in making call scheduling decisions, minimizes dependence on hardcopy forms, and greatly reduces the

Table 5. Generic Types of Monitoring Reports

1. *Queue listing.* A listing of the cases currently assigned to a queue. Displayed or printed by the supervisor, these reports are used to monitor, and in some systems, to assign the caseload in the various queues.

2. *Call record listing.* A listing of the call record for a case. In reviewing a case, the supervisor may find it more convenient to print a listing of the call record instead of reviewing this information online. Also, the listing can be used as a cover sheet when assigning refusals or other problem cases to interviewers.

3. *Current status report.* A report that summarizes the current status of all cases in the survey sample. Separate columns may show cumulative results through the preceding day, cumulative results at the end of the current day, and the net change, indicating the effect of the current day's production.

4. *Historical status report.* Similar to the current status report except that it shows the results of past processing steps in addition to the current status. For example, if a case has passed through a tracing operation and is currently assigned to the interviewing mode, only its current status related to interviewing operations will be shown in the current status report, whereas the historical status report will also include tracing final results for all cases that have passed through that process.

5. *Interviewer performance report.* Shows relevant production and performance data for each interviewer.

6. *Duration report.* Identifies cases that have remained in a particular status category beyond a prescribed duration.

7. *Exception report.* Identifies cases that have deviated from a prescribed processing path.

8. *Miscellaneous diagnostic reports.* Reports used to monitor various other aspects of the call scheduling system, such as number of missed appointments, average queue size across time, average number of calls to complete a case, etc.

9. *Cost control reports.* Reports produced by the CATI system for fiscal management purposes. These reports may include such data as average interviewing time per interview, labor cost per interview, toll charges per interview, computer charges per interview, and total cost per interview. These data are compared with budgeted assumptions and significant variances investigated.

considerable labor associated with case management in a conventional system. It facilitates the use of complex and sophisticated calling protocols, permits implementation of a priority system, and provides the telephone supervisor and survey management with a variety of utility programs and system produced reports that are useful in forecasting workloads, scheduling interviewer labor, monitoring caseloads, evaluating interviewer performance, and managing and controlling the overall progress of the CATI survey.

Although substantial progress has been made in CATI call scheduling in recent years, there remains ample opportunity for further research and development. In spite of the advantages of automated call scheduling, questions about its efficiency still remain. Although results reported by Catlin and Ingram (this volume, Chapter 27) are mixed on the relative efficiency of automated versus manual call scheduling, Nicholls and Groves (1986) cite evidence of CATI's superior efficiency and suggest that this may be attributable to the use of automated call scheduling rather than to the use of a computerized questionnaire. More research is needed in this area.

Other critical areas for future research are the scheduler, calling protocols, and the priority system. Scheduler software systems are still at a relatively early stage of development, and questions abound as to the optimal system configuration for various types of applications. Calling protocols are still dominated by intuition and folklore, leaving unresolved such issues as the optimal timing of initial calls in an RDD survey, when cold callbacks are best scheduled, and the most effective methods of classifying appointments and converting refusals. Finally, the key issue for priority systems is the development of optimal formulas for the priority score approach. Serious methodological research in these areas would make a significant contribution to the sponsoring organization as well as to the larger CATI community.

CHAPTER 26

QUESTIONNAIRE DESIGN FOR CATI: DESIGN OBJECTIVES AND METHODS

Carol C. House
U. S. Department of Agriculture

William L. Nicholls II
U. S. Bureau of the Census

This chapter presents objectives and methods of computer-assisted telephone interviewing (CATI) questionnaire design. It focuses on authoring questionnaires for item and screen based CATI systems which permit both forward and backward movement, as described in the general introduction to this section. However, much of the discussion may be relevant to other forms of CATI and to computer-assisted personal interviewing.

Although many design principles for paper questionnaires apply to computerized data collection, there are important differences. A change to computer-assisted methods alters the way some questions are asked and most answers are recorded. In addition it modifies the roles of both survey interviewer and questionnaire designer. The interviewer becomes more dependent on the work of the designer. The designer must view a CATI instrument as a complex computer program as well as a survey questionnaire and adapt the design to meet common programming objectives.

We propose six basic objectives for CATI questionnaire design, and list them in order of importance. They are to:

1. Collect survey data, including responses to questions, interviewer comments, and other relevant information

2. Meet interviewer needs by making displays quickly comprehensible, providing interviewers access to information, and expediting movement through the questionnaire

3. Ensure program correctness so that the instrument functions as intended under all circumstances

4. Function efficiently so that the CATI instrument makes minimum demands on hardware resources while maintaining rapid response times

5. Provide usable documentation thereby ensuring that its structure, coding, and operation are understandable to others

6. Be portable to other surveys collecting similar information.

This chapter covers five of these objectives but focuses on the first two, obtaining needed survey data and meeting interviewer needs. Program efficiency is not addressed at all because it is largely system dependent. This chapter draws in part on two of the authors' earlier papers. The first paper (House, 1985) presented a general introduction to the design of CATI questionnaires both as survey instruments and as computer programs. The second paper (Nicholls and House, 1987) introduced the six objectives listed above and focused on methods of ensuring program correctness.[1]

1. COLLECTING SURVEY DATA

The primary objective of any questionnaire is to collect data while minimizing error, bias, and respondent burden. To achieve this, CATI questionnaire designers must expand traditional methods of design in the areas of developing specifications, writing survey questions, and assuring data quality.

---

[1] The authors thank the editors of the *Journal of Official Statistics* and the *Proceedings of the U.S. Census Bureau's Third Annual Research Conference* for permission to reproduce selected parts of these earlier papers in this chapter.

1.1 Specifications

The first task is to obtain detailed survey specifications. Those new to CATI often assume that a paper and pencil version of a questionnaire constitutes a sufficient set of specifications. It's a good place to start, but it is insufficient for several reasons.

First, a CATI instrument controls the flow of the interview with branching instructions provided by the questionnaire author. The interviewer cannot arbitrarily omit blocks of questions or revise their order to meet unusual situations. If more flexible forms of movement are required, they must be specified in advance and built into the design. Therefore a CATI questionnaire requires unusually thorough specifications on likely sequences of interviewer movement and permissible variations.

Second, a CATI instrument typically employs computer-supported quality enhancement features not available in paper and pencil questionnaires. These include: (1) "fills," which modify the wording of questions and instructions; (2) online range and consistency edits; and (3) "help screens" containing such aids as definitions of key terms and summaries of entered data. It is not necessarily evident from a paper form where these features can be most effectively used in the questionnaire. The designer must work with the client to incorporate those decisions into the total specifications package.

Finally, a CATI instrument is a complex system for data capture and display. The widths, variable types, and formats of all data items must be specified. For example, textual material from open question responses may be either preserved in full in machine readable form, converted to codes, or discarded after an edit review. When input data (such as last known address and telephone number) are used in the interview, requirements for their display, updating, and inclusion in the output file must be developed. Unlike traditional surveys, CATI surveys can rarely separate decisions on data capture and data preparation from those on questionnaire design.

Table 1 provides a checklist of topics requiring specifications for a CATI questionnaire. As discussed above, a designer must go beyond the paper questionnaire to develop these. Possible additional sources of information include: interviewer manuals; procedures manuals; interviewer aids such as flash cards; input file specifications and layouts; key entry specifications; editing specifications; and output file requirements. Unusual situations are often omitted from formal documentation and are handled on an ad hoc basis. Key survey personnel may be the best source for documenting these situations and for developing fills, online edits, and help screens. Sometimes the CATI

Table 1. Checklist of Specification Topics

* Format of input variables from record systems or previous contact, with instructions for updating, displaying and outputting this data.
* Format of output variables and acceptable values.
* Procedures for handling responses to open questions, "other, specify" replies, and interviewer notes.
* Edit specifications and wording to reconcile edit failures.
* Audit trail requirements.
* Use of fills.
* Use of help or reference screens.
* Missing data options, including branching logic.
* Forms of nonstandard movement between items.
* Introductions and explanations of the survey to the respondent.
* Respondent selection and substitution procedures.
* Procedures for refusals, inaccessibles, and other field outcomes.
* Criteria which define a usable interview.

author is left with no option but to provide his or her own specifications, illustrated by demonstration instruments for client review.

1.2 Question Design

Mechanical differences between CATI and traditional interviewing procedures necessitate changes in the structure of many survey questions. In this section, we explore some of these differences and suggest ways to design several generic types of survey questions for CATI.

The restricted size of a video display screen accounts for many of the differences between paper and CATI questions. A CATI screen can rarely display more than two or three scripted questions at one time. This makes designing a CATI questionnaire similar to preparing a paper and pencil questionnaire on a set of "3 x 5" index cards. Questions with long introductions or a large number of precoded responses often must be redesigned or divided into subquestions on consecutive screens. Large tables or grids must either be split up into repeated sequences of individual items or programmed using form based CATI with brief labels rather than full question text.

Character sets also are limited on many video screens. Italic and bold type or different font sizes are not typically available; and not all systems have such replacements as reverse video, highlighting, and flashing characters. Color screens are currently uncommon in CATI. At the same time, the need for alternate character sets is somewhat reduced because automatic branching and computer supplied fills replace many interviewer instructions.

The following paragraphs suggest working procedures for preparing several common types of survey questions.

Check answer questions with single answer precoded responses are the simplest to format in CATI. Their screen layout parallels that of a paper form. The primary design extensions needed to prepare for CATI include the use of fills and handling missing data.

When using a fill, the CATI author must ensure that: (a) the filled value is nonempty and correct when the screen is reached; (b) the value is appropriately right or left adjusted; and (c) its length, when added to the fixed text, will not overflow the line. CATI systems differ in their support for these tasks.

Explicit missing data precodes are required with CATI. However, some systems permit "blind precodes" which do not display on the screen, or permit function keys to serve as alternatives to displayed precodes for "missing data." These features provide a closer analog to many paper forms because they omit the actual display of missing data precodes to avoid their overuse. Once missing data codes are established by one of these methods, the CATI author must next specify the branching path from each missing data outcome to the rest of the questionnaire.

Multiple answer precoded questions are easily accommodated in screen based CATI systems and in item based systems that permit multiple answer spaces per screen. The CATI interviewer enters precode numbers in as many consecutive answer spaces as necessary. The CATI author need only: (a) determine the maximum number of answers permitted; (b) add a code for "no further answers" which will exit the screen; and (c) establish edits to prevent the same code from being entered more than once. Multiple answer questions are more difficult to design for CATI systems, which permit only one answer space per screen. A series of YES-NO questions could be used, one for each alternative. Another approach is to redisplay the multiple answer question on consecutive screens, each with a single answer space, but with fills which indicate responses already chosen.

Questions requiring a *numeric* response are easily handled in CATI. The CATI author need only determine: (a) the acceptable range for editing; (b) the variable length for storage; and (c) the appropriate missing data codes. If the number is large, such as a 10 digit telephone number, additional steps to punctuate the value may be needed to increase readability, such as: (301) 415–1212; 14,345,234; or $3,494.50.

On paper forms, the answer space for *open questions* can be set for the average length of reply. If more space is needed, the interviewer can write smaller or continue up the side of the page. Decisions on the use and processing of open question responses may be deferred until interviewing is complete.

With CATI, additional planning is required, since the proposed use of an open question often determines how it is stored. The item may be entered as a variable with an unusually large length. As such, it may be stored in the output data record and used in fills later in the interview. However, such variables may contribute to inefficiently large output records. When this type of question is displayed, the screen should show the allowable answer length to help the interviewer avoid unintentionally truncating a response. If the system does not do this automatically, the author should mark it with dashes or brackets. The author also should develop procedures to handle responses longer than the fixed variable.

The second approach to open question responses is to record them in a separate file which stores lengthy textual material consecutively by case number and item label rather than in variables. Entries can be as long as needed. In most systems, an entry made to this file cannot be displayed in fills and can be reviewed only by special interviewer actions.

1.3 Increasing Data Quality

When used effectively, CATI provides new tools that control response errors and selected forms of item nonresponse. Merely emulating a paper and pencil survey with CATI will not necessarily increase data quality (Nicholls and Groves, 1986; Groves and Nicholls, 1986). Such improvements are likely to occur only when the questionnaire design capitalizes on CATI's special features.

Some quality enhancement features are standard. Automatic branching between items controls skip errors; and built-in editing of check answer and numeric items can prevent out of range entries. Open question responses can be required to have some minimum number of characters to increase the likelihood of a full response.

Further edits may check an item's consistency with prior entries, input data from records, or a response from a previous interview. These checks must be prepared by the CATI author. They involve four steps. (1) A logical or arithmetic comparison is made as a background process. If the edit fails, then step 2 is executed. (2) A display informs the interviewer of the discrepancy and supplies wording to elicit the respondent's help in resolving it. This screen should provide a summary of the values found to be inconsistent. (3) The interviewer corrects the entry or a previous entry, or provides a note to explain the inconsistency. (4) The interviewer returns to the mainstream of the questionnaire. This must be possible whether the inconsistency is resolved or not.

Item nonresponse may be reduced by building in probes to follow a "no answer" or "don't know" response. For example, an initial question may ask for the exact date of an event, but then branch to another question that asks for a range of days if the respondent cannot answer. For items of special importance, a "don't know" might be followed by a probe of the form: "This is very important to the survey. Would you think about it for a minute and see if you can recall that information."

An audit trail of the interview can provide valuable feedback to the designer on the quality of the survey data and the questionnaire. It can be an indispensable part of the pretesting and quality control process by identifying troublesome questions. Some CATI systems provide these functions automatically. With others, the questionnaire designer must take special steps to retain needed information.

2. MEETING INTERVIEWERS' NEEDS

Designing a CATI instrument to meet interviewer needs is a demanding and multifaceted task. Its complexity stems from a basic objective of CATI — to remove certain interviewer tasks and options and make them computer controlled. Responsibility for these tasks is transferred to the instrument designer. This makes it critical that the designer has a full understanding of the CATI interviewer's situation and then uses this knowledge to recognize his or her own increased responsibilities in the interviewing process.

The CATI interviewer's primary job is interviewing, not operating a CATI system. He must answer respondent questions, clarify ambiguous concepts, recognize and correct response errors, and encourage reluctant respondents to provide data. When a CATI instrument is well designed, the interviewer learns to rely on it while devoting little attention to it. He expects the CATI screens to tell him what to do in virtually every situation. If a problem occurs, the interviewer wants to be able to correct it quickly before the relationship with the respondent is jeopardized.

Complex solutions which require multiple interviewer actions are unlikely to be used or will be used incorrectly.

The instrument designer must design the questionnaire to minimize the interviewer's problems. To accomplish these goals the designer should reduce the negative effects of segmenting the questionnaire into individual screens and provide the interviewer with controlled flexibility.

2.1 Addressing the Effects of Segmentation

Because a CATI display screen is small, the interviewer generally sees only small segments of the questionnaire at a time. This is called the *segmentation effect*. Segmentation has its benefits: it focuses the interviewer's attention on the current question and makes it more difficult to omit questions or ask them out of sequence. However, it can be disorienting and frustrating unless the instrument designer takes steps to counteract its "isolating" effects. There are two general ways to do this: provide screens that the interviewer can quickly comprehend, and provide access to all information the interviewer will need.

Segmentation means a CATI interviewer cannot glance ahead to the next question, and thus needs each screen to be immediately comprehensible when it is displayed. Comprehensibility can be enhanced by:

1. Using a task oriented design that places the first thing the interviewer is to do at the top of each screen. If a question is to be read, it should appear first. If a special instruction such as "ASK ONLY IF NECESSARY" is to be followed, it should appear first.

2. Standardizing formats so that question text, response categories, and answer spaces are in approximately the same place on each screen. Use different character sets or highlighting to distinguish instructions from questions to be read verbatim.

3. Provide uncluttered screens with adequate "white space" to improve readability and aid interviewers in locating the appropriate response categories.

Segmentation also limits the interviewer's access to information. Some common examples include the answer to the previous question and the respondent's name. The instrument designer can provide this type of information in a variety of ways.

1. Fills may build information directly into the question to be asked, such as "You said JOHN was 19 years old." Or fills may provide supplementary information intended only for the interviewer, such as "ASKING ABOUT JOHN, AGE 19," which the interviewer does not read aloud. Fills are the best method of displaying information which requires limited space.

2. Putting consecutive, related questions on the same screen provides continuity and reduces segmentation effects. Screen based CATI systems and some item based systems allow the designer to display several questions on the same screen, either concurrently or consecutively.

3. "Reference or help" screens are separate displays an interviewer can access through a simple command or function key. They may contain information needed less frequently or which require more space than can be handled through "fills." For example, reference screens may display names of household members or definitions of key terms used in the survey.

4. A fourth method is to assist the interviewer in returning to a previous question. The designer can provide a list of item labels or names of key, previously answered questions to be used in "backup commands." The list may appear as a fill on the current display or in special reference screens.

2.2 Designing in Flexibility

One of the greatest strengths in CATI data collection is its control of the interview. This control leads to many of the advantages attributed to CATI, but is also CATI's greatest weakness. It strips the interviewer of the ability to improvise when unusual situations occur. The entire burden of anticipating and planning for what will transpire in an interview is placed on the shoulders of the questionnaire designer.

Some restrictions imposed by a CATI system are the result of expedient solutions for the systems designer and may present major problems in the field. For example, an interviewer might not be allowed to interview individual household members in the order most convenient for them if the CATI system is structured to accept the data only in the sequence their names were added to the roster matrix.

Controlled flexibility is needed to enable the interviewer to cope with different interviewing situations without sacrificing the special advantages of CATI. For example, a designer can employ "skipping" to access reference screens or to allow an interviewer to skip out of an interrupted interview and return to the place he or she left off on a different day. The designer may also allow an interviewer the option of skipping over a section of questions (perhaps of a sensitive nature) and returning to them at the end of the questionnaire.

There will always be unusual occurrences in interviews that can not be predicted. For these, the questionnaire author can provide the interviewer with a way to record the answer to a question, no matter how unexpected that answer might be. With precoded answers, this means an "other, specify" code. For fixed length open questions, interviewers must be taught to deal with an answer that is too large. Missing data codes are also important, with an "escape" that allows the interview to continue. Finally, whenever possible the design should facilitate nonstandard movement, such as backing up and reviewing answers.

3. PROGRAM CORRECTNESS, PORTABILITY, AND DOCUMENTATION

3.1 Program Correctness and Portability

Program correctness is a critical objective of CATI questionnaire design. An instrument that does not function correctly typically defeats other critical objectives by failing to collect needed survey data and by hindering rather than assisting interviewers at their work. First, the instrument must function properly when the interviewer moves through it from beginning to end in a consistently forward direction. This is a fairly straightforward requirement and can be achieved through the familiar logic of standard computer programming. There is, however, the more complex requirement that the instrument function properly after backing up, changing answers, or taking any other action the system permits. Modern CATI systems accommodate this unusual goal, but only when appropriate care is taken in questionnaire design.

Portability is another important objective of instrument design. To minimize the time and costs of future instrument development, CATI questionnaire sections should be transferable with minimum change to other surveys collecting similar information.

Following the dictum of modern quality control, correctness and portability should be designed in, not tested out. In particular, designing a complex instrument with the limited goal of correctness during standard movement alone, and then attempting to reach correctness under other types of movement by testing and debugging is a virtually hopeless task. Similarly, using the code from an older instrument can produce serious debugging problems if it was not designed to accommodate the objective of portability. CATI instruments should be designed from the start with the goals of consistent correctness and portability in mind.

3.2 Modular Design

Modular design, a key concept in structured programming (Yourdon, 1975), provides the building blocks for program correctness and portability for CATI instrument design. From a questionnaire point of view, a module is a self-contained group of questions, or items, which have a clearly defined function, a single lead item in standard forward movement, and a place in a modular hierarchy. A module may consist, for example, of the group of items that deal with dialing instructions, or a set of interrelated questions on the same substantive topic. We shall also see that a module may consist of only a single displayed item or question.

Programming principles of modular design should be followed in constructing CATI instruments. The designer "formally" identifies the major functions of the program first, and then proceeds to identify the lesser functions that extend from the major ones. For example, a major function might be "obtaining an acceptable respondent" or "obtaining a health profile of the respondent." Obtaining an acceptable respondent could be broken down into the functions of "locating a household," "obtaining permission to conduct the survey," and "randomizing the respondent selection." Each of these can be broken down into smaller and smaller pieces.

By using modules the designer has broken a large task down into smaller blocks that are easier to design, program, debug, document, modify, reuse, and review with clients. The designer can be sure that each individual part is working correctly before combining it with another. Once combined, changes in one module do not affect the others, except in very carefully constructed interfaces. This allows a module to be lifted and reused in another instrument with minimal debugging problems.

3.3 Virtual Items and Correctness

Use of the design techniques discussed above will produce a modular hierarchy, at the bottom of which we find the smallest modules, which may be single items or very "simple" combinations of items called "virtual items." The manner in which virtual items are combined to form the overall modular hierarchy will have real impacts on whether the instrument can be expected to function correctly under nonstandard movement. This chapter cannot go into a great deal of detail on programming logic, but will examine one sample problem that can arise when an interviewer backs up and changes an answer during an interview. It will use this example to demonstrate how the structural relationship between virtual items within modules, in addition to modular design itself, can be employed to assure correct functioning.

Consider the common situation when an interviewer backs to an earlier question and changes its answer. The branching path may remain the same or be redirected to a different sequence of questions. The result is two series or paths of answered items, one based on the former value of the changed entry and another based on its current value. We say items on the current path are applicable, while those no longer on the current path are inapplicable. When a designer wishes to perform a logic check based on the answer to an earlier question, he wants to be assured that the earlier question is on the current branching path. This is where the structural relationship between items can help.

In Figure 26.1, assume that each of the boxes represents an individual virtual item or question within a larger module. Next assume that the designer wishes to use the answer to Item B within a logic check to be executed at Item D. Does he need to worry whether Item B is on the current branching path? The answer is "No." These items are connected in a linear structure, meaning that each item directly follows the previous one with no branching or contingent paths. An interviewer cannot reach Item D (and execute the logic check) without passing through Item B. Thus when a logic check at Item D is executed, Item B will always be both answered and applicable. Linear structures between virtual items eliminate the concern of inapplicability.

Frequently a designer needs a more complex structure between items, such as the tree structure shown in Figure 26.2. A tree structure is one in which the branching paths diverge after the lead item, somewhat like the branches of a tree, but do not converge again at any item within the structure. Notice that there are three distinct branching paths in this structure: ABDG, ACEH, ACFI. An interviewer could enter answers that proceed down path ACFI, then back up to Item C and change the answer so that he is taken down path ACEH. This would create an answered but inapplicable status for Items F and I.

The key to working with tree structures is to realize that the different paths, once diverged, never converge again within that module. If a logic check is executed at Item H, the current interviewing path through the instrument is ACEH (no other paths to Item H exist). Items F and H, for example, can never be on a single branching path. This means that a designer *will not* use the answer to Item F in a logic check at Item H, because if the interview proceeds along path ACEH (so that the logic

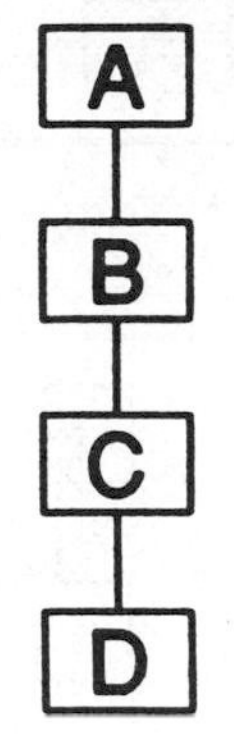

Figure 26.1 Linear structure.

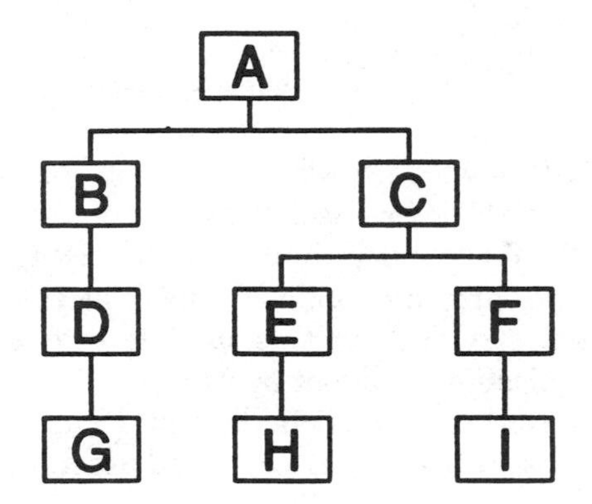

Figure 26.2 Tree structure.

check is executed), he knows that Item F is either unanswered or answered/inapplicable. Further he knows that all prior items on the path to H (Items A, C, and E) are both answered and applicable. Thus in tree structures, a designer can use values entered in any previous items on the branching path in commands at any subsequent item on that path, without concern for applicability.

The final illustration of this problem appears in Figure 26.3. Here the structure between items is much more complex than in either the linear or the tree structures discussed above. It is called a net structure because when diagrammed it resembles a fisherman's net. The defining properties of a net structure are the same as those of a tree structure, except that its branching paths converge on at least one item within the structure. In the figure, two paths converge at Item E, and two others converge at Item K.

A designer may wish to use the value of the entry to Item C in a logic check to be executed at Item K. Consider this scenario: an interview first proceeds down path ACEH to Item H, backs to Item A, changes the

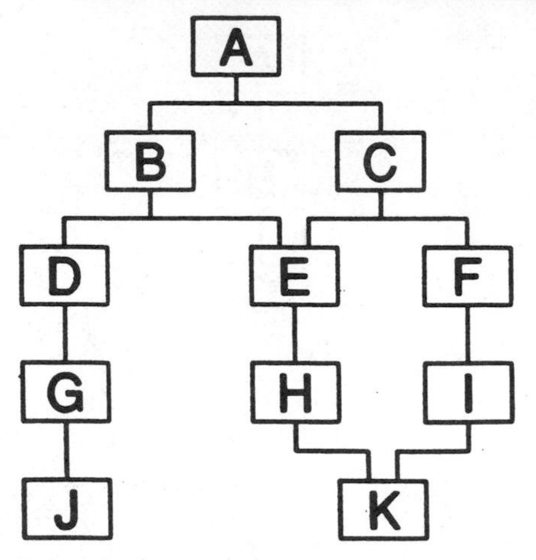

Figure 26.3 Net structure.

answer, and proceeds down path ABEHK. Then Item C will be answered but inapplicable when the logic check is executed at Item K. The designer must, therefore, plan for this possibility when designing the logic check at Item K. He may perform an "onpath" check at this point, or perhaps reformulate the net structure in this module into a more simple tree structure where applicability will not be a problem.

The reader is referred to Nicholls and House (1987) for a more comprehensive discussion of virtual items and programming logic.

3.4 Documentation

A CATI questionnaire author has not fully completed his or her work until that instrument is adequately documented. There are many reasons to document. First, by describing how the survey specifications were met, written documentation is valuable for supervisory and client review. Second, documentation is a valuable aid in testing, debugging, and changing a complex CATI instrument. Third, good documentation, like good design, increases the value of a CATI instrument by making its components more portable to other surveys.

General standards of documentation for CATI instruments do not currently exist. Documentation needs will vary with the CATI system employed, the type of CATI surveys undertaken, and local experiences and practices. The authors can offer only general recommendations reflecting good programming practice and their own experiences.

First, the documentation of CATI instruments is greatly facilitated by following modular design principles. The instrument's major components are documented first, then the submodules within them, and lastly the virtual items. For a larger module, the documentation should include a statement of its function, a listing of its submodules, and explanations of unusual procedures or coding employed. It may also be useful to list input or output variables that are used, created, or modified in the module. For a smaller module, comments interspersed in the program file may provide similar information.

Second, a set of flowcharts showing the relationships among the principal modules and their submodules should be a major component of documentation. Labeled flowcharts provide references for written descriptions and a table of contents for more detailed documentation on specific modules and submodules. Flowcharts are especially valuable in trouble shooting instruments or changing those that have been in use for a long time.

Third, standardized programming techniques can reduce the volume of documentation because their use means that procedures need be documented only once or even can be left undocumented when they follow standardized procedures.

Fourth, documentation should include a record of changes made to an instrument over time. Dated annotations of change within the programs, periodic hardcopy printouts, or change logbooks may all be helpful. File copies of dated prior versions are convenient to keep and can be compared to current versions easily through any of a variety of computer utilities.

4. SUMMARY AND CONCLUSIONS

Questionnaire design with CATI, as on paper, is far from an exact science. Readers who may have sought a road map to quality design in this chapter must settle for the few scattered directional signs the authors provide.

What do these signs say? First, the basic objectives of CATI design, such as collecting quality data, supporting interviewer functions, and documenting the process are basic to all questionnaire design, but may require different approaches to be workable under CATI. Survey specifications must be more detailed with CATI, and the questionnaire author may require considerable tenacity and imagination to pull together an acceptable package. Questions may have to look different, be grouped differently, or have their answers recorded in different manners.

Second, the survey instrument is the mechanism by which the interviewer's needs and expectations, often changeable and unexpected, mix with the rigid controls of a CATI system. Whether the result of this encounter produces a powerful data collection partnership, or merely the grinding of gears, is largely dependent on the work of the instrument's authors. Their responsibilities in the interviewing process are to counteract many of the negative segmentation effects, and to provide controlled flexibility to handle different interviewing situations. The designer also initiates the use of CATI's quality enhancement features to control response and recording errors and minimize nonresponse.

Third, a CATI questionnaire is also a complex computer program. This fact generates the need for additional design objectives that are specific to computerized data collection. Modular design techniques used in programming, along with thorough documentation, can provide the structure to build a program code that functions correctly and will be portable to other surveys.

CHAPTER 27

THE EFFECTS OF CATI ON COSTS AND DATA QUALITY: A COMPARISON OF CATI AND PAPER METHODS IN CENTRALIZED INTERVIEWING

Gary Catlin and Susan Ingram[1]
Statistics Canada

The rapid growth of computer-assisted telephone interviewing (CATI) in both private and public sector survey organizations in recent years is based, in part, on the expectations that CATI will enhance the quality of telephone survey data, reduce the costs of its collection, and increase its timeliness. The early literature on CATI appeared to promise all these benefits and perhaps the increased use of CATI confirms them. But solid published evidence supporting these claims is lacking. In a recent two article series reviewing research on the effects of CATI, Nicholls and Groves (1986) concluded that while many of CATI's potential benefits remain viable, few have been sufficiently demonstrated.

Previous studies comparing CATI and paper and pencil telephone methods have been published by Coulter (1985), Groves and Mathiowetz (1984), Harlow, Rosenthal, and Ziegler (1985), and House (1984). Nicholls and Groves (1986) and Groves and Nicholls (1986) have summarized additional unpublished comparisons and impressionistic evidence. Much of the comparative data have been based on surveys with small sample sizes, first attempts at CATI by organizations inexperienced with it, and studies with inadequate experimental designs.

The objective of the present study was to evaluate the effects of CATI on survey costs and data quality in a centralized interviewing facility

[1]The authors would like to thank William Nicholls and Lecily Hunter for their useful comments on earlier drafts of this chapter. We would also like to thank the members of the CATI project team for their contributions to this research.

under relatively controlled conditions, and with a study of sufficient size and duration to permit more generalizable results. To control possibly confounding factors, the same survey was conducted by CATI and paper and pencil methods in alternate weeks from the same site and with the same interviewers. The scale of the test was large. Between February and September 1987, more than 10,000 household interviews were completed, approximately evenly divided between the two methods. Since analysis was still in progress as this chapter was written, some results are based on less than the full eight months of data. Further reports on this research are planned after publication of the present volume.

This test was a joint Statistics Canada/U.S. Bureau of the Census (USBC) research project. Statistics Canada used the USBC Census CATI System and evaluated the impact of CATI using the Canadian Labour Force Survey (LFS) as a vehicle. The LFS is similar to the Current Population Survey conducted by the USBC with questions concerning employment and job search activities relating to a one week reference period.

This chapter does not address all the many steps required to undertake a CATI survey. We have not attempted to investigate the costs of installing a CATI system, developing a CATI questionnaire, training CATI interviewers, or processing CATI data. These activities have limited generalizability to other surveys or organizations because of the many ways in which they could be completed. Rather we will focus on the effect of CATI on data quality and survey costs during the data collection period. Before discussing these results, we will first review the study's methodology.

1. METHODOLOGY

The target population was the civilian noninstitutionalized population, largely paralleling that of the regular Canadian Labour Force Survey, in the province of Ontario. Nontelephone households and collective dwellings with 10 or more people were, however, excluded. Nontelephone households constitute approximately two percent of households and collective dwellings less than 1 percent of households in Ontario (Catlin, Choudhry, and Hofmann, 1984).

The sample was selected using random digit dialing (RDD). A systematic sample of working banks was chosen, based on information obtained from Bell Canada and other sources, and a two digit randomly generated number was added to form a complete telephone number. Screening of the RDD numbers to identify households was conducted one month prior to data collection, and households were assigned randomly to either CATI or paper and pencil interviewing. Households remained in the same interviewing method for the duration of their time in the survey.

A sample size of approximately 1,000 households per month was the target for each interviewing method. In both February and March, approximately 333 households were introduced in each method. During April and May, 167 households were added, bringing each sample size to the required 1,000. From June to September, one-sixth of the sample was replaced each month following the practice in the LFS.

Interviewing was conducted from a centralized telephone facility at Statistics Canada's head office in Ottawa. All interviewers worked on both CATI and paper and pencil interviewing, to reduce the effect of interviewer variability on the test results. Interviewers were trained for one week covering labour force concepts and both CATI and paper and pencil procedures. Two supervisors and 8 to 11 interviewers were employed each month.

Two one-week interviewing periods were required each month, one for CATI and one for paper and pencil interviewing. Each month, the interviewing took place in the same two weeks; however the interviewing methods assigned to each week alternated from month to month. The hours of interviewing were from 8:30 a.m. to 9:30 p.m. on weekdays and as required on Saturday.

The Labour Force interview consists of two parts: (1) the Household Record Docket, also called the control card or household roster, which contains items on household membership and demographics; and (2) the Labour Force questionnaire with items concerning labour force activity during the previous week asked for each eligible household member. Although both interviewing methods used the standard question wording of the regular LFS, the CATI instrument automatically personalized the question wording based on previous answers. The CATI interviews also employed automatic branching, enforced probing, editing, call scheduling, and sample management. In this test, interviews assigned to CATI were not conducted by paper and pencil even during system or hardware problems. Respondents were informed that there was a technical problem and that they would be called back as soon as possible.

Both the CATI and paper and pencil samples were processed using the existing LFS system. The same edit and imputation, automated industry and occupation coding, and tabulation modules were employed for both methods.

2. RESPONSE RATES

CATI may influence response rates either through its ability to contact eligible survey households or through its impact on the interviewer-respondent interaction. Response rates were based on all eligible households identified during the RDD screening in the month prior to their first month in the survey. Overall, CATI had a lower response rate, 84.2 compared to 86.3 percent for the paper and pencil method. The difference was almost exclusively a factor of noncontact rather than refusals. Each is examined separately below.

2.1 Household Contact

The literature suggests that call scheduling in a CATI system should improve a survey's ability to contact households by imposing a logic to the frequency and timing of calls during a fixed survey period (Groves and Nicholls, 1986). In Chapter 25, Weeks states that automated call scheduling offers numerous and significant advantages over traditional manual systems by eliminating interviewer error and allowing for programmed calling protocols that would be impossible to implement in a paper and pencil system.

The call scheduling algorithm for the CATI sample employed four factors to assign priorities to calls for a specific time period: (1) the best time to call reported in prior months; (2) the results of unsuccessful calls; (3) the time at which prior calls were made; and (4) appointment information. In the paper and pencil approach, supervisors made assignments of sample cases to interviewers who determined which households to call based on the same information. This was a manual operation and no ordering based on combined priorities was feasible.

The noncontacts included households that were never contacted as well as those reached who could not complete the interview at that time. As shown in Table 1, the noncontact component was significantly higher for CATI. This was consistent over all months, although the difference was not statistically significant in the last three survey months. This unexpected finding seems to have been the result of the longer CATI interview and, to a lesser extent, computer downtime.

The CATI interviews were approximately 20 percent longer than the comparable paper and pencil interviews. This permitted the paper and pencil interviewers to attempt more calls in the same time period, especially in the most productive interviewing days early in the week. For example, on Mondays the two methods completed approximately the same percent of calls made, as shown in Figure 27.1; but 13 percent more

Table 1. Response Rates by Interviewing Method (February to September)

	Paper and Pencil	CATI	Percentage Difference	SD of Difference
Complete	86.3	84.2	2.1*	1.02
Refusal	8.0	7.8	0.2	0.84
No Contact/Absent	4.8	7.6	2.8*	0.53
Other	0.9	0.3	0.6*	0.19

*Statistically significant at the 5% level.

paper and pencil interviews were completed because more calls could be made with the shorter paper interviews. Later in the week, as productivity for both methods declined, the CATI sample therefore had more households remaining to be called.

The effects of the longer CATI interview on the contact rate diminished in later months when, as shown in Figure 27.2, the average length of interview declined for both methods. In later months, the difference between CATI and paper noncontact rates was negligible. Across all eight months, the number of calls required to complete an interview was virtually identical, 2.7 for CATI and 2.6 for paper and pencil. The longer CATI interview may require more interviewers to obtain the same response rate as the paper and pencil interview.

Computer downtime may have been a minor but contributing factor. Traditionally, Monday to Wednesday were the most productive interviewing days. If the number of calls made during this period was limited for any reason, it was difficult to recover. Our CATI system ran on a minicomputer shared with other users. Although the system was reliable, it experienced some downtime in the first few months of interviewing. Losing just a few hours of interviewing from Monday to Wednesday had a far greater impact than losing hours on a Thursday or Friday. In a typical month, approximately 300 interviewer hours were required to complete the sample. In one month, we lost four interviewing hours on a Monday when eight interviewers were on staff. Although this was a loss of only 10 percent of the total interviewing time, numerous calls on Friday could not compensate for the loss of calls on Monday. The impact over the eight months of interviewing was small, but when downtime did occur, recovery from it was difficult.

The longer CATI interview resulted in fewer calls being made to active cases at the beginning of the week. Later in the week, the CATI scheduler seemed to ensure that a high proportion of CATI cases were called each day. As shown in Figure 27.3, the paper and pencil

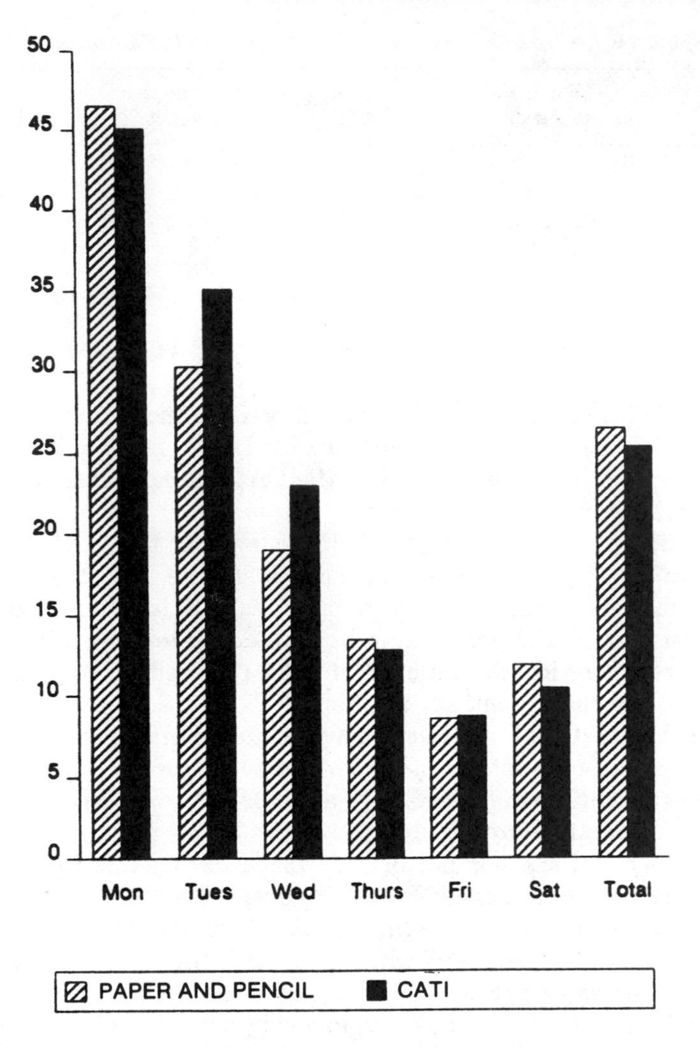

Figure 27.1 Completed interviews as percentage of calls made, by day and method.

interviewers tended to call a subset of the active cases more frequently. The CATI households that were not contacted in a given month were not the same households from month to month. Overall, 92.1 percent of households responded at least once during their time in the sample for both the CATI and the paper methods, even though in any given month the response rate for CATI was lower.

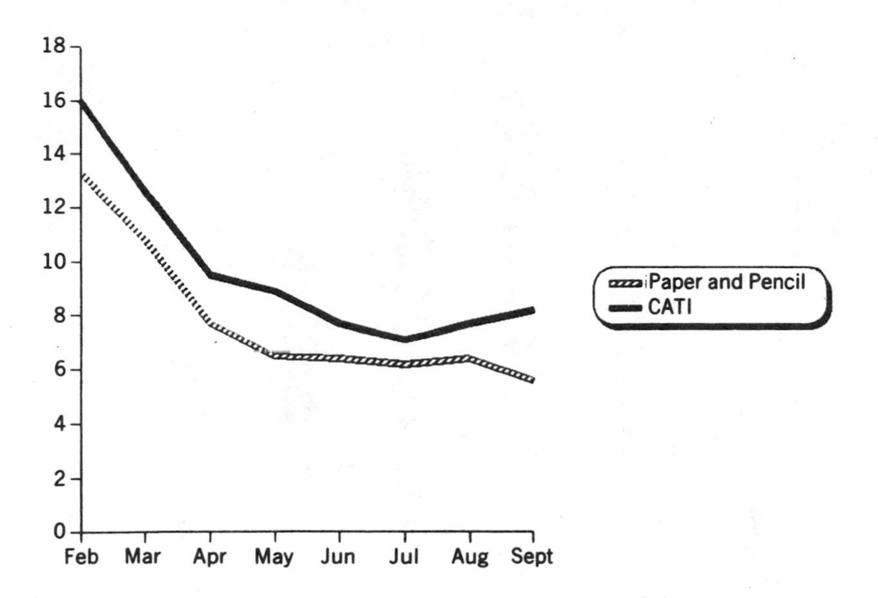

Figure 27.2 Average length in minutes of completed interviews, by month and method.

2.2 Refusals

Previous studies are inconclusive about the impact of CATI on refusals (Nicholls and Groves, 1986). The interviewers' lack of familiarity and confidence with CATI may have some negative impact (Nicholls, 1978), but no data are available on whether CATI affects the refusal rate when interviewers are equally adept at both conventional and CATI interviewing. Table 1 shows no significant difference between the two collection methods. Although we speculated that the more structured CATI approach might provoke more refusals during an interview's first few critical minutes, this did not occur.

Interviewer comments to respondents that they were entering the information into a computer and the sound of the keyboard had no apparent effect on the respondents' willingness to cooperate. Although some month to month variations occurred in the refusal rates, there were no significant differences over the eight months of data. It seems that CATI had no impact on the survey's refusal rate.

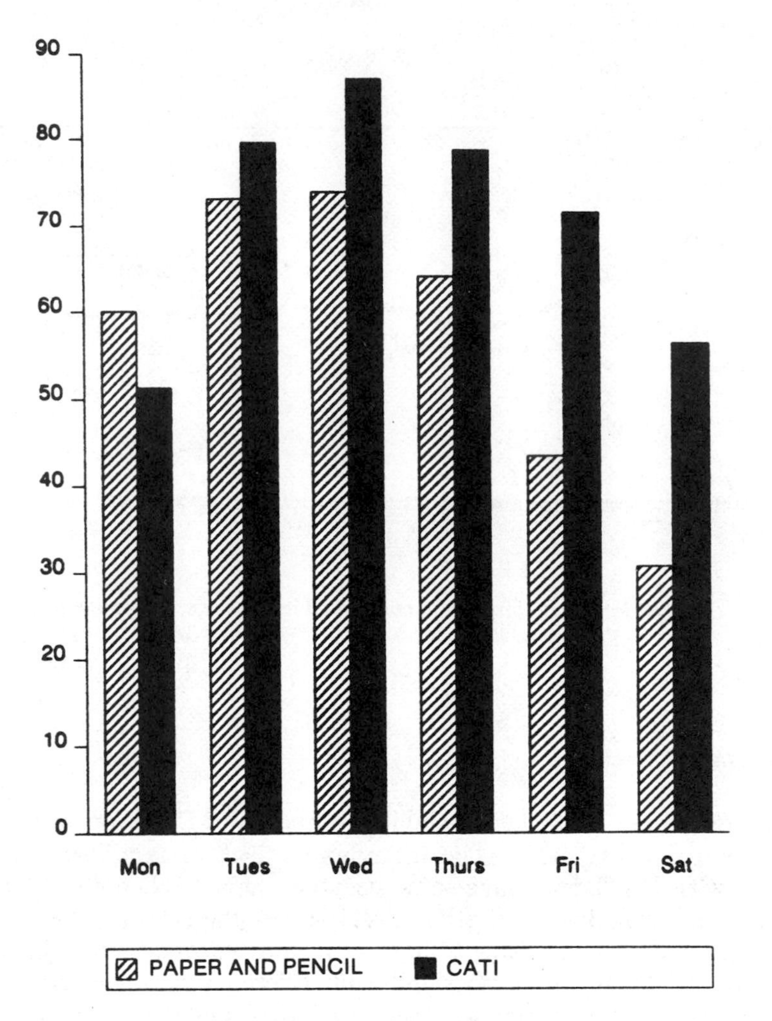

Figure 27.3 Percentage of active cases called, by day and method (February to September).

3. INTERVIEW QUALITY

Two components of data quality will be examined: (1) item nonresponse and data inconsistency as measured by item error rates; and (2) the codability of responses to a series of industry and occupation open questions.

3.1 Item Nonresponse and Consistency Errors

An advantage frequently attributed to CATI is its rigid control of the interview process (Groves and Nicholls, 1986). The interviewer is forced through the appropriate question sequence and required to make an entry for each question applicable to the respondent. While this theoretically keeps interviewers from missing questions, it does not ensure that a meaningful answer can be determined by the interviewer and respondent. Although a CATI system could exclude "don't know" and "refused" from acceptable entries, in practice these responses are allowed in CATI just as in paper and pencil interviews. In a study, using the same questionnaire for CATI and paper and pencil, Groves and Mathiowetz (1984) found that the levels of item nonresponse (combined "don't know," "refused," and "not ascertained") for a series of demographic and income questions were about the same.

The largest effect of CATI's rigid control of the interview process may be its automatic control of the question sequences. This should be especially true for applications with complicated branching where CATI removes the burden of implementing the questionnaire logic from the interviewer. In the same study noted above, Groves and Mathiowetz compared CATI and paper interviewing for a series of 28 questions involving complicated branching. They found that 1.8 percent of the CATI entries were inconsistent compared with 8.8 percent of the non-CATI entries.

The LFS processing system edits responses for both validity and consistency. An edit report is produced providing the discrepancy or edit failure rate for each item, that is, the percent of its entries that are inconsistent or blank, where "don't know" and "refused" count as blank. These discrepancy rates reflect the total error from all sources during collection and data entry. While many sources are common to both modes, such as misunderstanding questions and misrecording answers, different types of errors are possible in the two methodologies. CATI may have instrument or programming errors, while the paper and pencil method may have errors in its separate data entry stage. As discussed earlier, CATI's use of the computer should reduce or eliminate several types of error.

Table 2 compares the CATI and paper and pencil discrepancy rates by month for the Household Record Docket and the Labour Force Questionnaire.

The similar discrepancy rates by mode for the Household Record Docket reflect similar proportions of "don't know" and "refused" in the CATI and paper and pencil methods. The Household Record Docket does not have a complicated branching structure. Household membership and

Table 2. Discrepancy Rates by Interviewing Method (February to July)

Month	Method	Household Record Docket	Labour Force Questionnaire	Overall
February:	CATI	0.7	2.0	1.5
	Paper and pencil	1.0	4.0	2.7
March:	CATI	0.5	2.6	1.7
	Paper and pencil	0.9	7.7	4.7
April:	CATI	0.4	1.7	1.2
	Paper and pencil	0.7	6.3	3.8
May:	CATI	0.5	1.8	1.2
	Paper and pencil	0.6	3.2	2.0
June:	CATI	0.8	1.4	1.2
	Paper and pencil	0.7	3.6	2.3
July:	CATI	0.7	1.6	1.2
	Paper and pencil	0.6	3.5	2.2
Total:	CATI	0.6	1.8	1.3
	Paper and pencil	0.7	4.6	2.9

demographics are collected in a tabular format for paper and pencil, giving interviewers flexibility in the order in which the data are collected. CATI necessitated the structuring of this portion of the interview, but this involved only simple question sequencing. Although some consistency edits were built into the CATI instrument, the more complex consistency checks, such as on family composition, were not included.

In contrast, the Labour Force questionnaire has a very complicated branching logic. The differences in the discrepancy rates by mode for this form largely reflect the effect of CATI's automatic control of the question sequences. The paper and pencil error rate is at least double the CATI rate in most months and is three times as large in some months. In one series of Labour Force questions, 22.2 percent of the paper and pencil entries, but only 2.0 percent of the CATI entries, were in error. The residual CATI discrepancy rate is not a result of skip pattern errors. The discrepancy rates for both interviewing methods include failures to obtain a meaningful answer.

3.2 Recording Responses to Open Questions

While CATI's control over the question sequence may reduce some types of response errors, there is concern that it may increase other types of error. Most speculations about increased response errors in CATI deal with open questions, where interviewers must type in the responses (Groves and Nicholls, 1986). Because interviewers are unable to type the entries quickly enough, typographical errors may occur or the entries may be incomplete or unintelligible. Paper and pencil interviewers often review the responses to open questions for their completeness and legibility after the respondent is no longer on the telephone; but our CATI instrument required a complete entry as each question was asked. No previous studies have compared CATI and paper and pencil textual entries to open questions in the same survey.

In this study, a measure of recording completeness is available for a series of industry and occupation open questions used in the assignment of industry and occupation codes. As a first step, the files are run through a computer module which attempts to assign the appropriate codes automatically. Table 3 compares the rates of cases automatically coded by interviewing method.

The comparisons show no particular pattern. The rates are equal in some months, while one method or the other shows some advantage in other months. Neither the total nor the individual month differences are statistically significant. The number of words used to describe the industry or occupation was also studied. Over the five months of data thus far examined, the average number of words recorded by CATI and paper and pencil was virtually identical. It would appear there is no systematic mode difference in recording answers to open questions.

Table 3. Coding Rates by Interviewing Method (February to July)

Coded Automatically	CATI	Paper and Pencil
February	31.7	29.2
March	33.5	31.6
April	29.7	32.4
May	34.2	29.9
June	28.1	35.2
July	32.2	33.8
Total	31.8	31.5

4. COST

At least partial justification for CATI has been that it combines data capture and interviewing and, therefore, should be cheaper. The problem, however, is more complicated than this.

Some researchers speculate that CATI will increase the length and, therefore, the cost of interviews. Harlow, Rosenthal, and Ziegler (1985) reported that interviews were approximately 14 percent longer with CATI. Other authors have suggested that experienced paper and pencil interviewers begin asking the next question while recording the last and may enter responses to open questions more quickly (Nicholls and Groves, 1986; Groves and Mathiowetz, 1984). On the other hand, CATI call scheduling decreases paperwork and supervision (Coulter, 1985). The net effect of CATI is not clear.

This evaluation included only those costs incurred once the survey infrastructure was present. These include interviewers' and supervisors' salaries, long distance telephone charges, computer costs for CATI interviewing, and data capture costs for paper and pencil interviewing. The costs were analyzed both as a function of the number of households contacted and the number of interviews completed. As shown in Table 4, the overall differences are small, but CATI appears to be slightly less expensive per household contacted and slightly more expensive per interview completed.

Table 4. Cost (in $ Canadian) per Unit by Interviewing Method (February to September)

	CATI	Paper and Pencil	Percentage Difference
Interviewers	2.87	2.74	+5
Supervision	.63	.73	−16
Data Capture/Computer	.54	.81	−33
Telephones*	1.52	1.40	+9
Total per household	5.55	5.68	
N	(7,076)	(6,934)	−2
Total per completed interview	6.63	6.58	
N	(5,963)	(5,986)	+1

*Estimates for February, July, August, and September.

As suggested in the literature, supervision is less costly in a CATI environment. Supervisors spend less time with the scheduling of calls and the associated managing of documents. This was confirmed by debriefing of the supervisors who found their workload during a week of CATI interviewing much less onerous. The relatively small size of each monthly sample may underestimate the saving in supervision that can be achieved.

The interviewer component of costs in Table 4 is less straightforward. The average length of a completed interview was two minutes longer for CATI. This is consistent with results obtained by Harlow et al. (1985). Despite the longer CATI interview the costs for CATI and paper and pencil interviewing were virtually equal per household contacted. The paper and pencil interviewers spent more time between calls editing the completed interviews and determining which household to call next.

We did not find that paper and pencil interviewers recorded open question responses more quickly than CATI interviewers, although some writers have speculated that this could affect the length and cost of CATI interviews (Nicholls, 1978). In an informal attempt to measure this, we monitored interviews and handwrote responses on a questionnaire as the CATI interviewer entered the names of household members and industry and occupation descriptions on a keyboard. It was rarely possible to match the speed of the CATI entry unless the interviewer probed for clarification or spelling. Most of the interviewers hired for the project had some keyboard experience, and this certainly was a factor in their speed of entering information. Others were encouraged to use a microcomputer package that taught typing skills.

We also observed that the interviewers soon became comfortable with the flow of the questions, even when using CATI, and began asking the next question while entering the answer to the last. This may not occur in surveys of brief duration, but it was certainly a factor in this application.

Although CATI has no separate data entry cost, there were charges for its shared use of the minicomputer. When these costs were compared to traditional data entry for paper and pencil, the CATI costs were considerably less. Telephone charges were higher for CATI as would be expected with its the longer interview.

Based on total cost per completed interview, CATI is slightly more expensive. The total cost per sample household is approximately equal due to the larger proportion of CATI noninterviews.

5. CONCLUSION

Our data have generally confirmed the results of other CATI investigations and the optimism that has encouraged the proliferation of

CATI systems in survey research. The major impact of CATI will be in terms of data quality. In this evaluation a paper questionnaire was translated to a CATI instrument with little alteration. Even in this situation, the item based error rates were reduced with CATI by 50 percent overall and by as much as two thirds in the more complex of the two questionnaires included in the survey. In the LFS, each month of collection is independent, not relying on previous months' data except as an aid to interviewers. Surveys that employ previous data or edit against historical information should show even greater benefits.

The response rate for CATI in our evaluation was lower. There was no difference in the refusal rate but the noncontact rate was higher until the last three months of interviewing. This would seem to be an impact on cost rather than quality. During the last three months, as the average length of interview decreased for both methods, the difference in the noncontact rate was no longer evident. More interviewers may be required due to the slightly longer interview.

The data obtained to date leave little doubt that CATI interviewing will be somewhat more expensive due to slightly longer interviews. However, savings are realized in reduced time between interviews, in reduced supervision, and in the combining of collection and data capture in CATI. Additional cost information will have to be collected and analyzed before reaching final conclusions. The overall impact on individual surveys may depend on their size, duration, and the ratio of supervision to interviewing staff.

SECTION F

ADMINISTRATION OF TELEPHONE SURVEYS

CHAPTER 28

INTRODUCTION: ADMINISTRATION OF TELEPHONE SURVEYS

Lars E. Lyberg
Statistics Sweden

Every telephone survey should follow a step by step process that includes problem formulation, sample size calculation, questionnaire development, data collection, data processing (coding and data capture), methods of estimation and other analyses, and a survey report. To successfully carry out the process an administrative plan is needed. When the authors of Section F met and discussed the contents of the section, it was suggested that an alternative heading of the section might be: administration and operation of telephone surveys. This suggestion perhaps reflects the fact that often the administrative issues are not easily defined or easily distinguished from other survey issues. By and large one might say that administration is planning *and* implementation.

The literature on the administration of telephone surveys (or other modes for that matter) is scarce. This is so, partly because of a confusing terminology (some administrative issues are, in fact, dealt with under various labels or the administrative issues are part of an overall telephone survey plan), and partly because some administrative issues are not considered scholarly material by journal editors and researchers. However, in his book on telephone survey methods, Frey (1983) has devoted a chapter to administration. There he discusses issues like timing of callbacks, assignments, interviewing, supervision, technical monitoring, and budgeting. To illustrate his approach he provides a flowchart of telephone survey tasks. Another important reference is Groves and Lepkowski (1985) where four optional administrative structures for conducting large scale mixed mode surveys are discussed. Lavrakas

(1987) provides advice on how to run a telephone survey, including the not so glamorous but important details of preparing interviewers' workspace so that work can be started up efficiently, and advising supervisors about how to deal with unexpected problems.

The crucial administrative issues are those which can have an important impact on errors and costs. The design of the data collection system is such a crucial issue. Here, the basic choice is between a centralized and a decentralized facility. A common opinion is that centralized facilities offer advantages regarding costs, quality control, and balancing interviewer workloads. But centralization can have drawbacks, too. Too much specialization can have negative effects on morale and data quality; in mixed mode surveys centralization of only some of the interviewing may increase costs; and "long-distance" telephone interviewing may emphasize differences in cultural background and dialect between interviewers and respondents, leading to increased nonresponse rates.

Another crucial administrative issue is error control. Here the recruitment, training, and monitoring of interviewers are imperative. Large agencies and organizations, which have more or less permanently employed interviewers, are in a much better position for error control than smaller organizations, which usually have to rely on temporary interviewers. Organizations with temporarily employed interviewers must recruit more, and more of the training has to be basic rather than study-specific.

An interesting administrative problem is the choice of calling times to save money and to keep response rates at a reasonable level. Several organizations have conducted studies to find out when respondents can most easily be contacted. Some general guidelines can undoubtedly be provided. However, population characteristics (e.g., subpopulations with distinctive patterns of being away from home) and survey designs (e.g., permitting proxy response) can play a role. Contact patterns can also exhibit temporal variation (over weeks, months, years). Attempts to find optimal contact times are found in U. S. Bureau of the Census (1972), Vidgerhous (1979), Weeks et al. (1980), and Weeks (this volume, Chapter 25).

Administration also includes issues like coordinating the schedules of different parts of the survey activities (e.g., sampling, questionnaire development), identifying the personnel required, assigning staff responsibilities, preparing progress reports (sometimes required on a daily basis), and monitoring costs of the data collection.

The transition to CATI presents new administrative problems, and since the technology is new to many environments, organizations have adopted a variety of solutions to them. CATI demands new skills associated with the computer, new staff relationships, and new interaction patterns. The coordination of design and timing has become more crucial with the advent of CATI. New staff members have sometimes been added for CATI studies. As Berry and O'Rourke (this volume, Chapter 29) note, two basic models have emerged, "project" and "department," but in some organizations hybrid models exist. No single model works best in all survey situations.

It is not uncommon that policy decisions made by management for other reasons affect or limit an organization's administrative possibilities and its likely operational efficiency. For example, at Statistics Sweden a policy decision was made to have a decentralized interviewer organization supplemented by a relatively small centralized telephone interviewing group. The decentralized interviewers conduct telephone interviews from their homes, while the centralized group mainly deals with small special surveys and follow-ups on nonresponse cases. All development work and choice of administrative models was done within this restricted framework. Because data collection organizations are often part of larger organizations with other goals, survey researchers are often not free to optimize the administrative structure of the survey activities.

Chapter 29, by Berry and O'Rourke, gives an account of how various survey organizations in a number of countries operate. The information is based on a survey of 27 organizations conducted by the authors. Most of these organizations used CATI, and the chapter concentrates on administrative issues that are the result of the introduction of the new technology. Chapter 30, by Cannell and Oksenberg, deals with administrative issues in monitoring. These issues include: selection of monitors, what to monitor, frequency of monitoring, costs of coding interviewer behaviors, training of monitors, and feedback to interviewers. Chapter 31, by Bass and Tortora, presents the results of a case study of a special administrative structure (in fact a modification of structure IV discussed in Groves and Lepkowski, 1985). Bass and Tortora include in their study a comparison of response rates, survey statistics, and variances by facility. Chapter 32, by Dillman and Tarnai, deals with administrative issues in mixed mode surveys, where one of the modes is telephone. Mixed mode surveys present administrative demands that go well beyond those of procedures for single modes. The authors identify and discuss eight administrative issues important to a variety of mixed mode surveys. These issues are data comparability, questionnaire construction and pretesting, personnel requirements, sample design and respondent selection, costs, centralization vs. decentralization, interfacing the modes, and data processing and analysis.

Research on and documentation of administrative issues have been neglected over the years. This section does not cover all issues relevant to the administration of telephone surveys. The most obvious omission is a treatment of survey costs. We have not been able to present research or documentation on the costs of telephone data collection, budget monitoring, the balancing of errors and costs, etc. It is hoped that the chapters of this section, however, will serve as a stimulus for future efforts on these and other administrative issues in telephone survey methodology.

CHAPTER 29

ADMINISTRATIVE DESIGNS FOR CENTRALIZED TELEPHONE SURVEY CENTERS: IMPLICATIONS OF THE TRANSITION TO CATI

Sandra H. Berry
The Rand Corporation

Diane O'Rourke
The University of Illinois

Administrative designs for survey research centers have changed, both as a result of the use of telephone methods in general and also as a result of the use of computer-assisted telephone interviewing (CATI). As telephone interviewing became a more feasible and popular alternative to face to face interviewing, survey organizations were able to centralize more of their data collection staff. The advantages of centralization are numerous. In addition to increased quality control (Dillman, 1978; Frey, 1983) come all of the improvements in communication from having the research, interviewing, and data processing staffs in close proximity (Groves and Kahn, 1979). In a 1979 survey of survey research organizations (Spaeth, 1979), over 80 percent of the responding organizations that conducted telephone interviewing had centralized telephone facilities. Administratively, this centralization increased the number and layers of personnel at the central office (field supervisors and interviewers), but most relationships between the data collection section and other sections could remain the same as they were during the era when face to face interviewing was the dominant survey method.

In the 1970s and 1980s, the use of CATI has become common. Among those responding to the Spaeth (1979) survey, 10 percent of the academic organizations and 30 percent of the commercial organizations were using CATI in 1979. By 1987 over 40 percent of the academic organizations responding to a similar survey were using CATI (Spaeth, 1987) and, given the lead that they took in its inception, it is likely that a vast majority of commercial firms are now using it.

The introduction of CATI has created some organizational problems for centralized telephone facilities. Traditionally, interviewers represented the main resource, and such facilities were organized and staffed to perform a primarily interpersonal or people management task. The introduction of CATI meant that a quite different, technical resource had to be managed — hardware and software used in a complicated interaction with interviewers and other staff. In most cases, the managers of pencil and paper surveys were not familiar with CATI and were not adept at dealing with the hardware and software problems that emerged. On the other hand, the programmers were not accustomed to dealing with interpersonal issues such as interviewer training and management.

Such issues have been dealt with only briefly in the survey methodology literature. One of the earliest contributions was made by Groves (1983), who discussed how the new technology should be injected into an existing survey unit. Particularly in large organizations where there are departments devoted to specific phases of the survey, problems arise regarding what unit should be responsible for the machine and how the existing units can best make the transition to their greater interdependence due to the CATI system.

Despite the fact that staff organization and administration can be central to the success or failure of a technological change such as CATI, very little has been written on the topic. Most of the literature on CATI relates to its technical aspects — how it could change procedures or products — with little or no mention of people and roles (e.g., staffing requirements, task assignments, and role delineation). Where "people" have been discussed in the CATI literature, they have mostly been the interviewers, interviewer supervisors, and occasionally the respondents (Groves et al., 1980; Morton and House, 1983; Sudman, 1983; Groves and Mathiowetz, 1984).

CATI requires different skills. All persons who work on a CATI project should be "computer friendly" (more or less friendly depending on how friendly the software is). The need for this additional skill is most evident in the area of questionnaire design. Questionnaire designers now need to understand the characteristics of a well designed CATI questionnaire — its setup, edit checks, data records, etc. (House, 1985; House and Nicholls, this volume, Chapter 26; Jabine, 1985; Nicholls and Groves, 1986).

Organizational response to this change can vary: the questionnaire designer can learn the computer skills necessary to program the questionnaire, the questionnaire designer can turn the draft instrument over to a programmer, or, if the software is sufficiently friendly, a clerical intermediary can computerize the appropriate translations. For example, Westat, Inc. (1983) has used what it calls "authors" or nonprogrammers who "translate a survey questionnaire into the command language and enter the code into the computer." In addition, field and sampling staff must learn to relate to CATI, since CATI affects case management functions such as sampling, call scheduling, and progress report production (Nicholls and Groves, 1986).

Another impact of CATI on organization is that CATI requires more interaction among staff, who must now relate more often and more intensively than in the past (Groves, 1983; House, 1985; Nicholls and Groves, 1986). In a non-CATI environment, it is possible for discrete staffs involved in sampling, field operations, coding, and data processing to maintain fairly narrow and well defined relationships. CATI now defines new relationships and reorders some tasks. With everyone dependent on the machine, coordination of design, as well as timing, is more important than ever (Groves, 1983). This increased interaction and interdependence can result in administrative problems beyond those that were normally experienced in survey organizations.

Of course, problems associated with changes introduced specifically by CATI can be seen as a variant of the more general problems of diffusion of technology within an organization, and research findings from other settings can be instructive. For example, in a study of 26 office settings where computerized procedures were introduced, a major finding was that the implementation problems were not inherently computer problems. Rather, the problems were system characteristics that management could readily fix, such as unhelpful technical manuals and unaccommodating office environments (Bikson and Gutek, 1983, p. 326). In addition, the authors found that two features of organizational characteristics — variety in individual's work and the organization's orientation toward change —predicted how completely users integrated computer systems into their work and how happy they were with them. White collar workers in organizations that viewed change as a positive, problem solving, and achievable goal used their systems more and indicated greater satisfaction with the systems. Interestingly, Bikson and Gutek did not find a consistent relationship between satisfaction and either system use or productivity.

1. SURVEY OF TELEPHONE SURVEY CENTERS

In order to determine how telephone centers operate, with and without CATI, we conducted a small scale survey of telephone centers. In an effort to capture variation in CATI usage, two sample lists were used. We sent questionnaires to the main U.S. organizations reporting CATI use in the Spaeth survey (1987) and a list of primarily European organizations. Characteristics of the sample and the responding organizations appear in Table 1. Of course, the sample is not a probability sample of all survey organizations. For example, commercial organizations are poorly represented. Consequently, any inferences from the results should be made only to this sample.

1.1 Questionnaire

Information sought in the questionnaire appears in Table 2 and includes background information about the CATI facility, an organization chart of the telephone survey center, a description of skills required in positions listed in the organization chart, a breakdown of who performs specific tasks, and a description of changes due to the introduction of CATI.

1.2 Characteristics of the Telephone Centers

As shown in Table 3, the responding survey organizations represent a wide range of variation in characteristics that relate to CATI — including years of CATI experience, number of CATI studies and interviews completed, size in terms of number of stations, and kinds of hardware and software.

Table 1. Survey Sample and Responding Organizations

Location and Type	Sample (n=53)	Responding Organizations (n=27)	Responding Organizations That Use CATI (n=23)
Location			
United States	24	18	16
Non-U.S.	29	9	7
Type			
University based	21	13	12
Government	18	4	3
Private research	14	10	8

Table 2. Questionnaire Information

(1) Background data
- When the organizations began using CATI
- The hardware, software, and operating systems utilized
- The tasks that CATI is used for (RDD sample generation, call scheduling, etc.)
- The typical and largest number of interviewing stations utilized
- In 1986, the number of CATI interviews completed, the percentage of telephone interviews done on CATI, and the typical and maximum number of items and minutes in a CATI interview

(2) An organizational chart
- Indicating levels of relationships from the director of the center to the interviewers
- Indicating who reports to whom

(3) Position skills required
- For each position listed in the organizational chart, whether skills were required in the areas of (a) management and supervision, (b) administration, (c) survey methodology, (d) sampling or statistics, (e) general programming, (f) CATI programming, and (g) survey operations
- In addition, respondents were asked to categorize the personnel in each position within broad salary ranges, as another indicator of level of expertise

(4) Tasks
- Who is responsible for, spends the most time at, and participates in the following tasks on CATI and pencil and paper studies:

	CATI	Pencil and Paper
Deciding whether or not to use CATI (vs. pencil and paper)		x
Developing a cost estimate	x	x
Scheduling the project	x	x
Monitoring costs and schedule	x	x
Designing and pretesting the instrument	x	x
Putting the instrument on CATI	x	
Testing/debugging the instrument on CATI	x	
Selecting interviewers	x	x
Training interviewers	x	x
Supervising interviewers	x	x
Maintaining quality control	x	x
Sample management	x	x
Assigning cases and retiring cases	x	x
Setting decision rules for the scheduler	x	
Running the scheduler	x	
Performing data reduction		x

(5) Changes due to CATI in
- Where and how telephone center staff are recruited
- Need for management or supervisory staff retraining or professional development
- How interviewers are selected and trained
- Salary levels for telephone center staff
- Career patterns for telephone center staff

(6) Miscellaneous problems and comments

1.3 Survey Center Organization

Respondents were asked to draw an organizational chart, starting at the top with the director of the center, showing positions through the interviewer and clerical level, and indicating with connecting lines who reports to whom. The latitude and longitude of departments and positions vary greatly among centers, matched only by an equally diverse range of job titles in relation to level in organization, responsibilities, skills, and salary.

From the 25 organizational charts submitted by the responding centers, we discovered that these organizations were layered in from four to eight tiers. The configuration of those layers varies in the level at which general survey management and administration for the center as a whole becomes the positions concerned with the field department or data collection project management. Table 4 shows nine examples of such configurations.

Table 3. Characteristics of Responding Centers that Use CATI ($n = 23$)

First CATI study	
Prior to 1980	3
1980 – 1984	13
1985 – 1987	7
Number of CATI studies conducted	
Fewer than 6	6
6 – 20	6
26 – 50	4
51 or more	7
Number of CATI interviews in 1986	
Fewer than 1,000 interviews	4
1,000 – 9,999	10
10,000 +	9
Number of CATI stations "typically" used	
10 or fewer	13
11 – 39	6
40 or more	4
Hardware	
Mainframe or minicomputer	13
Super microcomputer	3
Personal computer (PC)	7
Software[a]	
Berkeley "CASES"	5
Wisconsin "CASS"	5
Amrigon-based	3
Quantime	3
Developed in-house	7
Other	2

[a]Two organizations use two kinds of software.

Among the 25 organizations, eight contain four layers from the director to the interviewers, nine contain five layers, and the other eight are comprised of six or more layers.

The number of layers appears to be only somewhat related to the size of the organization. The responding centers were asked to number each position of their organizational charts and then indicate the number of individuals (not full-time equivalents) in those positions. Counting only the number of individuals between the director and the interviewers (those in that vertical segment of the organization), one can view the shape of that segment of the hierarchy. Clearly, many of the centers are not organized in pyramid fashion. For example, while one organization has 28 people in four layers in positions between the director and interviewers, another organization has eight people in six layers. Larger organizations do not necessarily have more layers than small or medium-sized organizations; in fact, some of the larger organizations have fewer layers.

1.4 Frequency of Types of Skills

The 27 centers provided descriptions of the skills required for a total of 296 positions in their organization charts and we examined the distribution of skills required for those positions. Interestingly, the most common skills required in those positions were management and supervisory, which were required in 63 percent of the positions. About half the positions required administrative or survey skills, including survey methodology (study and instrument design) or survey operations (interviewing, coding, etc.). Sampling, CATI programming, and general programming skills were required in slightly over a fifth of the positions, as shown in the first column of Table 5.

Of the 296 positions, 277 could be associated with a count of individuals who filled those positions, which ranged from one to 800. We recalculated the distribution of skills, weighting by the number of individuals in each position. Not surprisingly, survey operations skills are most frequently represented among individuals occupying survey center positions, about 83 percent of the individuals. The remaining skills are required by 20 to 29 percent of the individuals, as shown in the second column of Table 5.

Table 4. Layers of Organization by Title: Nine Examples

FOUR-LAYER ORGANIZATIONS

1.	Director	Director
2.	Field Director	Assistant Director or Manager
3.	Field Supervisors	Supervisors
4.	Interviewers	Interviewers

FIVE-LAYER ORGANIZATIONS

1.	Director	Director	Director	Director
2.	Operations Manager	Operations Director	Survey Research Director	Data Collection Manager
3.	Head Supervisor	Study Directors	Project Managers	Operations Supervisor
4.	Supervisors	Shift Supervisors	Supervisors	Shift Supervisors
5.	Interviewers	Interviewers	Interviewers	Interviewers

SIX-LAYER ORGANIZATIONS

1.	Director	Director	Director
2.	Field Manager	Site Manager	Assistant Director
3.	Telephone Center Manager	Field Head	Project Coordinators
4.	Head Supervisor	Site Field Manager	Field Supervisor
5.	Shift Supervisors	Supervisors	Telephone Supervisors
6.	Interviewers	Interviewers	Interviewers

Table 5. Frequency of Skills by Position and Individuals in Telephone Survey Centers[a]

Type of Skill	Percent of Positions	Percent of Individuals
Management and supervisory	63.8	28.5
Administration	48.6	25.1
Survey methodology	42.9	26.9
Survey operations	54.4	82.7
Sampling statistics	23.3	21.0
General programming	21.2	23.3
CATI programming	23.0	20.5
Number of positions	296	277
Number of centers	27	26[b]

[a]Includes CATI and non-CATI centers.
[b]One center did not indicate number of individuals in each position.

1.5 Differentiation of Skills

Positions in telephone survey centers often require multiple skills. In our sample, the average number of skills required per position was 2.8 (std. dev. = 1.3). The distribution of these skills is not random. Table 6 shows the percentage of positions requiring each pair of skills. For example, 44 percent of the positions require management and supervision combined with administrative skills. Both management and administration are combined with survey methods skills in slightly less than a third of the positions. Other combinations occur less frequently, but have interesting patterns. Sampling and statistical skills are associated with all but the survey operations skills, indicating that this position requires familiarity with all parts of the survey process. All of the statistical and programming skills are associated with each other and with survey method skills.

Based on the descriptions of which positions were responsible for various tasks in the telephone centers, we looked for an explanation of the variation we observed.

1.6 Project Versus Department Models of Assigning Tasks

One source of variation in the assignment of tasks seems to be in the "department" compared with the "project" model of directing work on

Table 6. Percent of Positions Requiring Pairs of Skills in Telephone Survey Centers

	Admin-istration	Survey Methods	Survey Operations	Sampling and Statistical	Programming	
					General	CATI
Management and supervision	44	31	33	18	11	13
Administration		32	20	15	11	12
Survey methods			21	20	13	15
Survey operations				13	9	13
Sampling statistics					9	9
General programming						15

telephone survey projects. In some centers the work is organized around studies or "projects" and is managed by study or project directors, who take complete responsibility for coordinating the tasks on a project by project basis. In an extreme example of that model, there is no permanently staffed telephone center, and telephone surveys are carried out by groups of staff members put together specifically for each survey project that is carried out. Under the "department" model, work on projects is divided by functional area, e.g., instrument design, interviewing, and data reduction. The extreme of the department model is a permanent telephone center that manages all telephone surveys carried out at that center, with little or no involvement from the survey design and management staff once the questionnaire is designed and basic specifications are agreed on.

Each of these models has advantages and disadvantages. The project model allows for flexibility in carrying out different kinds of projects and ensures that work in each phase is carried out in a way that is compatible with other phases. The department model allows more efficiency and more uniformity across projects and makes it easier to build a body of skills in a staff over time. To some degree these models are associated with size; sustaining the department model requires a fairly large and stable work flow. It also works best when projects are fairly similar in character.

Only a few centers appeared to be at either extreme; most operated a permanent facility at some level and also involved project design staff in the interviewer training and the management of the telephone survey operations. As indicators of this division of tasks, we examined two key tasks, the decision of whether or not to use CATI for a particular survey (relevant for centers that used both methods), and budgeting of the project (for CATI projects). Focusing on the position designated as "responsible for" these tasks, we found the distribution of responsibility shown in Table 7.

While directors and associate directors in the centers we surveyed are responsible for much of the shaping of the projects they manage, this responsibility is not located exclusively in those positions. In about a quarter of the centers study staff are responsible for the decision about the use of CATI and the project budget. In a quarter of the centers, field staff make the CATI decision, and in one center they are also responsible for the development of the project budget.

1.7 Instrument Design for Pencil and Paper and CATI Surveys

The introduction of CATI to a pencil and paper facility usually requires the incorporation of different skills and more interaction among staff. As an example of how CATI affects a center's organization, we looked at instrument design and CATI programming tasks.

Table 7. Location of Responsibility for Key Tasks

	Number with Responsibility for	
Locus of Responsibility	CATI Use Decision	Budget for Survey
Director or associate director	10	10
Administrator	0	8
Study staff	5	5
Operations staff	5	1
Total	20[a]	24[b]

[a]Includes centers that do both CATI and pencil and paper surveys.
[b]Includes centers that do CATI and/or pencil and paper surveys.

House and Nicholls point out elsewhere in this volume (Chapter 26) that CATI instrument design is significantly different from pencil and paper instrument design. However, as shown in Table 8, many centers have not changed their basic approach to who performs these tasks.

In most centers that do either pencil and paper interviewing, CATI interviewing, or both, the same staff designs the initial instrument. In slightly over half the centers the director or assistant director of the center is responsible for this task. In most of the remaining centers, the

Table 8. Location of Responsibility for Instrument Design and Programming[a]

	Design Task		
Locus of Responsibility	Pencil and Paper Questionnaire Design	CATI Questionnaire Design	CATI Questionnaire Programming
Center director or assistant director	12	12	5
Project or study director	5	7	4
Field director and staff	1	2	3
General or CATI programmer	0	0	9
Total	18	21	21

[a]Based on four centers that do only pencil and paper telephone surveys, 7 that do only CATI, and 14 that do both. Two centers provided no usable data.

person with responsibility for the study design and coordination designs the instrument. The field director or field staff take responsibility for instrument design in only a few centers. However, only half the center directors and study directors who design instruments also do the actual CATI programming. In the remaining centers, the instrument design is turned over to a programmer, or a member of the field department, to be turned into an actual CATI instrument. In some cases this transfer of responsibility probably represents a more efficient use of time by center directors and study coordinators who are thoroughly familiar with CATI, in others it may represent a lack of direct knowledge about the technical possibilities and requirements of CATI instrument design.

2. CHANGES BROUGHT ABOUT BY THE TRANSITION TO CATI

Our questionnaire included open questions about changes brought about entirely or in part by the use of CATI. Table 9 shows the frequency of responses.

Most of the changes discussed under the recruitment question dealt with changes in interviewer recruitment, specifically the need to recruit interviewers with the ability to use the keyboard and the ability to learn to use CATI, which we will discuss below. However, one center commented that they now recruit supervisory staff who already have programming or CATI skills.

Most of the centers had had a need for retraining of the existing management and supervisory staff. One center commented:

> The staff needs to know basic computer programs for controlling sample, reviewing data, monitoring, editing, processing data, etc. Also, how to deal with problem machines. We have minimal formal training for them and a "manual" with formulas.

Others commented that more supervisory staff was needed for CATI problem solving. Most centers indicated that they handled this by training existing staff; one mentioned that long term supervisors were having problems becoming comfortable with CATI. Only a few mentioned that they had solved the problem by hiring new staff who already had these skills.

Smaller centers commented that developing and maintaining the technical skills needed to use CATI was a problem. Technical training for their secretaries was one method they had used. Presumably secretaries are assured longevity within these organizations due to the broad usefulness of their nonsurvey skills.

Interviewer recruitment and training was the area where the most consistent types of changes were described. For example, one center commented that they had recruited an entirely new pool of interviewers to replace existing interviewers who did not wish to use CATI. Eight of the centers now look for typing skills and a few look for general computer experience or aptitude. At least one center now routinely tests applicants for such skills. A few mentioned more emphasis on educational background as an interviewer selection criterion.

The need for recruitment of interviewers with keyboard experience may be derived more from problems of strain in using CATI for response

Table 9. Frequency of Mention of Changes Dues to the Use of CATI

Type of Change	Number Indicating a Change[a]
Recruitment of telephone center staff	7
Need for management or supervisory staff retraining	16
How interviewers are selected and trained	16
Salary levels for telephone center staff	4
Career patterns for telephone center staff	4

[a]Base refers to 22 centers that had changed from pencil and paper to CATI.

entry than the need for extensive typing as part of conducting interviews. Typically, the typing demands of CATI systems are not great. However, Malt (1987, pp. 322–323) in a "holistic" analysis of the human-computer interface, draws a distinction between the use of computers for "copy-keying" (e.g., of text) and "think-keying" which seems relevant to CATI:

> 'Think-keying' is a term applied to authoring directly in the keyboard. Thinking in words is a function of the linguistic, the...Left Hemisphere [of the brain]. The words are then transferred to the Right Hemisphere for musculo-spacial translation into keystrokes. It is logical that if the Left Hemisphere is involved in originating the copy to be keyed, it will operate more efficiently if all finger actions are controlled by the Right Hemisphere... That so many executive PCs remain unused may well be due to the strain felt by those who are unable to touch key. If they cannot find characters on a keyboard without involving the left brain their thinking process is handicapped.

Interviewers' use of the keyboard for CATI falls into the "think-keying" category since the keyboard entries are being generated from conversations with respondents. Thinking about the keyboard entry process at the same time places a strain on their left brain resources; they simply cannot do the two things at the same time. Failure to understand this problem may lead the interviewers and their supervisors to conclude that they "just cannot do CATI," when their need is really keyboard practice for which there are several good self-study programs. One center termed it "a need to learn motor skills as well as computer skills."

Extra interviewer training was also reported by seven of the centers. They mentioned a time period of from one half a day to a full day of additional training and use of one tutorial (CASS) as an extra feature of training.

Four of the centers mentioned paying the staff more for work on CATI. The amount varied from "significantly higher, when [the supervisors] are good enough to compete on the programmer market" to "increased somewhat" for good interviewers. One mentioned conversion of CATI supervisors from part-time to full-time staff with benefits.

The same themes are echoed in the responses to the question on career patterns. One center commented that their supervisors "get computer experience and move out into better paying jobs." Another noted that familiarity with computers has enhanced everyone's career possibilities. A third said that supervisors now can "move into programming rather than just survey design and management."

These comments reflect a general need among telephone survey centers using CATI for more investment in the technical training of their staffs, either through recruiting staff with these (highly paid) skills or through funding additional training of existing staff. This investment makes each staff member more valuable, potentially raising the costs of retaining good staff through the need for higher pay, more job stability, and/or benefits that are not now standard in the field, at least not at the interviewer supervisor level. This puts an additional strain on centers whose workload is sporadic, as stated in several of the general comments. For example:

> Few years go by without a slack period of several weeks or months. It is becoming increasingly difficult to maintain experienced staff from which to draw our experts. Most long term employees want 'benefits' we cannot provide, i.e., health insurance, vacation, etc.

3. GENERAL OBSERVATIONS ON THE TRANSITION TO CATI

Based on the survey results and the authors' observations of the transition from pencil and paper interviewing to CATI, there are many points at which the interpersonal and technical sides of telephone survey management must work effectively together to produce an efficient and high quality survey effort. Bad collaborations can result in strained relationships, as well as quality control failures, increasing costs, and schedule slippages as each person assumes the other one is doing what needs to be done or systems are designed in one sphere making them difficult or impossible to operate in the other. Points of collaboration may become points of strain for organizations transitioning to CATI as well as for organizations running side-by-side operations. Some points requiring close collaboration are described below, along with the reasons for potential strain.

3.1 Deciding When to Use CATI

Making the decision about when to use CATI requires balancing issues such as the difficulty of the interviewing task, the cost of CATI setup, hardware constraints, and the ability to do the survey as a pencil and paper task (for example, the number and complexity of skips). The decision requires both technical and interpersonal judgment. Moreover, the critical issue of how resources will be divided within an organization, in the short and long term, must also be considered.

3.2 Budgeting

CATI survey budgets must be estimated by technical and interviewer supervision staffs working together. Changes in the placement of effort has implications for the relative sizes of staffs. If CATI replaces traditional operations, for example, instrument design and programming staffs may increase, while the traditional data reduction staff may face shrinking support.

3.3 Questionnaire Design

CATI questionnaires must be designed to deal with difficulties that are likely to arise during the interview, such as the need to go back to verify information and to correct interviewer and respondent errors. This requires close coordination of technical and interpersonal staffs.

Management of instrument design is, in part, a social process of communication and decisionmaking. This process requires that specifications be developed and reviewed by persons representing expertise in the substantive research area, the craft of formulating questions, the training and supervision of interviewers, and the technical requirements and possibilities of CATI. Each may raise objections, request clarifications, or suggest additional possibilities. The people involved may have difficulty communicating because they come from different disciplines, use different technical terms, and, in some cases, lack interpersonal skills.

Written documentation can be very helpful in providing and communicating a record of what decisions were made and how. Flowcharts can be useful in making specifications clear and in testing for errors (Jabine, 1985). Further, time should be allowed for iterations as decisions are made, evaluated, respecified, and tested.

3.4 Questionnaire Testing

Testing CATI questionnaires is a technical problem that usually requires the cooperation of an alert and patient field staff. On one hand, errors in the questionnaire can affect the data and cause considerable frustration, especially for interviewers. On the other, since a CATI questionnaire, especially one operating off a central processor, can be changed quite easily it is tempting to polish and polish until it is "perfect."

3.5 Interviewer Selection, Training, and Management

Many organizations have found that technical people must be involved in the traditionally interpersonal domain of interviewer selection, training, and management. Interviewers must be able to deal with computers to function in a CATI environment. This may require revisions in recruiting criteria, training approach, and the availability of technically trained supervisors to sort out problems of interfacing between the computer, the questionnaire, and the interviewer. Traditional survey managers may perceive this as undermining their authority in the area that is most central to their traditional skills. However, the needs of interviewers and supervisors are now reflected in more user friendly systems, including help screens, data tables, and other aids to orientation, and in the development of self-study modules for training.

3.6 Sample Management

Many supervisors complain that with a CATI survey that uses a scheduler, they lose the "feel" of the progress of a survey. Instead of piles of cards to sort, they must deal with more abstract queues and reports. Further, they must specify in detail decisions about prioritizing calls, scheduling cases, and retiring cases that may have been made intuitively in the past. They must work together with programmers to be sure that their decisions are implemented correctly and have the desired effects. Again, as systems become more sophisticated, it seems feasible to incorporate computer graphics to describe the progress of a survey. For example, pie charts or histograms could summarize the status of the sample or the age of cases in the infamous "supervisor hold" category.

3.7 Cost and Productivity Monitoring

CATI systems are capable of producing detailed information on interviewer productivity and costs on a daily, or even hourly, basis. This can be very valuable, but once such information exists, it creates its own pressure to use it. Interviewer supervisors must deal with questions from survey cost monitors which may be difficult and time consuming to answer. Deciding on the content and frequency of monitoring must be a joint technical and interpersonal management decision.

4. CONCLUSIONS

It is clear from the literature, the survey data we collected, and from conversations with many colleagues that the organization of survey centers is important. It affects the way work is done and how well work is done. Further, it has a profound effect on the working environment. Yet, from the variation we observed, it is also clear that there is no single method that works best in all situations. The existing literature offers very little guidance about what the choices are.

In this chapter, we have tried to describe how a variety of telephone survey centers are organized. We have also described some of the organizational issues that arise when a technological change, CATI, is introduced. Since the technology for doing survey research will continue to develop, we should think carefully about how to organize, not only for the way we do work today, but also for the flexibility to adapt to the future.

CHAPTER 30

OBSERVATION OF BEHAVIOR IN TELEPHONE INTERVIEWS

Charles Cannell and Lois Oksenberg
The University of Michigan

Researchers who are responsible for survey data collection have frequently recognized the need for evaluating the activities of the interviewing staff. The usual assessment, especially for face to face interviews, is to review completed questionnaires and examine distributions of each interviewer's work for nonresponses, uncodable information, etc. These techniques are inexpensive and easily conducted, but are inadequate to identify most interviewing errors. For example, Wilcox (1963), comparing an analysis of tape recordings of interviews with reviews of completed questionnaires, found that only 12 percent of the interviewer errors could be identified through an examination of the questionnaires. The other 88 percent were "invisible errors" — failure to read the question correctly, errors in probing, errors in recording responses, etc. Such invisible errors include the main component of the interviewer's task — the use of prescribed techniques to obtain responses. Adequate evaluation of the interviewer's performance requires some method of directly observing the interviewers themselves.

In face to face interviews, direct observation has usually been limited to having a supervisor accompany interviewers to observe their work. In telephone interviews, the supervisor may listen to the interview on an extension telephone. Most often, observations are comments about the interviewer's performance for discussion following the interview. The comments are usually subjective and unsystematic, varying with the interest and orientation of the observer.

Standardized coding of interviewer behavior provides an objective method for evaluating interviewer performance. Data from behavior coding for respondents as well as for interviewers also have other benefits for survey methodology. Studying interviewer and respondent behaviors often reveals problems that interviewers or respondents have with survey questions, signaling deficiencies in wording or design of the survey questions. Studying the behavior patterns may indicate problem questions to the study director and may suggest how to improve them.

In addition to these immediate applications, behavioral data are useful to survey methodologists in analyses of the effects of various interview procedures on respondents. For example, standardized behavior coding can be used to study effects of our interviewer techniques on respondent feelings about the interview, to discover what interactive style is productive of positive respondent efforts, to discover what techniques produce better quality responses, etc.

With the growth of interest in methodological issues of response bias, interaction between interviewers and respondents, effects of variation in question design, and interviewing technique, there is a demand for detailed, objective data on what transpired in the interview. A systematic procedure for observing the behaviors of interviewers and respondents in survey interviews is needed.

1. OBSERVATION TECHNIQUES

Social scientists have long used observation methods for studies of communication within dyads and groups, and for studies of interactive patterns and role performance. For example, Chapple (1953) and, later, Matarazzo and Saslow (1961) measured speech duration and duration of pauses by members of dyads.

Students of group dynamics have made extensive use of behavioral analyses in studies of communication patterns, group cohesiveness, etc. (Lippitt and Zander, 1943). Flanders (1960) and others developed systems for studying student-teacher interactions in classrooms. Coding of the behavior of counselors and clients was done by Carl Rogers and his students (1944) to study processes in counseling.

Probably the best known and most completely developed system for measuring interaction in groups is Bales's *Interaction Process Analysis* (1950). His method focuses on an evaluation of member participation, conformity to norms, and affect related to performance of group tasks. Categories in the Bales system include: shows tension, suggests a solution, gives opinion, shows aggression, etc.

As these few examples suggest, systems of observation vary in complexity and in the level of inference required of the observer. Behavior can be observed at a detailed, specific level or at a broader level. Some systems require only a simple recording of specific behavior, which often can be done by a relatively untrained observer. Other systems, especially those requiring inferences about meanings of behaviors, may require highly trained observers.

Observation procedures have not been widely applied to survey interviews, and the history of their use is brief. Face to face interviews were the first to be studied (Cannell et al., 1968). The early studies used an observer who accompanied the interviewer and coded interview behavior live. Now, observation of behavior in face to face interviews is usually done from tape recordings. Observation of telephone interviews can be done live (on an extension telephone) or from tape recordings. Whether observation is of face to face or telephone interviews, and whether it is live or from tape recordings makes little difference in the system that is used. The techniques for coding behavior are the same. For this reason, this chapter discusses observation of both telephone and face to face interviews.

Studies of survey interviewing have used a variety of systems, ranging from simply categorizing verbal behavior at an elementary level to those requiring a high level of inference and interpretation. The nature of code categories used depends on the research objective. Three main objectives of interview observation are the following.

- *To monitor interviewer performance.* The purpose is to provide information for use by supervisors in correcting errors in interviewer performance. Information can be used for feedback to interviewers on a single interview or can be accumulated over all interviews taken by that interviewer in a survey, or codes for all interviewers can be summed for evaluation of total staff performance.

- *To identify survey questions that cause problems for the interviewer or respondent.* Summing interviewer behavior codes by question indicates those that caused problems. Such problem indicators include repeats of the question, use of several probes, giving of explanations, etc. If codes for respondent behavior are included, problems often become more clearly specified. Such potential respondent problem behaviors include failures to give an adequate response, requests for interviewer clarification, and "don't know" responses.

- *To provide basic data for methodological studies.* Behavioral observation data are particularly useful in investigations of the effects of interviewing techniques, question designs, etc. on responses. Experimental studies that manipulate an "input" variable, such as interviewing style, are more valuable if they include a behavioral analysis demonstrating how effects on responses are produced.

In this chapter we will give examples of the use of observation systems for each of these three purposes. The major attention will be given to observation for the purpose of monitoring interviewer performance.

2. MONITORING INTERVIEWERS AS A SUPERVISORY TECHNIQUE

Each survey organization specifies rules of conduct that identify acceptable and unacceptable interviewer behavior. The rules vary among survey organizations, depending on their concept of interviewing. The rules of conduct cover various types of interviewer behavior — how and when to ask questions, how and when to probe for more information, how and when to define or clarify questions, etc.

It is the supervisory function to monitor how well the staff adheres to the rules and to effect greater conformity. The supervisory function is not to question whether or not the established rules are correct but only to see how adequately they are followed.

Since interviewing rules are statements of explicit behaviors, observation systems for monitoring interviewer performance are quite straightforward. Depending on the level of detail desired, observation usually can be conducted reliably with a minimal amount of training.

A centralized telephone survey facility is an excellent setting for monitoring interviewer performance, since both monitoring and feedback to interviewers can be arranged efficiently, and feedback can follow the conclusion of the interview for maximum effect. Telephone interviews can be monitored live or from tape recordings. The advantage of recordings is that passages can be replayed for clarity when needed. Another advantage is that the coding can be expanded to cover more behaviors, or more refined coding can be used. Many state laws in the United States permit recording telephone calls provided the respondent gives permission. A further advantage is that coding from tapes can be arranged efficiently. To monitor live telephone interviews, both interviewer and monitor must be available at the same time. Since much of an interviewer's time is spent in activities other than actual interviewing, a monitor may have to waste time waiting for an interview to begin. Since prompt feedback is important, the value of monitoring is lessened if coding of the tapes is delayed.

Regardless of whether monitoring is live or from tapes, the system used must meet the following criteria:

- Provide objective and reliable measures of interviewer behaviors
- Be cost-efficient, not too expensive to operate and maintain

- Focus on behaviors central to task performance
- Provide the basis for feedback to interviewers
- Be unobtrusive, not interfering with the interviewer's activities
- Be nonthreatening to interviewers, and perceived as having positive value.

Table 1 is an example of a code system that was used for live monitoring of a telephone survey [half using computer-assisted telephone interviewing (CATI), half using paper and pencil] (Mathiowetz and Cannell, 1980). It is a simple system that includes the major categories of interviewers' verbal behavior. It also includes a few ratings of characteristics, such as pace and manner of reading. Interviewers in this study had received special training in feedback, pace, and delivery. The categories in the code system reflect a particular interest in evaluating these aspects of performance. Coding live interviews with such a detailed system requires monitors who are well trained in keeping up with the interview pace.

Table 2 reproduces the coding form used with a simpler coding system. The system is easier to use, yet still can provide the basis for evaluating interviewer performance. The study for which this system was designed involved both monitoring of interviewer performance and evaluation of questions (U.S. National Center for Health Statistics, 1987).

Coding systems are flexible and can easily be adapted to special situations. For example, for CATI, codes for accuracy of recording responses can be included. The more elaborate the code system, however, the harder it is for monitors to keep up with the pace of the interview. If the coding is complex, tape recordings may be used to achieve satisfactory coding reliability.

The process of coding is for a monitor (often a supervisor) to listen to an interview and code each interviewer behavior for each question that is monitored. Extension phones are used; for CATI interviews, slave screens may be used. Several kinds of recording forms have been used, the format depending on the complexity of the code system and the questionnaire characteristics (i.e., extensiveness of skip patterns). Whatever the format, the goal is to make the recording of codes easy and free of recording errors.

2.1 Administration of Monitoring for Interviewer Supervision

Procedures, frequency, and methods of monitoring will vary depending on the experience level of the interviewers, and with the complexity of the

Table 1. Codes for Monitoring Interviewer Behavior

QUESTION-ASKING	
11	Reads question exactly as printed
12	Reads question incorrectly — minor changes
16	Reads question incorrectly — major changes
17	Fails to read question
REPEATING QUESTIONS	
21	Repeats question correctly
25	Repeats question — unnecessarily
26	Repeats question — incorrectly
27	Fails to repeat question
DEFINING/CLARIFYING	
31	Clarifies or defines correctly
35	Defines or clarifies — unnecessarily
36	Defines or clarifies — incorrectly
37	Fails to define or clarify
FEEDBACK	
41	Delivers feedback — correctly
45	Delivers feedback — inappropriately
46	Delivers feedback — incorrectly
47	Fails to deliver feedback
PACE/TIMING	
55	Reads item too fast or too slow
56	Timing between items — too fast
57	Timing between items — too slow
OVERALL CLARITY	
65	"Unnatural" manner of reading item (poor inflection, exaggerated or inadequate emphasis, "wooden" or monotone expression)
66	Mispronunciation leading to (possible) misinterpretation

Source: Mathiowetz and Cannell (1980, pp. 525–528). This is a slight revision of the system described.

questionnaire. What is described here is based on our experiences with monitoring. There is no research, to our knowledge, to guide decisions.

• *Selection of monitors.* It is essential that monitors know interviewing techniques and what constitutes acceptable and unacceptable behavior. Since the monitoring is used to evaluate interviewers' performance, it is appropriate for the coding to be done by interviewing supervisors.

• *What to monitor.* Monitoring can cover the entire interview or be limited to small segments. Since the objective is to identify general interviewing problems, the segments need be only long enough to obtain an adequate sample of work. The sample should contain instances of the varieties of types of questions in the questionnaire.

Table 2. Interviewer Monitoring Form

Question Number	Asks Question 1=Correctly 2=Minor Changes 3=Major Changes 4=In Error	Probes or Clarifies 1=Correctly 3=Incorrectly	Other Behaviors 1=Appropriate 3=Inappropriate	Omissions 3=Fails to Probe 4=Fails to Clarify	Notes

- *Frequency of monitoring.* Frequency depends on several variables. Monitoring needs to be heavy at the beginning of the survey to avoid consistent problems that threaten the response validity. Interviewers who are inexperienced or have demonstrated problems need more frequent attention. Overall, perhaps 10 to 15 percent of the interviews should be monitored, with weaker interviewers having a disproportionately higher percentage. The idea is not to overburden the supervisors, but to ensure that each interviewer is monitored regularly.

- *Costs of coding interview behaviors.* For supervisory monitoring, the codes can be designed in categories that are sufficiently detailed for adequate evaluation yet broad enough to permit keeping up with coding live telephone interviews. If coding from recordings, this means the tapes do not need to be stopped for the coder to keep up. In taped interviews, coders can use the "fast forward" button for efficiently selecting and coding various segments of the interview.

2.2 Training Monitors

Monitors should undergo the same training as the interviewers, thereby eliminating the possibility of a discrepancy between the interviewers' and monitors' basic knowledge of the questionnaire and purpose of the study. For monitoring training, several practice interviews should be recorded, representing a broad distribution of interviewing problems. Illustrations of both correct and incorrect behavior of each of the code categories should be included.

The monitoring training consists of three parts: (1) initial exposure to the codes, (2) coding of several taped interviews, and (3) simulated production coding. Introduction to the codes involves detailed explanations of each code and examples of the more difficult categories. Several taped interviews should then be coded by each monitor. In coding these interviews, emphasis should be on learning the use of the codes, without worrying about the pace of coding. Next, the monitors' and the instructors' coding of these interviews should be compared and discussed. Finally, a few interviews, different from those already coded, should be coded in a manner simulating production monitoring. This type of practice is particularly important if monitoring is to be done live. Each monitor must feel comfortable enough with the codes to keep up with the interview pace.

With adequate training and well-defined code categories, agreement among coders as to the codes to assign to interviewer behaviors can be quite high. The level of coding agreement in the study in which the codes in Table 1 were used varied for different types of codes. Agreement among monitors reached an 88 percent level for codes for concrete behaviors (codes 11–47). For codes involving ratings (codes 55–66), agreement was somewhat lower, at a 79 percent level.

2.3 Feedback of Results of Monitoring to Interviewers

The immediate purpose of monitoring is to correct interviewing errors. For this objective, the codes show acceptable and unacceptable behavior for each question asked. The monitoring results are reviewed with the interviewer soon after the interview. For the trained interviewer, a high proportion of behaviors are acceptable and good performance can be rewarded while errors can be identified objectively, problems discussed, and solutions proposed.

In order for such feedback to be effective in improving performance, the interviewer must know why he or she performed inadequately. There are at least three major reasons for inadequate performances (Cannell et al., 1975):

1. The interviewer does not know what constitutes adequate performance. The training has not been successful in communicating the theory or concepts that can be used to evaluate his or her performance.

2. The interviewer may understand the *principles* of good performance but not be able to determine whether or not a particular behavior conforms to those principles. For example, he or she may not be able to differentiate directive (inappropriate) from nondirective (appropriate) probes, or the interviewer may fail to read the response alternatives in the question properly because he or she did not recognize them as part of the question. In both of these instances, the interviewer knows the correct principle but cannot distinguish between behavior that does or does not conform to that principle.

3. The interviewer knows the principle, and can distinguish between adequate and inadequate performance, but lacks *skill* in the application of approved interviewing techniques. This situation is, of course, characteristic of new interviewers, but it also sometimes plagues experienced interviewers who have not been active for some time and those who lack poise or are ill at ease and feel pressured during the interview.

Through discussion of observed errors with the interviewer, the underlying problem can usually be identified, and steps can be taken to solve it.

The idea of monitoring begins in training, when the codes are introduced for use in role-played interviews. Role-playing groups consist of an interviewer, a respondent, and an observer. The observer uses the coded interviewer behaviors as the basis for discussion of the techniques following the interview. This procedure serves two goals. First, it organizes the observer's role and provides a systematic basis for feedback. And, since trainees play all three roles, familiarity with the codes focuses attention on acceptable and unacceptable techniques of interviewing.

As interviewers start regular assignments, they are told that their work will be monitored periodically, as the monitor's time schedule permits and without notice. The procedure is readily accepted as an extension of training. It is acceptable because interviewers are treated as professionals, interested in performing well. Especially since the monitors are experienced interviewers who can empathize with interviewing problems, the procedures usually are viewed positively.

2.4 Monitoring to Evaluate an Entire Interviewing Staff

In addition to its use as a supervisory technique, monitoring can provide a basis for evaluating overall performance of the interviewing staff and identifying common weaknesses. This is done by summing codes over all interviews by all interviewers. The overall results indicate specific topics where remedial training is needed, or may indicate more general areas in which the interviewer training program needs improvement. We include here a few examples of how monitoring can be used for these purposes.

Table 3 compares monitoring results for question-asking and probing behavior from four studies. As easily seen, percentages of acceptable and unacceptable interviewer behavior varied widely among the four groups of interviewers represented. There were very high levels of major changes in question wording as well as unacceptable probes in the Marquis (1971) and the Brenner (1982) studies. From descriptions of field procedures for these studies it is apparent that the poor interviewer performances reflect insufficient training in interviewing techniques as well as inadequate field supervision.

Differing results for the four studies also reflect, to some extent, differences in definitions of the code categories. For example, in the Kjøller (1975) study questions turned into statements by interviewers based on information previously obtained were coded as asked with minor changes. In the other three studies, this would have been a major wording

Table 3. Quality of Question-Asking and Probing Behavior in Four Monitoring Studies

	Percent of Interviewer Behaviors			
	Marquis Study	Kjøller Study	Oksenberg Study	Brenner Study
Question-asking				
Exact wording	63.3	99.3[a]	84.7	70.0
Minor wording change	19.4	na	9.9	7.7
Major wording change	17.3	0.7	5.4	22.3
	100.0	100.0	100.0	100.0
Probes				
Acceptable	68.5	98.5	89.5	44.3
Unacceptable	31.5	1.5	10.5	55.7
	100.0	100.0	100.0	100.0

[a]Includes questions asked with minor wording changes.

Note: Figures in these tables are calculated from information appearing in the published reports. The Oksenberg study is based on live monitoring of telephone interviews. The others are based on tape recordings of face to face interviews.

Sources: Marquis, 1971; Kjøller, 1975; Oksenberg, 1981; Brenner, 1982.

change. The Kjøller study also coded a wider range of other wording changes as minor rather than major.

The probing results reflect different behavior patterns among the four sets of interviewers. For example, over a third of the unacceptable probing in the Brenner study involved seeking information not required by the question. With a well-trained staff, this behavior is so unusual that there is no point in establishing a separate code for it.

The Oksenberg (1981) study (Table 4) illustrates the range of performance that can occur within a single interviewing staff. In that study there was considerable difference between the four best and four worst interviewers on nearly all performance measures. A study of 60 tape-recorded interviews taken by 30 interviewers by Cannell et al., (1975) also showed a wide range of behavior. Unacceptable question-asking behavior ranged from 2.5 percent for the best interviewer to 86.8 percent for the worst interviewer. For unacceptable probing, comparable figures were 2.0 percent and 38.3 percent, while for other unacceptable behavior, they were 0 percent and 76.0 percent.

A study by Prüfer and Rexroth (1985) illustrates behavior changes from the first pretest to the second pretest. For example, the percent of all probing and clarification that was unacceptable declined from 38 percent in the first pretest to 8 percent in the second. The percent of all "other behavior" that was unacceptable declined from 85 to 55 percent.

Table 4. Average Percent of Unacceptable Behaviors by Behavior Type for Four Best Interviewers and Four Worst Interviewers

	Average Percent	
Interviewer Behavior Type	Four Best Interviewers	Four Worst Interviewers
Unacceptable question-asking		
Minor changes in wording	9	67
Major changes in wording	2	16
Unacceptable clarification	0	2
Unacceptable probing		
No probing when needed	1	3
Use of unacceptable probes	3	12
Unacceptable answer repetition	0	4
Unacceptable feedback	1	10
Unacceptable comments	1	6
Unacceptable pace	1	35
Any unacceptable behavior	25	88

Note: The 17 interviewers who took four or more interviews were ranked in terms of average proportion of questions they asked that exhibited each of the types of unacceptable behaviors shown in the table. The four highest and four lowest averages for each measure then were averaged to produce the values in this table.
Source: Oksenberg, 1981.

Unacceptable question asking (major wording changes) was low in both pretests. Improvements in probing, clarification, and other behavior probably reflect the effects of training following review of behavior from the first pretest.

3. IDENTIFICATION OF PROBLEM QUESTIONS

As investigators examine behavior data, it has become apparent that the particular question being asked has a large influence on the interviewer behavior that ensues. At the simplest level, questions vary greatly in the amount of interviewer activity involved. Overall interviewer activity, in turn, reflects the amount of question repeating, clarifying, probing, feedback, and so on. The amount and types of interviewer behavior associated with a question provide useful information. Not only can behavior coding provide information about interviewer performances, but it also can reveal certain kinds of problems with the questions themselves. For example, "yes-no" questions should involve only one or two interviewer behaviors — asking the question and possibly a feedback statement. Behavior coding has revealed that some such questions involved considerably more interviewer behavior. Interviewers repeated the question, used one or more probes, explained or defined terms, etc. This suggested that the higher level of behavior signals that either the interviewer or the respondent was having difficulty with the question.

Examining the distributions of types and amount of interviewer behavior will not diagnose the precise nature of the problem, but it may indicate that a problem exists. Examining the kinds of behaviors often will lead to a hypothesis as to the nature of the difficulty. Examining the question itself may convince the analyst that his or her hypothesis is supported, indicating the redesign or rewording required to correct the problem.

Including codes for respondent behavior as well as for interviewer behavior sharpens the indicators of problems. Respondent problems may be signaled by inadequate answers, "don't know" responses, requests for clarification, irrelevant comments, or refusals to answer. Codes may indicate various types of problems such as:

- *Problems with comprehension of the question.* Difficulties arise when the respondent fails to understand the question because of the vocabulary level, the lack of clarity of the concept, or the ambiguity of question wording.

- *Problems with cognitive processing of information.* Some questions require considerable respondent effort in retrieving and organizing information requested. At times the information is simply not accessible or is unknown.

- *Problems with question wording.* Questions may be complex or awkwardly worded, making them difficult for the interviewer to ask.

Morton-Williams and Sykes (1984) and Sykes and Morton-Williams (1987) conducted a study that gives some credence to the use of behavior coding to identify problems with questions. They coded 89 tape-recorded household interviews, including behaviors of both interviewer and respondent. These interviews were followed by a second interview designed to discover respondents' comprehension and interpretation of specific questions, including understanding particular words and phrases.

The followup interviews revealed that responses to some questions, apparently answered adequately, were based on misunderstanding or misinterpretation. However, it was often possible to infer from respondent behavior codes that the particular questions caused difficulty.

For example, for the following question, 18 percent of the respondents asked for clarification.

> Q.3. Would you say that the gap in wealth between the richer countries and the poorer countries is, on the whole, getting wider, getting narrower? Even if you aren't sure I'd like you to tell me what you think.

The followup interviews revealed many respondent difficulties of interpretation and meaning. The authors concluded,

> The concepts used in this question were thus seen as complex, imprecise, and confusing; many respondents clearly were not able to handle them at the level of broad abstraction perhaps hoped for by the researcher. A large number of those in the followup interviews said that they did not feel that they were competent to answer, yet very few said, "Don't know."

In another question, "How do you feel about the amount of noise, in your area, from cars and lorries or other road traffic?," 10 percent of the respondents requested clarification and 17 percent gave inadequate responses. The followup interview verified confusion and a lack of understanding.

A study by Oksenberg (National Center for Health Statistics, forthcoming) coded interviewer behavior for interviewer monitoring purposes and also tabulated those behaviors for individual questions. (Code categories are shown in Table 2.) Several behavior patterns were observed that suggest problems with questions. An examination of the particular questions exhibiting these patterns gives strong inference of the validity of the problem indicators. Three questions, with potential problem patterns, are shown in Table 5.

Behavior coding indicated that interviewers had difficulty reading question 101 as worded. Respondents often interrupted the interviewer before the final sentence of the question was completed. Other times the interviewer simply neglected to read the final sentence. The structure of question 101 either made respondents impatient to reply or confused them as to when a reply was expected. Question 394 was repeated and explained frequently. Many respondents did not know what "dental sealants" were. Question 117 showed a different pattern of especially frequent probes. Interviewers were required to record an exact figure. Respondents either did not understand the degree of precision expected or were uncertain about the exact number.

Table 5. Questions with a High Proportion of Interviewer Behavior Indicating Problems

Question	# of Times Q Was Asked	Percent of Times			
		Q Asked With Major or Minor Changes	Q was repeated	Q was explained	Q was probed
Q101 During the last two weeks, did you (or anyone in the family living there) go to a dentist? Include all types of dentists, such as orthodontists, oral surgeons, and all other dental specialists, as well as dental hygienists.	185	19.5	9.7	4.9	1.1
Q394 (Have you/Has anyone in the family) had dental sealants placed on (your/their) teeth?	39	2.6	28.2	25.6	—
Q117 During the past 12 months, that is since (MONTH), 1985, how many visits did (you/_) make to a dentist? (Include the _ visits you already told me about.)	180	5.6	6.7	2.8	10.0

Source: National Center for Health Statistics (forthcoming).

These investigations suggest the potentiality of identifying problems with questions by examining interviewer and respondent behavior. Extensions of the present coding systems may make it possible to objectively identify questions that cause difficulties and to diagnose the nature of the difficulties.

The least scientifically rigorous aspect of survey research is the development and testing of questions. It is ironic that the creation of the measuring instrument is based primarily on past experience, with only a few "common sense" principles as guidance. The usual procedure for testing questions is to complete a few interviews, using experienced interviewers, and then evaluate the questions based on interviewers' reports that questions either did or did not "work." The "did not work" covers a wide variety of factors, including respondent hostility to a question and difficulty either in understanding what was wanted or in organizing the information needed to respond adequately. There is often a lack of agreement among interviewers as to whether or not a question is a problem and, if it is, the nature of the difficulty. The critical task of creating a scientific measuring instrument is left to the subjective evaluation of the researcher, with little objective information from the pretest experience.

These descriptions illustrate that present methods of evaluating questions are based largely on subjective evaluation. Developing coding systems for evaluation may help to improve survey questions.

4. MONITORING BEHAVIOR IN METHODOLOGICAL STUDIES

Another use for behavior coding is to provide data to the methodologist about factors affecting response quality. Analysis of interviewer and respondent behavior in regular (nonexperimental) survey interviews provides a valuable basis for generating hypotheses about the likely effects of different interviewing techniques, question designs, and other behavior on responses. In experimental tests of hypotheses — in which, for example, the effects of different types of interviewer training are compared — behavior coding demonstrates how effects on responses are produced.

Morton-Williams and Sykes (1984) used behavior coding of regular interviews to study the effects of "good" and "bad" interviewer behaviors on response quality. An example of their findings is that when interviewers ask questions correctly, only 12 percent of immediately following respondent behaviors were undesirable (mainly consisting of irrelevant or uncodable responses); but when questions were asked poorly, undesirable respondent behavior increased to 27 percent. With regard to interviewer behavior after initial question reading, a comparison of respondent behavior following "good" and "bad" interviewer behavior dramatically illustrates the results of poor technique. When interviewers used "good" techniques (conforming to interviewing rules), only 13 percent of the following respondent behaviors were undesirable; but when interviewers used "bad" techniques, undesirable respondent behavior increased to 74 percent. "Bad" technique consisted mainly of digressions, interpreting the question, evaluative comments, and leading, misleading, or incomplete probes. Undesirable respondent behavior included digressions and irrelevant or inadequate responses.

Marquis and Cannell (1969) also analyzed coded behavior from regular interviews to investigate the effects of various interviewer behaviors on respondent behavior. Table 6, for example, shows the effect on respondent behavior of repeating the question, used to follow up responses that in one way or another did not meet the question objectives. As the table indicates, if the question was repeated, the probability of an adequate response was .38. Repeating the question was an effective technique for eliciting an adequate response.

Table 7, from the same study, illustrates a probable weakness in the application of interviewing technique. The table gives the probability of interviewer feedback following various types of respondent behavior. Since virtually all feedback consisted of a positive, supportive statement, the figures show that positive feedback is as likely to follow poor respondent behavior as good. This suggests that interviewers may be misusing a potentially useful technique, in that positive feedback following poor respondent behavior is likely to encourage poor behavior rather than promote good behavior.

Table 6. Probabilities of Various Respondent Behaviors Following Question Repetition

Respondent Behavior	Probability
Gives adequate answer	.38
Gives inadequate answer	.14
Elaborates answer	.12
Repeats previous answer	.05
Asks for clarification	.09
Talks to third person	.05

Source: Marquis and Cannell (1969).

Table 7. Probabilities of Interviewer Feedback Following Various Types of Respondent Behavior

Type of Respondent Behavior	Probability that Interviewer Feedback Follows
Gives adequate answer	.28
Gives inadequate answer	.24
Gives "don't know" answer	.18
Refuses to answer	.55
Elaborates answer	.30
Repeats answer	.32
Gives suggestion	.33

Source: Marquis and Cannell (1969).

The Morton-Williams and Sykes and the Marquis and Cannell studies illustrate the importance of good interviewing techniques and suggest ways in which the application of good techniques might be fine tuned to particular circumstances. Dijkstra et al. (1985) used behavior coding in an experimental study to investigate the impact of training in two different styles of interviewing on both use of good techniques by interviewers and adequacy of respondent behavior. The investigators trained one group of interviewers in a socio-emotional interviewing style, in which they were encouraged to react in a personal, sympathetic, understanding manner and to initiate conversation independent of the interviewing. The second group of interviewers were trained in a style of interviewing in which they were to employ such person-oriented behaviors only minimally, at a minimal level of social acceptability. Both groups received the same training in task-oriented interviewing techniques.

Behavior coding of subsequent interviews showed that, in terms of task-oriented techniques, performance for the socio-emotional interviewers was less adequate than that of the formal interviewers. For formal interviewers, 13.7 percent of the task-oriented behaviors were inadequate, while for socio-emotional interviewers the figure was 20.5 percent. For example, socio-emotional interviewers were more likely to use inadequate clarifications and interpretations — typically directive actions that may bias responses. While respondents interviewed in the socio-emotional style revealed more personal information, interviewing style did not affect the amounts of adequate or inadequate task-oriented respondent behaviors that were the overall measures of response quality used in the study. However, additional analysis of responses to seven open questions indicated that respondents interviewed in the socio-emotional style reported significantly more relevant information than did respondents interviewed in the formal style (Dijkstra, 1987). Studies like this one help to evaluate the effects of various interviewing practices and may lead to recommendations as to how current interviewing guidelines might be improved.

Effects of interviewer behavior on respondent behavior may be immediately evident or delayed. For example, the interviewer uses a suggestive probe, gets a partial response, repeats the original question, and gets another inadequate response. The inadequate response is not the result of the repeated question but of the earlier suggestive probe. Brenner (1982) has made a beginning attempt to study immediate and delayed effects by an analysis of sequences of interviewer and respondent behaviors. Figure 30.1 illustrates his method. The figure describes interviewer-respondent interaction following the asking of closed questions.

Coding of behaviors indicating affective states such as irritation, boredom, or enthusiasm may reveal important influences on interviewer and respondent behavior. Affective states are likely to be generated over the course of the interview. Coding of affect indicators, and when they occur, can reveal the course of development of affective states as well as their effects on responses. To our knowledge, no study of survey interviews has yet applied behavior coding for this purpose.

5. CONCLUSION

There are many reasons for wanting to know what transpired in the interaction between interviewer and respondent. Methodologists want to know the effects of various techniques in producing responses, learn what problems respondents have with particular questions or types of questions, study the affective responses to particular techniques, learn what underlies variability in responses obtained by different interviewers, or learn how adequately interviewers use techniques. While one can manipulate or identify interviewing techniques, interviewer characteristics, question characteristics, etc., and measure the effects on responses, one must make inferences about what transpired during the interview to produce the effects. Since several factors frequently are involved, these inferences are tenuous.

Van der Zouwen (1974) proposed the "black box" model of the interview, with the questions, characteristics of the interviewer, interview techniques, and situational variables as inputs and the answers as outputs. As the model suggests, knowing only input parameters provides limited insights into how the outputs are produced. Behavior coding is a technique for revealing some of the content of the black box. The data

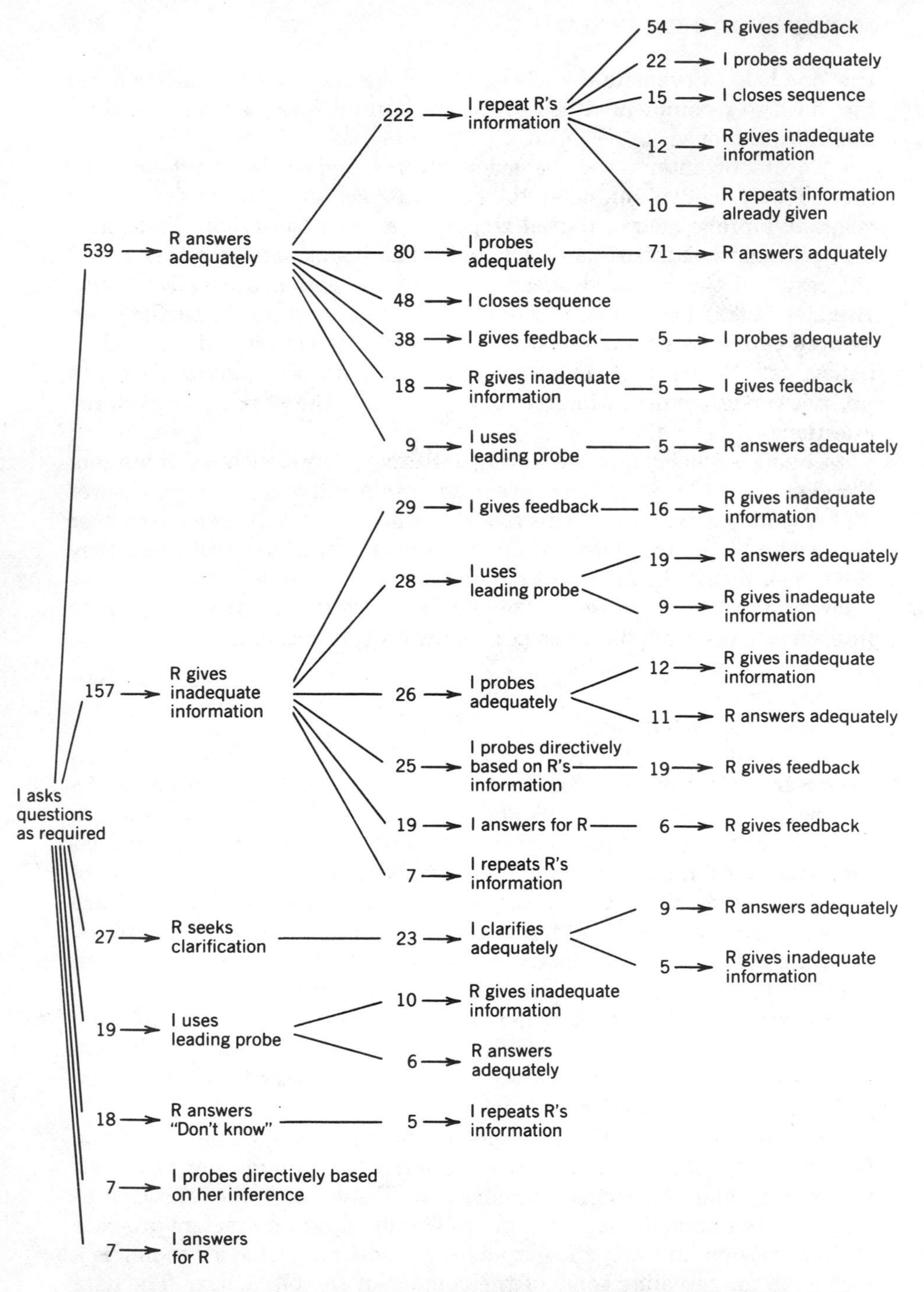

Figure 30.1 Interviewer-respondent interactions related to closed questions when question asked as required. This tree represents sequences of interviewer and respondent behavior following the asking of the original question. The numbers are the frequencies of the behaviors. From Brenner, M., "Response Effects of 'Role-Restricted' Characteristics of the Interviewer." In W. Dijkstra and H. van der Zouwen (eds.), Response Behavior in the Survey Interview. London: Academic Press, 1982, p. 154. Reproduced by permission of the publisher.

begin to help us understand factors that intervene between the questionnaire and the response.

We have summarized uses of behavior coding. The first is to monitor interviewer performance during an interview. The results are used as a supervisory technique for enforcing rule following behavior. The coded behavior for each interviewer can be summarized over an entire study as an evaluation technique, and it can be summarized over all interviewers to assess overall staff performance, evaluate training methods, and identify techniques for which training needs improvement.

The second use of analyses of behaviors is to evaluate survey questions. Evidence is clear that respondent and interviewer problems with questions are revealed in the amount and types of behaviors. Examination of the problem indicators often suggests hypotheses about the nature of the problems, which frequently can be confirmed by examining the question itself. Behavior coding has not been used extensively in survey research for this purpose. It is a procedure that holds considerable promise for improving the quality of survey responses and reducing unwanted biases.

The third potential use is to provide data to the methodologist about factors affecting response quality. Such data can provide valuable insights into the interview and the factors that intervene between input and output variables.

Behavior coding has not been widely used in survey research. It has been our purpose in this chapter to describe techniques and illustrate uses of the methods to stimulate the interest of fellow methodologists.

CHAPTER 31

A COMPARISON OF CENTRALIZED CATI FACILITIES FOR AN AGRICULTURAL LABOR SURVEY

Robert T. Bass and Robert D. Tortora[1]
U. S. Department of Agriculture

When survey organizations conduct large scale telephone surveys, several options for administrative structures exist. Groves and Lepkowski (1985) study four structures. Structure I is a completely decentralized structure, where face to face interviewers also conduct telephone interviews in their region. Structure II has a single telephone interviewing facility conducting interviews in all regions. In structure III, telephone interviews are conducted from multiple centralized telephone facilities. These facilities are not administratively attached to the regional offices. In structure IV, each regional office has its own telephone facility. Other options exist for structuring centralized telephone interviewing facilities. Centralized versus decentralized telephone data collection is an important administrative issue that survey organizations need to address (this volume, Chapter 32).

It is important to consider various combinations of administrative structures for centralized telephone interviewing. Dillman and Tarnai note that the increasing use of computer-assisted telephone interviewing (CATI) systems encourages centralization. One factor is the possible lowering of interviewer variance (Groves and Magilavy, 1983). Interviewer efficiency may also increase if interviewers can call over multiple time zones. This extends the optimum calling period each day so a higher number of contacts can be expected (this volume, Chapter 15; Warde, 1986). A final factor to consider is the optimum use of scarce staff

[1]The authors wish to express appreciation to all the National Agricultural Statistics Service offices involved in this project.

resources, particularly the number of computer specialists needed to operate centralized CATI facilities. With respect to these resources, reducing the number of CATI facilities (in administrative structures III and IV) is an option if the remaining CATI facilities can call into the regions (or call across all regions) without CATI facilities. This chapter presents the results of a case study where this type of modification to structure IV is made. Specifically, 6 of 18 state data collection offices (here the region corresponds to state boundaries in the United States) have administratively attached CATI facilities. Each CATI facility telephones into all 18 states.

The purposes of this chapter are to compare response rates by facility and state, compare survey statistics and variances by facility, compute intraclass correlation coefficients and design effects, and compare estimates by experience level of facilities.

1. MOTIVATION FOR THE STUDY

The current administrative structure of the National Agricultural Statistics Service (NASS) consists of 44 data collection or field offices, covering all 50 states, and a headquarters staff. Each of the data collection offices has a telephone interviewing facility. Forty-two field offices collect data entirely within the boundaries of the states in which they are located. Presently, 15 of the 44 data collection offices have CATI systems (Tortora, 1985). Each CATI facility has a supermicrocomputer with 12 to 22 interviewing stations and uses the University of California at Berkeley CATI software (Shanks et al., 1983).

Before expanding CATI to cover all 50 states, NASS has decided to evaluate an alternative administrative structure, where the number of CATI facilities that are administratively attached to data collection offices is reduced. Under this structure, all field offices would have responsibility for face to face interviewing, but in only a few would there be an administratively attached CATI facility.

As noted by Marquis and Blass (1985) there are some possible disadvantages of centralization. They include extreme interviewer job specialization and possible negative effects on interviewer morale. In addition, the structure under study requires a reliable electronic communications network not only between the CATI facilities, but between the field offices and the CATI facilities. This communications network would allow for the transmission of samples between CATI facilities in the case of hardware failure. It would also allow for the necessary operational activities that are associated with dual frame surveys. Examples of these activities include duplication detection in list frame samples, domain determination for dual frame surveys, and name, address, telephone number, and auxiliary data updates for later surveys. These types of activities require a communication link between CATI facilities and field offices. A final disadvantage may be the resistance within selected field offices to losing a well established telephone interviewing facility. Assuming these disadvantages can be overcome, we decided to evaluate this new administrative structure for data collection using the Farm Labor Survey.

2. DESCRIPTION OF THE QUARTERLY FARM LABOR SURVEY

The Quarterly Farm Labor Survey is a dual frame survey having an area and a list sample. The stratified simple random sample from each frame is selected by state. Sampled farm operators are requested to report the number and type of workers and wage rates paid for the one week period containing the twelfth day of the month. Data collection begins on the Monday of the week after the week containing the twelfth and continues for about 12 days. Data from the survey are used to produce estimates of the number of workers, average wages for hired workers by type of work performed, and wage rates by method of payment. Estimates are published at selected state levels, for 15 regions, and for the entire United States.

The main reason for choosing the Quarterly Farm Labor Survey for this study is that it minimizes the variation in cultural farm practices and terminology that an interviewer must learn. One version of the questionnaire is usable in all states.

In this case study, we selected six CATI facilities to call across 18 nonadjoining states. Thus, interviewers in each facility called into all 18 states. The six states where the CATI facilities were located were included in the 18 states. Final telephone noncontacts were returned to the originating state for face to face follow up.

3. RESEARCH DESIGN

The research design necessary to fully investigate the stated objectives presents complicating factors. In order to construct a variance component design to measure the effect of facilities on a statistic's total variance, a random sample of centralized facilities is necessary. A sample of this type is impossible to achieve, as no telephone interviewing is currently conducted by NASS using any centralized administrative structure.

A second complicating factor in the design concerns the estimation of the effect on survey results due to the experience level of the interviewing facility. Administrative constraints allowed a maximum of six facilities to conduct centralized regional interviewing for this project. Sample size, training requirements, transmission of data, length of data collection period, and costs were important factors in limiting the total number of CATI facilities to six. The six centers were chosen to get a geographic spread and to include two experienced facilities (more than two years' experience in conducting CATI surveys). The chosen facilities were in Georgia, Pennsylvania, Nebraska, Tennessee, Texas, and Washington. The two experienced centers were Georgia and Nebraska. Measurement of differences between facilities by experience is difficult based on six cases.

Stratified simple random samples from list frames[2] of 18 states[3] were selected for inclusion in the "pool" of samples to be contacted by the six facilities. After withholding sample units without a telephone number, the sample pool added up to 4,866 units. We then randomly allocated the sample pool to each selected facility. Each facility began the study with 811 sample units.

Training was a major part of the study. A survey statistician, farm labor subject matter statistician, and a telephone supervisor from each center attended a 1-1/2-day training session covering topics pertinent to the study such as survey concepts, interviewing techniques, and data handling. A workshop format was used to provide extensive hands-on training. In the same way that NASS trains for its other surveys, each team returned to its facility and trained its interviewer staff. This facility training used the same topics and format used at the team training workshop.

Prior to actual data collection, each of the 18 states briefed the six facilities on localized conditions and terminology that might have an effect on responses. This step was another precaution to ensure standardized training of telephone interviewers prior to data collection.

The telephone interviewing period covered six days. Interviewers began contacting farmers and ranchers on Monday, April 20, 1987, and concluded data collection the following Saturday. Respondents were requested to report number of workers, hours worked, and wage rates paid

---

[2] Area sample units were not used in this study.

[3] Georgia, Illinois, Missouri, Montana, Nebraska, Nevada, New Jersey, New Mexico, New York, North Carolina, North Dakota, Oklahoma, Pennsylvania, South Carolina, Tennessee, Texas, Washington, and Wisconsin.

for the seven day period of April 12–18. The length of interview varied by operation but averaged 10 minutes.

4. DATA AND ANALYSES

Completion of the assigned sample units varied by facility. Table 1 gives the overall completion of cases for each facility. The Tennessee facility, with a 46.7 percent completion rate and a 48.0 percent noncontact rate, encountered problems in getting interviewers trained and available for work. All noncontacts were attempted at least five times (with the exception of Tennessee), but due to the unavailability of the farm operator or no answer, telephone interviewers were unable to complete an interview.

Response rates for the individual states called by each center remained fairly constant. The Georgia center averaged a 70.0 percent completion rate over the 18 states while maintaining a 68.2 percent completion rate for the sample within its own state boundary. Tennessee's completion rate averaged 46.7 percent over the 18 states and 52.5 percent within its state boundary. The remaining four centers exhibited a similar pattern. Response rate analysis does not show any difficulty for a facility to conduct telephone interviews in any of the 18 states. Refusal rates by facilities are not significantly different.

Telephone interviewers introduced themselves as calling for the (State) Agricultural Statistics Service, where (State) was filled by the CATI instrument with the name of the state where the sample unit originated. An hypothesis held by some is that an interviewer's accent, pronunciation of names and towns, or handling of unfamiliar farming practices may have an effect on response rates. Table 2 gives a summary by facility of the number of contacted operators inquiring as to the geographic location of the telephone interviewer. This total is labeled as respondent inquiry.

Interviewers commented that geographical location had no apparent effect on their ability to secure the cooperation of respondents. Other interviewers reported that in some cases, respondents actually welcomed the chance to converse with an interviewer in another part of the country and inquire about local weather and crop conditions. Considering the constant response rates, the small amount of respondent inquiry, and the nonsignificant refusal rates, these data show that centralized telephone data collection facilities can produce response rates at the same level as individual state data collection efforts. Using the location of the facility as a proxy measure of the interviewer's accent, no evidence can be found to

Table 1. Sample Unit Results by CATI Facility

Facility	Cases Assigned	Interviews Completed (%)	Refusals (%)	Noncontacts (%)
Georgia	811	70.0	10.6	19.4
Nebraska	811	73.5	10.2	16.3
Pennsylvania	811	69.4	9.7	20.8
Tennessee	811	46.7	5.3	48.0
Texas	811	81.9	8.6	9.5
Washington	811	76.6	8.3	15.1
Total	4866	69.7	8.8	21.5

Table 2. Respondent Inquiry of Interviewer Location

Center	Total Number of Respondents	Respondent Inquiries	Percent
Georgia	654	34	5.2
Nebraska	679	42	6.2
Pennsylvania	642	45	7.0
Tennessee	422	44	10.4
Texas	734	33	4.5
Washington	688	28	4.1
Total	3819	226	5.9

support the hypothesis of response rate differences based on the interviewer's accent.

The next analysis, a one-way analysis of variance with unequal treatment group sample sizes, compares differences in reported data between facilities. The dependent variable in each analysis is a ratio level variable. The low response rate at the Tennessee center presented some concern for data analysis. An examination of completed interviews by this facility shows no violation of randomization rules for attempted calls. The mean value for all statistics calculated for this facility is neither the smallest nor the largest value when compared in a range with the comparable statistic calculated for the other five facilities. For these reasons, we decided to include the data collected by the Tennessee center in the analysis.

Thirteen data items were examined, and as Table 3 shows, at the 0.05 level no significant differences were found between the data collected by

Table 3. Analysis of Variance for Data Items

Item	Mean	Sampling Error	Range of Means	Pr>F
Type of worker				
Self-employed	1.04	.02	1.01–1.08	.57
Unpaid	1.03	.05	0.95–1.12	.38
Paid	3.54	.36	3.17–4.16	.76
Hours worked				
Self-employed	57.50	.84	56.56–59.23	.70
Unpaid	41.57	.91	39.96–49.28	.83
Paid	36.86	1.26	33.98–40.37	.05
Wage rate				
Field worker	4.78	.16	4.39–5.12	.09
Livestock worker	4.39	.18	4.02–4.48	.50
Supervisor/manager	7.09	.54	5.88–8.51	.11
Other worker	6.50	.87	5.67–7.82	.79
All hourly workers	4.28	.13	4.05–4.40	.76
All piece workers	3.78	.73	2.58–4.92	.59
All other workers	4.88	.31	4.53–5.05	.34

each facility. The range of means shows the large variability in the data collected by each facility. Examination of the number of paid workers reveals that one facility produced an average per operation of 3.17 while another produced an average response almost one worker higher. The overall difference in paid workers, however, is not significant.

To examine facility effects for all data items in the study, an intraclass correlation based on the previously discussed one-way analysis of variance was used. The facilities are considered factors. The intraclass correlation ρ_{FAC} is given by

$$S_b - S_w \,/\, [S_b + (k-1)S_w]$$

where S_b is the between-facility variance, S_w is the within-facility variance, and k is average number of usable reports over facilities.

Table 4 summarizes the results of the intraclass correlation analysis. Eight of the 13 data items exhibit a negative ρ_{FAC}. This is the result of a small between-facility variance and a large within-facility variance. The largest ρ_{FAC} value of 0.0021 for average hours worked by paid workers does, however, show a facility effect.

To evaluate the effect of the facility on the total variance of the survey statistic, it is necessary to calculate an additional component, the design

Table 4. Intraclass Correlation Coefficients and Design Effects

Item	ρ_{FAC}	$deff_{FAC}$
Type of worker		
Self-employed	−.0004	.9980
Unpaid	.0001	1.0005
Paid	−.0008	.9960
Hours worked		
Self-employed	−.0007	.9965
Unpaid	−.0010	.9950
Paid	.0021	1.0105
Wage rate		
Field worker	.0016	1.0080
Livestock worker	−.0002	.9990
Supervisor/manager	.0015	1.0075
Other worker	−.0009	.9955
All hourly workers	−.0009	.9955
All piece workers	−.0004	.9980
All other workers	.0002	1.0010

effect $deff_{FAC}$. The number of facilities collecting data is a component in the variance inflation factor

$$deff_{FAC} = 1 + \rho_{FAC}(m - 1)$$

where m is the number of facilities. In this analysis m equals 6 and the variance is inflated primarily due to large values of ρ_{FAC}. Table 4 shows that the variance of the 13 survey statistics both decreased and increased depending on the value of ρ_{FAC}. For the data items with a positive ρ_{FAC}, the variance is inflated no more than 1 percent.

An additional analysis of the intraclass correlation values involves all combinations of facilities. This analysis offers some insight into the possible intraclass correlation coefficients that might be expected if fewer than six facilities were used. The intraclass correlation coefficient was calculated for all combinations of two through five facilities. Figure 31.1 shows the low, high, and mean ρ_{FAC} values for the variable paid workers.[4] The values stabilize as more facilities and therefore more cases are brought into the calculation. The mean ρ_{FAC} value remains fairly constant over all combinations of centers. All values of the mean

[4]The variable paid workers was chosen because the most important estimate in this survey is based on that variable.

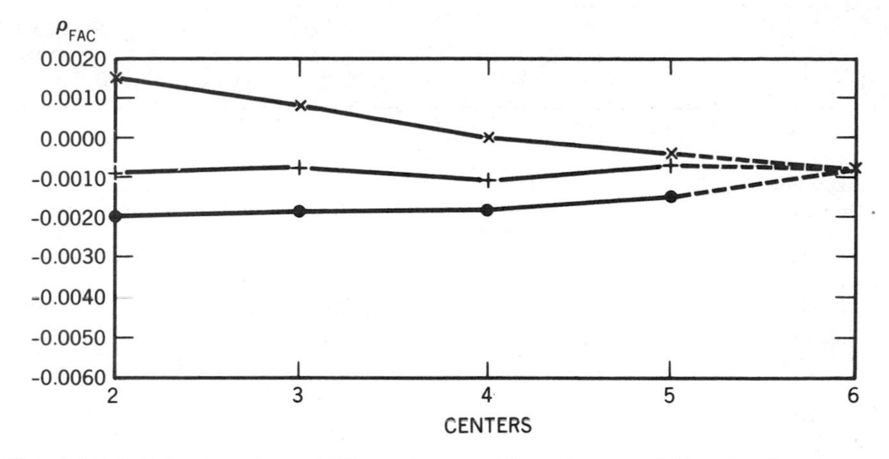

Figure 31.1 Intraclass correlation values: paid workers variable. • = low ρ_{FAC} values; + = mean ρ_{FAC} values; * = high ρ_{FAC} values; •• = the graph if an extremely low value were obtained.

remain negative. Within-facility variance continues to be larger than between-facility variance, showing a not so important facility effect. This analysis shows a general increase in the within-facility variance for the variable paid workers as the number of data collection facilities increases.

The final analysis of data involves the comparison of major data items collected by experience in conducting CATI interviews. Georgia and Nebraska were considered experienced facilities. Table 5 compares the differences in collected data by facility.

Once again, no differences are large enough to be considered statistically significant. These data support the continuation of the intensive training given to each state as CATI hardware is installed and the increased training given to interviewers in each facility before CATI data collection begins. Much has been learned from the experienced facilities. This information has been communicated in the training given to each facility prior to initial data collection using CATI.

A detailed cost analysis of regional data collection compared to data collection on an individual state basis was planned. Examination of these data shows that an improper accounting code was used by most states. Overall cost comparisons, however, point to a savings of approximately $1.30 per sample unit when comparing the regional CATI approach to the current NASS operational structure.

Table 5. Comparison of Data by CATI Experience

Item	Experienced Facility Mean n=1164[a]	Inexperienced Facility Mean n=2227[b]	Difference	Standard Error of Difference
Type of worker				
Self-employed	1.06	1.03	0.03	0.023
Unpaid	1.00	1.04	−0.04	0.054
Paid	3.33	3.64	−0.31	0.495
Hours worked				
Self-employed	57.29	57.61	−0.32	1.025
Unpaid	42.54	41.09	1.45	1.594
Paid	36.07	37.26	−1.19	1.305
Wage rate				
Field worker	4.99	4.67	−0.32	0.174
Livestock worker	4.34	4.42	−0.08	0.190
Supervisor/manager	7.25	7.01	0.24	0.645
Other worker	6.48	6.51	−0.03	1.115
All hourly workers	4.23	4.30	−0.07	0.144
All piece workers	3.35	3.93	−0.58	0.817
All other workers	4.88	4.87	0.01	0.392

[a]Includes Georgia and Nebraska.
[b]Includes Pennsylvania, Tennessee, Texas, and Washington.

5. SUMMARY AND CONCLUSIONS

Interviewers calling from six regional data collection facilities can produce response rates on the same level as those produced by individual state centers. No evidence was found in these data to suggest that a center experienced difficulties in securing response cooperation from farm operators in any particular state included in the study. Although the rate at which respondents inquired as to the interviewer's location varied greatly, no significant effects were found which might influence response rates.

Analysis of the reported data shows no differences in reported data between centers. The analysis of variance for 13 data items shows no differences between data collection facilities at the 0.05 level of significance. Further analyses of the data show an increase in the total variance of five survey statistics due to a facility effect. These increases, although small, are cause for concern and must be overcome by increased training and supervision. The encouraging results, however, are that facility effects for eight data items are negative and produce variance deflation.

A comparison of intraclass correlation coefficients for two through six facilities for the most important item of the survey showed almost no difference in the correlation coefficients. Differences in reported data between facilities by experience were all nonsignificant. Location of calling facility and level of experience do not appear to affect data quality.

One operational problem was encountered in this project. Electronic mail transmission of names between facilities and states for domain determination presented some inconveniences, but all facilities overcame this problem through patience and hard work. This problem will be solved in the near future as NASS offices begin using an online system for list and area sampling frames.

CHAPTER 32

ADMINISTRATIVE ISSUES IN MIXED MODE SURVEYS

Don A. Dillman and John Tarnai[1]
Washington State University

Traditionally the telephone has been viewed as a competing alternative to face to face and mail surveys. This view is particularly evident in research methods texts which typically, and often in laborious detail, describe the relative advantages and disadvantages of telephone surveys in relation to the other methods (e.g., Babbie, 1986; Simon and Burstein, 1985).

That situation is now changing. "Mixed mode" surveys, that is, surveys that combine the use of telephone, mail, and/or face to face interview procedures to collect data for a single survey project are occurring with increasing frequency. A second, or in some cases even a third, method to collect data for a single survey is being used throughout the world, including Australia, Canada, Finland, Denmark, Israel, West Germany, Sweden, the United Kingdom, the Netherlands, and the United States. Indeed, "mixed mode" is becoming one of the survey buzz words of the late 20th century.

There exists a strong intuitive appeal for designing and implementing mixed mode surveys. Each of the basic survey methods has certain limitations that defy solution, e.g., the enormous cost of face to face interviews, the lack of telephones in many households, and the lack of acceptable sample frames for household mail surveys. The joint use of

[1] The authors wish to acknowledge with thanks the help of many people, identified here by country, who responded to our inquiries about their use of mixed mode surveys. Their responses have frequently been reported in the text without further citation: J. A. Noordhoek (Denmark); I. Thomsen (Norway); E. Sudefeld (West Germany); P. Pietila (Finland); W. Sykes (United Kingdom); E. H. J. Vrancken (The Netherlands); D. J. Trewin (Australia); M. Kantorowitz (Israel); W. Nicholls and R. Tortora (United States).

more than one method offers promise for mitigating, if not overcoming, certain limitations of individual methods. Mixed mode survey designs offer the potential of using each method to its greatest advantage while avoiding its onerous weaknesses. The potential offered by mixed mode surveys is to obtain data sets of better quality at a lower cost than can be produced by any single method, at any cost.

Although interest in mixed mode surveys is now substantial, that interest is not entirely new. In 1964 Stanley Payne reported the combining of mail, telephone, and face to face methods for surveying the same respondents and advocated their joint use in other ways. Four years earlier, the U.S. Census Bureau used a combination of mail delivery and personal pick-up as well as the reverse, that is, personal delivery and mail return, for the 1960 decennial census. Also in the 1960s, the Australian Bureau of Statistics used a two phase mail and face to face interview method for the agricultural finance survey. Hochstim (1967) reported a comparison of mail, telephone, and face to face strategies for collecting health information which relied on the remaining two modes to obtain data from those who failed to respond to the initial mode of data collection. Nearly a decade later, Thornberry (1977) reported a similar analysis of all three methods, and Herman (1977) reported an analysis of voting behavior in which telephone and face to face modes had been used as the primary procedures, with followups being implemented by the remaining modes.

In the United States, a report on the role of telephone data collection methods and federal statistics included an analysis of federal surveys approved by the Office of Management and Budget and considered active as of August 22, 1981 (U.S. Office of Management and Budget, 1984). Of 2,137 approved surveys, 80.4 percent used a single method—68.8 percent self-administered, 2.2 percent telephone, and 9.4 percent face to face interviews. An additional 10 percent used two methods, 2.9 percent used all three methods, and procedures were unclear for the remaining 6.7 percent. Although little deliberate analysis of mixed mode procedures appeared in the literature before 1980, this lack of research attention does not appear to be the result of few, if any, mixed mode surveys being done. As a matter of fact, the number of mixed mode surveys exceeds those done by telephone or face to face methods.

Since 1980, deliberate research on mixed mode procedures has expanded significantly (Marquis and Blass, 1985; Massey et al., 1982; Woltman et al., 1980; Hall and Ford, 1979; Groves and Lepkowski, 1985; Carroll et al., 1986; and Thomas, 1987). In addition, mixed mode procedures have been adopted by many countries for major governmental surveys, including the U.S. Current Population Survey, the British Labour Force Survey, the Danish Omnibus and Labour Force Surveys, the

Finnish Labour Force Survey, the U.S. Residential Energy Consumption Surveys, and the U.S. National Agricultural Statistics Service Surveys.

The purpose of this chapter is to review the available research on mixed mode surveys and to identify administrative issues that must be resolved to assure the success of such surveys. The emergent nature of this area of research means we are reviewing work based more on the general reporting of mixed mode experiences, rather than formal methodological inquiries. Our conclusions are, therefore, somewhat tentative.

The chapter is divided into three sections. First, we distinguish six different types of mixed mode surveys, a necessary step for organizing and understanding the wide variety of mixed mode experiences reported in the literature. Second, five distinct forces that encourage greater use of mixed mode designs are discussed, followed by a brief comment on countervailing forces that have slowed down the acceptance of mixed mode procedures. We then discuss eight administrative issues — data comparability, questionnaire construction, personnel requirements, costs, sample design, and respondent selection, centralization versus decentralization, interfacing the modes, and data processing and analysis. Individually and collectively, these administrative issues stand as potential barriers to the design and implementation of successful mixed mode surveys.

1. TYPES OF MIXED MODE SURVEYS

Simply defined, a mixed mode survey is one that uses two or more methods to collect data for a single data set. The most common methods are telephone, face to face, and mail questionnaires. However, it is sometimes useful to distinguish self-administered questionnaires without mail delivery and/or return from other types of self-administered surveys, and to separate computer-assisted interviewing from surveys that use paper and pencil methods. By single data set, we mean data that are analyzed as a unit, which in practice may mean either cross-sectional data or longitudinal data from panel studies.

Based upon the above definition, at least six types of mixed mode surveys can be distinguished from one another. These types are identified below, independent of the particular combinations of methods being used. Each of these six types, and a generic example, is provided in Table 1.

Type A, the use of different modes to collect the same data from different respondents within a sample is perhaps the most common type of mixed mode survey. An example is use of the telephone to obtain data

Table 1. Six Types of Mixed Mode Surveys

Use of Mode	Example
A. To collect same data from different respondents within a sample	Mail for early respondents; telephone and/or face to face interviews for respondents who could not be contacted
B. To collect followup panel data from same respondent at a later time	Face to face interview for initial survey; telephone for panel study followup interview
C. To collect same data from different sample frames in same population for combined analysis	Dual frame survey using area probability sample for face to face interviews and RDD sample for telephone
D. To collect different data from same respondents during a single data collection period	Mail questionnaire administered to respondents immediately after completing a telephone interview
E. To collect data from different populations for comparison purposes	Telephone used to collect data from national household sample for comparison with mail questionnaires from a list frame sample
F. To increase response rate for another mode	Telephone contact used to encourage response to mail survey

from respondents who could not be located for a face to face interview. Telephone and/or address information obtained from neighbors or from other household members is essential for switching from one mode to another. Type A mixed mode surveys are encouraged by efforts to cut cost and decrease the time required for conducting a survey, and respondent preferences or aversion to being surveyed by the primary mode. Works by Hochstim (1967), Thornberry (1976), and Carroll et al. (1986) illustrate various combinations of Type A surveys.

Type B, the use of different modes to collect followup data from the same respondents, is illustrated by using face to face interviews for an initial survey with telephone interviews for the followup. Examples of this increasingly common procedure are the labor force survey in Great Britain, which utilizes an initial face to face interview followed by four telephone contacts on a quarterly basis, and the Current Population Survey in the United States, which calls for an initial face to face interview followed by four telephone followups, another face to face interview, and additional telephone followups. Type B surveys have also been conducted by the U.S. National Center for Health Statistics and the U.S. National Agricultural Statistics Service (Massey et al., 1982). In surveys of this type, the initial contact is typically used to obtain information that will facilitate later followups by the alternative method. Type B surveys appear to be driven in large part by cost considerations, with less expensive methods following initial use of the more expensive method.

Type C, the use of mixed modes to collect the same data from different sample frames in the same population for combined analysis, is relatively new. It is illustrated by dual frame surveys that use face to face interviews for an area probability sample frame and telephone interviews for a random digit dialing (RDD) sample frame. The usual objective appears to be to find an optimal mix which significantly reduces cost without a corresponding increase in sample bias. Research issues underlying the development of Type C dual frame surveys have been explored by Massey et al. (1982), Marquis and Blass (1985), Groves and Lepkowski (1985), and Thomas (1987).

Type D, the use of a different method to collect different data from the same respondents during a single data collection, represents a quite different use of mixed mode procedures, the use of which is typically motivated by quite different concerns. An example is a mail questionnaire administered to respondents immediately after their completion of a telephone interview. Such a questionnaire might be mailed to the telephone respondents in order to obtain information that is difficult to obtain over the telephone, for example, questions for which a visual display is helpful (e.g., ranking a large number of choices) or the completion of diaries that require several days of effort. The telephone interview can be used to obtain eligibility and background information which is combined with data obtained by the second mode into a single cross-sectional data set. This type of mixed mode survey was described and advocated by Payne (1964) more than 20 years ago in his call for greater use of mixed mode procedures. Another example of a Type D survey is the use of self-administered questionnaires during a face to face interview to obtain answers to extremely sensitive questions that respondents might not be willing to verbalize, especially when other members of the household are within hearing distance of an interview.

Type E, the use of different modes to collect data from different populations for comparison purposes, is distinctively different from other mixed mode procedures. Frequently researchers wish to compare the responses of different populations on certain attitudes or behaviors, e.g., attitudes on nuclear energy. An example is the use of the telephone to collect data from a national sample of the general public for comparison with mail questionnaires from various lists of organizational members (e.g., scientists, environmentalists, advocates of nuclear energy) for whom the unavailability of telephone numbers prohibits the use of the telephone. This type of mixed mode survey is being encouraged by increased use of both mail and telephone procedures in recent years which permits conducting surveys at a much lower cost than in the past (Dillman, 1977, 1978).

Type F, the use of a different mode to increase response rates for another mode, differs from all the preceding types in that the data themselves are collected by only one mode. The second and third modes are used to encourage response, for example, a telephone call to encourage response to a previously delivered mail questionnaire (Dillman et al., 1984; Marquis and Blass, 1985). Major factors driving this type of mixed mode effort are time constraints and the belief that a second mode will encourage response to the former.

Differentiation of these six types of mixed mode surveys makes it clear that what may appear on the surface as simply combining together two or more survey methods is in reality substantially more complex. Recognition that each type may involve various combinations of the three methods used in different orders, and that a variety of motivations drive the decision to conduct a mixed mode survey enlarges that complexity. Thus, it should come as no surprise that the range of administrative issues that must be considered is also complex, ranging from how to staff or conduct two or more modes to being sure that the same data are actually being obtained by each method.

2. FORCES ENCOURAGING THE USE OF MIXED MODE SURVEYS

The current high interest in mixed mode surveys is by no means a matter of chance. Several forces, including improvements in the response capabilities of mail and telephone alternatives to face to face interviews, rising survey costs, the vulnerability of methods to different sources of errors, and differential advancement in the application of new information technologies to each of the methods, have coalesced to encourage experimentation with mixed mode surveys. In opposition to these forces, the difficulty of changing long-established practices of many survey organizations, the establishment of new ones with limited survey capabilities, and certain administrative issues have acted to retard the rate of change toward mixed mode survey designs. Understanding the nature of these conflicting pressures helps to identify administrative issues in need of attention by researchers and prospective users.

2.1 Improvements in Capabilities of Mail and Telephone Survey Methods

One of the chief factors retarding the development of serious interest in mixed mode surveys prior to the mid-1970s, particularly among those responsible for governmental surveys, was the inferiority ascribed to telephone and mail surveys (Dillman, 1978). This barrier was recognized by Payne (1964) in his original call for consideration of mixed mode survey procedures.

Research done in the last decade has demonstrated that both mail and telephone surveys have far greater capabilities than they were generally accepted as having prior to that time (Dillman, 1978; Heberlein and Baumgartner, 1978; Frey, 1983; Goyder, 1985; Groves and Kahn, 1979). Improvements in sample frame coverage and expected response rates for questionnaires of considerable length in both mail and telephone surveys led to a case for mixed mode surveys again being made, fourteen years after Payne's original call for mixed mode research (Dillman, 1978).

The improvement in telephone survey methods described by Groves and Kahn (1979) as one of the two most important survey developments in the 20th century (the other one being probability sampling) has been particularly dramatic. Several of the most recent and useful discussions of issues involved in doing mixed mode surveys are in fact discussions focused on switching from surveying entirely by face to face methods to partial use of telephone methods (Marquis and Blass, 1985; Massey et al., 1982; Woltman et al., 1980; Groves and Lepkowski, 1985; Thomas, 1987; Sykes and Collins, 1987).

2.2 Rising Survey Costs

Precise cost comparisons for doing mail, telephone, and face to face surveys are virtually impossible to make. Relative costs differ substantially for various populations, sample designs, and the organizations that conduct surveys. However, it is usually the case that telephone surveys are significantly less costly to implement than are face to face interviews, and mail surveys are usually less costly than telephone surveys. A second mode is often used in Type A surveys when it is judged that additional interview attempts by the first mode are likely to be fruitless. Thus, interview costs for a second mode, regardless of which mode is used, are sometimes (but by no means always) less than would be additional effort by the first mode. The development of Type C, dual frame area probability—RDD surveys, is mostly motivated by cost considerations, as is the use of an alternative followup method in panel studies (Type B). Thus, it is not surprising that rising costs are among the most important forces encouraging the use of mixed mode procedures (e.g., Massey et al., 1982; Thomas, 1987; Hall and Ford, 1979).

2.3 Survey Time Constraints

Many surveys, ranging from completion of television viewing diaries to livestock on hand surveys, must be done within a limited time period in order to avoid recall error and to meet organizational reporting deadlines. This trend is becoming increasingly important as advanced industrial nations make the transition to an information age, which places a premium on reporting timely information for precisely defined populations (Dillman et al., 1985).

Switching from one mode to another sometimes provides an opportunity for speeding the completion of a survey. For example, the speed of telephone surveys is limited by the number of interviewing stations. In large-scale surveys of several thousand respondents, greater speed can often be achieved by starting with mail questionnaires, all of which can be sent out at the same time, saving telephone and face to face contacts for nonrespondents. Thus, the mixed mode time advantage stems from both time sensitive characteristics intrinsic to each method and characteristics of the survey situation.

2.4 Vulnerability of Individual Methods to Different Sources of Errors

Four major types of potential error encountered in doing sample surveys include: noncoverage by the sample frame, sampling bias, measurement errors, and nonresponse. Each of the three survey methods is subject to each type of error to different degrees and in different ways. For example, some households do not have telephones, thus creating a noncoverage problem. Use of area probability methods for selecting households and face to face methods for interviews in homes without telephones overcomes that problem. Although in the United States virtually all households can be reached by mail, no adequate list exists for doing household surveys of the general public. However, names and telephone numbers can sometimes be obtained through initial screening to facilitate the use of less expensive mail and/or telephone contacts. Cost considerations force the use of multi-stage cluster sampling methods or face to face interviews, whereas simpler, more efficient designs can be used for telephone methods.

This differential vulnerability to error of various types gives rise to dual sampling frame approaches as a means of optimizing cost and the extent of error from different sources. For example, Carroll et al. (1986) demonstrate that the use of a mail questionnaire as a followup to a face to face interview corrected certain response biases by getting more responses from one person households, males, older respondents, and those living in multi-unit buildings in central metropolitan areas.

Each method may pose different measurement problems, for example, anonymity and allowable questionnaire complexity. RDD telephone surveys and certain mail methods can allow complete anonymity, whereas face to face interviews do not. The complete reliance on verbal communication restricts the complexity of some questions in telephone surveys. A mixed mode approach allows reliance on the best method from a measurement perspective using supplementation by a less desirable method.

Response rate considerations also vary by method. When cost is not a barrier, response to face to face interviews is usually highest, telephone next, and mail third (Dillman, 1978). Widespread concern appears to exist that response rates to all methods are lower than in the past. The response rate problems for telephones may soon become particularly problematic because of screening devices such as answering machines, number recognition devices (which are now being test-marketed), alienation from the use of telephones for product marketing, and changes in telephone behavior norms that make it acceptable to ignore ringing telephones and terminate unwanted calls without a sense of guilt.

It is well documented that people vary in their survey mode preferences (Groves and Kahn, 1979; Hall and Ford, 1979). If given a choice, some would prefer each method over the others. The mixed mode design provides an opportunity for the respondent to switch methods. The opportunity to change modes seems likely to be a powerful means for reducing nonresponse error, regardless of which method is used first. The advantages of mixed mode designs for reducing nonresponse error is one of the most frequently reported reasons for conducting surveys of this type (Hochstim, 1967; Carroll et al., 1986; Woltman et al., 1980; Whitmore et al., 1985; Herman, 1977; Thornberry, 1977).

2.5 Differential Advancement in Application of New Information Technologies

A fifth factor encouraging the use of mixed mode surveys is differential change in the survey technologies by means of which the various methods are implemented, especially the development of computer assisted telephone interviewing (CATI). Face to face interviewing runs counter to major societal trends, being both labor and energy intensive (Dillman et al., 1985). In contrast, mail surveying has been able to take advantage of new word processing capabilities for printing of quality questionnaires and personalization of correspondence, important factors that encourage high quality and increased response rate (Dillman, 1978). The task of sending followups to nonrespondents in mail surveys, which is the single most important stimulus to high response rate, has been greatly facilitated by using word processing equipment.

Telephone surveying has utilized information technologies to even greater advantage with networked CATI systems. In addition to direct computer input of data by the interviewer, thus bypassing the usual coding and data entry process, CATI systems provide for selection of random digits, automatic dialing, scheduling of initial contacts, scheduling needed return calls, and data base management. These same technologies make possible the development of telephone questionnaires with very complex skip patterns that would be difficult and more time-consuming to handle on paper and pencil face to face interviews and prohibitive on mail questionnaires. Although face to face interviews are now being done by computer-assisted methods, these applications are much less prevalent than for the telephone. Technology also exists for displaying information from previous questions that inform the interviewer whether and how later questions get asked. In sum, improvements in survey technologies, which disproportionately benefit mail and in particular telephone interviewing, have been important forces for encouraging the use of mixed mode surveys.

2.6 Countervailing Forces

The forces which encourage the use of mixed mode surveys are quite strong and seem likely to become stronger. There is little doubt that mixed mode surveys can significantly reduce costs, produce data in a more timely fashion, overcome error problems to a greater extent than can any single method, and gain efficiency from recent improvements in survey technology. These advantages are such that one may appropriately wonder why greater use of mixed mode surveys has not already been achieved. There are several reasons. Only a decade or so has passed since mail, and in particular telephone, surveys received general recognition as viable survey methodologies that might be used in concert with or in place of face to face procedures. As with most new practices, their adoption takes considerable time.

Ironically, the same technological advances that make mixed mode surveys practical also mitigate against them. Small local or regional organizations can easily implement national telephone surveys, but would have great difficulty implementing face to face interview surveys. The establishment of these small survey organizations, which appears to be occurring with greater frequency, may mitigate against the use of mixed mode designs, especially dual frame, face to face, and telephone surveys.

Adoption of mixed mode designs by most existing survey organizations requires changing well-established organizational arrangements, for example, field interviewing offices and data processing operations. It also requires the learning of new data collection skills and development of new organizational capabilities for coordinating their use. Another major obstacle to mixing modes in general population surveys is the sampling problem, that is, the difficulty of obtaining sample frames that contain both addresses and telephone numbers. Administrative issues such as these exist as perhaps the greatest barriers to the increased use of mixed mode surveys.

3. THE ADMINISTRATIVE ISSUES

Administrative issues vary greatly from one survey situation to another. Whereas use of the telephone to encourage response to a mail survey or to schedule a face to face interview (Type E) may be a rather simple matter, deciding to conduct a dual frame panel survey (combination of Types B and C) is not. Although a number of articles have identified administrative issues as major concerns in mixed mode design (e.g., Marquis and Blass, 1985; Massey et al., 1982; Groves and Lepkowski, 1985), surprisingly little research has been done on the best way to resolve them. In this section, we identify eight administrative issues important to a variety of mixed mode surveys.

3.1 Data Comparability

Mixed mode Types A, B, C, and in some instances D and E, involve combining data into a single data set. For this to be done, it is essential to know whether respondents provide the same answers regardless of which method is used. Data comparability establishes constraints and greatly influences when and how mixed mode surveys get done. In particular, it influences what can be concluded from raw data comparisons and the statistical treatment of those data for analysis.

Data comparability across methods is addressed in several other chapters in this book (de Leeuw and van der Zouwen; Körmendi; Sykes and Collins; and Bishop and Hippler), so here we will comment only briefly. Past research has reported a number of specific instances in which telephone and face to face results have differed (for example, Cannell et al., 1987; Thornberry, 1977; Hochstim, 1967; Groves and Kahn, 1979; Woltman et al., 1980). The general conclusion reached in these and other studies is that differences that exist tend to be minor. Very few comparisons of mail and the other methods have been made. However, a consistent finding across the available comparisons is that on questions using vague quantifiers (e.g., very favorable, somewhat favorable, etc.), telephone respondents are more likely to provide extreme responses than are mail respondents (Hochstim, 1967; Dillman and Mason, 1984).

A similar but less-pronounced tendency toward extremeness in telephone versus face to face responses has been observed in vaguely quantified questions reported by Jordan et al. (1980), Groves and Kahn (1979), and Hochstim (1967). Another comparison between telephone and self-administered questionnaires shows that the latter are less prone to response category order effects than are the former (this volume, Chapter 20). Most past research on mode comparisons has not focused on questions using vague quantifier answer choices, a type of question on which the potential for response differences seems greater than for most other questions (Dillman and Mason, 1984). If future research confirms such differences, switching modes for followup studies aimed at measuring attitudinal change could pose significant problems. A conservative approach would be not to mix modes, particularly mail and either of the other methods, when changes in beliefs or attitudes are being measured across time, as in Type B surveys. However, if the error properties of any such differences can be learned, then the combining together of data from different modes might be achieved through statistical adjustments. Thus, even if people do respond differently to questions posed via each of the survey modes, this situation alone does not preclude conducting mixed mode surveys.

Overall, when the wide variety of question structures typically used in surveys is juxtaposed with the small amount of past research on mode differences, there remain many unanswered questions. The need for research on this issue is particularly urgent.

3.2 Questionnaire Construction and Pretesting

Reliance on a single mode has allowed, if not strongly encouraged, questionnaire designers to do what works best for that single method.

Writers of face to face questionnaires incorporate response flashcards to help respondents answer long or complex questions, and use blank tables with appropriate headings for certain questions (e.g., household composition) that interviewers and respondents find easy to fill out in a cooperative manner. For telephone interviews, such implicit tables need to be made explicit and flashcards cannot be used (Marquis and Blass, 1985). The writers of telephone questionnaires tend to use fewer response choices and make maximum use of skip patterns. CATI encourages researchers to randomly order response choices to overcome possible order effects such as those identified by Schuman and Presser (1981). The writers of mail questionnaires typically try to minimize the use of screen questions, avoid open-ended questions, and are more reluctant to ask the same question again later in the survey as a consistency check.

Even seemingly simple issues such as the use of "don't know" categories can be troublesome in mixed mode surveys. Often in telephone surveys "don't know" is not presented as a choice, but accepted if it is volunteered. In mail surveys there is no similar way to accept a volunteered choice without explicitly presenting a category for it. Whether a "don't know" choice is explicitly presented may dramatically affect the proportions selecting that choice. Similarly, asking questions about sex and ethnic origin in the same way in all three modes poses interesting (if not embarrassing) difficulties (Thomas, 1987).

The problems of developing common questionnaires stem to a considerable degree from the reliance on different senses: hearing only for the telephone, touch and vision for the mail questionnaires, and all of these senses for the face to face interview. The problems of developing a common approach to questionnaire construction have been recognized by a number of researchers, for example, Marquis and Blass (1985), Dillman and Mason (1984), Massey et al. (1982), and Thomas (1987).

The problem of writing common questionnaires for all methods involves much more than the wording of questionnaires. A mail questionnaire should not be encumbered by extraneous information for coders or researchers, and a telephone or face to face questionnaire must have needed interviewer directions. A face to face questionnaire typically has instructions for the use of flashcards, and any information that should not be seen by the respondent is written in coded form or eliminated. Individual researchers, and to some extent organizations, tend to construct questionnaires in a format that works well for a specific mode, and when "adapted" for a second mode may not work as well. This problem appears to be greater between mail questionnaires and each of the other two methods (Dillman, 1978).

In some cases, it may not be possible to transfer a questionnaire from one mode to another without significantly altering the questions and ultimately the response distributions (Massey et al., 1982). In cases where face to face interviews have been used to collect government statistics over many years, this problem may be one of the main deterrents for changing to mixed mode designs.

In recent years, work has been done on the development of performance feedback techniques, i.e., providing positive feedback for desired respondent behavior (Marquis and Blass, 1985). Such techniques can be used in both face to face and telephone interviews, but might produce comparability problems if used in mixed mode surveys with mail questionnaires.

Pretesting also presents additional problems. Each mode must be tested separately. A question that works well in one method but not in the other(s) will most likely have to be changed in all methods. Thus, pretesting of mixed mode surveys poses complexities and additional efforts that go well beyond typical pretesting procedures for individual methods.

In sum, shifting to mixed mode data collection requires rethinking of questionnaire construction and pretesting principles so that procedures that produce comparable results are used across all of the modes employed in a single survey. Such questionnaire construction principles have not to date been systematically developed and published. Developing such principles represents a high priority for future research.

3.3 Personnel Requirements

Each mode has different personnel requirements. Developing a mixed mode capability inevitably means training existing personnel or adding new staff. Operating a CATI system requires computer keyboard and programming skills that may be completely lacking among personnel of an organization that engages only in the implementation of face to face and/ or mail surveys. Good face to face interviewers may not make good telephone interviewers, and vice versa.

Use of CATI could result in a geographic switch from where interviewers are located, resulting in not enough work to keep field interviewers busy (Tortora, 1985). Coders and data entry operators may also be underemployed, as much of the CATI work bypasses them. Changing organizational structures from centralized to decentralized, or vice versa, as described by Groves and Lepkowski (1985), may require terminating some people and hiring others.

The personnel problems encountered in developing a mixed mode capability can perhaps be divided into short and long term ones. In the short run, the concern may be with getting current employees to develop new skills, while avoiding the temptation to slightly adapt existing methods that will not work well for the new method. In the long term, the problem is developing a broader set of skills for all employees so that they can comfortably switch back and forth between modes as needed for the completion of a survey. For example, if one is combining an area probability frame using face to face interviews with an RDD frame that uses telephone interviews, it may be possible to administer each of the components using separate personnel. In other instances, and particularly in small organizations, that may not be possible. Also, if one is using the Type A mixed mode design in which one follows or switches modes part way through a survey, changing from one set of personnel to another may be quite cumbersome.

Thomas (1987) suggests that some mixed mode surveys may create morale problems for interviewers. A frequently employed procedure is to use the more expensive and time-consuming survey procedures for initial nonrespondents. Interviewers who are assigned the task of obtaining these interviews may come to feel that they are having to work harder than other interviewers. This is in fact a general problem in all survey research organizations. Survey supervisors tend to ask their most experienced and capable interviewers to conduct difficult interviews or to complete interviews with potential refusals. Unless such interviewers are provided with extra compensation or with other rewards, they may rightly come to view their jobs as undesirable in comparison to other interviewers. In sum, the extensive use of mixed mode surveys poses personnel requirements not typically faced by survey organizations in the past.

3.4 Sample Design and Respondent Selection

Whereas face to face household interviews generally use area probability sampling with clustering to lower costs, there is usually no reason to use these less efficient sampling designs for mail and telephone procedures. Switching to mixed mode designs requires decisions to be made about which sampling method to use and/or how different sample frames are to be combined into an overall design.

The problem is perhaps most acute in national dual frame surveys, which combine an RDD sample frame for telephone interviews with an area probability sampling frame for face to face interviews. Considerable research has been done on this topic, much of it directed toward developing rules for optimal allocation between the two modes (Casady et al., 1981; Groves and Lepkowski, 1985; Sirken and Casady, 1982). The appropriate solution will vary depending upon the survey topic and proportion of households without a telephone. The importance of this issue is sufficiently great that investigations will undoubtedly continue.

Changing even part of a sample design to allow the use of telephone interviewing and avoid travel costs associated with face to face interviews for a study in which many years, if not decades, of data have already been collected, is most likely to be greeted with a lack of enthusiasm (Thomas, 1987). The possibilities of using different administrative designs (Groves and Lepkowski, 1985) may also be greeted with resistance.

Closely related to sample design is the issue of respondent selection. Dual frame designs may require changes in who gets interviewed. For example, Massey et al. (1982) report that the U.S. National Health Interview Survey uses a family-style respondent rule, asking everyone sixteen and over who is at home to participate in the interview. Switching to the telephone would result in the rule being changed. For other mixed mode surveys, use of telephone or mail encourages going after a particular respondent rather than a knowledgeable household member. Issues of respondent selection require making accommodations to limitations of the different modes and necessitates careful consideration by those planning to do mixed mode surveys.

3.5 Costs

Inasmuch as cost is one of the most important factors encouraging the use of mixed mode surveys (Massey et al., 1982), calculating how to achieve cost-efficiency must be considered one of the most important administrative issues. Often an elusive consideration in single mode designs, in a mixed mode context costs become even more difficult to calculate. In addition to the usual costs of each method, necessary communication and coordination is likely to impose additional costs. Also, in a Type A survey, substituting one method for another that has already been tried means that the cost of the new method has to be compared to the likely cost of additional efforts by the original method, rather than the cost per respondent for the interviews already completed (Massey et al., 1982). Such calculations often require considerable speculation.

Ideally, in a Type A survey, one wants to phase in a second mode at an optimal time from the standpoint of cost. Calculations of this nature must usually be made on the basis of incomplete data. Groves and Lepkowski (1985) have calculated costs for one type of dual frame survey and have concluded that different administrative structures, ranging from centralized to decentralized, have little impact on cost efficiency. Such calculations might differ substantially from one organization to another.

To date, no systematic efforts to develop procedures for allocating costs in mixed mode designs of the various types listed in Table 1 have been done. Developing rules for allocating costs which will allow comparisons to be made across survey situations represents an important priority for future research.

3.6 Centralization Versus Decentralization

Conducting face to face interviews for national surveys tends to force decentralization of survey operations in order to have interviewers located near where people live. The conduct of telephone interviews can either be centralized or decentralized (Groves and Lepkowski, 1985; Massey et al., 1982; Marquis and Blass, 1985; Tortora, 1985). The development of sophisticated CATI systems with features such as automatic call scheduling, rescheduling, and dialing encourages centralization of interviewing facilities. The many advantages of such centralization for training, assigning, and supervising interviewers have been described by Marquis and Blass (1985). In addition, Groves and Lepkowski (1985) have noted the economies of scale and greater ease in balancing interviewer loads in centralized facilities.

However, Marquis and Blass (1985) also raise questions about possible negative effects on interviewer morale. They note, for example, that centralization also removes the opportunity for face to face social contact, which many face to face interviewers consider quite desirable. They also indicate how a large centralized interviewing center may result in extreme specialization of tasks, for example, the best interviewers only being assigned refusal conversions. Alternatively, personnel might be rotated among several jobs. These considerations raise at least two important administrative issues which need to be examined.

The organization of face to face interviewing on the one hand and telephone on the other creates opposing forces in an organization. What is efficient for one mode may be inefficient for the other. Dealing with these opposing forces represents a major management challenge for large survey organizations. For example, personnel who benefit from doing face to face interviews might pursue decisions to start a telephone followup somewhat later in the survey process in order to provide more work for face to face interviewers. Those who benefit from telephone interviewing might do the opposite, resulting in tensions that could lower organizational efficiency and affect data quality.

The second issue important to the centralization versus decentralization issue is the long term effect on interviewing quality. Centralization implies that respondents will increasingly be surveyed by interviewers who are culturally different from themselves, and that interviewers will not have background experiences (which would come from face to face interviews) that help them grasp the subtleties of respondent answers and correctly understand colloquial expressions. One result could be lower respondent cooperation.

It is interesting that the possibility of centralizing survey operations is developing at a time when most institutions of post-industrial societies are in the process of decentralizing. In developing centralized models it seems important that we not in effect develop assembly line production procedures which many other segments of society are now trying to shed.

3.7 Interfacing the Modes

Conduct of mixed mode surveys requires that methods be effectively and efficiently interfaced. Doing so requires knowledge, effort, and money, in addition to that needed for implementation of a single method used alone. These, in effect, are interfacing costs.

Work needs to be done to determine what principles should guide the decisions to switch methods. Switching from face to face to telephone interviews in a survey could be done after only one or two face to face attempts, or after five or six attempts. Wording of introductions or letters used in a later mode need to be done in a way that is consistent with earlier respondent contacts. If speed of completing a survey is a goal, the rules for switching modes might be very different than if cost is the main issue, and still different if response rate is the major concern. Thomas (1987) observes that coordinating information from different survey modes is limited by the slowest method, which will usually be the mail self-administered mode, or the personal interview mode. Information needed for a second mode must, therefore, await the outcome of the first mode. If the first mode is relatively slow in providing the information, this will delay the implementation of the second mode.

In sum, doing mixed mode surveys usually involves more than using different methods to collect data from different people. The activities associated with each method have to be linked together, and such coordination usually requires additional resources.

Another aspect of interfacing modes is gathering data from one mode to facilitate another. Such activities range from asking neighbors for a not-at-home neighbor's telephone number and/or mailing address. One of the exciting prospects of mixed mode surveying is for organizations to use screening efforts to get information that would allow, for example, mail procedures to be used on household samples for which an adequate sample frame has not in the past been available.

3.8 Data Processing and Analysis

Doing mixed mode surveys raises a host of intriguing processing and analysis issues. Data processors are likely to be confronted with handling both computerized data from CATI interviews and paper and pencil forms. Coders may need to interact with the interviewers and supervisors to see how certain problems that appear on face to face or mail forms have been handled in order to assure consistency with decisions made by the telephone interviewers. In the past, achieving consistent decisions has not required extensive communication between personnel involved in coding activities and those who do the interviewing.

Mixed mode survey operations also result in the need to add information to data files about how data were collected. The study by Carroll et al. (1986) demonstrates the kinds of data processing problems that may arise in mixed mode surveys. The authors indicate that a major problem with the mail followup procedure was the integration of the abbreviated mail data with the more complete personal interview data. The steps involved in accurately interpreting the data and accounting for missing information resulted in problems for the data analysis staff.

As the number of mixed mode surveys increases, it will be important for such information to be considered in data analysis, at least until the time that we are completely confident that people have responded in the same fashion to each mode. If differences persist, analysis and mode effects from many surveys will likely provide needed information for how to statistically adjust data to take into account mode effects.

4. CONCLUSION

In this chapter, we have identified eight administrative issues that must receive attention if the effective use of mixed mode surveys is to become a reality. Dealing with these issues is complicated when one considers the variety of mixed mode surveys which might be done, six of which were identified here. However, a number of powerful forces have coalesced to encourage the use of mixed mode surveys. The strength of these forces make it likely that attention will be focused on overcoming these administrative barriers for the conduct of mixed mode surveys.

It remains to be seen what the consequences of an increased emphasis on doing mixed mode surveys will be. On the one hand it may usher in an era in which there truly is an emphasis on total survey error. The development of mixed mode approaches could encourage practitioners who focused on only one mode to learn about and effectively use the others. In sum, the growing emphasis on mixed mode surveys could prove to be a stimulus for creating a single discipline or subdiscipline of survey research rather than an activity which has some components mostly of interest to statisticians, others of interest mostly to sociologists, others of main interest to psychologists, etc.

The development of mixed mode methods may also encourage the development of new cleavages and distinctions in the survey research community. During the past decade the development of viable mail and telephone survey methods technologies which can be handled by small survey organizations encouraged the development of many regional and local survey research centers. Development of mixed mode strategies which rely on expensive and sophisticated CATI systems requires more resources than those needed to do a single method survey. One result may be increased distinctions among the survey capabilities of various organizations, at a time when more and more organizations are in the business of doing surveys.

Reliance on a mixed mode strategy could in fact encourage the centralization of survey research at a time when more use of surveys is being encouraged by major processes of societal decentralization. We look forward with much interest to seeing how these tensions will be resolved during the coming decades.

BIBLIOGRAPHY OF TELEPHONE SURVEY METHODOLOGY

This bibliography contains a comprehensive list of the literature in telephone survey methodology as well as other references made by authors of this volume. References by authors to articles that are not specifically relevant to telephone surveys are marked with an (*). This classification system is somewhat subjective but is more informative than an unclassified list of papers and books.

The bibliography began with a listing compiled by the Survey Research Center at The University of Michigan, updated in September, 1985. Staff at Statistics Sweden added entries from other bibliographies and reference lists (e.g., *Current Index of Statistics*). To that set, references made in individual chapters of the volume were added, if not already present.

The bibliography is extensive but no bibliography can be fully complete. It includes a number of unpublished reports that were judged important even if they were difficult to obtain. Some but not all of the contributed papers presented at the November, 1987, International Conference on Telephone Survey Methodology, are included. Further, it has not been possible to check all of the entries, and errors no doubt remain. The user should be cautious.

Adams, W.C., and Smith, D.J., "Effects of Telephone Canvassing on Turnout and Preferences: A Field Experiment," *Public Opinion Quarterly*, Vol. 44, Fall 1980, pp. 389-395.

Adler, M.K., "Types of Surveys," in M.K. Adler, *Modern Market Research: A Guide for Business Executives*, New York, Philosophical Library, 1957, pp. 78-92.

Adler, M.K., "The Use of the Telephone in Industrial Market Research," *Scientific Business*, Vol. 1, No. 4, 1964, pp. 336-342.

Alexander, C., Sebold, J., and Pfaff, P., "Some Results with an Experiment With Telephone Sampling for the US National Crime Survey," *Proceedings of the Section on Survey Research Methods, American Statistical Association*, 1986.

Alwin, D.F., Finney, J.M., and Otto, L.B., "An Investigation of Completion Rates and Response Differences in Telephone and Mail Methods of Administering a Sample Survey Instrument," Survey Research Center, Institute for Social Research, The University of Michigan, Ann Arbor, 1980.

*Andersen, B.H., and Christoffersen, M.N., Om Sporgeskemaer. Socialforsknings-instituttets studie 46, Copenhagen, Denmark, 1982. (In Danish)

Andersen, R., "Experience of the Center for Health Administration Studies (CHAS) With Telephone Survey Methods/Potential Uses of Computer Assisted Methods in Studies of Health and Well Being," paper presented at the conference on Computer-Assisted Telephone Interview Technology, Berkeley, CA, March 1980.

Anderson, C.L., and Halford, L.J., "A Four State Comparison of Variable Sampling and Data Collecting Procedures," *Pacific Sociological Review*, Vol. 13, No. 3, 1970, pp. 149-150.

*Anderson, D., and Aitkin, M., "Variance Component Models With Binary Response: Interviewer Variability," *Journal of the Royal Statistical Society*, Series A, Vol. 47, No. 2, 1985, pp. 203-210.

*Aneshensel, C.S., Frerichs, R.R., Clark, V.A., and Yokopenic, P.A., "Measuring Depression in the Community: A Comparison of Telephone and Personal Interviews," *Public Opinion Quarterly*, Vol. 46, No. 1, Spring 1982, pp. 110-121.

Aneshensel, C.S., Frerichs, R.R., Clark, V.A., and Yokopenic, P.A., "Telephone Versus In-Person Surveys of Community Health Status," *American Journal of Public Health*, Vol. 72, No. 9, September 1982, pp. 1071-1102.

*Apple, W., Streeter, L.A., and Krauss, R. M., "Effects of Pitch and Speech Rate on Personal Attributions," *Journal of Personality and Social Psychology*, Vol. 37, 1979, pp. 715-726.

*Argyle, M., *Social Interaction*, London, Methuen, 1969.

Aronson, S.H., "The Sociology of the Telephone," *International Journal of Comparative Sociology*, Vol. 12, 1971, pp. 153-167.

Ashikaga, T., Greene, C.J., Young, M.R., and MacPherson, B.V., "Results of Random Digit Dialing Survey Methods in a Rural State," *Proceedings of the Section on Survey Research Methods, American Statistical Association*, 1982, pp. 577–580.

Assael, H., "Comparison of Brand Share Data by Three Reporting Systems," *Journal of Marketing Research*, Vol. 4, November 1967, pp. 400–401.

*Assael, H., and Keon, J., "Non Sampling Versus Sampling Errors in Survey Research," *Journal of Marketing*, 1982, Vol. 46, pp. 114–123.

AT&T, *The World's Telephones*, Atlanta, R.H. Donnelley, 1982.

*Ayidiya, S.A., "Unanticipated Similarity of Response Effects in Mail and Interview Surveys," paper presented at the 42nd Annual Conference of the American Association for Public Opinion Research, Hershey, PA, May 14–17, 1987.

*Babbie, E., *The Practice of Social Research*, 3rd ed., Belmont, CA, Wadsworth Publishing Co., 1986.

*Bailar, B.A., "The Quality of Survey Data," *Proceedings of the Section on Survey Research Methods, American Statistical Association*, 1984.

Bailar, B.A., "Nonresponse: What It Is and What We Do About It," *Statistical Journal of the United Nations*, Vol. 2, 1984, pp. 381–392.

*Bailar, B.A., "Information Needs, Surveys, and Measurement Methods," paper presented at the Conference on Panel Surveys, Washington, DC, 1986.

Bailar, B.A., and Lanphier, C., *Development of Survey Methods to Assess Survey Practices*, American Statistical Association, 1978.

*Bailar, B.A., Bailey, L., and Stevens, J., "Measures of Interviewer Bias and Variance," *Journal of Marketing Research*, Vol. 14, 1977, pp. 337–343.

Bailar, B.A., Bailey, L., and Corby, C., "A Comparison of Some Adjustment of Weighting Procedures for Survey Data,", in N.K. Namboodiri (ed.), *Survey Sampling and Measurement*, New York, Academic Press, 1978.

*Bales, R.F., *Interaction Process Analysis*, Cambridge, MA, Wesley Press, 1950.

Ball, D.W., "Toward a Sociology of Telephones and Telephoners," in M. Truzzi (ed.), *Sociology and Everyday Life*, Englewood Cliffs, NJ, Prentice-Hall, 1968, pp. 59–75.

*Bangert-Drowns, R.L., "Review of Developments in Meta-analytic Method," *Psychological Bulletin*, Vol. 99, No. 3, 1986, pp. 388–399.

Banks, M.J., "Comparing Health and Medical Care Estimates of the Phone and Nonphone Populations," *Proceedings of the Section on Survey Research Methods, American Statistical Association*, 1983, pp. 569–574.

Banks, M.J., and Andersen, R.M., "Estimating and Adjusting for Nonphone Coverage Bias Using Center for Health Administration Studies Data," in National Center for Health Services Research, Health Survey Research Methods: Proceedings of the 4th Biannual Conference, *National Center for Health Services Research Proceedings Series*, Department of Health and Human Services Publication No. (PHS) 84-3346, 1982.

Banks, M.J., and Hagan, D.E., "Reducing Interviewer Screening and Controlling Sample Size in a Local-Area Telephone Survey," *Proceedings of the Section on Survey Research Methods, American Statistical Association*, 1984, pp. 271-273.

Bean, J.A., "Estimate and Sampling Variability in the Health Interview Survey," *Vital and Health Statistics* Series 2, No. 38, Department of Health, Education, and Welfare Publication No. (HRA) 74-1288, 1974.

Becker, H.J., *An Alternative to RDD Sampling in Small Area Studies*, Johns Hopkins University, Center for Social Organization of Schools, Baltimore, 1979.

Bemelmans-Spork, M.E.J., and Sikkel, D., "Data Collection With Hand-held Computers," *Proceedings of the 45th Session, International Statistical Institute*, Book 3, Topic 18.3, 1985, pp. 1-16.

Bemelmans-Spork, M.E.J., and Sikkel, D., "Observation of Prices With Hand-held Computers," *Statistical Journal of the United Nations Economic Commission for Europe*, Vol. 3, 1985, pp. 153-160.

Bemelmans-Spork, M.E.J., Keller, W., and Sikkel, D., "Use of Handheld Microcomputers in the Netherlands Central Bureau of Statistics," *Proceedings of the First Annual Research Conference of the U.S. Census Bureau*, U.S. Bureau of the Census, 1985, pp. 47-52.

Bennett, C.T., "A Telephone Interview: A Method for Conducting a Follow-up Study," *Mental Hygiene*, Vol. 45, April 1961, pp. 216-220.

Bercini, D.H., and Massey, J.T., "Obtaining the Household Roster in a Telephone Survey: The Impact of Names and Placement on Response Rates," *Proceedings of the Social Statistics Section, American Statistical Association*, 1979, pp. 136-140.

Bergsten, J.W., "Some Methodological Results from Four Statewide Telephone Surveys Using Random Digit Dialing," *Proceedings of the Section on Survey Research Methods, American Statistical Association*, 1979, pp. 239-244.

Bergsten, J.W., and Pierson, S.A., "Telephone Screening for Rare Characteristics Using Multiplicity Counting Rules," *Proceedings of the Section on Survey Research Methods, American Statistical Association*, 1982, pp. 145-150.

Bergsten, J.W., Weeks, M.F., and Bryan, F.A., "Effects on an Advance Telephone Call in a Personal Interview Survey," *Public Opinion Quarterly*, Vol. 48, No. 3, Fall 1984, pp. 650-657.

Berry, M., and Miller, P.V., "Reducing Response Effects in Telephone Interviews," *Proceedings of the Section on Survey Research Methods, American Statistical Association*, 1980, pp. 294-298.

Beza, A., "Study Design and Telephone Surveys," paper presented at the Field Directors Conference, Santa Monica, CA, 1981.

Biel, A.L., "Abuses of Survey Research Techniques: The Phony Interview," *Public Opinion Quarterly*, Vol. 31, No. 2, Summer 1967, p. 298.

Biemer, P.P., "Optimal Dual Frame Sample Design: Results of a Simulation Study," *Proceedings of the Section on Survey Research Methods, American Statistical Association*, 1983, pp. 630–635.

Biemer, P.P., "Review of 'Surveys by Telephone: A National Comparison With Personal Interviews'" (by R.M. Groves and R.L. Kahn), *Journal of the American Statistical Association*, Vol. 78, 1983, pp. 996–997.

*Biemer, P.P., and Stokes, S.L., "An Improved Procedure for Estimating the Components of Response Variance in Complex Surveys," August 1983, U.S. Bureau of the Census, unpublished manuscript.

Biemer, P.P., and Wolfgang, G., "Estimates of the Index of Inconsistency from the RDD-1 Study," U.S. Bureau of the Census, internal memorandum for record, 1985.

Biemer, P.P., Chapman, D.W., and Alexander, C.F., "Some Research Issues in Random-Digit-Dialing Sampling and Estimation," *Proceedings of the First Annual Research Conference of the U.S. Census Bureau*, U.S. Bureau of the Census, March 1985, pp. 71–86.

*Bikson, T.K., and Gutek, B.A., "Advanced Office Systems: An Empirical Look at Use and Satisfaction," N-1970-NSF, The RAND Corporation, Santa Monica, CA, 1983.

*Bishop, G.F., "Experiments With the Middle Response Alternative in Survey Questions," *Public Opinion Quarterly*, Vol. 51, NO. 2, 1987, pp. 220–232.

*Bishop, G.F., Oldendick, R.W., Tuchfarber, A.J., and Bennett, S.E., "Pseudo-Opinions on Public Affairs," *Public Opinion Quarterly*, Vol. 44, No. 2, 1980, pp. 198–209.

*Bishop, G.F., Oldendick, R.W., and Tuchfarber, A.J., "Effects of Presenting One Versus Two Sides of an Issue in Survey Questions," *Public Opinion Quarterly*, Vol. 46, No. 1, 1982, pp. 69–85.

*Bishop, G.F., Oldendick, R.W., and Tuchfarber, A.J., "Effects of Filter Questions in Public Opinion Surveys," *Public Opinion Quarterly*, Vol. 47, No. 4, 1983, pp. 528–546.

*Bishop, G.F., Oldendick, R.W., and Tuchfarber, A.J., "What Must My Interest in Politics Be If I Just Told You 'I Don't Know'?" *Public Opinion Quarterly*, Vol. 48, No. 2, 1984, pp. 510–519.

*Bishop, G.F., Oldendick, R.W., and Tuchfarber, A.J., "The Importance of Replicating a Failure to Replicate: Order Effects on Abortion Items," *Public Opinion Quarterly*, Vol. 49, No. 1, 1985, pp. 105–114.

*Bishop, G.F., Tuchfarber, A.J., and Oldendick, R.W., "Opinions on Fictitious Issues: The Pressure to Answer Survey Questions," *Public Opinion Quarterly*, Vol. 50, No. 2, 1986, pp. 240–250.

Blair, J., and Czaja, R., "Locating a Special Population Using Random Digit Dialing," *Public Opinion Quarterly*, Vol. 46, Winter 1982, pp. 585–590.

Blankenship, A.B., "Listed Versus Unlisted Numbers in Telephone Survey Samples," *Journal of Advertising Research*, Vol. 17, February 1977, pp. 39–46.

Blankenship, A.B., *Professional Telephone Surveys*, McGraw Hill, 1977.

Blankenship, A.B., and Pearson, M.M., "Guidelines for Telephone Group Interviews," *Journal of the Academy of Marketing Science*, Vol. 5, 1977, pp. 1–8.

Blankenship, A.B., Crossley, A., Heidingsfield, M.S., Herzog, H., and Kornhauser, A., "Questionnaire Preparation and Interviewer Technique," *Journal of Marketing*, Vol. 14, 1949, pp. 399–433.

Boettinger, H., "Our Sixth-and-a-Half Sense," in I. Pool (ed.), *The Social Impact of the Telephone*, Cambridge, MA, MIT Press, 1977.

Booth, A., and Johnson, D.R., "Tracking Respondents in a Telephone Interview Panel Selected by Random Digit Dialing," *Sociological Methods & Research*, Vol. 14, No. 1, August 1985, pp. 53–64.

Bortner, B.Z., "The Use of Cathode Ray Tube Interviewing on Your Research Projects," paper presented at the 18th ARF Conference, New York, 1972.

Bortner, B.Z., and Assael, H., "Continuous Tracking Studies via WATS Lines and Personal Interviewing," presented at the Annual Conference of the Advertising Research Foundation, New York, 1967.

Botman, S.L., Massey, J.T., and Shimizu, I.M., "Effect of Weighting Adjustments on Estimates from a Random-Digit-Dialed Telephone Survey," *Proceedings of the Section on Survey Research Methods, American Statistical Association*, 1982, pp. 139–144.

*Bradburn, N.M., "Response Effects," in P.H. Rossi, J.D. Wright, and A.B. Anderson (eds.), *Handbook of Survey Research*, Orlando, Academic Press, 1983.

*Bradburn, N.M., Sudman, S., et al., *Improving Interviewing Method and Questionnaire Design: Response Effects to Threatening Questions in Survey Research*, San Francisco, Jossey-Bass, 1979.

Brandon, B.B., *The Effect of the Demographics of Individual Households on Their Telephone Usage*, Cambridge, MA, Ballinger Publishing Co, 1981.

Brehm, J., "Analysis of Result Code Disposition for Continuous Monitoring by Time in Field: Report to the Board of Overseers, National Election Studies," Working Paper #7, Ann Arbor, National Election Studies, Institute for Social Research, 1985.

*Brenner, M., "Response Effects of 'Role-Restricted' Characteristics of the Interviewer," in W. Dijkstra and H. van der Zouwen (eds.), *Response Behaviour in the Survey Interview*, London, Academic Press, 1982.

*Brown, B.L., Strong, W.L., and Rencher, A.C., "Perceptions of Personality From Speech: Effects of Manipulations of Acoustical Parameters," *Journal of the Acoustical Society of America*, Vol. 54, 1973, pp. 29–35.

Brown, J.L., and Wilson, R.R., "Summary Report — 1984 NHIS-RDD Feasibility Study," Bureau of the Census, internal memorandum, October 1984.

Brown, P.R., and Bishop, G.F., "Who Refuses and Resists in Telephone Surveys? Some New Evidence," University of Cincinnati, 1982.

Brunner, G.A., and Carroll, S.J., Jr., "The Effect of Prior Telephone Appointments on Completion Rates and Response Content," *Public Opinion Quarterly*, Vol. 31, No. 4, Winter 1967–68, pp. 652–654.

Brunner, J.A., and Brunner, G.A., "Are Voluntarily Unlisted Telephone Subscribers Really Different?" *Journal of Marketing Research*, Vol. 8, February 1971, pp. 121–124.

Bryant, B.E., "Respondent Selection in a Time of Changing Household Composition," *Journal of Marketing Research*, Vol. 12, May 1975, pp. 129–135.

Burke, J., Morganstein, D., and Schwartz, S., "Toward the Design of an Optimal Telephone Sample," *Proceedings of the Section on Survey Research Methods, American Statistical Association*, 1981, pp. 448–453.

Burt, C.W., "Random Digit Dialing Experiments: An Investigation Into the Criterion Validity of a Telephone Interviewer Selection Procedure," U.S. Bureau of the Census, unpublished report, March 1983.

Buse, R.C., "Increasing Response Rates in Mailed Questionnaires," *American Journal of Agricultural Economics*, Vol. 55, No. 3, August 1973, pp. 503–508.

Bushery, J.M., *National Crime Survey — Evaluation of Increased Telephone Interviewing Procedure*, Washington, DC, U.S. Bureau of the Census, 1980.

Bushery, J.M., Cowan, C.D., and Murphy, L.R., "Experiments in Telephone-Personal Visit Surveys," *Proceedings of the Section on Survey Research Methods, American Statistical Association*, August 1978, pp. 564–569.

Buss, T.F., "Sampling Bias in Statewide Telephone Surveys of High School Dropouts," *Proceedings of the Section on Survey Research Methods, American Statistical Association*, 1979, pp. 234–238.

Cahalan, D., "Measuring Newspaper Readership by Telephone: Two Comparisons With Face to Face Interviews," *Journal of Advertising Research*, Vol. 1, No. 2, December 1960, pp. 1–6.

Cahalan, D., and Treiman, B., "Drinking Behavior, Attitudes, and Problems in Marin County, California," Social Research Group, School of Public Health, University of California, Berkeley, unpublished paper, January 1976.

*Campbell, B., "Race-of-Interviewer Effects among Southern Adolescents," *Public Opinion Quarterly*, Vol. 45, 1981, pp. 231–244.

Cannell, C.F., *The Telephone Interview: Progress and Prospects*, Survey Research Center, Institute for Social Research, The University of Michigan, Ann Arbor, 1978.

Cannell, C.F., and Fowler, F.J., "Interviewers and Interviewing Techniques," *Advances in Health Survey Research: Proceedings of a National Conference, Airlie House, Airlie, VA, May 1–2, 1975*, Washington, DC, Health Resources Administration, 1977, pp. 13-23.

Cannell, C.F., and Groves, R.M., "Development and Experimentation With Telephone Data Collection Techniques (Machine-Readable Data File)," Survey Research Center, Institute for Social Research, The University of Michigan, Ann Arbor, 1979.

*Cannell, C.F., Fowler, F.J., and Marquis, K.H., "The Influence of Interviewer and Respondent Psychological and Behavioral Variables on the Reporting in Household Interviews," *Vital and Health Statistics*, Series 2, Number 26, Public Health Service, 1968.

*Cannell, C.F., Lawson, S.A., and Hausser, D.L., *A Technique for Evaluating Interviewer Performance*, Ann Arbor, MI, Survey Research Center, Institute for Social Research, The University of Michigan, 1975.

Cannell, C.F., Groves, R.M., and Miller, P.V., "The Effects of Mode of Data Collection on Health Survey Data," *Proceedings of the Social Statistics Section, American Statistical Association*, 1981, pp. 1-6.

Cannell, C.F., Miller, P.V., and Oksenberg, L.F., "Research on Interviewing Techniques," in S. Leinhardt (ed.), *Sociological Methodology*, San Francisco, Jossey-Bass, 1981, pp. 389–437.

Cannell, C., Groves, R., Magilavy, L., Mathiowetz, N., and Miller, P., "An Experimental Comparison of Telephone and Personal Health Surveys," *Vital and Health Statistics*, Series 2, No. 106, Public Health Service, 1987, (PHS), pp. 87-1380.

Carroll, D., Cohen, R., Slider, C., and Thompson, W., "Use of a Mailed Questionnaire to Augment Response Rates for a Personal Interview Survey," presented at the annual meeting of the American Association for Public Opinion Research, St. Petersburg, FL, May 15-18, 1986.

Carter, R.E., Jr., "Field Methods in Communication Research," in R.O. Nafziger and D.M. White (eds.), *Introduction to Mass Communication Research*, revised edition, Baton Rouge, Louisiana State University Press, 1963, pp. 78-127.

Carter, R.E., Jr., and Troldahl, V.C., "Use of a Recall Criterion in Measuring the Educational Television Audience," *Public Opinion Quarterly*, Vol. 26, 1962, pp. 114-121.

Casady, R.J., and Sirken, M.G., "A Multiplicity Estimator for Multiple Frame Sampling," *Proceedings of the Section on Survey Research Methods, American Statistical Association*, 1980, pp. 601-609.

Casady, R.J., Snowden, C.B., and Sirken, M.G., "A Study of Dual Frame Estimators for the National Health Interview Survey," *Proceedings of the Section on Survey Research Methods, American Statistical Association*, 1981, pp. 444-447.

CASRO (Council of American Survey Research Organizations), Report of the CASRO Completion Rates Task Force, New York, Audits & Surveys Company, Inc., unpublished report, 1982.

Catlin, G., and Murray, S., "Canadian Victimization Surveys: A Report on Pretests in Edmonton and Hamilton, *Survey Methodology*, Vol. 5, No. 2, December 1979, pp. 200-237.

Catlin, G., Choudhry, H., and Hofmann, H., "Telephone Ownership in Canada," Statistics Canada, unpublished report, 1984.

Champion, E., *Use of Telephone Interviewing in SORAR*, U.S. Bureau of the Census, March 30, 1976.

Champion, E., "Supplement to 'Use of Telephone Interviewing in SORAR,'" March 30, 1976, U.S. Bureau of the Census, July 20, 1976.

Champion, E., "Update of 'Use of Telephone Interviewing in SORAR,'" March 30, 1976, U.S. Bureau of the Census, April 5, 1978.

*Chapanis, S., Ochsman, R., Parish, R., and Weeks, G., "Studies in Interactive Communication: I. The Effects of Four Communication Modes on the Behaviour of Teams During Co-operative Problem Solving," *Human Factors*, No. 14, 1972, pp. 487-509.

*Chapple, E.D., "The Standard Experimental (Stress) Interview as Used in Interaction Chronograph Investigations," *Human Organization*, Vol. 12, 1953, pp. 23-32.

Chilton Research Services, *The History and Development of CATI*, Chilton Research Services, Radnor, PA (undated).

Chilton Research Services, *A National Probability Sample of Telephone Households Using Computerized Sampling Techniques*, Chilton Research Services, Radnor, PA (undated).

Choudhry, H., "Results From Telephone Interviewing Experiment in the Non Self Representing Areas of the Labour Force Survey," Statistics Canada, internal memorandum, 1984.

*Chromy, J.R., "Design Optimization With Multiple Objective," *Proceedings of the Section on Survey Research Methods, American Statistical Association*, 1987.

Cialdini, R.B., Vincent, J.E., Lewis, S.K. Catalan, J., Wheeler, D., and Darby, B.L., "Reciprocal Concessions Procedure for Inducing Compliance: The Door-in-the-Face Technique," *Journal of Personality and Social Psychology*, Vol. 31, 1975, pp. 206-215.

*Clarke, L., Phibbs, M., Klepez, A., and Griffiths, D., "General Household Advance Letter Experiment," *Survey Methodology Bulletin*, Office of Population Censuses and Surveys, 1987.

*Cohen, J., *Statistical Power Analysis for the Behavioral Sciences*, New York, Academic Press, 1969.

*Cohen, R., Slider, C., and Thompson, W., "Use of a Mailed Questionnaire to Augment Response Rates for a Personal Interview Survey," presented at the annual meeting of the American Association for Public Opinion Research, St. Petersburg, FL, May 15–18, 1986.

Collins, M., "Computer-Assisted Telephone Interviewing in the U.K.," *Proceedings of the Section on Survey Research Methods, American Statistical Association*, 1983, pp. 636-641.

Collins, M., "Telephone Interviewing in Consumer Surveys," *Market Research Society Newsletter*, No. 11, October 1983.

Collins, M., and Sykes, W., "The Problems of Non-Coverage and Unlisted Numbers in Telephone Surveys in Britain," *Journal of the Royal Statistical Society*, Series A 150, Vol. 3, 1987, pp. 241–253.

Colombotos, J., "The Effects of Personal vs. Telephone Interviews on Socially Acceptable Responses," *Public Opinion Quarterly*, Vol. 29, No. 3, Fall 1965, pp. 457–458.

Colombotos, J., "Personal Versus Telephone Interviews: Effect on Responses," *Public Health Reports*, Vol. 84, No. 9, September 1969, pp. 773–782.

Computer-Assisted Survey Methods Program, *Computer-Assisted Survey Methods at the Office of Computing Affairs*, University of California, Berkeley, 1984.

Computer-Assisted Survey Methods Program, *A Demonstration Questionnaire Written in the Q Language*, University of California, Berkeley, 1984.

Cook, W.A., Nelson, R., and Oldach, W.H., "Supplier Image: Application of CRT's in Trade-Off Analysis," paper presented at the 21st Annual Conference of the Advertising Research Foundation, 1975.

Coombs, L., and Freedman, R., "Use of Telephone Interviews in a Longitudinal Fertility Study," *Public Opinion Quarterly*, Vol. 28, No. 1, Spring 1964, pp. 112–117.

*Cooper, H.M., "Statistically Combining Independent Studies: A Meta-analysis of Sex Differences in Conformity Research," *Journal of Personality and Social Psychology*, Vol. 37, 1979, pp. 131–146.

Cooper, S.L., "Random Sampling by Telephone: An Improved Method," *Journal of Marketing Research*, Vol. 1, No. 4, November 1964, pp. 45–48.

Cotter, P.R., Cohen, J., and Coulter, P.B., "Race-of-Interviewer Effects in Telephone Interviews," *Public Opinion Quarterly*, Vol. 46, No. 2, Summer 1982, pp. 278–284.

Coulter, R., "A Comparison of CATI and non-CATI on a Nebraska Hog Survey," Staff Report No. 85, Statistical Research Division, Statistical Reporting Service, U.S. Department of Agriculture, Washington, DC, April 1985.

Cowan, C.D., "Review of 'Surveys by Telephone: A National Comparison With Personal Interviews'" (by R.M. Groves and R.L. Kahn), *Public Opinion Quarterly*, Vol. 46, 1982, pp. 139–141.

Cowan, C.D., Roman, A.M., Wolter, K.M., and Woltman, H.F., "A Test of Data Collection Methodologies: The Methods Test," *Proceedings of the Section on Survey Research Methods, American Statistical Association*, 1979, pp. 141–146.

*Crespi, I., and Morris, D., "Question Order Effect and the Measurement of Candidate Preference in the 1982 Connecticut Elections," *Public Opinion Quarterly*, Vol. 48, No. 3, 1984, pp. 578–591.

Cummings, K.M., "Random Digit Dialing: A Sampling Technique for Telephone Surveys," *Public Opinion Quarterly*, Vol. 43, No. 2, Summer 1979, pp. 233–244.

Cunningham, J.M., Westerman, H.H., and Fischhoff, J., "A Follow-up Study of Patients Seen in a Psychiatric Clinic for Children," *American Journal of Orthopsychiatry*, Vol. 26, July 1956, pp. 602–610.

Curry, J.L., "Bell Ringing: Survey Sampling and the Telephone Company," New York, American Telephone and Telegraph Company, paper presented at the American Association of Public Opinion Research, May 22, 1977.

Curtin, R.T., "Indicators of Consumer Behavior: The University of Michigan Surveys of Consumers," *Public Opinion Quarterly*, Vol. 46, No. 3, Fall 1982, pp. 340–352.

Czaja, R., Blair, J., and Sebestik, J.P., "Respondent Selection in a Telephone Survey: A Comparison of Three Techniques," *Journal of Marketing Research*, Vol. 19, August 1982, pp. 381–385.

Dalenius, T.E., "Privacy and Telephone Surveys," *Proceedings of the Business and Economics Statistics Section, American Statistical Association*, 1980, pp. 83–85.

Danbury, T., "Alternative Sampling Models for Random Digit Dialing Surveys," paper presented at the 21st Annual Conference of the Advertising Research Foundation, November 1975.

*Davies, M., "Co-operative Problem Solving: An Exploratory Study," Communications Studies Group Paper, No. 3/71159/DV, 1971.

*De Leeuw, E.D., and Hox, J.J., "Artifacts in Mail Surveys: The Influence of Dillman's Total Design Method on the Quality of Responses," in W.E. Saris and I.N. Gallhofer (eds.), *Sociometric Research*, Vol. I, London, Macmillan, 1987.

De Leeuw, E.D., and van der Zouwen, J., *A Methodological Comparison of the Data Quality in Telephone and Face to Face Surveys: A Meta Analysis of the Research Literature*, University of Amsterdam, Amsterdam, Publication No. 36, 1987.

de Sola Pool, I., *The Social Impact of the Telephone*, Cambridge, MA, MIT Press, 1977.

Dekker, F., and Doorn, P.K., "Computer Assisted Telephonic Interviewing: A Research Project in the Netherlands," paper presented for the Conference of the Institute of British Geographers, Durham, United Kingdom, January 1984.

DeMaio, T.J., "Response Rates in RDD," U.S. Bureau of the Census, internal memorandum, December 1981.

DeMaio, T.J., "Refusals in Telephone Surveys: When Do They Occur?" paper presented at the 1984 meeting of the American Association for Public Opinion Research, 1984.

*DeMaio, T.J., "Social Desirability and Survey Measurement: A Review," in C.F. Turner and E. Martin (eds)., *Surveying Subjective Phenomena*, Vol. II, Russel Sage Foundation, New York, 1984.

Denney, W.M., and Shanks, J.M., "Using CATI for Direct Data Entry: A Feasibility and Cost Comparison Study," Survey Research Center Working Paper, University of California, Berkeley, 1981.

Denteneer, D., Bethlehem, J.G., Hundepool, A.J., and Keller, W.J., "The BLAISE System for Computer-Assisted Survey Processing," *Proceedings of the Third Annual Research Conference of the U.S. Bureau of the Census*, U.S. Bureau of the Census, 1987, pp. 112–127.

Dickson, G.M., "Data Capture and Validation Using Portable Terminals," COMPSTAT, Physica Verlag, Vienna, 1984.

Dijkstra, W., "Interviewing Style and Respondent Behavior: An Experimental Study of the Survey Interview," *Sociological Methods and Research*, Vol. 16, No. 2, November 1987, pp. 309–334.

*Dijkstra, W., and van der Zouwen, J. (eds.), *Response Behaviour in the Survey Interview*, London/New York, Academic Press, 1982.

*Dijkstra, W., Van der Veen, L., and van der Zouwen, J., "A Field Experiment on Interviewer-Respondent Interaction," in M. Brenner, J. Brown, and D. Canter (eds.), *The Research Interview*, London, Academic Press, 1985.

*Dillman, D. A., "Our New Tools Need Not Be Used in the Same Old Way," *Journal of Community Development Society*, Vol. 8, No. 1, 1977, pp. 32–43.

Dillman, D.A., *Mail and Telephone Surveys: The Total Design Method*, New York, John Wiley & Sons, 1978.

*Dillman, D.A., "You Have Been Randomly Selected....Survey Methods for the Information Age," 52nd Distinguished Faculty Address, Washington State University, Department of Rural Sociology, Pullman, WA, 1985.

Dillman, D.A., and Frey, J.H., "Coming of Age: Interviews by Telephone," paper presented at the annual meetings of the Pacific Sociological Association, San José, CA, March 1974.

Dillman, D.A., and Frey, J.H., "Coming of Age: Interviews by Telephone," paper presented at the annual meetings of Survey Research, New York, Academic Press, 1983, pp. 359–378.

*Dillman, D.A., and Mason, R.G., "The Influence of Survey Method on Question Response," paper presented at the annual meeting of the American Association for Public Opinion Research, Delavan, WI, 1984.

Dillman, D.A., Gallegos, J.G., and Frey, J.H., "Reducing Refusal Rates for Telephone Interviews," *Public Opinion Quarterly*, Vol. 40, No. 1, Spring 1976, pp. 66–78.

*Dillman, D.A., Dillman, J.J., and Makela, C.J., "The Importance of Adhering to Details of the Total Design Method (TDM) for Mail Surveys," in D.C. Lockhart (ed.), Source Book 22, *New Directions for Program Evaluation*, San Francisco, Jossey-Bass, 1985.

Donnelley Marketing, *Donnelley Marketing Advantages*, Stamford, CT, R.H. Donnelley Corp., 1986.

Doorn, P.K., and Dekker, F., "Computer-Assisted Telephone Interviewing: An Application in Planning Research," *Environment and Planning A*, Vol. 17, 1985, pp. 795–813.

Drew, D., and Jaworski, R., "Telephone Survey Development on the Canadian Labour Force Survey," internal report to Statistics Canada, 1987.

*Drew, D., Armstrong, J., van Baaren, A., and Deguire, Y., "Methodology of Address Register Construction Using Several Administrative Sources," Statistics Canada, *Proceedings of the International Symposium on Statistical Uses of Administrative Data*, New York, John Wiley & Sons, forthcoming, 1988.

Duncanson, J.P., "The Average Telephone Call Is Better Than the Average Telephone Call," *Public Opinion Quarterly*, Vol. 33, 1969/1970, pp. 112–116.

Durako, S.J., and McKenna, T.W., "Collecting Health Interview Survey Data by Telephone: A Mailout Experiment," *Proceedings of the Business and Economic Statistics Section, American Statistical Association*, 1980, pp. 77–82.

Dutka, S., and Frankel, L.R., "Sequential Survey Design Through the Use of Computer-Assisted Telephone Interviewing," *Proceedings of the Section on Business and Economic Statistics, American Statistical Association*, 1980, pp. 73–76.

Eastlack, J.O., Jr., "Recall of Advertising by Two Telephone Samples," *Journal of Advertising Research*, Vol. 4, 1964, pp. 25–29.

Eastlack, J.O., Jr., and Assael, H., "Better Telephone Surveys Through Centralized Interviewing," *Journal of Advertising Research*, Vol. 6, No. 1, March 1966, pp. 2–7.

*Efron, B., and Morris, C., "Stein's Estimation Rule and Its Competitors — An Empirical Bayes Approach," *Journal of the American Statistical Association*, Vol. 68, 1973, pp. 117–130.

Falthzik, A.M., "When to Make Telephone Interviews," *Journal of Marketing Research*, Vol. 9, November 1972, pp. 451–452.

Fay, R., "Discussion of 'Use of Historical Data in a Current Interview Situation: Response Error Analyses and Applications to Computer-Assisted Telephone Interviewing,'" (by Pafford and Coulter), *Proceedings of the Third Annual Research Conference of the U.S. Bureau of the Census*, U.S. Bureau of the Census, 1987, pp. 320–323.

Ferrari, P.W., "Preliminary Results from the Evaluation of the CATI Test for the 1982 National Survey of Natural and Social Scientists and Engineers," U.S. Bureau of the Census, internal draft manuscript, February 1984.

Ferrari, P.W., "An Evaluation of Computer-Assisted Telephone Interviewing Used During the 1982 Census of Agriculture," unpublished evaluation report, Agriculture Division, U.S. Bureau of the Census, Washington, DC, 1986.

Ferrari, P.W., and Bailey, L., "Preliminary Results of 1980 Decennial Census Telephone Followup Nonresponse," *Proceedings of the Section on Survey Research Methods, American Statistical Association*, 1981, pp. 264–269.

Ferrari, P.W., Storm, R.R., and Tolson, F.D., "Computer-Assisted Telephone Interviewing," *Proceedings of the Section on Survey Research Methods, American Statistical Association*, 1984, pp. 594–599.

Field, D.R., "The Telephone Interview in Leisure Research," *Journal of Leisure Research*, Vol. 5, 1973, pp. 51–59.

Fielder, E.P., "Computer-Assisted Telephone Interviewing: Some Implications for Policy Research," *International Journal of Policy Analysis and Information Systems*, Knowledge Systems Laboratory, University of Illinois at Chicago Circle, 1979, pp. 662–668.

Fink, J.C., "Quality Improvement and Time Savings Attributed to CATI — Reflections on 11 Years of Experience," paper presented at the 36th Annual Meetings of the American Association of Public Opinion Research, Buck Hill Falls, PA, 1981.

Fink, J.C., "Quality Improvement and Time Savings Attributed to CATI — Reflections on 11 Years of Experience," paper presented at the 1982 American Educational Research Association Meeting, New York, 1982.

Fink, J.C., "CATI's First Decade: The Chilton Experience," *Sociological Methods & Research*, Vol. 12, No. 2, November 1983, pp. 153–168.

Fitti, J.E., "Some Results From the Telephone Health Interview System," *Proceedings of the Section on Survey Research Methods, American Statistical Association*, 1979, pp. 244–249.

*Flanders, N.A., "Teacher Influence, Pupil Attitudes and Achievement," Minneapolis, University of Minnesota, mimeo, 1960.

Fleishman, E., and Berk, M., "Survey of Interviewer Attitudes Toward Selected Methodological Issues in the National Medical Care Expenditure Survey," paper prepared for presentation at the Third Biennial Conference on Health Survey Research and Methods, Reston, VA, May 1979.

Flemion, J.A., "NCS-RDD Extended Follow-Up," U.S. Bureau of the Census, internal memorandum, July 1984.

Fletcher, J., and Thompson, H., "Telephone Directory Samples and Random Telephone Number Generation," *Journal of Broadcasting*, Vol. 18.2, 1974, pp. 187–191.

*Folsom, R.E., Jr., Chromy, J.R., and Williams, R.L., "Optimum Allocation of a Medical Care Provider Record Check Survey: An Application of Survey Costs Minimization Subject to Multiple Variance Constraints," paper presented at the Joint National Meetings of the Institute of Management Science and the Operations Research Society of America, New Orleans, April 30-May 2, 1979.

Forsythe, J.B., "Obtaining Cooperation in a Survey of Business Executives," *Journal of Marketing Research*, Vol. 14, August 1977, pp. 370–373.

Frankel, J., and Sharp, L., "Measurement of Respondent Burden," *Statistical Reporter*, January 1981, pp. 105–111.

Frankel, L.R., "On the Definition of Response Rates," a special report of the CASRO Task Force on Completion Rates, 1982.

Frankel, M.R., and Frankel, L.R., "Some Recent Developments in Sample Survey Design," *Journal of Marketing Research*, Vol. 14, August 1977, pp. 280–293.

Freeman, H.E., "Research Opportunities Related to CATI," *Sociological Methods & Research*, Vol. 12, No. 2, November 1983, pp. 143–152.

Freeman, H.E., and Shanks, J.M., "The Emergence of Computer-Assisted Survey Research," *Sociological Methods & Research*, Vol. 12, No. 2, November 1983, pp. 115–118.

Freeman, H.E., Kiecolt, K.J., Nicholls, W.L. II, and Shanks, J. M., "Telephone Sampling Bias in Surveying Disability," *Public Opinion Quarterly*, Vol. 46, No. 3, Fall 1982, pp. 392–407.

Freeman, J., and Butler, E.W., "Some Sources of Interviewer Variance in Surveys," *Public Opinion Quarterly*, Vol. 40, Spring 1976, pp. 79–91.

Frey, J.H., "Characteristics of Listed Vs. Unlisted Households in a Rapidly Growing Population," paper presented at the annual meetings of the Pacific Sociological Association, Anaheim, CA, 1979.

Frey, J.H., *Survey Research by Telephone*, Beverly Hills, Sage Publications, 1983.

Frey, J.H., "An Experiment With a Confidentiality Reminder in a Telephone Survey," *Public Opinion Quarterly*, Vol. 50, Summer 1986, pp. 267–269.

Friedman, L., and Friedman, F.H., "Does the Perceived Race of a Telephone Interviewer Affect the Responses of White Subjects to an Attitude Toward Negroes Scale?" *Proceedings of the Section on Survey Research Methods, American Statistical Association*, 1978, pp. 556–558.

Friedman, S.B., "People and Places: A Report on a Survey of Peoples' Locational Preferences," draft, State Planning Office, Wisconsin Department of Administration, September 1973.

Fry, H.G., and McNaire, S., "Data Gathering by Long Distance Telephone," *Public Health Reports*, Vol. 73, No. 9, September 1958, pp. 831–835.

Ghosh, D., "Improving the Plus 1 Method of Random Digit Dialing," *Proceedings of the Section on Survey Research Methods, American Statistical Association*, 1984, p. 285.

*Glass, G.V., McGaw, B., and Smith, M.L., *Meta-Analysis in Social Research*, Sage, Beverly Hills, CA, 1981.

Glasser, G.J., and Metzger, G.D., "Random-Digit Dialing as a Method of Telephone Sampling," *Journal of Marketing Research*, Vol. 9, No. 1, February 1972, pp. 59–64.

Glasser, G.J., and Metzger, G.D., "National Estimates of Nonlisted Telephone Households and Their Characteristics," *Journal of Marketing Research*, Vol. 12, August 1975, pp. 359–361.

Glenn, N.D., "Aging, Disengagement, and Opinionation," *Public Opinion Quarterly*, Vol. 33, No. 1, 1969, pp. 17–33.

Goldberg, D., Sharp, H., and Freedman, R., "The Stability and Reliability of Expected Family Size Data," *Milbank Memorial Fund Quarterly*, Vol. 37, 1959, pp. 369–385.

Gower, A.R., "Non-Response in the Canadian Labour Force Survey," *Survey Methodology*, Vol. 5, No. 1, June 1979, pp. 29–58.

*Goyder, J., "Face-to-Face Interviews and Mailed Questionnaires: The Net Difference in Response Rate," *Public Opinion Quarterly*, Vol. 49, 1985, pp. 234–252.

*Goyder, J., *The Silent Majority: Non-Respondents on Sample Surveys*, Polity Press, New York, 1986.

Grimes, M.D., and Pinhey, T.K., *Response Bias in Cross-Sex Telephone Interviews: An Empirical Assessment and Commentary*, Baton Rouge, Louisiana, State University, paper presented at the annual meeting of the Southwestern Sociological Association, April 1976.

Gross, K.H., "Random-Digit Dialing Sampling Methods," *Encyclopedia of Statistical Sciences*, Vol. 7, New York, John Wiley & Sons, 1986, pp. 506–508

Grossman, R., and Weiland, D., "The Use of Telephone Directories as a Sample Frame: Patterns of Bias Revisited," *Journal of Advertising*, Vol. 7, No. 14, Summer 1978, pp. 31–35.

Groves, R.M., "An Experimental Comparison of National Telephone and Personal Interview Surveys," *Proceedings of the Social Statistics Section, American Statistical Association*, Part I, 1977, pp. 232–241.

Groves, R.M., "An Empirical Comparison of Two Telephone Sample Designs," *Journal of Marketing Research*, Vol. 15, November 1978, pp. 622–631.

Groves, R.M., "On the Mode of Administering a Questionnaire and Response to Open-ended Items," *Social Science Research*, Vol. 7, 1978, pp. 257-271.

Groves, R.M., "Actors and Questions in Telephone and Personal Interview Surveys," *Public Opinion Quarterly*, Vol. 43, No. 2, Summer 1979, pp. 190–205.

Groves, R.M., "A Researcher's View of the SRC Computer-Based Interviewing System: Measurement of Some Sources of Error in Telephone Survey Data," paper presented at the Third Biennial Conference on Health Survey Research Methods, 1979.

Groves, R.M., "Computer Assisted Telephone Interviewing and the Future of Survey Research," presentation for the Conference on Computer-Assisted Telephone Interviewing, Berkeley, CA, March 1980.

Groves, R.M., "Implications of CATI: Costs, Errors, and Organization of Telephone Survey Research," *Sociological Methods & Research*, Vol. 12, No. 2, November 1983, pp. 199-215.

Groves, R.M., "Research on Survey Data Quality," *Public Opinion Quarterly*, 50th Anniversary Issue, Winter 1987, pp. S156-S172.

Groves, R.M., *Survey Costs and Survey Errors*, John Wiley & Sons, New York, forthcoming.

Groves, R.M., and Fultz, N.H., "Gender Effects Among Telephone Interviewers in a Survey of Economic Attitudes," *Sociological Methods & Research*, Vol. 14, No. 1, August 1985, pp. 31-52.

Groves, R.M., and Kahn, R.L., *Surveys by Telephone: A National Comparison With Personal Interviews*, New York, Academic Press, 1979.

Groves, R.M., and Lepkowski, J.M., "Alternative Dual Frame Mixed Mode Survey Designs," *Proceedings of the Section on Survey Research Methods, American Statistical Association*, 1982, pp. 154-159.

Groves, R.M., and Lepkowski, J.M., "Cost and Error Modelling for Large-Scale Telephone Surveys," *Proceedings of the First Annual Research Conference of the U.S. Bureau of the Census*, U.S. Bureau of the Census, 1985, pp. 330-357.

Groves, R.M., and Lepkowski, J.M., "Dual Frame, Mixed Mode Survey Designs," *Journal of Official Statistics*, Vol. 1, No. 3, 1985, pp. 263-286.

Groves, R.M., and Lepkowski, J.M., "An Experimental Implementation of a Dual Frame Telephone Sample Design," *Proceedings of the Section on Survey Research Methods, American Statistical Association*, 1986, pp. 340-345.

Groves, R.M., and Magilavy, L.J., "Estimates of Interviewer Variance in Telephone Surveys," *Proceedings of the Section on Survey Research Methods, American Statistical Association*, 1980, pp. 622-627.

Groves, R.M., and Magilavy, L.J., "Increasing Response Rates to Telephone Surveys: A Door in the Face for Foot-in-the-Door?" *Public Opinion Quarterly*, Vol. 45, No. 3, Fall 1981, pp. 346-358.

Groves, R.M., and Magilavy, L.J., "Investigations of the Magnitude and the Sources of Interviewer Effects in Telephone Surveys," paper presented at the International Statistical Institute, Madrid, 1983.

Groves, R.M., and Magilavy, L.J., "Measuring and Explaining Interviewer Effects in Centralized Telephone Surveys," *Public Opinion Quarterly*, Vol. 50, No. 2, Summer 1986, pp. 251–266.

Groves, R.M., and Mathiowetz, N.A., "Computer Assisted Telephone Interviewing: Effects on Interviewers and Respondents," *Public Opinion Quarterly*, Vol. 48, No. 1B, 1984, pp. 356–369.

Groves, R.M., and Nicholls, W.L., II, "The Status of Computer-Assisted Telephone Interviewing: Part II — Data Quality Issues," *Journal of Official Statistics*, Vol. 2, No. 2, 1986, pp. 117–134.

Groves, R.M., and Robinson, D., "Final Report on Callback Algorithms on CATI Systems," The University of Michigan, Survey Research Center, prepared under a Joint Statistical Agreement with the U.S. Bureau of the Census, unpublished memorandum, 1982.

Groves, R.M., and Scott, J., "An Attempt to Measure the Relative Efficiency of Telephone Surveys of Social Science Data Collection," paper presented at the American Association for Public Opinion Research annual conference, 1976.

Groves, R.M., and Snowden, C., "The Effects of Advanced Letters on Response Rates in Linked Telephone Surveys," *Proceedings of the Section on Survey Research Methods, American Statistical Association*, 1987.

Groves, R.M., Berry, M., and Mathiowetz, N., "Some Impacts of Computer Assisted Telephone Interviewing on Survey Methods," *Proceedings of the Section on Survey Research Methods, American Statistical Association*, 1980, pp. 519–524.

Groves, R.M., Cannell, C., and Miller, P., "A Methodological Study of Telephone and Face-to-Face Interviewing," paper presented at the 36th Annual AAPOR Conference, Buck Hill Falls, PA, 1981.

Groves, R.M., Magilavy, L.J., and Mathiowetz, N.A., "The Process of Interviewer Variability: Evidence from Telephone Surveys," *Proceedings of the Section on Survey Research Methods, American Statistical Association*, 1981, pp. 438–443.

Groves, R.M., Lepkowski, J.M., and Landenberger, B.D., "Matching Census Data to Telephone Exchanges in the State of Michigan," U.S. Bureau of the Census, internal draft report, June 1984.

Guenzel, P.J., Berckmans, T.R., and Cannell, C.F., "General Interviewing Techniques: A Self-Instructional Workbook for Telephone and Personal Interviewer Training," Institute for Social Research, The University of Michigan, 1983.

Gunn, W.J., and Rhodes, I.N., "Physician Response Rates to a Telephone Survey: Effects of Monetary Incentive Level," *Public Opinion Quarterly*, Vol. 45, No. 1, Spring 1981, pp. 109–115.

Hagan, D.E., and Collier, C.M., "Respondent Selection Procedures for Telephone Surveys: Must They Be Intrusive?" paper presented at the conference of the American Association for Public Opinion Research, Baltimore, MD, 1982.

Hagan, D.E., and Collier, C.M., "Must Respondent Selection Procedures for Telephone Surveys Be Invasive?" *Public Opinion Quarterly*, Vol. 47, No. 4, Winter 1983, pp. 547–556.

*Hall, K.B., and Ford, F., "The Effects of Data Collection Methods," Statistical Research Division, Economics, Statistics, and Cooperative Service, U.S. Department of Agriculture, Washington, DC, December 1979.

*Hansen, M., Hurwitz, W., and Madow, W., *Sample Survey Methods and Theory*, Vol. 1, New York, John Wiley, 1953.

Hansen, M.H., Hurwitz, W.N., and Bershad, M.A., "Measurement Errors in Censuses and Surveys," *Bulletin of the International Statistical Institute*, Vol. 38, No. 2, 1961, pp. 359–374.

Hansen, M.H., Hurwitz, W.N., and Pritzker, L., "The Estimation and Interpretation of Gross Differences and the Simple Response Variance," in C.R. Rao (ed.), *Contributions to Statistics*, Oxford, Pergamon Press, 1964.

Harlow, B.L., and Hartge, P., "Telephone Household Screening and Interviewing," *American Journal of Epidemiology*, Vol. 117, No. 5, 1983, pp. 632–633.

Harlow, B.L., Rosenthal, J.F., and Ziegler, R.G., "A Comparison of Computer-Assisted and Hard Copy Telephone Interviewing," *American Journal of Epidemiology*, Vol. 122, 1985, pp. 335–340.

Harris, M., "Documents Used During the Selection and Training of Social Survey Interviewers and Selected Papers on Interviewers and Interviewing," Social Survey Division, Central Office of Information, Great Britain, 1952.

Hartge, P., Cahill, J.I., West, D., Hauck, M., Austin, D., Silverman, D., and Hoover, R., "Design and Methods in a Multi-Center Case-Control Interview Study," *American Journal of Public Health*, Vol. 74, No. 1, January 1984, pp. 52–56.

Hartge, P., Brinton, L.A., Rosenthal, J.F., Cahill, J.I., Hoover, R.N., and Waksberg, J., "Random Digit Dialing in Selecting a Population-Based Control Group," *American Journal of Epidemiology*, Vol. 120, 1984, pp. 825–833.

*Hartley, H., "Multiple Frame Surveys," *Proceedings of the Social Statistics Section, American Statistical Association*, 1962, pp. 203–206.

*Hartley, H., "Multiple Frame Methodology and Selected Applications," *Sankhyā*, Series C, Vol. 36, 1974, pp. 99–118.

*Hatchett, S., and Schuman, H., "White Respondents and Race-of-Interviewer Effects," *Public Opinion Quarterly*, Vol. 39, 1975–76, pp. 523–28.

Hauck, M., "Paper Users Methodology—Results of the Pilot Study," Survey Research Laboratory, University of Illinois, Urbana, May 1973.

Hauck, M., "Use of the Telephone for Technical Surveys," Survey Research Laboratory, University of Illinois, Urbana, 1974.

Hauck, M., and Cox, M., "Locating a Sample by Random Digit Dialing," *Public Opinion Quarterly*, Vol. 38, No. 2, Summer 1974, pp. 253-260.

Hauck, M., and Goldberg, J., "Telephone Interviewing on the NLRB Election Study," *Survey Research Newsletter*, Survey Research Laboratory, University of Illinois, Urbana, Vol. 5, No. 1, January 1973, pp. 15-16.

*Health and Welfare Canada, Technical Report on the Health Promotions Survey, 1985.

*Heberlein, T.A., and Baumgartner, R., "Factors Affecting Response Rates to Mailed Questionnaires: A Quantitative Analysis of the Published Literature," *American Sociological Review*, Vol. 43, No. 4, 1978, pp. 447-462.

Heckman, J.J., "Sample Selection Bias as a Specification Error," *Econometrica*, Vol. 47, No. 1, January 1979, pp. 153-161.

Hedges, L., and Olkin, I., *Statistical Methods for Meta-Analysis*, Orlando, FL, Academic Press, 1985.

Henson, R., Roth, A., and Cannell, C.F., "Personal vs. Telephone Interviews and the Effects of Telephone Reinterviews on Reporting of Psychiatric Symptomatology," Survey Research Center, Institute for Social Research, The University of Michigan, 1974.

Herman, J.B., "Mixed Mode Data Collection: Telephone and Personal Interviewing," *Journal of Applied Psychology*, Vol. 62, 1977, pp. 399-404.

Herzog, A.R., and Kulka, R.A., "Interviewing Older Adults: A Comparison of Telephone and Face-to-Face Modalities," unpublished paper, June 1981; revised and abbreviated version of three papers presented at the 33rd Annual Meeting of the Gerontological Society, San Diego, November 1980.

Herzog, A.R., Rodgers, W.L., and Kulka, R.A., "Interviewing Older Adults: A Comparison of Telephone and Face-to-Face Modalities," *Public Opinion Quarterly*, Vol. 47, No. 3, Fall 1983, pp. 405-418.

Hildum, D.C., and Brown, R.W., "Verbal Reinforcement and Interviewer Bias," *Journal of Abnormal and Social Psychology*, Vol. 55, No. 1, July 1956, pp. 108-111.

*Hippler, H.J., and Hippler, G., "Reducing Refusal Rates in the Case of Threatening Questions: The 'Door-in-the-Face' Technique," *Journal of Official Statistics*, Vol. 2, No. 7, 1986, pp. 25-33.

*Hippler, H.J., and Schwarz, N., "Not Forbidding Isn't Allowing: The Cognitive Basis of the Forbid-Allow Asymmetry," *Public Opinion Quarterly*, Vol. 50, No. 1, 1986, pp. 87-96.

Hochstim, J.R., "Comparison of Three Information-Gathering Strategies in a Population Study of Sociomedical Variables," *Proceedings of the Social Statistics Section, American Statistical Association*, 1962, pp. 154-159.

Hochstim, J.R., "Alternatives to Personal Interviewing," *Public Opinion Quarterly*, Vol. 27, No. 4, Winter 1963, pp. 629-630.

Hochstim, J.R., "A Critical Comparison of Three Strategies of Collecting Data from Households," *Journal of the American Statistical Association*, Vol. 62, No. 319, September 1967, pp. 976-989.

Hogan, H.W., "Some Effects of 'Social Desirability' in Survey Studies, An Extended Replication," Cookeville, Tennessee Technical University, April 1976, paper presented at the annual meeting of the Southern Sociological Society in Miami, FL.

Hogue, C.R., and Chapman, D.W., "An Investigation of PSU Cutoff Points for a Random Digit Dialing Survey," *Proceedings of the Section on Survey Research Methods, American Statistical Association*, 1984, pp. 286-291.

*Hoinville, G., "Carrying Out Surveys Among the Elderly, Some Problems of Sampling and Interviewing," *Journal of the Market Research Society*, Vol. 25, No. 3, 1983, pp. 223-237.

*Holt, D., Scott, A., and Ewings, P., "Chi-Squared Tests with Survey Data," *Journal of the Royal Statistical Society*, Series A, Vol. 143, Part 3, 1980, pp. 303-320.

Hooper, C.E., "The Coincidental Method of Measuring Radio Audience Size," in A.B. Blankenship (ed.), *How to Conduct Consumer and Opinion Research*, New York, Harper & Brothers, 1946, pp. 156-171.

Horn, D., "Smoking Among Teenagers," *Public Health Reports*, Vol. 83, No. 6, June 1968, pp. 458-460.

Horton, R.L., "A New Look at Telephone Interviewing Methodology," *Pacific Sociological Review*, Vol. 21, 1978, p. 3.

Horvitz, D.G., "Discussion of Two Articles" (by A. Rustemeyer, G.H. Shure, M. Rogers, and R.J. Meeker; and W. L. Nicholls II), pp. 1-17, *Proceedings of the Section on Survey Research Methods, American Statistical Association*, 1978, pp. 18-19.

Horvitz, D.G., "Discussion of 'Beyond CATI: Generalized and Distributed Systems for Computer-Assisted Surveys'" (by Shanks and Tortora), *Proceedings of the First Annual Research Conference of the U.S. Bureau of the Census*, U.S. Bureau of the Census, 1985, pp. 372-375.

House, C.C., "The Department of Agriculture's Experience With Computer-Assisted Telephone Interviewing: An Overview of the First Test Year," unpublished memorandum, Statistical Research Division, Statistical Reporting Service, U.S. Department of Agriculture, 1981.

House, C.C., "Computer-Assisted Telephone Interviewing on Cattle Multiple Frame Survey," Staff Report No. 82, Statistical Research Division, Statistical Reporting Service, U.S. Department of Agriculture, Washington, DC, 1984.

House, C.C., "Questionnaire Design With Computer-Assisted Telephone Interviewing," *Journal of Official Statistics*, Vol. 1, No. 2, 1985, pp. 209-219.

House, C.C., and Morton, B.T., "Measuring CATI Effects on Numerical Data," *Proceedings of the Section on Survey Research Methods, American Statistical Association*, 1983, pp. 135-138.

House, C.C., and Morton, B.T., "Training Interviewers for Computer-Assisted Telephone Interviewing," *Proceedings of the Section on Survey Research Methods, American Statistical Association*, 1983, pp. 129-134.

Ibsen, C.A., and Ballweg, J.A., "Telephone Interviews in Social Research: Some Methodological Considerations," *Quality and Quantity*, Vol. 8, No. 2, June 1974, pp. 181-192.

*Immerman, F.W., and Mason, R.E., "Sampling Design and Sample Selection Procedures, Youth Attitude Tracking Study, 1986," Report RTI/3624/01-01W, Research Triangle Park, NC, Research Triangle Institute, 1986.

Inderfurth, G.P., "Interviewer-Respondent Interaction in a Telephone Survey of Physicians' Offices," paper presented at the 30th Annual Meeting of the American Association for Public Opinion Research, 1975.

Inglis, K., Groves, R.M., and Heeringa, S., "Alternative Telephone Sample Designs for the Black Household Population," *Proceedings of the Section on Survey Research Methods, American Statistical Association*, 1985, pp. 203-208.

Inglis, K., Groves, R.M., and Heeringa, S., "Telephone Sample Designs for the U.S. Black Household Population," *Survey Methodology*, Vol. 13, No. 1, 1987, pp. 1-14.

*Jabine, T., "Flow Charts: A Tool for Developing and Understanding Survey Questionnaires," *Journal of Official Statistics*, Vol. 1, No. 2, 1985, pp. 189-207.

*Jackson, G.B., "Methods for Integrative Reviews," *Review of Educational Research*, Vol. 50, 1980, pp. 428-460.

Janofsky, A.I., "Affective Self-Disclosure in Telephone Versus Face-to-Face Interviews," *Journal of Humanistic Psychology*, Vol. 11, 1971, pp. 93-103.

Janusz, J., "Hand-Held Terminals Carry Wide Range of Applications and Capabilities," *Data Communications*, October 1982, pp. 123-131.

Jarvis, I., "Practical Experience With Central Telephone Interviewing," in *Valve of Money in Market and Social Research*, ESOMAR Congress, Bristol, 1978.

Johnson, M.A., "The Feasibility of Telephone Surveys for Research on Urban Travel Behavior," Working Paper No. 7530 of the Travel Demand Forecasting Project, Institute of Transportation and Traffic Engineering, University of California, Berkeley, 1975.

Johnson, T.P., Hougland, J.G., and Clayton, R.R., "Obtaining Reports of Sensitive Behavior: A Comparison from Telephone and Face to Face Interviews," paper presented at the International Conference on Telephone Survey Methodology, Charlotte, November 1987.

Jones, G.K., Massey, J.T., and Tenebaum, M., "Considerations for Including Special Places in Telephone Surveys," *Proceedings of the Section on Survey Research Methods, American Statistical Association*, 1984, pp. 274-279.

Jones, R., "Variations in Household Telephone Access: Implications for Telephone Surveys," *The Australian Journal of Statistics*, Vol. 24, 1982, pp. 18-32.

Jones, S.M., and Chromy, J.R., "Improved Variance Estimators Using Weighting Class Adjustments for Sample Survey Nonresponse," *Proceedings of the Section on Survey Research Methods, American Statistical Association*, 1982, pp. 105-110.

*Jones, W.H., and Lang, J.R., "Reliability and Validity Effects Under Mail Survey Conditions," *Journal of Business Research*, Vol. 10, 1982, pp. 339-353.

Jordan, L.A., Marcus, A.C., and Reeder, L.G., "Response Styles in Telephone and Household Interviewing: A Field Experiment," *Public Opinion Quarterly*, Vol. 44, No. 2, Summer 1980, pp. 210-222.

Joyce, E., "Telephone Dialing by Computer," *BYTE*, Vol. 5, No. 1, 1980, pp. 122-128.

Judd, R.C., "Telephone Usage and Survey Research," *Journal of Advertising Research*, Vol. 6, No. 4, December 1966, pp. 38-39.

*Kalton, G., and Anderson, D.W., "Sampling Rare Populations," *The Journal of the Royal Statistical Society*, Series A, Vol. 149, Part 1, 1986, pp. 65-82.

*Kalton, G., Collins, M., and Brook, L., "Experiments in Wording Opinion Questions," *Applied Statistics*, Vol. 27, No. 2, 1978, pp. 149-161.

*Kalton, G., Roberts, J., and Holt, D., "The Effects of Offering a Middle Response Option With Opinion Questions," *The Statistician*, Vol. 29, No. 1, 1980, pp. 65-78.

Karweit, N., and Meyers, E.D., Jr., "Computers in Survey Research," in P.H. Rossi, J.D. Wright, and A.B. Anderson (eds.), *Handbook of Survey Research*, Academic Press, 1983, pp. 379-414.

Katz, D., and Cantril, H., "Public Opinion Polls," *Sociometry*, Vol. 1, 1937, pp. 155-179.

Kegeles, S.S., Fink, C.F., and Kirscht, J.P., "Interviewing a National Sample by Long-Distance Telephone," *Public Opinion Quarterly*, Vol. 33, No. 3, Fall 1969, pp. 412-419.

Kerin, R., and Peterson, R., "Scheduling Telephone Interviews: Lessons From 250,000 Dialings," *Journal of Advertising Research*, Vol. 23, 1983, pp. 41-47.

Kerssemakers, F.A.M., "An Empirical Comparison of Two Modes of Data-Collection," United Nations Economic and Social Council, CES/AC, 48/44, 1983.

Kildegaard, I.C., "Telephone Trends," *Journal of Advertising Research*, Vol. 6, No. 2, June 1966, pp. 56-60.

Kildegaard, I.C., "Rejoinder, Comments to Robert Judd's article, 'Telephone Usage and Survey Research,'" *Journal of Advertising Research*, Vol. 6, No. 4, December 1966, pp. 40-41.

Kincannon, C.L., "The Census Role in Scientific and Technical Manpower Programs for the 70's," *Proceedings of the Social Statistics Section, American Statistical Association*, 1972, pp. 181–189.

Kish, L., "A Procedure for Objective Respondent Selection Within the Household," *Journal of the American Statistical Association*, Vol. 44, September 1949,pp. 380–387.

*Kish, L., "Studies of Interviewer Variance for Attitudinal Variables," *Journal of the American Statistical Association*, Vol. 57, 1962, pp. 92–115.

*Kish, L., *Survey Sampling*, New York, John Wiley & Sons, 1965.

*Kjøller, M. *Interaktionen mellem interviewer og respondent*, Socialforsknings-instituttet (The Danish National Institute of Social Research), Copenhagen, Denmark, 1975 (In Danish).

Klecka, W.R., "Structured Design Considerations for Questionnaires Used in a Computer-Assisted Survey Environment," paper presented at the annual Field Directors' Conference, Santa Monica, CA, 1981.

Klecka, W.R., and Tuchfarber, A.J., Jr., "Random Digit Dialing as an Efficient Method for Political Polling," *Georgia Political Science Association Journal*, Vol. 2, No. 1, Spring 1974, pp. 133–151.

Klecka, W.R., and Tuchfarber, A.J., Jr., "Random Digit Dialing: A Comparison to Personal Surveys," *Public Opinion Quarterly*, Vol. 42, 1978, pp. 105–114.

Kofron, J.H., and Nelson, R.O., "Centralized Long-Distance Interviewing — A Revolution in Data Collection," paper presented to the Advertising Research Foundation, 12th Annual Conference, October 1966.

Kofron, J.H., Bayton, J.A., and Bortner, B.Z., "Guidelines for Choosing Between Long-Distance Telephone and Personal Interviewing," paper presented at the Advertising Research Foundation, 15th Annual Conference, October 1969.

Kofron, J.H., Kilpatrick, D.J., and Brown, A.J., "Cathode Ray Tube (CRT) WATS Line Interviewing," Philadelphia, Chilton Research Services, paper presented at the 29th Advertising Research Foundation, 1974.

Koo, H.P., Ridley, J.C., Piserchia, P.V., Dawson, D.A., Bachrach, C.A., Holt, M.I., and Horvitz, D.G., "An Experiment on Improving Response Rates and Reducing Call-Backs in Household Surveys," *Proceedings of the Social Statistics Section, American Statistical Association*, Part II, 1976, pp. 491–494.

*Koolwijk, J. van, "Unangenehme Fragen, Paradigma für die Reaktionen der Befragten im Interview," *Kölner Zeitschrift für Soziologie und Sozialpsy-chologie*, Vol. 21, 1969, p. 4.

Koons, D.A., "Current Medicare Survey, Telephone Interviewing Compared With Personal Interviews," Response Research Staff Report No. 74-4, U.S. Bureau of the Census, Washington, DC, October 1974.

Kooyman, C.A., "Towards the Design of a CATI-Based Telephone Tracing Operation," presentation summary for meetings of the American Association for Public Opinion Research, St. Petersburg Beach, FL, 1986.

Körmendi, E., Egsmose, L., and Noordhoek, J., "Datakvalitet ved telefon-interview," Socialforskingsinstituttet, Studie 52, Copenhagen, Denmark, 1986.

Kovar, M.G., and Fitti, J.E., "A Linked Follow-up Study of Older People," *Proceedings of the Section on Survey Research Methods, American Statistical Association*, 1985, pp. 179–184.

*Krosnick, J.A., and Alwin, D.F., "An Evaluation of a Cognitive Theory of Response-Order Effects in Survey Measurement," *Public Opinion Quarterly*, Vol. 51, No. 2, 1987, pp. 201–219.

Kulka, R.A., Weeks, M.F., Lessler, J.T., and Whitmore, R.W., "A Comparison of the Telephone and Personal Interview Modes for Conducting Local Household Health Surveys," *Proceedings of the Fourth Conference on Health Survey Research Methods*, 1982, pp. 116–127.

*Kviz, F.J., "Towards a Standard Definition of Response Rate," *Public Opinion Quarterly*, Vol. 41, 1977, pp. 265–267.

Lacey, B.H., "Random Digit Dialing Experiments: An Analysis of Job Requirements for Telephone Interviewers," *Proceedings of the Fourth Conference on Health Survey Research Methods*, 1982, pp. 279–294.

*Laird, N., and Lewis, T., "Empirical Bayes Confidence Intervals Based on Bootstrap Samples," *Journal of the American Statistical Association*, Vol. 82, 1987, pp. 739–750.

Landenberger, B.D., Groves, R.M., and Lepkowski, J.M., "A Comparison of Listed and Randomly Dialed Telephone Numbers," *Proceedings of the Section on Survey Research Methods, American Statistical Association*, 1984, pp. 280–284.

Landon, E.L., Jr., and Banks, S.K., "Relative Efficiency and Bias of Plus-One Telephone Sampling," *Journal of Marketing Research*, Vol. 14, August 1977, pp. 294–299.

Larson, O.N., "The Comparative Validity of Telephone and Face-to-Face Interviews in the Measurement of Message Diffusion From Leaflets," *American Sociological Review*, Vol. 17, No. 4, August 1952, pp. 471–476.

Larson, R.F., and Catton, W.R., Jr., "Can the Mail-Back Bias Contribute to a Study's Validity?" *American Sociological Review*, Vol. 24, 1959, pp. 243–245.

Lavrakas, P.J., *Telephone Survey Methods: Sampling, Selection, and Supervision*, Newbury Park, CA, Sage Publications, 1987.

Lavrakas, P.J., and Maier, R.A., Jr., "The Magnitude and Nature of RDD Panel Attrition," Northwestern University Survey Laboratory, Evanston, IL, mimeo, 1984.

Lavrakas, P.J., and Tyler, T.R., "Low Cost Telephone Surveys," paper presented at Evaluation '83, Chicago, 1983.

LeBaily, R.K., "Method Artifacts in Telephone and In-Person Rape Victimization Surveys," Center for Urban Affairs, Northwestern University, Evanston, IL, January 1979.

LeBaily, R.K., and Lavrakas, P.J., "Generating a Random Digit Dialing Sample for Telephone Surveys," paper presented at Issue '81, The Annual SPSS Convention, San Francisco, 1981.

Lebby, D.E., "On-Line Inter-Active Interviewing: New Dimensions for Survey Research," paper presented at the Conference of the American Association for Public Opinion Research, 1978.

Lebby, D.E., "CATI's First Decade: The Chilton Experience," paper presented at the conference on Computer-Assisted Telephone Interview Technology, Berkeley, CA, March 1980.

Lepkowski, J.M., and Groves, R.M., "The Impact of Bias on Dual Frame Survey Design," *Proceedings of the Section on Survey Research Methods, American Statistical Association*, 1984, pp. 265–270.

Lepkowski, J.M., and Groves, R.M., "A Two Phase Probability Proportional to Size Design for Telephone Sampling," *Proceedings of the Section on Survey Research Methods, American Statistical Association*, 1986, pp. 357–362.

Lepkowski, J.M., and Groves, R.M., "A Mean Squared Error Model for Dual Frame, Mixed Mode Survey Design," *Journal of the American Statistical Association*, Vol. 81, 1986, pp. 930–937.

Leuthold, D.A., and Scheele, R., "Patterns of Bias in Samples Based on Telephone Directories," *Public Opinion Quarterly*, Vol. 35, No. 2, Summer 1971, pp. 249–257.

*Lievesley, D., "Unit Nonresponse in Interviewing Surveys," Social and Community Planning Research, London, 1987.

*Light, R.J., and Pillemer, D.B., *Summing Up: The Science of Reviewing Research*, Cambridge, MA, Harvard University Press, 1984.

Linebarger, J., "Analysis of NCS Recheck Results," unpublished memorandum, Washington, DC, U.S. Bureau of the Census, January 18, 1978.

*Lippitt, R., and Zander, A. "Observation and Interview Methods for the Leadership Training Study," mimeo, New York, Boy Scouts of America, 1943.

Little, R.J.A., and Rubin, D.B., *Statistical Analysis With Missing Data*, New York, John Wiley & Sons, 1987.

Locander, W.B., and Burton, J.P., "The Effect of Question Form on Gathering Income Data by Telephone," *Journal of Marketing Research*, Vol. 13, No. 2, May 1976, pp. 189–192.

Locander, W., Sudman, S., and Bradburn, N., "An Investigation of Interview Method, Threat, and Response Distortion," *Journal of the American Statistical Association*, Vol. 71, June 1976, pp. 269–275.

Love, L.T., and Kusch, G.L., "Efficiency of Telephone Follow-up of Questionnaire Coverage Item A," Dane County, Washington, DC, U.S. Bureau of the Census, unpublished memorandum, May 23, 1969.

Lucas, W.A., and Adams, W.C., "An Assessment of Telephone Survey Methods," (R-2135-NSF), Santa Monica, CA, Rand Corporation, 1977.

Lund, R.E., "Estimators in Multiple Frame Surveys," *Proceedings of the Social Statistics Section, American Statistical Association*, 1968, pp. 282–288.

Lyberg, L., "Plans for Computer-Assisted Data Collection at Statistics Sweden," *Proceedings of the 45th Session, International Statistical Institute*, Book 3, Topic 18.2, 1985, pp. 1–11.

Lyons, W., and Durant, R.F., "Interviewer Costs Associated With the Use of Random Digit Dialing in Large Area Sample," *Journal of Marketing*, Vol. 44, 1980. pp. 65–69.

Madow, W.G., Olkin, I., and Rubin, D.B. (eds.), *Incomplete Data in Sample Surveys*, Vol. I-III, New York, Academic Press, 1983.

*Mahalanobis, P., "Recent Experiments in Statistical Sampling in the Indian Statistical Institute," *Journal of the Royal Statistical Society*, Vol. 109, 1946, pp. 325–370.

Maier, N.R.F., and Thurber, J., "Accuracy of Judgments of Deception When an Interview Is Watched, Heard or Read," *Personnel Psychology*, Vol. 21, 1968, pp. 23–30.

*Malt, L.G., "Skills Analysis in Human-Computer Interface — A Holistic Approach," in G. Salvendy, S.L. Sauter, and J.J. Hurrell, Jr. (eds.), *Social, Ergonomic and Stress Aspects of Work with Computers*, Amsterdam, Elsevier Science Publishers, 1987.

Mangione, T.W., Hingson, R., and Barrett, J., "Collecting Sensitive Data: A Comparison of Three Survey Strategies," *Sociological Methods & Research*, Vol. 10, No. 3, 1982, pp. 337–346.

Marcus, A.C., and Crane, L.A., "Telephone Interviewing in Health Survey Research," paper presented at the American Public Health Association Meeting, November 16, 1983.

Marcus, A.C., and Crane, L.A., "A Review of Telephone Interviewing in Health Survey Research," UCLA Jonsson Comprehensive Cancer Center, Division of Cancer Control, Los Angeles, CA, 1983.

Marcus, A.C., and Crane, L.A., "Telephone Surveys in Public Health Research," *Medical Care*, Vol. 24, No. 2, February 1986, pp. 97–112.

Marcus, A.C., and Telesky, C.W., "Non-Participation in Telephone Follow-Up Interviews," *Proceedings of the Fourth Conference on Health Survey Research Methods*, 1982, pp. 128–134.

Marcus, A.C., and Telesky, C.W., "Non-Participation in Telephone Follow-Up Interviews," *American Journal of Public Health*, Vol. 73, No. 1, January 1983, pp. 72–77.

Market Research Development Fund, "Comparing Telephone and Face-to-Face Surveys," Marplan Ltd., 1985.

*Marquis, K.H., "Effects of Race, Residence, and Selection of Respondent on the Conduct of the Interview," in J.B. Lansing et al. (eds.), *Working Papers on Survey Research in Poverty Areas*, Survey Research Center, Institute for Social Research, The University of Michigan, Ann Arbor, 1971.

Marquis, K.H., and Blass, R., "Nonsampling Error Considerations in the Design and Operation of Telephone Surveys," *Proceedings of the First Annual Research Conference of the U.S. Bureau of the Census*, U.S. Bureau of the Census, 1985, pp. 301–329.

*Marquis, K.H., and Cannell, C.F., "A Study of Interviewer-Respondent Interaction in the Urban Employment Survey," Research Report to U.S. Department of Labor (Contract No. 81-24-68-26), Survey Research Center, Institute for Social Research, The University of Michigan, 1969.

*Marquis, K.H., and Cannell, C.F., "Effect of Some Experimental Interviewing Techniques on Reporting in the Health Interview Survey," *Vital and Health Statistics*, Series 2, Number 41, Public Health Service, 1971.

Massey, J.T., "New Initiatives Involving the Health Interview Survey," *Proceedings of the Section on Survey Research Methods, American Statistical Association*, 1978, pp. 589–593.

Massey, J.T., "The Emergence of Telephone Interviewing in Survey Research," Policy Forum for *Medical Care*, Vol. 24, No. 2, February 1986, pp. 95–96.

Massey, J.T., Barker, P.R., and Moss, A.J., "Comparative Results of Face-to-Face and Telephone Interviews in a Survey on Cigarette Smoking," paper presented at the American Public Health Association Meeting, November 1979.

Massey, J.T., Barker, P.R., and Hsiung, S., "An Investigation of Response in a Telephone Survey," *Proceedings of the Section on Survey Research Methods, American Statistical Association*, 1981, pp. 426–431.

Massey, J.T., Marquis, K.H., and Tortora, R.D., "Methodological Issues Related to Telephone Surveys by Federal Agencies," *Proceedings of the Social Statistics Section, American Statistical Association*, 1982, pp. 63–72.

*Matarazzo, J.D., and Saslow, G., "Differences in Interview Interaction Behavior Among Normal and Deviant Groups," in I.A. Berg and B.M. Bass (eds.), *Conformity and Deviation*, New York, Harper and Row, 1961.

Mathiowetz, N.A., and Cannell, C.F., "Coding Interviewer Behavior as a Method of Evaluating Performance," *Proceedings of the Section on Survey Research Methods, American Statistical Association*, 1980, pp. 525–528.

Mathiowetz, N.A., and Groves, R.M., "The Effects of Respondent Rules on Health Survey Reports," *Journal of the American Public Health Association*, Vol. 75, No. 6, June 1985, pp. 639–644.

Mathiowetz, N.A., Ward, E.P., White, A.A., and Northrup, N., "Mode of Initial Contact for Personal Interviews: Findings from Two Experiments," *Proceedings of the Section on Survey Research Methods, American Statistical Association*, 1986, pp. 287–292.

McCarthy, P.J., "Replication: An Approach to the Analysis of Data from Complex Surveys," *Vital and Health Statistics*, U.S. Department of Health, Education, and Welfare Publication (PHS) No. 1000, Series 2, No. 14, 1966.

McCarthy, P.J., "Pseudoreplication: Further Evaluation and Application of the Balanced Half-Sample Technique," *Vital and Health Statistics Series*, U.S. Department of Health, Education, and Welfare Publication (PHS) No. 1000, Series 2, No. 31, 1969.

McCormick, K.J., "The Characteristics of Nontelephone Households in the United States," contributed paper to the International Conference on Telephone Survey Methodology, 1987.

McDonald, C., "Telephone Surveys: A Review of Research Findings," The Market Research Development Fund, The Market Research Society, 1981.

McDonald, C., "Comparison of the Telephone and Face-to-Face Modes," *Proceedings of the MRDF Seminar on Telephone Surveys*, London, The Market Research Development Fund, September 1985.

McGowan, H., "Telephone Ownership in the National Crime Survey," unpublished memorandum, U.S. Bureau of the Census, 1982.

McKenzie, J., "The Accuracy of Telephone Call Data Collected by Diary Methods," *Journal of Marketing Research*, Vol. 20, November 1983, pp. 417–427.

McMillan, R.K., "Telecentral Communication—An Innovation in Survey Research," paper presented to the Advertising Research Foundation, 11th Annual Conference, New York, October 1965.

McQuire, B., and Leroy, D.J., "Comparison of Mail and Telephone Methods of Studying Media Contractors," *Journal of Broadcasting*, Vol. 21, No. 4, Fall 1977, pp. 391–400.

Meeker, R.J., "A Demonstration System for Computer-Assisted Telephone Surveys; Final Report: Prototype Development," Technical Report, Center for Computer-Based Behavioral Science, Institute for Social Science Research, University of California at Los Angeles, 1976.

*Mehrabian, A., and Williams, M., "Nonverbal Concomitants of Perceived and Intended Persuasiveness," *Journal of Personality and Social Psychology*, Vol. 13, 1969, pp. 37–58.

Miller, P.V., "Alternative Questioning Procedures for Attitude Measurement in Telephone Surveys," *Proceedings of the Section on Survey Research Methods, American Statistical Association*, 1981, pp. 432–437.

Miller, P.V., "A Comparison of Telephone and Personal Interviews in the Health Interview Survey," *Proceedings of the Fourth Conference on Health Survey Research Methods*, 1982, pp. 135–145.

Miller, P.V., "Alternative Question Forms for Attitude Scale Questions in Telephone Interviews," *Public Opinion Quarterly*, Vol. 48, No. 4, Winter 1984, pp. 766–778.

Miller, P.V., and Cannell, C.F., "A Study of Experimental Techniques for Telephone Interviewing," *Public Opinion Quarterly*, Vol. 46, No. 2, Summer 1982, pp. 250–269.

Mitchell, G.H., and Rogers, E.M., "Telephone Interviewing," *Journal of Farm Economics*, Vol. 40, August 1958, pp. 743–747.

Mitofsky, W., "Sampling of Telephone Households," unpublished CBS memorandum, 1970.

Monsees, M.L., and Massey, J.T., "Adapting Procedures for Collecting Demographic Data in a Personal Interview to a Telephone Interview," *Proceedings of the Section on Survey Research Methods, American Statistical Association*, 1979, pp. 130–135.

Mooney, H.W., Pollack, B.R., and Corsa, L., Jr., "Use of Telephone Interviewing to Study Human Reproduction," *Public Health Reports*, Vol. 83, No. 12, December 1968, pp. 1049–1060.

Morganstein, D., and Burke, J., "Use of a Telephone Survey and Log-Linear Models to Evaluate a Federal Motor Vehicle Safety Standard," *Proceedings of the Section on Survey Research Methods, American Statistical Association*, 1980, pp. 429–434.

*Morris, C., "Parametric Empirical Bayes Inference: Theory and Applications," *Journal of the American Statistical Association*, Vol. 78, 1983, pp. 47–54. 1170

Morton, B.T., and House, C.C., "Training Interviewers for Computer-Assisted Telephone Interviewing," *Proceedings of the Section on Survey Research Methods, American Statistical Association*, 1983, pp. 129–134.

*Morton-Williams, J., and Sykes, W., "The Use of Interaction Coding and Follow-up Interviews to Investigate Comprehension of Survey Questions," *Journal of the Market Research Society*, Vol. 26, No. 2, 1984, pp. 109–127.

*Morton-Williams, J., and Young, P., "Obtaining the Survey Interview — An Analysis of Tape-Recorded Doorstep Introductions," *Journal of the Market Research Society*, Vol. 29, No. 1, 1986, pp. 35–53.

Muirhead, R.C., Gower, A.R., and Newton, F.T., "The Telephone Experiment in the Canadian Labour Force Survey," *Survey Methodology*, Vol. 1, No. 2, December 1975, pp. 158–179.

Mullet, G.M., "Telephone Dialing Systems — Some Issues and Answers," *Proceedings of the Section on Survey Research Methods, American Statistical Association*, 1979, pp. 231–233.

Mullet, G.M., "The Efficacy of Plus-One Dialing: Self-Reported Status," *Proceedings of the Section on Survey Research Methods, American Statistical Association*, 1982, pp. 575–576.

Mulry-Liggan, M.H., "A Comparison of a Random Digit Dialing Survey and the Current Population Survey," *Proceedings of the Section on Survey Research Methods, American Statistical Association*, 1983, pp. 214–219.

Mulry-Liggan, M.H., and Chapman, D.W., "The Design and Selection of a Sample for the Bureau of the Census Random Digit Dialing Experiment," *Proceedings of the Section on Survey Research Methods, American Statistical Association*, 1982, pp. 133–138.

Nealon, J., "The Effects of Male vs. Female Telephone Interviewers," *Proceedings of the Section on Survey Research Methods, American Statistical Association*, 1983, pp. 139-141.

Nelson, R.O., "The McMillan System of Random Digit Dialing as Used by Chilton Research Services," Chilton Research Services, Philadelphia, paper presented at the meetings of the American Statistical Association, March 1977.

Nelson, R.O., Peyton, B.L., and Bortner, B.Z., "Use of an On-Line Interactive System: Its Effects on the Speed, Accuracy, and Cost of Survey Results," paper presented at the 18th ARF Conference, New York, 1972.

Nicholls, W.L., II, "Sampling and Field Work Methods of a Telephone Survey of Urban Users of California Wildland Areas," a report of the Survey Research Center of the University of California at Berkeley (undated).

Nicholls, W.L., II, "Sampling and Field Work Methods of the Travel Demand Forecasting Pilot Telephone Interview Survey," a report of the Survey Research Center of the University of California at Berkeley, 1975.

Nicholls, W.L., II, "Designing Telephone Surveys for the Greater Bay Area," Survey Research Center, The University of California at Berkeley, 1977.

Nicholls, W.L., II, "Experiences With CATI in a Large Scale Survey," *Proceedings of the Section on Survey Research Methods, American Statistical Association*, 1978, pp. 9-17.

Nicholls, W.L., II, "California Disability Survey: Technical Report," Survey Research Center, The University of California at Berkeley, 1979.

Nicholls, W.L., II, "A Project Plan for a Program of Research Into and Development of a Computer-Assisted Data Collection System and Assistance in Implementation Planning of a CATI System," U.S. Bureau of the Census, unpublished memorandum, 1980.

Nicholls, W.L., II, "Progress Toward Computer-Assisted Telephone Interviewing: A Report to the ASA Census Advisory Committee," U.S. Bureau of the Census, unpublished memorandum, 1981.

Nicholls, W.L., II, "Discussion of 'Telephone Surveys,'" *Proceedings of the Section on Survey Research Methods, American Statistical Association*, 1981, pp. 454-456.

Nicholls, W.L., II, "Development of CATI at the U.S. Census Bureau," *Proceedings of the Section on Survey Research Methods, American Statistical Association*, 1983, pp. 642-647.

Nicholls, W.L., II, "CATI Research and Development at the Census Bureau," *Sociological Methods & Research*, Vol. 12, No. 2, November 1983, pp. 191-197.

Nicholls, W.L., II, "Discussion of 'Use of Telephones in Surveys,'" *Proceedings of the Section on Survey Research Methods, American Statistical Association*, 1984, pp. 292-294.

Nicholls, W.L., II, and Groves, R.M., "The Status of Computer Assisted Telephone Interviewing: 1985," *Proceedings of the 45th Session, International Statistical Institute*, Book 3, Topic 18.1, 1985, pp. 1-16.

Nicholls, W.L., II, and Groves, R.M., "The Status of Computer-Assisted Telephone Interviewing: Part I – Introduction and Impact on Cost and Timeliness of Survey Data," *Journal of Official Statistics*, Vol. 2, No. 2, 1986, pp. 93-115.

Nicholls, W.L., II, and Groves, R.M., "The Status of Computer-Assisted Telephone Interviewing," *Proceedings of the 18th Symposium on the Interface, Computer Science and Statistics*, 1986, pp. 130-137.

Nicholls, W.L., II, and House, C.C., "Designing Questionnaires for Computer-Assisted Interviewing: A Focus on Program Correctness," *Proceedings of the Third Annual Research Conference of the U.S. Bureau of the Census*, U.S. Bureau of the Census, 1987, pp. 95-111.

Nicholls, W.L., II, and Lavender, G.A., "Berkeley SRC CATI Preliminary Interviewer Manual for Random Adult Studies," The University of California at Berkeley, Survey Research Center, Working Paper 32 (SMIS 10124), December 1979.

Nicholls, W.L., II, Lavender, G.A., and Shanks, J.M., "Berkeley SRC CATI: An Overview of Berkeley SRC CATI Version I," The University of California at Berkeley, Survey Research Center, 1980.

Nisselson, H., "Discussion of 'An Experimental Comparison of National Telephone and Personal Interview Surveys,'" (by Groves), *Proceedings of the Social Statistics Section, American Statistical Association*, 1977, p. 242.

Northrup, R.M., and Deniston, O.L., "Comparison of Mail and Telephone Methods to Collect Program Evaluation Data," *Public Health Reports*, Vol. 82, No. 8, August 1967, pp. 739-745.

*Nowakowska, M., "A Model of Answering to a Questionnaire Item," *Acta Psychologica*, Vol. 34, 1970, pp. 420-439.

O'Neil, M.J., "Estimating the Nonresponse Bias Due to Refusals in Telephone Surveys," *Public Opinion Quarterly*, Vol. 43, No. 2, Summer 1979, pp. 218-232.

O'Neil, M.J., "Varieties in Random Digit Dialing Sampling Designs," paper presented at the annual meetings of the American Association of Public Opinion Research, Buck Hill Falls, PA, 1979.

O'Neil, M.J., Groves, R.M., and Cannell, C.F., "Telephone Interview Introductions and Refusal Rates: Experiments in Increasing Respondent Cooperation," *Proceedings of the Section on Survey Research Methods, American Statistical Association*, 1979, pp. 252-255.

O'Rourke, D., and Blair, J., "Improving Random Respondent Selection in Telephone Surveys," *Journal of Marketing Research*, Vol. 20, November 1983, pp. 428-432.

O'Toole, B.I., Battistuta, D., Long, A., and Crouch, K., "A Comparison of Costs and Data Quality of Three Health Survey Methods: Mail, Telephone, and Personal Home Interview," *American Journal of Epidemiology*, Vol. 124, 1986, pp. 317–328.

Oakes, R.H., "Differences in Responsiveness in Telephone Versus Personal Interviews," *Journal of Marketing*, Vol. 19, No. 2, October 1954, p. 169.

Ognibene, P., "Traits Affecting Questionnaire Response," *Journal of Advertising Research*, Vol. 10, 1970, pp. 18–20.

Oksenberg, L., "Analysis of Monitored Telephone Interviews," Research Report to Bureau of the Census (JSA 80-23), Survey Research Center, Institute for Social Research, The University of Michigan, 1981.

Oksenberg, L., Coleman, L., and Cannell, C., "Personal vs. Telephone Interviewers' Voices and Refusal Rates in Telephone Surveys," Survey Research Center, Institute for Social Research, The University of Michigan, 1974.

Oksenberg, L., Coleman, L., and Cannell, C., "Interviewers' Voices and Refusal Rates in Telephone Surveys," *Public Opinion Quarterly*, Vol. 50, 1986, pp. 97–111.

Oldendick, R.W., Sorenson, S.B., Tuchfarber, A.J., and Bishop, G.F., "Last Birthday Respondent Selection in Telephone Surveys: A Further Test," paper presented at the Midwest Association of Public Opinion Research meetings, Chicago, 1985.

Ono, M., "Income Nonresponse Rates by Personal and Telephone Interviews by Regional Office — March 1970 CPS," Washington, DC, U.S. Bureau of the Census, unpublished memorandum, 1979.

Orwin, R.G., and Boruch, R.F., "RRT Meets RDD: Statistical Strategies for Assuring Response Privacy in Telephone Surveys," *Public Opinion Quarterly*, Vol. 46, No. 4, Winter 1982, pp. 560–571.

Ott, L.E., "An Empirical Investigation of Telephone Sampling Bias," *Proceedings of the Business and Economics Statistics Section, American Statistical Association*, 1975, pp. 461–471.

*Packwood, W.T., "Loudness as a Variable in Persuasion," *Journal of Counseling Psychology*, Vol. 21, 1974, pp. 1–2.

Pafford, B.V., and Coulter, D., "Use of Historical Data in a Current Interview Situation: Response Error Analyses and Applications to Computer-Assisted Telephone Interviewing," *Proceedings of the Third Annual Research Conference of the U.S. Bureau of the Census*, U.S. Bureau of the Census, 1987, pp. 281–298.

Palit, C.D., "Discussion of 'An Experimental Comparison of National Telephone and Personal Interview Surveys'" (by Groves), *Proceedings of the Social Statistics Section, American Statistical Association*, 1977, pp. 243–244.

Palit, C.D., "A Microcomputer Based Computer Assisted Interviewing System," *Proceedings of the Section on Survey Research Methods, American Statistical Association*, 1980, pp. 516–518.

Palit, C.D., "Design Strategies in RDD Sampling," *Proceedings of the Section on Survey Research Methods, American Statistical Association*, 1983, pp. 627–629.

Palit, C.D., "Computer Assisted Survey Systems: Questionnaire Development Module," Madison, WI, 1985.

Palit, C.D., "Discussion of 'Designing Questionnaires for Computer-Assisted Interviewing: A Focus on Program Correctness' (by Nicholls and House) and 'The BLAISE System for Computer-Assisted Survey Processing'" (by Denteneer, Bethlehem, Hundepool and Keller), *Proceedings of the Third Annual Research Conference of the U.S. Bureau of the Census*, U.S. Bureau of the Census, 1987, pp. 128–129.

Palit, C.D., and Sharp, H., "Microcomputer Assisted Telephone Interviewing," unpublished paper, Wisconsin Survey Research Laboratory, University of Wisconsin-Extension, February 9, 1981.

Palit, C.D., and Sharp, H., "Microcomputer-Assisted Telephone Interviewing," *Sociological Methods & Research*, Vol. 12, No. 2, November 1983, pp. 169–189.

Pannekoek, J., "Interviewer Variance in a Telephone Survey," *Journal of Official Statistics*, Vol. 4, forthcoming, 1988.

Pavalko, R.M., and Lutterman, K.G., "Characteristics of Willing and Reluctant Respondents," *Pacific Sociological Review*, Vol. 16, No. 4, 1973, pp. 463–476.

Payne, S.L., *The Art of Asking Questions*, Princeton, Princeton University Press, 1951.

Payne, S.L., "Some Advantages of Telephone Surveys," *Journal of Marketing*, Vol. 20, No. 3, January 1956, pp. 278–28l.

*Payne, S.L., "Combination of Survey Methods," *Journal of Marketing Research*, Vol. 1, 1964, pp. 61–62.

Payne, S.L., "Data Collection Methods: Telephone Surveys," in R. Ferber (ed.), *Handbook of Marketing Research*, 1974, New York, McGraw-Hill, pp. 2–105 – 2–123.

*Pearce, W.B., and Conklin, F. "Nonverbal Vocalic Communication and Perceptions of a Speaker," *Speech Monographs*, Vol. 38, 1971, pp. 235–241.

Perry, J.B., Jr., "A Note on the Use of Telephone Directories as a Sample Source," *Public Opinion Quarterly*, Vol. 32, No. 4, Winter 1968–1969, pp. 691–695.

Philipp, S.F., and Cicciarella, C.F., "An Apple II Package for Computer-Assisted Telephone Interviewing," *Behavior Research Methods and Instrumentation*, Vol. 15, No. 4, 1983, pp. 456–458.

*Phillips, D.L., *Knowledge From What? Theories and Methods in Social Research*, Chicago, Rand McNally, 1971.

Pietilä, P. "The Effect of Personal Interviews in an Additional Survey of the Finnish Labor Force Survey," contributed paper for the Conference on Telephone Survey Methodology, Charlotte, 1987.

Poe, G.S., "Methodological Issues in a Telephone Panel Survey on Smoking Practices," paper presented by Owen Thornberry, Jr., at the American Public Health Association Meeting in Anaheim, CA, 1984.

Potthoff, R.F., "Some Generalizations of the Mitofsky-Waksberg Technique for Random Digit Dialing," *Journal of the American Statistical Association*, Vol. 82, No. 398, June 1987, pp. 409–418.

Potthoff, R.F., "Generalizations of the Mitofsky-Waksberg Technique for Random Digit Dialing: Some Added Topics," *Proceedings of the Section on Survey Research Methods, American Statistical Association*, 1987, pp. 615–620.

Powell, A.E., and Klecka, W.R., "Does It Matter Who Is Missed in Telephone Surveys?" paper presented at the Annual Conference of the American Association for Public Opinion Research, May 1976.

Presser, S., "Discussion" of papers by House and Morton and Morton and House, *Proceedings of the Section on Survey Research Methods, American Statistical Association*, 1983, pp. 142–143.

*Prüfer, P., and Rexroth, M., "On the Use of the Interaction Coding Technique," *Zumanachrichten*, No. 17, November 1985.

Pyle, W.C., "Monitoring Human Resources 'On Line,'" *Michigan Business Review*, Vol. 22, No. 4, 1970, pp. 19–32.

Quinn, R.P., Gutek, B.A., and Walsh, J.T., "Telephone Interviewing: A Reappraisal and a Field Experiment," *Basic and Applied Social Psychology*, Vol. 1, No. 2, 1980, pp. 127–153.

Ravenholt, R.T., and Nixon, M., "The Telephone in Epidemiology of Staphylococcal Disease," *American Journal of Nursing*, Vol. 61, 1961, pp. 60–64.

Reeder, L.G., "Recent Literature Concerning the Use of the Telephone in Survey Research," Institute for Social Science Research, University of California at Los Angeles, 1976.

Reese, S.D., Danielson, W.A., Shoemaker, P.J., Chang, T.-K., and Hsu, H.-L., "Ethnicity-of-Interviewer Effects Among Mexican-Americans and Anglos," *Public Opinion Quarterly*, Vol. 50, No. 4, Winter 1986, pp. 563–572.

Reid, A., "Comparing Telephone With Face to Face Contact," in I. Pool (ed.), *The Social Impact of the Telephone*, Cambridge, MA, MIT Press, 1977.

*Reingen, P.H., and Kernan, J.B., "Compliance With an Interview Request: A Foot-in-the-Door, Self-Perception Interpretation," *Journal of Marketing Research*, Vol. 14, August 1977, pp. 365–369.

Research Triangle Institute, "Excerpts From Screens Used in the Youth Attitude Tracking Study II," unpublished memorandum, 1984.

Research Triangle Institute, "The Youth Attitude Tracking Study II," Telephone Interviewer Manual, RTI/ 23U-2927-04DR, 1984.

Rich, C.L., "Is Random Digit Dialing Really Necessary?" *Journal of Marketing Research*, Vol. 14, August 1977, pp. 300–305.

Robinson, J.P., and Triplett, T., "Activity Pattern Differences Between Telephone and Non-Telephone Households," contributed paper to the International Conference on Telephone Survey Methodology, Charlotte, NC, 1987.

*Rogers, C., "The Nondirective Method as a Technique for Social Research," *The American Journal of Sociology*, Vol. 50, No. 3, 1944, pp. 279–283.

Rogers, C.M., "CATI 79 Release Notes," Center for Computer-Based Behavioral Studies, University of California, Los Angeles, 1979.

Rogers, C.M., Rogers, M.S., Seward, L., and Shure, G.H., "CATI (Computer-Assisted Telephone Interviewing) Programmer Documentation, Version II, Parts 1 & 2," Center for Computer-Based Behavioral Studies, University of California at Los Angeles, 1979.

Rogers, C.M., Rogers, M.S., Seward, L., and Shure, G.H., "CATI User Documentation," Institute for Social Science Research, University of California at Los Angeles, 1979.

Rogers, T.F., "Interviews by Telephone and in Person: Quality of Responses and Field Performance," *Public Opinion Quarterly*, Vol. 40, No. 1, Spring 1976, pp. 51–65.

Roman, A.M., and Woltman, H.F., *The Methods Test Panel: Analysis of the Unemployment Rate*, Washington, DC, U.S. Bureau of the Census, 1980.

Rorer, L.G., "The Great Response-Style Myth," *Psychological Bulletin*, Vol. 63, 1965, pp. 129–156.

*Rosenthal, R., "Combining Results of Independent Studies," *Psychological Bulletin*, Vol. 85, 1978, pp. 185–193.

*Rosenthal, R., and Rosnow, R.L., *Artifact in Behavioral Research*, New York, Academic Press, 1969.

*Rosenthal, R., and Rubin, D.B., "Meta-Analytic Procedures for Combining Studies With Multiple Effect Sizes," *Psychological Bulletin*, Vol. 99, 1986, pp. 400–406.

Roshwalb, I., Spector, L., and Madansky, A., "New Methods of Telephone Interviewing A&S/CATI," *Fieldwork, Sampling and Questionnaire Design, Proceedings*, 32nd Congress, ESOMAR, Brussels, 1979, pp. 189–200.

Roslow, S., and Roslow, L., "Unlisted Phone Subscribers Are Different," *Journal of Advertising Research*, Vol. 7, August 1972, pp. 35–38.

Roughmann, K.J., and Haggerty, R.J., "Measuring the Use of Health Services by Household Interviews: A Comparison of Procedures Used in Three Child Health Surveys," *International Journal of Epidemiology*, Vol. 3, No. 1, pp. 71–81.

Rustemeyer, A., "Measuring Interviewer Performance in Mock Interviews," *Proceedings of the Social Statistics Section, American Statistical Association*, 1977, pp. 341–346.

Rustemeyer, A., "Toward Development of a Computer Assisted Telephone Interviewing System," mimeograph, Statistical Research Division, U.S. Bureau of the Census, July 1977.

Rustemeyer, A., "Summary of the 1978 CATI Conference," U.S. Bureau of the Census, January 9, 1979.

Rustemeyer, A., and Heller, G., "Computer-Assisted Interviewing," paper presented to the Conference of European Statisticians – Seminar on "Statistics in the Coming Decade," Washington, DC, March 24, 1977 (SMIS 77208 007).

Rustemeyer, A., and Levin, A., "Report on a Telephone Survey Using Computer Assistance," Statistical Research Division, U.S. Bureau of the Census, November 1977.

Rustemeyer, A., and Levin, A., "Description of and Requirements Study for Computer-Assisted Data Collection," Washington, DC, U.S. Bureau of the Census, unpublished report, 1979.

Rustemeyer, A., Shure, G.H., Rogers, M., and Meeker, R.J., "Computer Assisted Telephone Interviewing: Design Considerations," *Proceedings of the Section on Survey Research Methods, American Statistical Association*, 1978, p. 8.

Salmon, C.T., and Nichols, J.S., "The Next-Birthday Method of Respondent Selection," *Public Opinion Quarterly*, Vol. 47, 1983, pp. 270 276.

San Augustine, A.J., and Friedman, H.H., "The Use of the Telephone Interview in Obtaining Information of a Sensitive Nature," *Proceedings of the Section on Survey Research Methods, American Statistical Association*, 1978, pp. 559–561.

*Schaeffer, N., "Evaluating Race-of-Interviewer Effects in a National Survey," *Sociological Methods and Research*, Vol. 8, 1980, pp. 400–419.

*Scherer, K.R., "Personality Markers in Speech," in K.R. Scherer and H. Giles (eds.), *Social Markers in Speech*, Cambridge, Cambridge University Press, 1979.

Schmiedeskamp, J.W., "Reinterviews by Telephone," *Journal of Marketing*, Vol. 26, No. 1, January 1962, pp. 28–34.

Schuman, E.A., and McCandles, B., "Who Answers Questionnaires," *Journal of Applied Psychology*, Vol. 24, 1940, pp. 758–769.

*Schuman, H., and Converse, J., "The Effects of Black and White Interviewers on Black Responses in 1968," *Public Opinion Quarterly*, Vol. 35, 1971, pp. 44–66.

*Schuman, H., and Ludwig, J., "The Norm of Even-Handedness in Surveys as in Life," *American Sociological Review*, Vol. 48, 1983, pp. 112–120.

*Schuman, H., and Presser, S., *Questions and Answers in Attitude Surveys: Experiments on Question Form, Wording, and Context*, New York, Academic Press, 1981.

*Schuman, H., Kalton, G., and Ludwig, J., "Context and Contiguity in Survey Questionnaires," *Public Opinion Quarterly*, Vol. 47, No. 1, 1983, pp. 112–115.

*Schuman, H., Ludwig, J., and Krosnick, J.A., "The Perceived Threat of Nuclear War, Salience, and Open Questions," *Public Opinion Quarterly*, Vol. 50, No. 4, 1986, pp. 519-536.

Scott, C.A., "The Effects of Trial and Incentives on Repeat Purchase Behavior," *Journal of Marketing Research*, Vol. 13, 1976, pp. 263-269.

Scott, S., and Yurachek, M., "Telephone Collection in an Establishment Survey of Job Openings," *Proceedings of the Section on Survey Research Methods, American Statistical Association*, 1981, pp. 91-96.

Seaman, J., "Review of 'The Effect of the Demographics of Individual Households on Their Telephone Usage'" (by B.B. Brandon), *Journal of the American Statistical Association*, Vol. 77, 1982, p. 497.

Sebestik, J.P., and Sudman, S., "What Makes a Good Telephone Interviewer?" Survey Research Laboratory, University of Illinois, Urbana-Champaign, 1977.

Sewell, W.H., and Shah, V.P., "Socioeconomic Status, Intelligence, and the Attainment of High Education," *Sociology of Education*, Vol. 40, 1967, pp. 1-23.

Shanks, J.M., "The Development of CATI Methodology," paper presented at the Conference on Computer-Assisted Telephone Interview Technology, Berkeley, CA, March 1980.

Shanks, J.M., "The Current Status of Computer-Assisted Telephone Interviewing: Recent Progress and Future Prospects," *Sociological Methods & Research*, Vol. 12, No. 2, November 1983, pp. 119-142.

Shanks, J.M., and Freeman, H.E., "The Emergence of Computer-Assisted Survey Research," a summary of the National Science Foundation-Supported Conference on Computer-Assisted Survey Methods, 1981.

Shanks, J.M., and Tortora, R., "Beyond CATI: Generalized and Distributed Systems for Computer-Assisted Surveys," *Proceedings of the First Annual Research Conference of the U.S. Bureau of the Census*, U.S. Bureau of the Census, 1985, pp. 358-371.

Shanks, J.M., Freeman, H.E., and Nicholls, W.L., II, "A First Report on the California Disability Survey: A Large Scale Telephone Survey to Locate and Describe the Disabled Population," paper presented at the Annual Conference of the American Association for Political Opinion Research, June 1978.

Shanks, J.M., Coleman, C., Lavender, G., Monsky, S.F., Morton, B.T., and Nicholls, W.L. II, "Introduction to Computer-Assisted Telephone Interviewing (CATI) and Demonstration of the SRC Berkeley CATI System," presented at the Conference on Computer-Assisted Survey Technology, Berkeley, CA, 1980.

Shanks, J.M., Lavender, G., and Nicholls, W.L., II, "Continuity and Change in Computer-Assisted Surveys: Development of Berkeley SRC CATI," *Proceedings of the Section on Survey Research Methods, American Statistical Association*, 1980, pp. 507-512.

Shanks, J.M., Denney, W.M., Hendricks, J.S., and Brody, R.A., "Citizen Reasoning About Public Issues and Policy Tradeoffs: A Progress Report on Computer-Assisted Political Surveys," The Russel Sage Foundation, unpublished memorandum, October 1981.

Shanks, J.M., Nicholls, W.L., II, and Freeman, H.E., "The California Disability Survey: Design and Execution of a Computer-Assisted Telephone Study," *Sociological Methods & Research*, Vol. 10, No. 1, 1981, pp. 123–140.

Shanks, J.M., Morton, B.T., McIntosh, R.A., and Lavender, G., "Computer-Assisted Surveys: A User's Guide," Computer-Assisted Methods Technical Report, University of California at Berkeley, 1983.

Shanks, J.M., Morton, B.T., McIntosh, R.A., and Walton, C., "The Berkeley System for Computer-Assisted Data Collection: Demonstration and Exhibits," paper prepared for meetings of the American Association for Public Opinion Research, Buck Hill Falls, PA, 1983.

Shanks, J.M., McIntosh, R.A., and Morton, B.T., "Computer-Assisted Surveys: Recent Activities of the Computer-Assisted Survey Methods Program," paper prepared for the Annual Meeting of the American Sociological Association, San Antonio, TX, 1984.

Shapiro, S., Yaffe, R., Fuchsberg, R.R., and Corpeño, H.C., "Medical Economics Survey-Methods Study: Design, Data Collection, and Analytical Plan," *Medical Care*, Vol. 14, No. 11, November 1976, pp. 893–912.

Sharf, D.J., and Lehman, M.E., "Relationship Between the Speech Characteristics and Effectiveness of Telephone Interviewers," *Journal of Phonetics*, Vol. 12, 1984, pp. 219–228.

Shih, W.-F.P., "An Evaluation of Random Digit Dialing Household Survey," *Proceedings of the Section on Survey Research Methods, American Statistical Association*, 1980, pp. 736–739.

Shih, W.-F.P., "Nonresponse to Income Questions in Telephone Surveys," *Proceedings of the Section on Survey Research Methods, American Statistical Association*, 1983, pp. 283–288.

Short, J., Williams, E., and Christie, B., *The Social Psychology of Telecommunications*, London, Wiley, 1976.

Shure, G.H., and Meeker, R.J., "A Computer-Based Experimental Laboratory," *American Psychologist*, Vol. 25, 1970, pp. 962–969.

Shure, G.H., and Meeker, R.J., "A Minicomputer System for Multiperson Computer-Assisted Telephone Interviewing," *Behavioral Research Methods and Instrumentation*, Vol. 12, No. 2, 1978, pp. 196–202.

Shure, G.H., Freeman, H.E., and Treiman, D.J., "Computer-Assisted Telephone Interviewing," unpublished manuscript, Institute for Social Science Research, The University of California at Los Angeles, 1976.

Siemiatycki, J., "A Comparison of Mail, Telephone, and Home Interview Strategies for Household Health Surveys," *American Journal of Public Health*, Vol. 69, No. 3, March 1979, pp. 238–245.

Siemiatycki, J., and Campbell, S., "Nonresponse Bias and Early Versus All Responders in Mail and Telephone Surveys," *American Journal of Epidemiology*, Vol. 120, 1984, pp. 291–301.

Siemiatycki, J., Campbell, S., Richardson, L., and Aubert, D., "Quality of Response in Different Population Groups in Mail and Telephone Surveys," *American Journal of Epidemiology*, Vol. 120, 1984, pp. 302–314.

Silver, J., "A Discussion of Telephone Interviewing, Respondent Weariness and Memory Bias in the Survey of Residential Alterations and Repairs," Washington, DC, U.S. Bureau of the Census, 1976.

*Simmons, D.M., *Nonlinear Programming for Operations Research*, Englewood Cliffs, NJ, Prentice-Hall, 1975.

Simmons, W.R., and Bean, J.A., "Impact of Design and Estimation Components on Inference," in N.L. Johnson and H. Smith, Jr. (eds.), *New Developments in Survey Sampling*, New York, Wiley-Interscience, 1969.

*Simon, J.L., and Burnstein, P., *Basic Research Methods in Social Science*, 3rd ed., New York, Random House, 1985.

Singer, E., "Telephone Interviewing as a Black Box," paper presented at the Third Biennial Conference on Health Survey Research Methods, Reston, VA, 1979.

*Singer, E., Frankel, M., and Glassman, M., "The Effect of Interviewer Characteristics and Expectations on Response," *Public Opinion Quarterly*, Vol. 47, 1983, pp. 68–83.

Sirken, M.G., and Casady, R.J., "Nonresponse in Dual Frame Surveys Based on Area/List and Telephone Frames," *Proceedings of the Section on Survey Research Methods, American Statistical Association*, 1982, pp. 151–153.

Sirken, M.G., Pifer, J.W., and Brown, M.L., "Survey Procedures for Supplementing Mortality Statistics," *American Journal of Public Health*, Vol. 50, No. 11, November 1961, pp. 1753–1764.

Slocum, W.L., Empey, L.T., and Swanson, H.S., "Increasing Response to Questionnaires and Structured Interviews," *American Sociological Review*, Vol. 21, 1956, pp. 221–225.

Smead, R.J., and Wilcox, J., "Ring Policy in Telephone Surveys," *Public Opinion Quarterly*, Vol. 44, No. 1, Spring 1980, pp. 115–116.

Smith, R., and Smith, R., "Evaluation and Enhancements of Computer Controlled Telephone Interviewing," *Proceedings of the Section on Survey Research Methods, American Statistical Association*, 1980, pp. 513–515.

Smith, T.W., "A Comparison of Telephone and Personal Interviewing," Chicago, National Opinion Research Center, 1984.

*Smith, T.W., "That Which We Call Welfare by Any Other Name Would Smell Sweeter: An Analysis of the Impact of Question Wording on Response Patterns," *Public Opinion Quarterly*, Vol. 51, No. 1, 1987, pp. 75–83.

Smith, T.W., "Phone Home? An Analysis of Household Telephone Ownership," contributed paper to the International Conference on Telephone Survey Methodology, Charlotte, NC, 1987.

Snowden, C.B., "The National Center for Health Statistics Linked Telephone Survey: A Potential Health Data Source," *Proceedings of the Section on Survey Research Methods, American Statistical Association*, 1985, pp. 238–241.

Snyder, M., and Cunningham, M.R., "To Comply or Not Comply: Testing the Self-Perception Explanation of the 'Foot-in-the-Door' Phenomenon," *Journal of Personality and Social Psychology*, Vol. 31, 1975, pp. 64–67.

Sobal, J., "What Should We Say After We Say Hello? Disclosing Information in Interview Introductions," paper presented at the annual meetings of the American Association of Public Opinion Research, Buck Hill Falls, PA, 1978.

*Sonquist, J.A., Baker, E.L., and Morgan, J.N., *Searching for Structure*, revised edition, Ann Arbor, MI, Institute for Social Research, 1973.

Spaeth, M.A., "Interviewing in Telephone Surveys," *Survey Research*, Vol. 5, 1973, pp. 9–10.

Spaeth, M.A., "Telephone Interviewing Facilities at Survey Research Organizations," *Survey Research*, Vol. 11, Nos. 3–4, Summer-Fall 1979, pp. 21–25.

Spaeth, M.A., "CATI Facilities at Survey Research Organizations," *Survey Research*, Vol. 18, Nos. 3–4, 1987, pp. 18–22.

Spitz, K.E., "Review of 'Surveys by Telephone: A National Comparison With Personal Interviews'" (by R.M. Groves and R.L. Kahn), *Journal of the Royal Statistical Society*, Series A, Vol. 143, Part 4, 1980, pp. 521–522.

"SRI Electric Interview System," Stanford Research Institute, Menlo Park, CA, 1974.

Stafford, J.E., "Influence of Preliminary Contact on Mail Returns," *Journal of Marketing Research*, Vol. 3, November 1966, pp. 410–411.

Steeh, C.G., "Trends in Nonresponse Rates, 1952–1979," *Public Opinion Quarterly*, Vol. 45, No. 1, Spring 1981, pp. 40–57.

Steel, D., and Boal, P., "Characteristics of Households and Persons Not Accessible by Telephone in Australia," contributed paper to the International Conference on Telephone Survey Methodology, Charlotte, NC, 1987.

Stephenson, G.M., Ayling, K., and Rutter, D.R., "The Role of Visual Communication in Social Exchange," *British Journal of Social and Clinical Psychology*, Vol. 15, 1976, pp. 114–120.

Stern, D.E., Jr., and Steinhorst, R.K., "Telephone Interview and Mail Questionnaire Applications of the Randomized Response Model," *Journal of the American Statistical Association*, Vol. 79, 1984, pp. 555–564.

Stock, J.S., "How to Improve Samples Based on Telephone Listings," *Journal of Advertising Research*, Vol. 2, No. 3, September 1962, pp. 50–51.

Stokes, S.L., "Interviewer Variability in RDD-1," U.S. Bureau of the Census and the University of Texas at Austin, December 1984.

Stokes, S.L., "Estimation of Interviewer Effects in Complex Surveys With Application to Random Digit Dialing," *Proceedings of the Second Annual Research Conference of the U.S. Bureau of the Census*, U.S. Bureau of the Census, 1986, pp. 21–31.

Stokes, S.L., "Estimation of Interviewer Effects for Categorical Items in a Random Digit Dial Telephone Survey," *Journal of the American Statistical Association*, in press.

Sudman, S., "New Uses of Telephone Methods in Survey Research," *Journal of Marketing Research*, Vol. 3, No. 2, May 1966, pp. 163–167.

Sudman, S., *Reducing the Cost of Surveys*, Chicago, Aldine, 1967.

Sudman, S., "The Uses of Telephone Directories for Survey Sampling," *Journal of Marketing Research*, Vol. 10, No. 2, May 1973, pp. 204–207.

Sudman, S., "Sample Surveys," *Annual Review of Sociology*, Vol. 2, Palo Alto, Annual Reviews, 1976, pp. 107–120.

Sudman, S., "Discussion of 'Telephone Interviewing Research,'" *Proceedings of the Section on Survey Research Methods, American Statistical Association*, 1978, pp. 562–563.

Sudman, S., "Optimum Cluster Designs Within a Primary Unit Using Combined Telephone Screening and Face-to-Face Interviewing," *Journal of the American Statistical Association*, Vol. 73, No. 362, June 1978, pp. 300–304.

Sudman, S., "Survey Research and Technological Change," *Sociological Methods & Research*, Vol. 12, No. 2, November 1983, pp. 217–230.

Sudman, S., and Bradburn, N., *Response Effects in Surveys, A Review and Synthesis*, Chicago, Aldine, 1974.

*Sudman, S., and Bradburn, N.M., *Asking Questions*, San Francisco, Jossey-Bass, 1982.

Sudman, S., and Ferber, R., "A Comparison of Alternative Procedures for Collecting Consumer Expenditure Data for Frequently Purchased Products," *Journal of Marketing Research*, Vol. 11, No. 2, May 1974, pp. 128–135.

Sudman, S., and Lannom, L.B., "A Comparison of Alternative Panel Procedures for Obtaining Health Data," *Proceedings of the Social Statistics Section, American Statistical Association*, Part I, 1977, pp. 511–516.

Sudman, S., Bradburn, N.M., Blair, E., and Stocking, C., "Modest Expectations," *Sociological Methods & Research*, Vol. 6, No. 2, November 1977, pp. 171–182.

Survey Research Center, "Motivation Analysis of Social Long Distance Telephone Calls," The University of Michigan, Ann Arbor, June 1955.

Survey Research Center, "Appraisal of Telephone Interviews for Obtaining Information on Consumer Attitudes," The University of Michigan, Ann Arbor, April 1960.

Survey Research Center, "Telephone Interviewing (/Reinterviewing)," in *Interviewer's Manual*, Revised Edition, The University of Michigan, Ann Arbor, 1976, pp. 33–34.

Survey Research Center, "CATI Application for Commercial Fleet Managers Survey," Institute for Social Research, The University of Michigan, Ann Arbor, 1983.

Survey Research Center, "Activities and Products of an SRC CATI Survey," Institute for Social Research, The University of Michigan, Ann Arbor, unpublished memorandum, 1984.

Survey Research Center, "Monitoring Report Form," Institute for Social Research, The University of Michigan, Ann Arbor, unpublished memorandum, 1984.

Survey Research Center, "CATI Application for National Association of Business Economists Survey," Institute for Social Research, The University of Michigan, Ann Arbor, 1985.

Survey Research Laboratory, "Interviewing in Telephone Surveys," *Survey Research Newsletter*, Vol. 5, Urbana, University of Illinois, 1975.

Survey Sampling, Inc., "Statistical Characteristics of Random Digit Telephone Samples Produced by Survey Sampling, Inc.," Westport, CT, Survey Sampling, Inc., 1986.

Swensson, B., "A Survey of Nonresponse Terms," *Statistical Journal of the United Nations*, Vol. 1, 1983, pp. 241–252.

Swint, A.G., and Powell, T.E., "CLUSFONE Computer-Generated Telephone Sampling Offers Efficiency and Minimal Bias," *Marketing News*, Vol. 17, May 13, Section 2, 1983, pp. 2–3.

Sykes, W., and Collins, M., "Comparing Telephone and Face-to-Face Interviewing in the UK," *Survey Methodology*, Vol. 13, No. 1, 1987, pp. 15–28.

Sykes, W., and Hoinville, G., "Telephone Interviewing on a Survey of Social Attitudes: A Comparison With Face-to-Face Procedures," London, Social and Community Planning Research, 1985.

*Sykes, W., and Morton-Williams, J., "Evaluating Survey Questions," *Journal of Official Statistics*, Vol. 3, No. 2, 1987, pp. 191–207.

Taylor, D.G., "Observations on the Behavior of Automated Telephone Interviewing," *Proceedings of the Third Biennial Conference on Health Survey Research Methods*, 1979, pp. 99–100.

Telser, E., "Telephone Versus Personal Interviews: Why Do It at All if We Can't Do It the Right Way?" unpublished paper presented at Midwestern American Association of Public Opinion Research, A.C. Nielsen Company (undated).

Telser, E., "Data Exorcises Bias in Telephone Versus Personal Interview Debate, but if You Can't Do It Right, Don't Do It at All," *Marketing News*, Vol. 10, 1976, pp. 6–7.

Thomas, R., "Telephone Interviewing at Social Survey Division," London, Social and Community Planning Research, 1987.

Thompson, D.J., and Tauber, J., "Household Survey, Individual Interview, and Clinical Examination to Determine the Prevalence of Heart Disease," *American Journal of Public Health*, Vol. 47, September 1957, pp. 1131–1140.

Thompson, N.R., "Nonresponse Bias From 'No Answer/Busy' Calls in a Telephone Survey," *Proceedings of the Section on Survey Research Methods, American Statistical Association*, 1979, pp. 250–251.

Thornberry, O.T., Jr., "An Evaluation of Three Strategies for the Collection of Health Interview Data From Households," unpublished doctoral dissertation, 1977.

Thornberry, O.T., Jr., "Review of 'Survey Research by Telephone'" (by J.H. Frey), *Journal of the American Statistical Association*, Vol. 80, 1985, pp. 483–484.

Thornberry, O.T., Jr., and Massey, J.T., "Correcting for Undercoverage Bias in Random Digit Dialed National Health Surveys," *Proceedings of the Section on Survey Research Methods, American Statistical Association*, 1978, pp. 224–229.

Thornberry, O.T., Jr., and Massey, J.T., "Coverage and Response in Random Digit Dialed National Surveys," *Proceedings of the Section on Survey Research Methods, American Statistical Association*, 1983, pp. 654–659.

Thornberry, O.T., Jr., and Massey, J.T., "Noninterview Bias in Random Digit Dialed Surveys," paper presented at the NCHS-sponsored Seminar on Methodological Needs for Behavioral Risk Factor Assessment Surveys, Houston, TX, 1984.

Thornberry, O.T., Jr., and Poe, G.S., "NCHS Research on the Telephone Interview: Some Observations," *Proceedings of the Social Statistics Section, American Statistical Association*, 1982, pp. 296–301.

Tigert, D.J., Barnes, J.G., and Bourgeois, J.C., "Research on Research: Mail Panel Versus Telephone Survey in Retail Image Analysis," *The Canadian Marketer*, Winter 1975, pp. 22–27.

Tomkins, L., and Massey, J.T. "Using a Most Knowledgeable Respondent Rule in a Household Telephone Survey," *Proceedings of the Section on Survey Research Methods, American Statistical Association*, 1986, pp. 281–286.

Tortora, R.D., "CATI in an Agricultural Statistics Agency," *Journal of Official Statistics*, Vol. 1, No. 2, 1985, pp. 301–314.

Traugott, M.W., "The Importance of Persistence in Respondent Selection for Preelection Surveys," *Public Opinion Quarterly*, Vol. 51, No. 1, Spring 1987, pp. 48–57.

Traugott, M.W., Groves, R.M., and Lepkowski, J., "Using Dual Frame Designs to Reduce Nonresponse in Telephone Surveys," *Public Opinion Quarterly*, Vol. 51, No. 4, Winter 1987, pp. 522–539.

Tremblay, K.R., and Dillman, D.A., "Research Ethics: Emerging Concerns From Increased Use of Mail and Telephone Survey Methods," *Humboldt Journal of Social Relations*, Vol. 5, 1977, pp. 64-89.

Trendex, Inc., "A Comparison of Phone Book Samples and Random Digit Dialing Samples," paper presented at the 22nd Annual Conference of the Advertising Research Foundation, 1976.

Troldahl, V.C., and Carter, R.E., Jr., "Random Selection of Respondents Within Households in Phone Surveys," *Journal of Marketing Research*, Vol. 1, No. 2, May 1964, pp. 71-76.

Trussel, R.E., Elinson, J., and Levin, M.L., "Comparisons of Various Methods of Estimating the Prevalence of Chronic Disease in a Community — The Hunterdon County Study," *American Journal of Public Health*, Vol. 46, No. 2, February 1956, pp. 82.

Tuchfarber, A.J., Jr., "Random Digit Dialing: A Test of Accuracy and Efficiency," Ph.D. dissertation, Cincinnati, University of Cincinnati, 1974.

Tuchfarber, A.J., Jr., "Random Digit Dialing: An Empirical Test of Reliability," paper presented at the annual conference of the American Association for Public Opinion Research, 1976.

Tuchfarber, A.J., Jr., and Klecka, W.R., "Demographic Similarities Between Samples Collected by Random Digit Dialing Versus Complex Sampling Designs," University of Cincinnati, Behavioral Sciences Laboratory, paper presented at the annual meeting of the American Association for Public Opinion Research, June 1975.

Tuchfarber, A.J., Jr., and Klecka, W.R., "Random Digit Dialing: Lowering the Cost of Victimization Surveys," Police Foundation, Washington, DC, 1976.

Tuchfarber, A.J., Jr., Klecka, W.R., Bardes, B.A., and Oldendick, R.W., 'Reducing the Cost of Victim Surveys," in W.G. Skogan (ed.), *Sample Surveys of the Victims of Crime*, Cambridge, MA, Ballinger Publishing Co., 1976, pp. 207-221.

Tucker, C., "Interviewer Effects in Telephone Surveys," *Public Opinion Quarterly*, Vol. 47, No. 1, Spring 1983, pp. 84-95.

Tull, D.S., and Albaum, G.S., "Bias in Random Digit Dialed Surveys," *Public Opinion Quarterly*, Vol. 41, No. 3, Fall 1977, pp. 389-395.

Turner, A.G., "A Specialized Analysis of the NCS Telephone Experiment," Washington, DC, U.S. Bureau of the Census, 1977.

Turner, A.G., "An Experiment to Compare Three Interviewing Procedures in the National Crime Survey," Washington, DC, U.S. Bureau of the Census, Statistical Research Division, 1977.

*Turner, C., and Martin, E.A., *Surveying Subjective Phenomena*, Vol. I, New York, Russel Sage Foundation, 1984.

Tyebjee, T.Z., "Telephone Survey Methods: The State of the Art," *Journal of Marketing*, Vol. 43, Summer 1979, pp. 68–78.

*United Nations, *Statistical Year Book*, New York, United Nations, 1986.

U.S. Bureau of the Census, "Characteristics of Households with Telephones, March 1965," *Current Population Reports* (Series P-20, No. 146), 1965.

U.S. Bureau of the Census, "Response Errors in Collection of Expenditure Data in Household Interviews: An Experimental Study," 1965, Technical Paper No. 11.

U.S. Bureau of the Census, *Who's Home When*, prepared by Dean Weber, Washington, DC, U.S. Government Printing Office, Working Paper 37, January 1972.

*U.S. Bureau of the Census, "The Current Population Survey, Design and Methodology," Technical Paper No. 40 of the U.S. Bureau of the Census, 1978.

U.S. Bureau of the Census, "Census of the Population: 1980, Detailed Housing Characteristics, Final Report," HC 80-1-B1, U.S. Summary, Washington, D.C., U.S. Government Printing Office, 1982.

*U.S. Bureau of the Census, Panel on Decennial Census Methodology, "The Bicentennial Census, New Directions for Methodology in 1990," Washington, DC, Academic Press, 1985.

U.S. Department of Justice, Law Enforcement Assistance Administration, National Criminal Justice Information and Statistics Service, "Criminal-Victimization Surveys in 13 American Cities," Report No. SD-NCP-C-4, 1975.

U.S. National Center for Health Statistics, "U.S. National Health Survey: The Statistical Design of the Health Household Interview Survey," Health Statistics, PHS Pub. No. 584-A2, Washington, D.C., U.S. Government Printing Office, 1958.

*U.S. National Center for Health Statistics, Kovar, M.G., and Poe, G.S., "The National Health Interview Survey Design, 1973–84, and Procedures 1975–83," *Vital and Health Statistics*, Series 1, No. 18, 1985.

U.S. National Center for Health Statistics, "An Experimental Comparison of Telephone and Personal Health Interview Surveys," *Vital and Health Statistics*, Series 2, No. 106, DHHS Pub. (PHS) 87-1380, August 1987.

U.S. National Center for Health Statistics, "Linked Telephone Surveys: A Test of a Methodology," *Vital and Health Statistics*, Series 2, No. 109, U.S. Department of Health and Human Services, forthcoming.

U.S. Office of Management and Budget, "The Role of Telephone Data Collection in Federal Statistics," Statistical Policy Working Paper 12, November 1984.

Van Bastelaer, A., and Leenders, P., "A Comparison of the Responses of a Restricted Sample of Telephone Owners With the Responses of a Total Sample," internal report to the Netherlands Central Bureau of Statistics, 1985.

*Van der Zouwen, J., "A Conceptual Model for the Auxiliary Hypotheses Behind the Interview," *Annals of Systems Research*, Vol. 4, 1974, pp. 21-37.

*Van der Zouwen, J., and De Leeuw, E.D., "Effects of the Method of Data Collection on the Outcome of Surveys; Towards a Cybernetic Model," paper presented at the Seventh International Congress of Cybernetics and Systems, London, September 1987. Published in abridged form in J. Rose (ed.), *Cybernetics and Systems: The Way Ahead*, Vol. 2, England, Lythan St. Annes, Thales Publications, 1987.

*Van der Zouwen, J., Dijkstra, W., and Van de Bovenkamp, J., "The Control of Interaction Processes in Survey Interviews," in F. Geyer and J. van der Zouwen (eds.), *Sociocybernetic Paradoxes; Observation, Control and Evolution of Self-Steering Systems*, London, Sage, 1986.

Vanderveer, R.B., "Training Interviewers by Telephone — Does It Work?" paper presented at the annual meeting of the American Association for Public Opinion Research, 1974.

Verbrugge, L.M., "Sex Differences in Illness and Health Actions," paper presented at the meetings of American Public Health Association, 1977.

Vigderhous, G., "Optimizing the Time Schedules and Response Rates in Telephone Interviews," paper presented at the annual meeting of the American Association of Public Opinion Research, June 1978.

Vigderhous, G., "Optimizing the Time Schedules and Response Rates in Telephone Interviews — A Study of the Seasonal Patterns," *Proceedings of the Section on Survey Research Methods, American Statistical Association*, 1979, pp. 256-261.

Vigderhous, G., "Scheduling Telephone Interviews: A Study of Seasonal Patterns," *Public Opinion Quarterly*, Vol. 45, No. 2, Summer 1981, pp. 250-259.

*Waelberg, H.J., and Haertel, E.H., "Research Integration: An Introduction and Overview," *Evaluation in Education*, Vol. 4, 1980.

Wage, J., *How to Use the Telephone in Selling*, Aldershot, England, Epping, Gower, 1973.

*Waksberg, J., "The Effects of Stratification With Differential Sampling Rates on Attributes of Subsets of the Population," *Proceedings of the Social Statistics Section, American Statistical Association*, 1973, pp. 429-434.

Waksberg, J., "Sampling Methods for Random Digit Dialing," *Journal of the American Statistical Association*, Vol. 73, No. 361, March 1978, pp. 40-46.

Waksberg, J., "A Note on Locating a Special Population Using Random Digit Dialing," *Public Opinion Quarterly*, Vol. 47, 1983, pp. 576-579.

Waksberg, J., "CLUSFONE and Waksberg: Let's Correct Misconceptions," *Marketing News*, Vol. 18, January 6, 1984, p. 2.

Waksberg, J., "Efficiency of Alternative Methods of Establishing Cluster Sizes in RDD Sampling," internal WESTAT memorandum for distribution, December 17, 1984.

Waksberg, J., "Discussion," *Proceedings of the First Annual Research Conference of the U.S. Bureau of the Census,* U.S. Bureau of the Census, 1985, pp. 87–92.

Waksberg, J., "Discussion of Nicholls and Groves Paper," *Proceedings of the 45th Session, International Statistical Institute,* Book 5, 1985, pp. 159–161.

Waksberg, J., "Discussion of 'Some Research Issues in Random-Digit-Dialing and Estimation'" (by Biemer, Chapman, and Alexander), *Proceedings of the First Annual Research Conference of the U.S. Census Bureau,* U.S. Bureau of the Census, 1985, pp. 87–92.

Waksberg, J., "Review of 'The Role of Telephone Data Collection in Federal Statistics'" (by Federal Committee on Statistical Methodology), *Journal of the American Statistical Association,* Vol. 80, 1985, p. 1077.

Waksberg, J., Berlin, M., and McKenna, T., "Selective Factors Affecting Success in Obtaining Mailing Addresses in Random Digit Dialing," paper presented at the American Statistical Association meetings, 1978.

Ward, E.M., Kramer, S., and Meadows, A.T., "The Efficacy of Random Digit Dialing in Selecting Matched Controls for a Case-Control Study of Pediatric Cancer," *American Journal of Epidemiology,* Vol. 120, 1984, pp. 582–591.

Warde, W., "Problems With Telephone Surveys," U.S. Department of Agriculture, NASS Staff Report No. SRB-NERS-86-01, Washington, DC, 1986.

Wasserman, I.M., and Friedman, M., "Characteristics of Respondents and Non-Respondents in a Telephone Survey Study of Elderly Consumers," paper presented at the Second Annual Conference of the Midwest Association for Public Opinion Research, November 18–20, 1976.

Weaver, C.N., Holmes, S.L., and Glenn, N.D., "Some Characteristics of Inaccessible Respondents in a Telephone Survey," *Journal of Applied Psychology,* Vol. 60, No. 2, 1975, pp. 260–262.

Weeks, M.F., Jones, B.L., Folsom, Jr., R.E., and Benrud, C.H., "Optimal Times to Contact Sample Households," *Public Opinion Quarterly,* Vol. 44, No. 1, 1980, pp. 101–114.

Weeks, M.F., Kulka, R.A., Lessler, J.T., and Whitmore, R.W., "Personal Versus Telephone Surveys for Collecting Household Health Data at the Local Level," *American Journal of Public Health,* Vol. 73, 1983, pp. 1389–1394.

Weeks, M.F., Kulka, R.A., and Pierson, S., "Optimal Scheduling of Calls for a Telephone Survey," paper presented at the 41st Annual Conference of the American Association for Public Opinion Research, St. Petersburg, FL, 1986.

Weeks, M.F., Kulka, R.A., and Pierson, S., "Optimal Call Scheduling for a Telephone Survey," *Public Opinion Quarterly,* Vol. 51, 1987, pp. 540–549.

Weinberg, E., "Data Collection: Planning and Management," in P.H. Rossi, J.D. Wright, and A.B. Anderson (eds.), *Handbook of Survey Research,* Academic Press, 1983, pp. 329–358.

Weller, T., "Telephone Interviewing Procedures," *Survey Research Newsletter*, Survey Research Laboratory, University of Illinois, Urbana, Vol. 5, No. 1, January 1973, pp. 13–14.

Westat, "An Introduction to the Westat FRAME System for Computer Assisted Telephone Interviewing (CATI/FRAME)," paper presented at the meetings of the American Association for Public Opinion Research, May 1983.

Wheatley, J.J., "Self-Administered Written Questionnaires or Telephone Interviews?" *Journal of Marketing Research*, Vol. 10, February 1973, pp. 94–96.

White, A.A., "Response Rate Calculation in RDD Telephone Health Surveys: Current Practices," *Proceedings of the Section on Survey Research Methods, American Statistical Association*, 1983, pp. 277–282.

Whitmore, R.W., Mason, R.E., Hartwell, T.D., and Rosenzweig, M.S., "Use of Geographically Classified Telephone Directory Listings in Multi-Mode Surveys," *Proceedings of the Section on Survey Research Methods, American Statistical Association*, 1983, pp. 721–727.

Whitmore, R.W., Mason, R.E., and Hartwell, T.D., "Use of Geographically Classified Telephone Directory Lists in Multi-Mode Surveys," *Journal of the American Statistical Association*, Vol. 80, 1985, pp. 842–844.

Wilcox, K., "Comparison of Three Methods for the Collection of Morbidity Data by Household Survey," Ph.D. dissertation, The University of Michigan, School of Public Health, 1963.

Willamowsky, Y., Friedman, H.H., Di Pietro, R., and Epstein, S., "Does the Age of a Telephone Interviewer Affect Subject's Responses?" *Proceedings of the Section on Survey Research Methods, American Statistical Association*, 1979, pp. 427–429.

Williams, E., "Experimental Comparisons of Face-to-Face and Mediated Communication: A Review," *Psychological Bulletin*, Vol. 84, No. 5, 1977, pp. 963–976.

*Williams, E., "Visual Interaction and Speech Patterns: An Extension of Previous Results," *British Journal of Social and Clinical Psychology*, No. 17, 1978, pp. 101–102.

Williams, G., "The Panasonic and Quasar Hand-Held Computers, Beginning a New Generation of Consumer Computers," *BYTE*, Vol. 6, No. 1, 1981, pp. 34–45.

Williams, L.E., and Chakrabarty, R.P., "The Michigan State Random Digit Dialing Survey of Sportsmen and Wildlife Associated Recreation," *Proceedings of the Section on Survey Research Methods, American Statistical Association*, 1983, pp. 648–653.

*Wilson, C., and Williams, E., "Watergate Words: A Naturalistic Study of Media and Communications," *Communication Research*, Vol. 4, No. 2, 1977, pp. 169–178.

*Winer, B., *Statistical Principles in Experimental Design*, second edition, New York, McGraw-Hill, 1971.

Wiseman, F., "Methodological Bias in Public Opinion Surveys," *Public Opinion Quarterly*, Vol. 36, No. 1, Spring 1972, pp. 105–108.

Wiseman, F., "Measurement Problems in the Calculation of Rates of Response and Nonresponse," Northeastern University, paper presented at the American Psychological Association meetings, September 1979.

Wiseman, F., "The Measurement and Magnitude of Nonresponse in U.S. Consumer Telephone Surveys," *Survey Methodology*, Vol. 6, special edition, 1980, pp. 133–150.

Wiseman, F., and McDonald, P., "Noncontact and Refusal Rates in Consumer Telephone Surveys," *Journal of Marketing Research*, Vol. 16, November 1979, pp. 478–484.

*Wolf, F.M., *Meta-Analysis; Quantitative Methods for Research Synthesis*, Beverly Hills, Sage, 1986.

Wolfle, L.M., "Characteristics of Persons With and Without Home Telephones," *Journal of Marketing Research*, Vol. 16, August 1979, pp. 421–425.

Woltman, H.F., and Bushery, J.M., "Results of the NCS Maximum Personal Visit-Minimum Telephone Interview Experiment," U.S. Bureau of the Census, Statistical Methods Division, 1977.

Woltman, H.F., Turner, A.G., and Bushery, J.M., "A Comparison of Three Mixed-Mode Interviewing Procedures in the National Crime Survey," *Journal of the American Statistical Association*, Vol. 75, 1980, pp. 534–543.

Yaffe, R., Shapiro, S., Fuchsberg, R.R., Rohde, C.A., and Corpeño, H.C., "Medical Economics Survey-Methods Study, Cost-Effectiveness of Alternative Survey Strategies," *Medical Care*, Vol. 16, No. 8, August 1978, pp. 641–659.

*Young, P., "The Survey Respondents' Experience of a Structured Interview," Survey Methods Centre, London, Social and Community Planning Research, 1987.

*Yourdon, E., *Techniques of Program Structure and Design*, Englewood Cliffs, NJ, Prentice-Hall, 1975.

Index

Applied Probability and Statistics (Continued)

HEIBERGER • Computation for the Analysis of Designed Experiments
HOAGLIN, MOSTELLER and TUKEY • Exploring Data Tables, Trends and Shapes
HOAGLIN, MOSTELLER, and TUKEY • Understanding Robust and Exploratory Data Analysis
HOCHBERG and TAMHANE • Multiple Comparison Procedures
HOEL • Elementary Statistics, *Fourth Edition*
HOEL and JESSEN • Basic Statistics for Business and Economics, *Third Edition*
HOGG and KLUGMAN • Loss Distributions
HOLLANDER and WOLFE • Nonparametric Statistical Methods
IMAN and CONOVER • Modern Business Statistics
JESSEN • Statistical Survey Techniques
JOHNSON • Multivariate Statistical Simulation
JOHNSON and KOTZ • Distributions in Statistics
Discrete Distributions
Continuous Univariate Distributions—1
Continuous Univariate Distributions—2
Continuous Multivariate Distributions
JUDGE, HILL, GRIFFITHS, LÜTKEPOHL and LEE • Introduction to the Theory and Practice of Econometrics, *Second Edition*
JUDGE, GRIFFITHS, HILL, LÜTKEPOHL and LEE • The Theory and Practice of Econometrics, *Second Edition*
KALBFLEISCH and PRENTICE • The Statistical Analysis of Failure Time Data
KISH • Statistical Design for Research
KISH • Survey Sampling
KUH, NEESE, and HOLLINGER • Structural Sensitivity in Econometric Models
KEENEY and RAIFFA • Decisions with Multiple Objectives
LAWLESS • Statistical Models and Methods for Lifetime Data
LEAMER • Specification Searches: Ad Hoc Inference with Nonexperimental Data
LEBART, MORINEAU, and WARWICK • Multivariate Descriptive Statistical Analysis: Correspondence Analysis and Related Techniques for Large Matrices
LINHART and ZUCCHINI • Model Selection
LITTLE and RUBIN • Statistical Analysis with Missing Data
McNEIL • Interactive Data Analysis
MAGNUS and NEUDECKER • Matrix Differential Calculus with Applications in Statistics and Econometrics
MAINDONALD • Statistical Computation
MALLOWS • Design, Data, and Analysis by Some Friends of Cuthbert Daniel
MANN, SCHAFER and SINGPURWALLA • Methods for Statistical Analysis of Reliability and Life Data
MARTZ and WALLER • Bayesian Reliability Analysis
MASON, GUNST, and HESS • Statistical Design and Analysis of Experiments with Applications to Engineering and Science
MIKÉ and STANLEY • Statistics in Medical Research: Methods and Issues with Applications in Cancer Research
MILLER • Beyond ANOVA, Basics of Applied Statistics
MILLER • Survival Analysis
MILLER, EFRON, BROWN, and MOSES • Biostatistics Casebook
MONTGOMERY and PECK • Introduction to Linear Regression Analysis
NELSON • Applied Life Data Analysis
OSBORNE • Finite Algorithms in Optimization and Data Analysis
OTNES and ENOCHSON • Applied Time Series Analysis: Volume I, Basic Techniques
OTNES and ENOCHSON • Digital Time Series Analysis